T0184456

Introduction to Statistics and Data Analysis

Christian Heumann · Michael Schomaker ·
Shalabh

Introduction to Statistics and Data Analysis

With Exercises, Solutions and Applications in R

Second Edition

 Springer

Christian Heumann
Department of Statistics
LMU Munich
Munich, Germany

Michael Schomaker
Department of Statistics
LMU Munich
Munich, Germany

Shalabh
Department of Mathematics and Statistics
Indian Institute of Technology Kanpur
Kanpur, India

ISBN 978-3-031-12025-1   ISBN 978-3-031-11833-3   (eBook)
https://doi.org/10.1007/978-3-031-11833-3

1st edition: © Springer International Publishing Switzerland 2016
2nd edition: © Springer Nature Switzerland AG 2022
This work is subject to copyright. All rights are reserved by the Publisher, whether the whole or part of the material is concerned, specifically the rights of translation, reprinting, reuse of illustrations, recitation, broadcasting, reproduction on microfilms or in any other physical way, and transmission or information storage and retrieval, electronic adaptation, computer software, or by similar or dissimilar methodology now known or hereafter developed.
The use of general descriptive names, registered names, trademarks, service marks, etc. in this publication does not imply, even in the absence of a specific statement, that such names are exempt from the relevant protective laws and regulations and therefore free for general use.
The publisher, the authors, and the editors are safe to assume that the advice and information in this book are believed to be true and accurate at the date of publication. Neither the publisher nor the authors or the editors give a warranty, expressed or implied, with respect to the material contained herein or for any errors or omissions that may have been made. The publisher remains neutral with regard to jurisdictional claims in published maps and institutional affiliations.

This Springer imprint is published by the registered company Springer Nature Switzerland AG
The registered company address is: Gewerbestrasse 11, 6330 Cham, Switzerland

Preface to the Second Edition

We are gratified by the success of the first edition of "Introduction to Statistics and Data Analysis". Our lecture materials have been in great demand, and our readers have conveyed to us numerous good suggestions and ideas.

We believe that modern approaches to statistics can be taught early in one's statistics education. The topics need to be presented clearly and they need to be well connected to both traditional methods and statistical software. Based on this belief and the proposals received from our readers, we decided to add three new chapters to the second edition. A newly added chapter on logistic regression extends our comprehensive treatment of regression analysis. We also added a chapter on simple random sampling which interconnects classical statistical results with the computing-based inference method of bootstrapping. Lastly, we developed a chapter on causal inference: we believe that an early formal treatment of the subject helps students and researchers to express their actual hypotheses accurately, analyze them appropriately and understand the requirements needed to engage with more sophisticated literature.

In Chap. 10, we now illustrate alternatives to making binary decisions with statistical tests: we show how the concepts of compatibility and functions of p- and S-values for varying hypotheses aid the interpretation of test results by focusing on gradations of evidence rather than binary decision rules.

We rely on continuous critical feedback from our readers to learn what can be improved and refined further. We welcome suggestions, which can be sent to christian.heumann@stat.uni-muenchen.de, shalab@iitk.ac.in, and michael.schomaker@stat.uni-muenchen.de.

As in the previous edition, our philosophy is to explain the most fundamental concepts comprehensively, relegate more technical material to the appendix, illustrate computations with interpretations and use exercises and software to engage in more detail with the challenges and implications of the presented methods.

The updated book homepage is now available at https://statsbook.github.io/. It contains all solutions to the R exercises, additional information, and a current list of errata.

We hope that our new edition proves to be valuable for teaching introductory statistics. We thank Veronika Rosteck from Springer for her tremendous support while preparing this updated manuscript and to Barnaby Sheppard for his help in editing the new content. We deeply appreciate and acknowledge the continuous caring support of our families while completing the book.

Munich, Germany   Christian Heumann
Munich, Germany   Michael Schomaker
Kanpur, India   Shalabh
April 2022

Preface to the First Edition

The success of the open-source statistical software "*R*" has made a significant impact on the teaching and research of statistics in the last decade. Analyzing data is now easier and more affordable than ever, but choosing the most appropriate statistical methods remains a challenge for many users. To understand and interpret software output, it is necessary to engage with the fundamentals of statistics.

However, many readers do not feel comfortable with complicated mathematics. In this book, we attempt to find a healthy balance between explaining statistical concepts comprehensively and showing their application and interpretation using *R*.

This book will benefit beginners and self-learners from various backgrounds as we complement each chapter with various exercises and detailed and comprehensible solutions. The results involving mathematics and rigorous proofs are separated from the main text, where possible, and are kept in an appendix for interested readers. Our textbook covers material that is generally taught in introductory level statistics courses to students from various backgrounds, including sociology, biology, economics, psychology, medicine, and others. Most often we introduce the statistical concepts using examples and illustrate the calculations both manually and using *R*.

However, while we provide a gentle introduction to *R* (in the appendix), this is not a software book. Our emphasis lies on explaining statistical concepts correctly and comprehensively, using exercises and software to delve deeper into the subject matter and learn about the conceptual challenges that the methods present.

The book homepage, https://statsbook.github.io/, contains additional material, most notably the software codes needed to answer the software exercises, and data sets. In the remainder of the book, we will use gray boxes

```
R-command()
```

to introduce the relevant *R* commands. In many cases, the code can be directly pasted into *R* to reproduce the results and graphs presented in the book; in others the code is abbreviated to improve readability and clarity, and the detailed code can be found online.

Many years of teaching experience, from undergraduate to post-graduate level, went into this book. The authors hope that the reader will enjoy reading it and find it a useful reference for learning. We welcome critical feedback to improve future editions of this book. Comments can be sent to christian.heumann@ stat.uni-muenchen.de, shalab@iitk.ac.in, and michael.schomaker@uct.ac.za who contributed equally to this book.

We thank Melanie Schomaker for producing some of the figures and giving graphical advice, Alice Blanck from Springer for her continuous help and support, and Lyn Imeson for her dedicated commitment which improved earlier versions of this book. We are grateful to our families who have supported us during the preparation of this book.

München, Germany   Christian Heumann
Cape Town, South Africa   Michael Schomaker
Kanpur, India   Shalabh
May 2016

Contents

About the Authors

Dr. Christian Heumann is a Professor at the Department of Statistics, LMU Munich, Germany, where he teaches students in both the Bachelor's and Master's programs. His research interests include statistical modeling, computational statistics and methods for missing data, also in connection with causal inference. Recently, he has begun exploring statistical methods in natural language processing.

Dr. Michael Schomaker is a Researcher and Heisenberg Fellow at the Department of Statistics, LMU Munich, Germany. He is an honorary Senior Lecturer at the University of Cape Town, South Africa and previously worked as an Associate Professor at UMIT—University for Health Sciences, Medical Informatics and Technology, Austria. For many years he has taught both undergraduate and post-graduate students from various disciplines, including the business and medical sciences, and has written contributions for various introductory textbooks. His research focuses on causal inference, missing data, model averaging, and HIV and public health.

Dr. Shalabh is a Professor at the Indian Institute of Technology Kanpur, India. As a post-doctoral researcher he worked at the University of Pittsburgh, USA and LMU Munich, Germany. He has over twenty years of experience in teaching and research. His main research areas are linear models, regression analysis, econometrics, error-measurement models, missing data models and sampling theory.

Part I
Descriptive Statistics

Introduction and Framework

1

Statistics is a collection of methods which help us to describe, summarize, interpret, and analyse data. Drawing conclusions from data is vital in research, administration, and business. Researchers are interested in understanding whether a medical intervention helps in reducing the burden of a disease, how personality relates to decision making, whether a new fertilizer increases the yield of crops, how a political system affects trade policy, who is going to vote for a political party in the next election, what are the long-term changes in the population of a fish species, and many more questions. Governments and organizations may be interested in the life expectancy of a population, the risk factors for infant mortality, geographical differences in energy usage, migration patterns, or reasons for unemployment. In business, identifying people who may be interested in a certain product, optimizing prices, and evaluating the satisfaction of customers are possible areas of interest.

No matter what the question of interest is, it is important to collect data in a way which allows its analysis. The representation of collected data in a **data set** or **data matrix** allows the application of a variety of statistical methods. In the first part of the book, we are going to introduce methods which help us in *describing* data, the second and third parts of the book focus on inferential statistics, which means *drawing conclusions* from data. In this chapter, we are going to introduce the framework of statistics which is needed to properly collect, administer, evaluate and analyse data.

1.1 Population, Sample and Observations

Let us first introduce some terminology and related notations used in this book. The **units** on which we measure data—such as persons, cars, animals, or plants—are called **observations**. These units/observations are represented by the Greek symbol

© Springer Nature Switzerland AG 2022
C. Heumann et al., *Introduction to Statistics and Data Analysis*,
https://doi.org/10.1007/978-3-031-11833-3_1

ω. The collection of all units is called **population** and is represented by Ω. When we refer to $\omega \in \Omega$, we mean a single unit out of all units, e.g. one person out of all persons of interest. If we consider a selection of observations $\omega_1, \omega_2, \ldots, \omega_n$, then these observations are called **sample**. A sample is always a subset of the population, $\{\omega_1, \omega_2, \ldots, \omega_n\} \subseteq \Omega$.

Example 1.1.1

- If we are interested in the social conditions under which Indian people live, then we would define all inhabitants of India as Ω and each of its inhabitants as ω. If we want to collect data from a few inhabitants then those would represent a sample from the total population.
- Investigating the economic power of Africa's platinum industry would require to treat each platinum related company as ω, whereas all platinum related companies would be collected in Ω. A few companies $\omega_1, \omega_2, \ldots, \omega_n$ comprise a sample of all companies.
- We may be interested in collecting information about those participating in a statistics course. All participants in the course constitute the population Ω, each participant refers to a unit or observation ω.

Remark 1.1.1 Sometimes the concept of a population is not applicable or difficult to imagine. As an example imagine that we measure the temperature in New Delhi every hour. A sample would then be the time series of temperatures in a specific time window, for example from January to March 2022. A population in the sense of observational units does not exist here. But now assume that we measure temperatures in several different cities; then all the cities form the population and a sample is any subset of the cities.

1.2   Variables

If we have specified the population of interest for a specific research question, we can think of what is of interest about our observations. A particular feature of these observations can be collected in a statistical **variable** X. Any information we are interested in may be captured in such a variable. For example, if our observations refer to human beings, X may describe marital status, gender, age, or anything else which may relate to a person. Of course, we can be interested in many different features, each of them collected in a different variable X_i, $i = 1, 2, \ldots, p$. Each observation ω takes a particular value for X. If X refers to gender, each observation, i.e. each person, has a particular value x which refers to either "male" or "female".

The formal definition of a variable is

$$X : \Omega \to S$$
$$\omega \mapsto x \tag{1.1}$$

This definition states that a variable X takes a value x for each observation $\omega \in \Omega$, whereby the number of possible values are contained in the set S.

Example 1.2.1

- If X refers to gender, possible x-values are contained in $S = \{\text{male, female}\}$. Each observation ω is either male or female and this information is summarized in X.
- Let X be the country of origin for a car. Possible values to be taken by an observation ω (i.e. a car) are $S = \{\text{Italy, South Korea, Germany, France, India, China, Japan, USA, ...}\}$.
- A variable X which refers to age may take any value between 1 and 125. Each person ω is assigned a value x which represents the age of this person.

1.2.1 Qualitative and Quantitative Variables

Qualitative variables are the variables which take values x that cannot be ordered in a logical or natural way. For example

- the colour of the eye,
- the name of a political party,
- the type of transport used to travel to work,

are all qualitative variables. Neither is there any reason to list blue eyes before brown eyes (or vice versa) nor does it make sense to list buses before trains (or vice versa).

Quantitative variables represent measurable quantities. The values which these variables can take can be ordered in a logical and natural way. Examples of quantitative variables are

- size of shoes,
- price for houses,
- number of semesters studied,
- weight of a person.

Remark 1.2.1 It is common to assign numbers to qualitative variables for practical purposes in data analyses, see Sect. 1.4 for more details. For instance, if we consider the variable "gender", then each observation can either take the "value" male or female. We may decide to assign 1 to female and 0 to male and use these numbers instead of the original categories. However, this is arbitrary, we could have also chosen "1" for male and "0" for female, or "2" for male and "10" for female. There is no logical and natural order on how to arrange male and female and thus the variable gender remains a qualitative variable, even after using numbers for coding the values that X can take.

1.2.2   Discrete and Continuous Variables

Discrete variables are variables which can only take a finite number of values. All qualitative variables are discrete, such as the colour of the eye or the region of a country. But also quantitative variables can be discrete: the size of shoes or the number of semesters studied would be discrete because the number of values these variables can take is limited.

Variables which can take an infinite number of values are called **continuous variables**. Examples are the time it takes to travel to university, the length of an antelope, and the distance between two planets. Sometimes it is said that continuous variables are variables which are "measured rather than counted". This is a rather informal definition which helps to understand the difference between discrete and continuous variables. The crucial point is that continuous variables can, in theory, take an infinite number of values; for instance, the height of a person may be recorded as 172 cm. However, the actual height on the measuring tape might be 172.3 cm which was rounded off to 172 cm. If one had a better measuring instrument we may have obtained 172.342 cm. But the real height of this person is a number with indefinitely many decimal places such as 172.342975328... cm. No matter what we eventually report or obtain, a variable which can take an infinite amount of values is defined to be a continuous variable.

1.2.3   Scales

The thoughts and considerations from above indicate that different variables contain different amounts of information. A useful classification of these considerations is given by the concept of the **scale** of a variable. This concept will help us in the remainder of this book to identify which methods are the appropriate ones to use in a particular setting.

Nominal scale.   The values of a *nominal variable* cannot be ordered. Examples are the gender of a person (male–female) or the status of an application (pending–not pending).

Ordinal scale.   The values of an *ordinal variable* can be ordered. However, the differences between these values cannot be interpreted in a meaningful way. For example, the possible values of education level (none–primary education–secondary education–university degree) can be ordered meaningfully but the differences between these values cannot be interpreted. Likewise, the satisfaction with a product (unsatisfied–satisfied–very satisfied) is an ordinal variable because the values this variable can take can be ordered but the differences between "unsatisfied–satisfied" and "satisfied–very satisfied" cannot be compared in a numerical way.

Metric scale.   The values of a *metric variable* can be ordered. Furthermore, the differences between these values can be interpreted in a meaningful way. For instance, the height of a person refers to a metric variable because the values can be ordered (170 cm, 171 cm, 172 cm,..) and differences between these values can be compared (the difference between 170 cm and 171 cm is the same as the difference between 171 cm and 172 cm). Sometimes the metric scale is divided further into sub-scales. While in the remainder of the book we typically don't need these classifications it is still useful to reflect on them:

Interval scale. Only differences between values, but not ratios, can be interpreted. An example for this scale would be temperature (measured in °C): the difference between $-2\,°C$ and $4\,°C$ is $6°C$ but the ratio of $4/-2 = -2$ doesn't mean that $-4\,°C$ is twice as cold as $2\,°C$.

Ratio scale. Both differences and ratios can be interpreted. An example is speed: 60 km/h is 40 km/h more than 20 km/h. Moreover, 60 km/h is three times faster than 20 km/h because the ratio between them is 3.

Absolute scale. The absolute scale is the same as the ratio scale, with the exception that the values are measured in "natural" units. An example is "number of semesters studied" where no artificial unit such as km/h or ° C is needed: the values are simply 1, 2, 3, ...

1.2.4 Grouped Data

Sometimes, data may be available only in a summarized form: instead of the original value one may only know the category or group the value belongs to. For example,

- it is often convenient in a survey to ask for the income (per year) by means of groups: [€0–€20,000), [€20,000–€30,000), . . . , > €100,000;
- if there are many political parties in an election, those with a low number of voters are often summarized in a new category "Other Parties";
- instead of capturing the number of claims made by an insurance company customer, the variable "claimed" may denote whether or not the customer claimed at all (yes–no).

If data are available in grouped form, we call the respective variable capturing this information a **grouped variable**. Sometimes these variables are also known as **categorical variables**. This is however not a complete definition because categorical variables refer to any type of variable which takes a finite, possibly small, number of values. Thus, any discrete and/or nominal and/or ordinal and/or qualitative variable may be regarded as a categorical variable. Any grouped or categorical variable which can only take two values is called a **binary variable**.

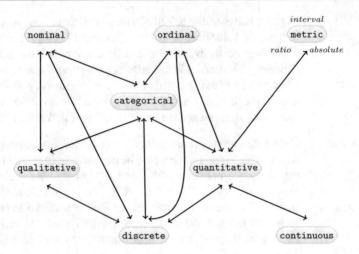

Fig. 1.1 Summary of variable classifications

To gain a better understanding on how the definitions from the above sections relate to each other, see Fig. 1.1: qualitative data is always discrete, but quantitative data can be both discrete (e.g. size of shoes or a grouped variable) or continuous (e.g. temperature). Nominal variables are always qualitative and discrete (e.g. colour of the eye), whereas metric variables are always quantitative (e.g. temperature). Categorical variables can be both qualitative (e.g. colour of the eye) or quantitative (satisfaction level on a scale from 1 to 5). Categorical variables are never continuous.

1.3   Data Collection

When collecting data, we may ask ourselves how to facilitate this in detail and how much data needs to be collected. The latter question will be partly answered in Sect. 9.5; but in general we can think of collecting data either on all subjects of interest, such as in a national census, or on a representative sample of the population. Most commonly, we gather data on a sample (described in the Part I of this book), and then draw conclusions about the population of interest (discussed in Part III of this book). A sample might either be chosen by us or obtained through third parties (hospitals, government agencies), or created during an experiment. This depends on the context as described below.

1.3.1   Survey

A survey typically (but not always) collects data by asking questions (in person or by phone) or providing questionnaires to study participants (as a print-out or online). For example, an opinion poll before a national election provides evidence about the future government: potential voters are asked by phone which party they are going to vote for in the next election; on the day of the election this information can be updated by asking the same question to a sample of voters who have just delivered their vote at the polling station (so-called exit poll). A behavioural research survey may ask members of a community about their knowledge and attitudes towards drug use. For this purpose, the study coordinators can send people with a questionnaire to this community and interview members of randomly selected households.

Ideally a survey is conducted in a way which makes the chosen sample representative of the population of interest. If a marketing company interviews people in a pedestrian zone to find their views about a new chocolate bar, then these people may not be representative of those who will potentially be interested in this product. Similarly, if students are asked to fill in an online survey to evaluate a lecture, it may turn out that those who participate are on average more satisfied than those who don't. Survey sampling is a complex topic on its own. The interested reader may consult Groves, Fowler, Couper, Lepkowski, Singer and Tourangeau (2009).

1.3.2 Experiment

Experimental data is obtained in "controlled" settings. This can mean many things, but essentially it is data which is generated by the researcher with full control over one or many variables of interest. For instance, suppose there are two competing toothpastes, both of which promise to reduce pain for people with sensitive teeth. If the researcher decided to randomly assign toothpaste A to half of the study participants, and toothpaste B to the other half, then this is an experiment because it is only the researcher who decides which toothpaste is to be used by any of the participants. It is not decided by the participant. The data of the variable toothpaste is controlled by the experimenter. Consider another example where the production process of a product can potentially be reduced by combining two processes. The management could decide to implement the new process in three production facilities, but leave it as it is in the other facilities. The production process for the different units (facilities) is therefore under control of the management. However, if each facility could decide for themselves if they wanted a change or not, it would not be an experiment because factors not directly controlled by the management, such as the leadership style of the facility manager, would determine which process is chosen. More details on experiments are given in Sect. 14.4.1.

1.3.3 Observational Data

Observational data is data which is collected routinely, without a researcher designing a survey or conducting an experiment. Suppose a blood sample is drawn from each patient with a particular acute infection when they arrive at a hospital. This data may be stored in the hospital's folders and later accessed by a researcher who is interested in studying this infection. Or suppose a government institution monitors where people live and move to. This data can later be used to explore migration patterns.

1.3.4 Primary and Secondary Data

Primary data is data we collect ourselves, i.e. via a survey or experiment. Secondary data, in contrast, is collected by someone else. For example, data from a national census, publicly available databases, previous research studies, government reports, historical data, data from the internet, among others, are secondary data.

1.4   Creating a Data Set

There is a unique way in which data is prepared and collected to utilize statistical analyses. The data is stored in a data matrix (= data set) with p columns and n rows, see Fig. 1.2. Each row corresponds to an observation/unit ω and each column to a variable X. This means that, for example, the entry in the fourth row and second column (x_{42}) describes the value of the fourth observation on the second variable. The examples below will illustrate the concept of a data set in more detail.

Example 1.4.1 Suppose five students take examinations in music, mathematics, biology, and geography. Their marks, measured on a scale between 0 and 100 (where 100 is the best mark), can be written down as illustrated in Fig. 1.3. Note that each row refers to a student and each column to a variable, i.e. a subject. We consider a larger data set in the next example.

Example 1.4.2 Consider the data set described in Appendix A.4.1. A pizza delivery service captures information related to each delivery, for example the delivery time, the temperature of the pizza, the name of the driver, the date of the delivery, the name of the branch, and many more. To capture the data of all deliveries during one month, we create a data matrix. Each row refers to a particular delivery, therefore

Fig. 1.2 Data set or data matrix

$$
\begin{array}{c|cccc}
\omega & \text{Variable 1} & \text{Variable 2} & \cdots & \text{Variable } p \\
\hline
1 & x_{11} & x_{12} & \cdots & x_{1p} \\
2 & x_{21} & x_{22} & \cdots & x_{2p} \\
\vdots & \vdots & \vdots & & \vdots \\
n & x_{n1} & x_{n2} & \cdots & x_{np}
\end{array}
$$

Fig. 1.3 Data set of marks of five students

$$
\begin{array}{c|cccc}
\omega & \text{Music} & \text{Mathematics} & \text{Biology} & \text{Geography} \\
\hline
\text{Student } A & 65 & 70 & 85 & 45 \\
\text{Student } B & 77 & 82 & 80 & 60 \\
\text{Student } C & 78 & 73 & 93 & 68 \\
\text{Student } D & 88 & 71 & 63 & 58 \\
\text{Student } E & 75 & 83 & 63 & 57
\end{array}
$$

Fig. 1.4 Pizza data set

$$\begin{pmatrix} \text{Delivery} & \text{Delivery Time} & \text{Temperature} & \cdots & \text{Branch} \\ 1 & 35.1 & 68.3 & \cdots & \text{East (1)} \\ 2 & 25.2 & 71.0 & \cdots & \text{East (1)} \\ \vdots & \vdots & \vdots & & \vdots \\ 1266 & 35.7 & 60.8 & \cdots & \text{West (2)} \end{pmatrix}$$

Table 1.1 Coding list for branch

Variable	Values	Code
Branch	East	1
	West	2
	Centre	3
	Missing	4

representing the observations of the data. Each column refers to a variable. In Fig. 1.4 the variables X_1 (delivery time in minutes), X_2 (temperature in $^\circ$ C), and X_{12} (name of branch) are listed.

The first row tells us about the features of the first pizza delivery: the delivery time was 35.1 minutes, the pizza arrived with a temperature of 68.3 $^\circ$C and the pizza was delivered from the branch in the East of the city. In total there were $n = 1266$ deliveries. For nominal variables, such as branch, we may decide to produce a coding list, as illustrated in Table 1.1: instead of referring to the branches as "East", "West", and "Centre" we may simply call them 1, 2, and 3. As we will see in Chap. 11 this has benefits for some analysis methods, though this is not needed in general.

If some values are missing, for example because they were never captured or even lost, then this requires special attention. In Table 1.1 we assign missing values the number "4" and therefore treat them as a separate category. If we work with statistical software (see below) we may need other coding such as NA in the statistical software R or . in Stata. More details can be found in Appendix A.

Another consideration when collecting data is that of **transformations**: We may have captured the velocity of cars in kilometres per hour, but may need to present the data in miles per hour; we have captured the temperature in degrees Celsius, whereas we need to communicate results in degrees Fahrenheit; or we have created a satisfaction score which we want to range from -5 to $+5$ while the score currently runs from 0 to 20. This is not a problem at all. We can simply create a new variable which reflects the required transformation. However, valid transformations depend on the scale of a variable. Variables on an interval scale can use transformations of the following kind:

$$g(x) = a + bx, \quad b > 0. \tag{1.2}$$

For ratio scales, only the following transformations are valid:

$$g(x) = bx, \quad b > 0. \tag{1.3}$$

In the above equation a is set to 0 because ratios only stay the same if we respect a variable's natural point of origin.

Example 1.4.3 The temperature in °F relates to the temperature in °C as follows

$$\text{Temperature in }°F = 32 + 1.8 \text{ Temperature in }°C$$
$$g(x) = a + b \qquad x$$

This means that 25 °C relates to $(32 + 1.8 \cdot 25)$ °F = 77 °F. If X_1 is a variable representing temperature by °C we can simply create a new variable X_2 which is temperature in °F. Since temperature is measured on an interval scale this transformation is valid.

Changing currencies is also possible. If we would like to represent the price of a product not in South African Rand but in € we simply apply the transformation

$$\text{Price in South African Rand} = b \cdot \text{Price in €}$$

where by b is the currency exchange rate.

1.4.1 Statistical Software

There are number of statistical software packages which allows data collection, management, and—most importantly—analysis. In this book, we focus on the statistical software R which is freely available at http://cran.r-project.org/. A gentle introduction to R is provided in Appendix A. A data matrix can be created manually using commands such as `matrix()`, `data.frame()`, and others. Any data can be edited using `edit()`. However, typically analysts have already typed their data into databases or spreadsheets, for example in Excel, Access, or MySQL. In most of these applications, it is possible to save the data as an ASCII file (*.dat*), a tab delimited file (*.txt*) or as a comma-separated values file (*.csv*). All of these formats allow easy switching between different software and database applications. Such data can easily be read into R by means of the following commands:

```
setwd('C:/directory')
read.table('pizza_delivery.dat')
read.table('pizza_delivery.txt')
read.csv('pizza_delivery.csv')
```

where `setwd` specifies the working directory where the data files are stored. Alternatively, loading the library `foreign` allows the import of data from many different statistical software packages, notably Stata, SAS, Minitab, SPSS, among others. A detailed description of data import and export can be found in the respective R manual available at http://cran.r-project.org/. Once the data is read into R it can be viewed with

```
View()
```

R File   Windows   Edit   Help

	day	date	time	operator	branch	driver	temperature	bill
1	Thursday	01-May-14	35.12837	Laura	East	Bruno	68.28772	58.4
2	Thursday	01-May-14	25.20307	Melissa	East	Salvatore	70.99779	26.4
3	Thursday	01-May-14	45.6434	Melissa	West	Salvatore	53.39415	58.1
4	Thursday	01-May-14	29.3743	Melissa	East	Salvatore	70.3066	35.2
5	Thursday	01-May-14	29.99461	Melissa	West	Salvatore	71.50169	38.4
6	Thursday	01-May-14	40.25432	Melissa	Centre	Bruno	60.7595	61.8
7	Thursday	01-May-14	48.72861	Laura	West	Bruno	58.2587	57.9
8	Thursday	01-May-14	34.02772	Melissa	West	Mario	68.12793	35.8
9	Thursday	01-May-14	28.20943	Laura	Centre	Mario	64.94661	36.6
10	Thursday	01-May-14	37.95479	Melissa	Centre	Bruno	60.00738	44.8
11	Thursday	01-May-14	42.07956	Melissa	Centre	Bruno	65.53872	49.5

Fig. 1.5 Viewing data in *R*

We can also can get an overview of the data directly in the *R*-console by displaying only the top lines of the data with head(). Both approaches are visualized in Fig. 1.5 for the pizza data introduced in Example 1.4.2.

1.5   Key Points and Further Issues

Note:

✓ The scale of variables is not only a formalism but an essential framework for choosing the correct analysis methods. This is particularly relevant for association analysis (Chap. 4), statistical tests (Chap. 10) and regression (Chaps. 11 and 12).

✓ Even if variables are measured on a nominal scale (i.e. if they are categorical/qualitative), we may choose to assign a number to each category of this variable. This eases the implementation of some analysis methods introduced later in this book.

✓ Data is usually stored in a data matrix where the rows represent the observations and the columns are variables. It can be analysed with statistical software. We use *R* (R Core Team, 2021) in this book. A gentle introduction is provided in Appendix A and throughout the book. A more comprehensive introduction can be found in other books, for example in Albert and Rizzo (2012) or Crawley (2013). Even advanced books, e.g. Adler (2012) or Everitt and Hothorn (2011), can offer insights to beginners.

1.6   Exercises

Exercise 1.1 Describe both the population and the observations for the following research questions:

(a) Evaluation of the satisfaction of employees from an airline.
(b) Description of the marks of students from an assignment.
(c) Comparison of two drugs which deal with high blood pressure.

Exercise 1.2 A national park conducts a study on the behaviour of their leopards. A few of the park's leopards are registered and receive a GPS device which allows measuring the position of the leopard. Use this example to describe the following concepts: population, sample, observation, value, variable.

Exercise 1.3 Which of the following variables are qualitative, and which are quantitative? Specify which of the quantitative variables are discrete and which are continuous:

Time to travel to work, shoe size, preferred political party, price for a canteen meal, eye colour, gender, wavelength of light, customer satisfaction on a scale from 1 to 10, delivery time for a parcel, blood type, number of goals in a hockey match, height of a child, subject line of an email.

Exercise 1.4 Identify the scale of the following variables:

(a) Political party voted for in an election.
(b) The difficulty of different levels in a computer game.
(c) Production time of a car.
(d) Age of turtles.
(e) Calender year.
(f) Price of a chocolate bar.
(g) Identification number of a student.
(h) Final ranking at a beauty contest.
(i) Intelligence quotient.

Exercise 1.5 Make yourself familiar with the pizza data set from Appendix A.4.1.

(a) First, browse through the introduction to *R* in Appendix A. Then, read in the data.
(b) View the data both in the *R* Data Editor and in the *R* Console.
(c) Create a new data matrix which consists of the first 5 rows and first 5 variables of the data. Print this data set on the *R* Console. Now, save this data set in your preferred format.
(d) Add a new variable "NewTemperature" to the data set which converts the temperature from °C to °F.

(e) Attach the data and list the values from the variable "NewTemperature".
(f) Use "?" to make yourself familiar with the following commands: `str`, `dim`, `colnames`, `names`, `nrow`, `ncol`, `head`, `tail`. Apply these commands to the data to get more information about it.

Exercise 1.6 Consider the research questions of describing parents' attitudes towards immunization, what proportion of them wants immunization against chicken-pox for their last-born child, and whether this proportion differs by gender and age.

(a) Which data collection method is the most suitable one to answer the above questions: survey or experiment?
(b) How would you capture the attitudes towards immunization in a single variable?
(c) Which variables are needed to answer all the above questions? Describe the scale of each of them.
(d) Reflect on what an appropriate data set would look like. Now, given this data set, try to write down the above research questions as precisely as possible.

→ Solutions to all exercises in this chapter can be found in Appendix B

Frequency Measures and Graphical Representation of Data

In Chap. 1, we highlighted that different variables contain different levels of information. When summarizing or visualizing one or more variable(s), it is this information which determines the appropriate statistical methods to use.

Suppose we are interested in studying the employment opportunities and starting salaries of university graduates with a Masters' degree. Let the variable X denote the starting salaries measured in €/year. Now suppose 100 graduate students provide their initial salaries. Let us write down the salary of the first student as x_1, the salary of the second student as x_2, and so on. We therefore have 100 observations $x_1, x_2, \ldots, x_{100}$. How can we summarize those 100 values best to extract meaningful information from them? The answer to this question depends upon several aspects like the nature of the recorded data, e.g. how many observations have been obtained (either small in number or large in number) or how the data was recorded (either exact values were obtained or the values were obtained in intervals). For example, the starting salaries may be obtained as exact values, say 51,500 €/year, 32,350 €/year etc. Alternatively, these values could have been summarized in categories like low income ($<30,000$ €/year), medium income ($30,000 - 50,000$ €/year), high income ($50,000 - 70,000$ €/year), very high income ($>70,000$ €/year). Another approach is to ask whether the students were employed or not after graduating and record the data in terms of "yes" or "no". It is evident that the latter classification is less detailed than the grouped income data which is less detailed than the exact data. Depending on which conceptualization of "starting salary" we use, we need to choose the approach to summarize the data, that is the 100 values relating to the 100 graduated students.

© Springer Nature Switzerland AG 2022
C. Heumann et al., *Introduction to Statistics and Data Analysis*,
https://doi.org/10.1007/978-3-031-11833-3_2

2.1 Absolute and Relative Frequencies

2.1.1 Discrete Data

Let us first consider a simple example to illustrate our notation.

Example 2.1.1 Suppose there are ten people in a supermarket queue. Each of them is either coded as "F" (if the person is female) or "M" (if the person is male). The collected data may look like

$$M, F, M, F, M, M, M, F, M, M.$$

There are now two categories in the data: male (M) and female (F). We use a_1 to refer to the male category and a_2 to refer to the female category. Since there are 7 male and 3 female students, we have 7 values in category a_1, denoted as $n_1 = 7$, and 3 values in category a_2, denoted as $n_2 = 3$. The number of observations in a particular category is called the **absolute frequency**. It follows that $n_1 = 7$ and $n_2 = 3$ are the absolute frequencies of a_1 and a_2, respectively. Note that $n_1 + n_2 = n = 10$ which is the same as the total number of collected observations. We can also calculate the **relative frequencies** of a_1 and a_2 as $f_1 = f(a_1) = \frac{n_1}{n} = \frac{7}{10} = 0.7 = 70\%$ and $f_2 = f(a_2) = \frac{n_2}{n} = \frac{3}{10} = 0.3 = 30\%$, respectively. This gives us information about the proportions of male and female customers in the queue.

We now extend these concepts to a general framework for the summary of **data on discrete variables**. Suppose there are k categories denoted as $a_1, a_2, \ldots, a_k$ with $n_j (j = 1, 2, \ldots, k)$ observations in category a_j. The **absolute frequency** n_j is defined as the number of units in the jth category a_j. The sum of absolute frequencies equals the total number of units in the data: $\sum_{j=1}^{k} n_j = n$. The **relative frequencies** of the jth class are defined as

$$f_j = f(a_j) = \frac{n_j}{n}, \quad j = 1, 2, \ldots, k. \tag{2.1}$$

The relative frequencies always lie between 0 and 1 and $\sum_{j=1}^{k} f_j = 1$.

2.1.2 Grouped Metric Data

Data on metric variables usually have a large number (k) of different values. Sometimes k may even be the same as n and in such a case the relative frequencies become $f_j = \frac{1}{n}$ for all j. However, it is possible to define intervals in which the observed values are contained.

Table 2.1 Frequency distribution for discrete data

Class intervals (a_j)	a_1	a_2	...	a_k
Absolute frequencies (n_j)	n_1	n_2	...	n_k
Relative frequencies (f_j)	f_1	f_2	...	f_k

Example 2.1.2 Consider the following $n = 20$ results of the written part of a driving licence examination (a maximum of 100 points could be achieved):

$$28, 35, 42, 90, 70, 56, 75, 66, 30, 89, 75, 64, 81, 69, 55, 83, 72, 68, 73, 16.$$

We can summarize the results in $k = 5$ class intervals such as $0 - 20, 21 - 40, 41 - 60, 61 - 80, 81 - 100$ and the data can be presented as follows:

Class intervals	$0 - 20$	$21 - 40$	$41 - 60$	$61 - 80$	$81 - 100$
Absolute freq.	$n_1 = 1$	$n_2 = 3$	$n_3 = 3$	$n_4 = 9$	$n_5 = 4$
Relative freq.	$f_1 = \frac{1}{20}$	$f_2 = \frac{3}{20}$	$f_3 = \frac{3}{20}$	$f_4 = \frac{9}{20}$	$f_5 = \frac{5}{20}$

We have $\sum_{j=1}^{5} n_j = 20 = n$ and $\sum_{j=1}^{5} f_j = 1$.

Now, suppose the n observations can be classified into k class intervals $a_1, a_2, \ldots, a_k$ where $a_j (j = 1, 2, \ldots, k)$ contains n_j observations with $\sum_{j=1}^{k} n_j = n$. The relative frequency of the jth class is $f_j = n_j/n$ and $\sum_{j=1}^{k} f_j = 1$. Table 2.1 displays the **frequency distribution** of a discrete variable X.

Example 2.1.3 Consider the pizza delivery service data (Example 1.4.2, Appendix A.4.1). We are interested in the pizza deliveries by branch and generate the respective frequency table, showing the distribution of the data, using the `table` command in R (after reading in and attaching the data) as:

```
table(branch)                    # abs. frequencies
table(branch)/length(branch)     # rel. frequencies
```

a_j	Centre	East	West
n_j	421	410	435
f_j	$\frac{421}{1266} \approx 0.333$	$\frac{410}{1266} \approx 0.323$	$\frac{435}{1266} \approx 0.344$

We have $n = \sum_j n_j = 1266$ deliveries and $\sum_j f_j = 1$. We can see from this table that each branch has a similar absolute number of pizza deliveries and each branch contributes to about one third of the total number of deliveries.

2.2   Empirical Cumulative Distribution Function

Another approach to summarize and visualize the (frequency) distribution of variables is the **empirical cumulative distribution function**, often abbreviated as "ECDF". As the name itself suggests, it gives us an idea about the cumulative relative frequencies up to a certain point. For example, say we want to know how many people scored up to 60 points in Example 2.1.2. Then this can be calculated by adding the number of people in the class intervals $0 - 20, 21 - 40$ and $41 - 60$ which corresponds to $n_1 + n_2 + n_3 = 1 + 3 + 3 = 7$ and is the **cumulative frequency**. If we want to know the relative frequency of people obtaining up to 60 points, we have to add the relative frequencies of the people in the class intervals $0 - 20, 21 - 40$ and $41 - 60$ as $f_1 + f_2 + f_3 = \frac{1}{20} + \frac{3}{20} + \frac{3}{20} = \frac{7}{20}$.

Before discussing the empirical cumulative distribution function in a more general framework, let us first understand the concept of ordered values. Suppose, the values of height of four people are observed as $x_1 = 180$ cm, $x_2 = 160$ cm, $x_3 = 175$ cm and $x_4 = 170$ cm. We arrange these values in an order, say ascending order, i.e. first the smallest value (denoted as $x_{(1)}$) and lastly the largest value (denoted as $x_{(4)}$). We obtain

$$x_{(1)} = x_2 = 160 \text{ cm}, \ x_{(2)} = x_4 = 170 \text{ cm},$$
$$x_{(3)} = x_3 = 175 \text{ cm}, \ x_{(4)} = x_1 = 180 \text{ cm}.$$

The values $x_{(1)}, x_{(2)}, x_{(3)}$ and $x_{(4)}$ are called **ordered values** for which $x_{(1)} < x_{(2)} < x_{(3)} < x_{(4)}$ holds. Note that x_1 is not necessarily the smallest value but $x_{(1)}$ is necessarily the smallest value. In general, if we have n observations $x_1, x_2, \ldots, x_n$ then the ordered data is $x_{(1)} \leq x_{(2)} \leq \ldots \leq x_{(n)}$.

Consider n observations $x_1, x_2, \ldots, x_n$ of a variable X which are arranged in ascending order as $x_{(1)} \leq x_{(2)} \leq \ldots \leq x_{(n)}$ (and are thus on an at least ordinal scale). The **empirical cumulative distribution function** $F(x)$ is defined as the cumulative relative frequencies of all values a_j which are smaller than, or equal to, x:

$$F(x) = \sum_{a_j \leq x} f(a_j). \tag{2.2}$$

This definition implies that $F(x)$ is a monotonically nondecreasing function, $0 \leq F(x) \leq 1$, $\lim_{x \to -\infty} F(x) = 0$ (the lower limit of F is 0), $\lim_{x \to +\infty} F(x) = 1$ (the upper limit of F is 1) and $F(x)$ is right continuous. Note that for metric variables F is defined with respect to all real values, not just the observed values. For ordinal variables looking at values other than those observed may however not make sense. Note also that ties are possible, for example it may happen that $x_{(3)} = x_{(4)}$.

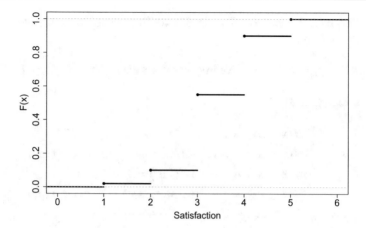

Fig. 2.1 ECDF for the satisfaction survey

2.2.1   ECDF for Ordinal Variables

The empirical cumulative distribution function of ordinal variables is a **step function**.

Example 2.2.1 Consider a customer satisfaction survey from a car service company. The 200 customers who had a car service done within the last 30 days were asked to respond regarding their overall level of satisfaction with the quality of the car service on a scale from 1 to 5 based on the following options: 1 = not satisfied at all, 2 = unsatisfied, 3 = satisfied, 4 = very satisfied, 5 = perfectly satisfied. Based on the frequency of each reply, we can calculate the relative frequencies and then the empirical cumulative distribution function, either manually (takes longer) or by using R (quick):

Satisfaction level (a_j)	$j = 1$	$j = 2$	$j = 3$	$j = 4$	$j = 5$
n_j	4	16	90	70	20
f_j	4/200	16/200	90/200	70/200	20/200
F_j	4/200	20/200	110/200	180/200	200/200

The F_j's are calculated as follows:

$$F_1 = f_1, \qquad F_3 = f_1 + f_2 + f_3,$$
$$F_2 = f_1 + f_2, \qquad F_4 = f_1 + f_2 + f_3 + f_4.$$

The ECDF for this data can be obtained by summarizing the data in a vector and using the `plot.ecdf()` function in R, see Fig. 2.1:

```
sv <- c(rep(1,4),rep(2,16),rep(3,90),rep(4,70),rep(5,20))
plot.ecdf(sv)
```

The ECDF can be used to obtain the relative frequencies for values contained in certain intervals as

$$H(c \leq x \leq d) = \text{relative frequency of values } x \text{ with } c \leq x \leq d.$$

It further follows that:

$$H(x \leq a_j) = F(a_j) \tag{2.3}$$
$$H(x < a_j) = H(x \leq a_j) - f(a_j) = F(a_j) - f(a_j) \tag{2.4}$$
$$H(x > a_j) = 1 - H(x \leq a_j) = 1 - F(a_j) \tag{2.5}$$
$$H(x \geq a_j) = 1 - H(X < a_j) = 1 - F(a_j) + f(a_j) \tag{2.6}$$
$$H(a_{j_1} \leq x \leq a_{j_2}) = F(a_{j_2}) - F(a_{j_1}) + f(a_{j_1}) \tag{2.7}$$
$$H(a_{j_1} < x \leq a_{j_2}) = F(a_{j_2}) - F(a_{j_1}) \tag{2.8}$$
$$H(a_{j_1} < x < a_{j_2}) = F(a_{j_2}) - F(a_{j_1}) - f(a_{j_2}) \tag{2.9}$$
$$H(a_{j_1} \leq x < a_{j_2}) = F(a_{j_2}) - F(a_{j_1}) - f(a_{j_2}) + f(a_{j_1}) \tag{2.10}$$

Example 2.2.2 Suppose, in Example 2.2.1, we want to know how many customers are not satisfied with their car service. Then using the data relating to the responses "1" *and* "2", we observe from the ECDF that $(16 + 4)/200\% = 10\%$ of the customers were not satisfied with the car service. This relates to using rule (2.3): $H(X \leq 2) = F(2) = 0.1$ or 10%. Similarly, the proportion of customers who are more than satisfied can be obtained using (2.5) as $H(X > 3) = 1 - H(x \leq 3) = 1 - 110/200 = 0.45$ or 45%.

2.2.2   ECDF for Metric Variables

In general, we can apply formulae (2.2)–(2.10) to metric data as well. However, before demonstrating their use, let us consider a somewhat different setting. Let us assume that a metric variable of interest is only available in the form of grouped data. We may assume that the observations within each group, i.e. each category or each interval, are distributed uniformly over the entire interval. The ECDF then consists of straight lines connecting the lower and upper values of the ECDF in each of the intervals. To understand this concept in more detail, we introduce the following notation:

e_{j-1}	lower limit of jth interval,
e_j	upper limit of jth interval,
$d_j = e_j - e_{j-1}$	width of the jth interval,
k	number of groups or intervals, e.g. $(e_0, e_1], (e_1, e_2], \ldots, (e_{k-1}, e_k]$,
n_j	number of observations in the jth interval.

Fig. 2.2 Illustration of the ECDF for metric data available in groups/intervals

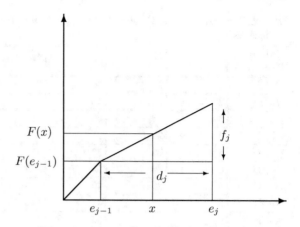

Under the assumption that all values in a particular interval are distributed uniformly within this interval, the empirical cumulative distribution function relates to a **polygonal chain** connecting the points $(e_0, 0)$, $(e_1, F(e_1))$, $(e_2, F(e_2))$, ..., $(e_k, 1)$. The ECDF can then be defined as

$$F(x) = \begin{cases} 0, & x < e_0 \\ F(e_{j-1}) + \dfrac{f_j}{d_j}(x - e_{j-1}), & x \in [e_{j-1}, e_j) \\ 1, & x \geq e_k \end{cases} \tag{2.11}$$

with $F(e_0) = 0$. The idea behind (2.11) is presented in Fig. 2.2. For any interval $[e_{j-1}, e_j)$ the respective lower and upper limits of the ECDF are $F(e_j)$ and $F(e_{j-1})$. If we assume values to be distributed uniformly over this interval, we can connect $F(e_j)$ and $F(e_{j-1})$ with a straight line. To obtain $F(x)$ with $x > e_{j-1}$ and $x < e_j$, we simply add the height of the ECDF between $F(e_{j-1})$ and $F(x)$ to $F(e_{j-1})$.

Example 2.2.3 Consider Example 2.1.3 of the pizza delivery service. Suppose we are interested in determining the distribution of the pizza delivery times. Using the function plot.ecdf() in R, we obtain the ECDF of the metric data, see Fig. 2.3a. Note that the structure of the curve is a step function but now almost looks like a continuous curve. The reason for this is that when the number of observations is large, then the lengths of class intervals become small. When these small lengths are joined together, they appear like a continuous curve. As the number of observations increases, the smoothness of the curve increases too. If the number of observations is not large, e.g. suppose the data is reported as a summary from the drivers, i.e. whether the delivery took <15 min, between 15 and 20 min, between 20 and 25 min, and so on, then we can construct the ECDF by creating a table summarizing the data features as in Table 2.2.

Figure 2.3b shows the ECDF based on the grouped data evaluated in Table 2.2. It is interesting to see that the graphs emerging from the use of the grouped data and un-grouped data are similar in this specific example.

Table 2.2 The values needed to calculate the ECDF for the grouped pizza delivery time data in Example 2.2.3

Delivery time	j	e_{j-1}	e_j	n_j	f_j	$F(e_j)$
[0; 10]	1	0	10	0	0.0000	0.0000
(10; 15]	2	10	15	3	0.0024	0.0024
(15; 20]	3	15	20	21	0.0166	0.0190
(20; 25]	4	20	25	75	0.0592	0.0782
(25; 30]	5	25	30	215	0.1698	0.2480
(30; 35]	6	30	35	373	0.2946	0.5426
(35; 40]	7	35	40	350	0.2765	0.8191
(40; 45]	8	40	45	171	0.1351	0.9542
(45; 50]	9	45	50	52	0.0411	0.9953
(50; 55]	10	50	55	6	0.0047	1.0000

(a) Metric data

(b) Grouped metric data

Fig. 2.3 Empirical cumulative distribution function for pizza delivery time

Suppose we are interested in calculating how many deliveries were completed within the desired time limit of 30 min, with a tolerance of maximum 10% deviation, i.e. a deviation of 3 min. We can evaluate the ECDF at $x = 33$ min. Based on (2.11), we calculate $H(X \le 33) = F(33) = F(30) + f(6)/5(33 - 30) = 0.2480 + 0.2946/5 \cdot 3 = 0.42476$. Thus we conclude, based on the grouped data, that only about 42% of the deliveries were completed in the desired time frame.

2.3 Graphical Representation of a Variable

Frequency tables and empirical cumulative distribution functions are useful in providing a numerical summary of a variable. Graphs are an alternative way to summarize a variable's information. In many situations, they have the advantage of conveying the information hidden in the data more compactly. Similarly, someone's mood can be more easily understood when looking at a smiley ☺ than by reading an essay about one's mood in a long paragraph.

2.3.1 Bar Chart

A simple tool to visualize the relative or absolute frequencies of observed values of a variable is a **bar chart**. A bar chart can be used for nominal and ordinal variables, as long as the number of categories is not very large. It consists of one bar for each category. The height of each bar is determined by either the absolute frequency or the relative frequency of the respective category and is shown on the y-axis. If the variable is measured on an ordinal level, then it is recommended to arrange the bars on the x-axis according to their ranks or values. If the number of categories is large then the number of bars will be large too and the bar chart, in turn, may not remain informative.

Example 2.3.1 Consider Example 2.1.1 in which ten people, queueing in a supermarket, were classified as being either male (M) or female (F). The absolute frequencies for males and females are $n_1 = 7$ and $n_2 = 3$, respectively. Since there are two categories, M and F, two bars are needed to construct the chart—one for the male category and another for the female category. The heights of the bars are determined as either $n_1 = 7$ and $n_2 = 3$ or $f_1 = 0.7$ and $f_2 = 0.3$. These graphs are shown in Fig. 2.4. Since the f_j are simply scaled versions of the n_j, the two bar charts look identical, only the scale on the y-axis differs between them.

Example 2.3.2 Consider the data in Example 2.1.3 where the pizza delivery times for each branch are recorded over a period of 1 month. The frequency table forms the basis for the bar chart, either using the absolute or relative frequencies on the y-axis. Figure 2.5 shows the bar charts for the number and proportion of pizza deliveries per branch. The graphs can be produced in R by applying the `barplot` command to a frequency table:

```
barplot(table(branch))
barplot(table(branch)/length(branch))
```

Remark 2.3.1 Instead of vertical bars, horizontal bars can be drawn using the optional argument `horiz=TRUE` in the `barplot` command.

(a) With absolute frequencies   (b) With relative frequencies

Fig. 2.4 Bar charts

(a) With absolute frequencies   (b) With relative frequencies

Fig. 2.5 Bar charts for the pizza deliveries per branch

2.3.2   Pie Chart

Pie charts are another option to visualize the absolute and relative frequencies of nominal and ordinal variables. A pie chart is a circle partitioned into segments where each of the segments represents a category. The size of each segment depends upon the relative frequency and is determined by the angle $f_j \cdot 360°$.

Example 2.3.3 To illustrate the construction of a pie chart, let us consider again Example 2.1.1 in which ten people in a supermarket queue were classified as being

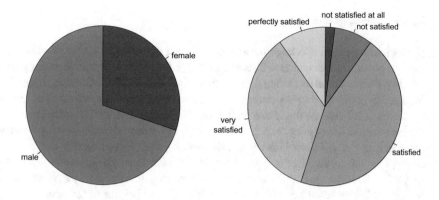

(a) For gender of people queueing   (b) For satisfaction with the car service

Fig. 2.6 Pie charts

either male (M) or female (F): M, F, M, F, M, M, M, F, M, M. The pie chart for this data will have two segments: one for males and another one for females. The relative frequencies are $f_1 = 7/10$ and $f_2 = 3/10$, respectively. The size of the segment for the first category (M) is $f_1 \cdot 360° = (7/10) \cdot 360° = 252°$ and the size of the segment for the second category (F) is $f_2 \cdot 360° = (3/10) \cdot 360° = 108°$. The pie chart is shown in Fig. 2.6a.

Example 2.3.4 Consider again Example 2.2.1 where 200 customers were asked about their level of satisfaction (5 categories) with their car service. The pie chart for this example consists of five segments representing the categories 1, 2, 3, 4 and 5. The size of the jth segment is $f_j \cdot 360°$, $j = 1, 2, 3, 4, 5$. For example, for category 1, there are 4 out of 200 customers which are not satisfied at all. The angle of the segment "not satisfied at all" therefore is $f_1 \cdot 360° = 4/200 \cdot 360° = 7.2°$. Similarly, we can calculate the angle of the other segments and obtain a pie chart as shown in Fig. 2.6b using the `pie` command in R.

```
pie(table(sv))
```

Remark Note that the area of a segment is *not* proportional to the absolute frequency of the respective category. Instead, the area of the segment is proportional to the angle $f_j \cdot 360°$ (and depends also on the radius of the whole circle). It has been argued that this may cause improper interpretations as the human eye may catch the segment's area more easily than the angle of a segment. Pie charts should therefore be used with caution.

2.3.3 Histogram

If a variable consists of a large number of different values, the number of categories used to construct bar charts will consequently be large too. A bar chart may thus not give a clear summary when applied to an ungrouped metric variable. Instead, a **histogram** is the appropriate choice to represent the distribution of values of metric variables. It is based on the idea to categorize the data into different groups and plot the bars for each category with height $h_j = f_j/d_j$ where $d_j = e_j - e_{j-1}$ denotes the width of the jth class interval or category. An important consideration for this concept is that the area of the bars (= height $\times$ width) is proportional to the relative frequency. This means that the widths of the bars need not necessarily to be the same because different widths can be adjusted with different heights of the bars.

Example 2.3.5 Consider Example 2.1.2 where the performance in a written driving licence examination was recorded for $n = 20$ people. The results were: 28, 35, 42, 90, 70, 56, 75, 66, 30, 89, 75, 64, 81, 69, 55, 83, 72, 68, 73, 16. Suppose we are interested in drawing a histogram for these data using 5 equally spaced class intervals. We then obtain:

Class intervals	(0, 20]	(20, 40]	(40, 60]	(60, 80]	(80, 100]
Absolute freq.	$n_1 = 1$	$n_2 = 3$	$n_3 = 3$	$n_4 = 9$	$n_5 = 4$
Relative freq.	$f_1 = \frac{1}{20}$	$f_2 = \frac{3}{20}$	$f_3 = \frac{3}{20}$	$f_4 = \frac{9}{20}$	$f_5 = \frac{5}{20}$
Height f_j/d_j	$h_1 = \frac{1}{400}$	$h_2 = \frac{3}{400}$	$h_3 = \frac{3}{400}$	$h_4 = \frac{9}{400}$	$h_5 = \frac{4}{400}$

Since the widths of class intervals are the same, the heights of the bars are proportional to the relative frequency of the respective category. The resulting histogram is displayed in Fig. 2.7.

Example 2.3.6 Recall Example 2.2.3 and the variable "pizza delivery time". Table 2.3 shows the summary of the grouped data and the values needed to calculate the histogram. Figure 2.8a shows the histogram with equal widths of delivery time intervals. We see a symmetric distribution of the pizza delivery times, but many delivery times exceeding the target time of 30 min. If the histogram is required to have different widths for different bars, i.e. different delivery time intervals for different categories, then it can also be constructed as shown in Fig. 2.8b. This representation is different from Fig. 2.8a. The following commands in R are used to construct histograms based on absolute and relative frequencies respectively:

```
hist(time)              # based on abs. frequencies
hist(time, freq=F)      # based on rel. frequencies
```

Fig. 2.7 Histogram for the scores of the people

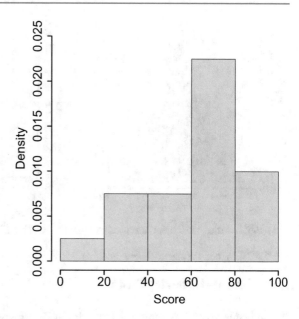

Table 2.3 Values needed to calculate the histogram for the grouped pizza delivery time data

Delivery time	j	e_{j-1}	e_j	d_j	f_j	h_j
[0; 10]	1	0	10	10	0.0000	0.00000
(10; 15]	2	10	15	5	0.0024	0.00047
(15; 20]	3	15	20	5	0.0166	0.00332
(20; 25]	4	20	25	5	0.0592	0.01185
(25; 30]	5	25	30	5	0.1698	0.03397
(30; 35]	6	30	35	5	0.2946	0.05893
(35; 40]	7	35	40	5	0.2765	0.05529
(40; 45]	8	40	45	5	0.1351	0.02701
(45; 50]	9	45	50	5	0.0411	0.00821
(50; 55]	10	50	55	5	0.0047	0.00094

Remark The R command `truehist()` from the library `MASS` presents an alternative to the `hist()` command. The default specifications are somewhat different and many users prefer it to the command `hist`.

(a) With same width for each category (b) With different widths per category

Fig. 2.8 Histogram for pizza delivery time

2.4   Kernel Density Plots

A disadvantage of histograms is that metric data is categorized artificially. The choice of the class intervals is crucial for the final look of the graph. A more elegant way to deal with this problem is to smooth the histogram in the sense that each observation may contribute to different classes with different weights and the distribution is represented by a continuous function rather than a step function. A **kernel density plot** can be produced by using the following function:

$$\hat{f}_n(x) = \frac{1}{nh} \sum_{i=1}^{n} K\left(\frac{x - x_i}{h}\right), \quad h > 0, \tag{2.12}$$

where n is the sample size, h is the bandwidth and K is a kernel function, for example

$$K(x) = \begin{cases} \frac{1}{2} & \text{if } -1 \le x \le 1 \\ 0 & \text{elsewhere} \end{cases} \quad \text{(rectangular kernel)}$$

$$K(x) = \begin{cases} \frac{3}{4}(1 - x^2) & \text{if } |x| < 1 \\ 0 & \text{else} \end{cases} \quad \text{(Epanechnikov kernel)}$$

To better understand this concept, consider Fig. 2.9a. The tick marks on the x-axis represent 5 observations: 3, 6, 7, 8, 10. On each observation x_i as well as its surrounding values, we apply a kernel function which is the Epanechnikov kernel in the figure. This means that we have 5 functions (grey, dashed lines) which refer to the five observations. These functions are largest at the observation itself and become gradually smaller as the distance from the observation increases. Summing up the functions, as described in Eq. (2.12), yields the solid black line which is the kernel density plot of the 5 observations. It is a smooth curve which represents the data distribution. The degree of smoothness can be controlled by the bandwidth h which is chosen as 2 in Fig. 2.9a.

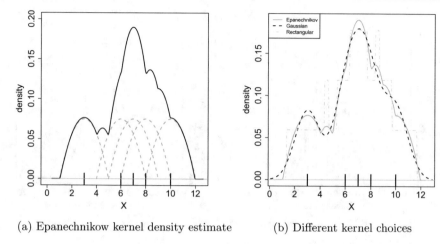

(a) Epanechnikow kernel density estimate   (b) Different kernel choices

Fig. 2.9 Construction of kernel density plots

The choice of the kernel may affect the overall look of the plot. Above, we have given the functions for the rectangular and Epanechnikov kernel. However, another common function for kernel density plots is the normal distribution function which is introduced in Sect. 8.2.2, see Fig. 2.9b for a comparison of different kernels. The kernel which is based on the normal distribution is called the "Gaussian kernel" and is the default in *R* where a kernel density plot can be produced combining the `plot` and `density` commands:

```
example <- c(3,6,7,8,10)
plot(density(example, kernel='gaussian'))
```

Please note that kernel functions are not defined arbitrarily and need to satisfy certain conditions, such as those required for probability density functions as explained in Chap. 7, Theorem 7.2.1.

Example 2.4.1 Let us consider the pizza data which we introduced earlier and in Appendix A.4.1. We can summarize the delivery time by using a kernel density plot using the *R* command `plot(density(time))` and compare it with a histogram, see Fig. 2.10a. We see that the delivery times are symmetric around 35 min.

If we shorten the bandwidth to a half of the default bandwidth (option `adjust=0.5`) the kernel density plot becomes more wiggly, which is illustrated in Fig. 2.10b.

(a) Default bandwidth   (b) Shorter bandwidth

Fig. 2.10 Kernel density plot for delivery time

2.5   Key Points and Further Issues

> **Note:**
>
> ✓ Bar charts and histograms are not the same graphical tools. Bar charts visualize the categories of nominal or ordinal variables whereas histograms visualize the distribution of metric variables. A bar chart does not require to have ordered values on the x-axis but a histogram always requires the values on the x-axis to be on a metric scale and to be ordered. The interpretation of a histogram is simplified if the class intervals are equally sized, since then the heights of the rectangles of the histogram are proportional to the absolute or relative frequencies.
>
> ✓ The ECDF can be used only for ordinal and metric variables, see Sect. 7.2 for the theoretical background of the cumulative distribution function.
>
> ✓ A pie chart summarizes observations from a discrete (nominal, ordinal or grouped metric) variable. It is only useful if the number of different values (categories) is small. It is to be kept in mind that the area of each segment is not proportional to the relative frequency of the respective category. The angle of the segment is proportional to the relative frequency of the respective category.
>
> ✓ Other possibilities to visualize the distribution of variables are, for example, box plots (Sect. 3.3) and stratified plots (Sects. 4.1.3, 4.3.1, and 4.4).

2.6   Exercises

Exercise 2.1 Consider the results of the national elections in South Africa in 2014 and 2009:

Party		Results 2014	Results 2009
ANC	(African National Congress)	62.15%	65.90%
DA	(Democratic Alliance)	22.23%	16.66%
EFF	(Economic Freedom Fighters)	6.35%	–
IFP	(Inkatha Freedom Party)	2.40%	4.55%
COPE	(Congress of the People)	0.67%	7.42%
Others		6.20%	5.47%

(a) Summarize the results of the 2014 elections in a bar chart. Do it manually and by using R.
(b) How would you compare the results of the 2009 and 2014 elections? Offer a simple solution that can be represented in a single plot. Construct this plot in R.

Exercise 2.2 Consider a variable X describing the time until the first goal was scored in the matches of the 2006 football world cup competition. Only matches with at least one goal are considered and goals during the xth minute of extra time are denoted as $90 + x$:

6	24	90+1	8	4	25	3	83	89	34	25	24	18	6
23	10	28	4	63	6	60	5	40	2	22	26	23	26
44	49	34	2	33	9	16	55	23	13	23	4	8	26
70	4	6	60	23	90+5	28	49	6	57	33	56	7	

(a) What is the scale of X?
(b) Write down the frequency table of X based on the following categories: $[0, 15)$, $[15, 30)$, $[30, 45)$, $[45, 60)$, $[60, 75)$, $[75, 90)$, $[90, 96)$.
(c) Draw the histogram for X with intervals relating to the groups from the frequency table.
(d) Now use R to reproduce the histogram. Compare the histogram to a kernel density plot of your choice.
(e) Calculate the empirical cumulative distribution function for the grouped data.
(f) Use R to plot the ECDF (via a step function) for

(i) the original data and
(ii) the grouped data.

(g) Consider the grouped data. Now assume that the values within each interval are distributed uniformly. Determine the proportion of first goals which occurred

  (i) in the first half, i.e. during the first 45 min,
 (ii) in the last ten minutes or during the extra time,
(iii) between the 20th and 65th minute, i.e. what is $H(20 \leq X \leq 65)$?

(h) Determine the time point at which in 80% of the matches the first goal was scored
    at or before this time point.

Exercise 2.3 Suppose we have the following information to construct a histogram
for a metric variable with 2000 observations:

j	e_{j-1}	e_j	d_j	h_j
1	0	1	1	0.125
2	1	4	3	0.125
3	4	7	3	0.125
4	7	8	1	0.125

(a) Determine the relative frequencies for each interval.
(b) Determine the absolute frequencies.

Exercise 2.4 A university survey was conducted on 500 first year students to obtain
knowledge about the size of their accommodation (in square meters).

j	size of accommodation (m^2) $e_{j-1} \leq x \leq e_j$	$F(x)$
1	8 – 14	0.25
2	14 – 22	0.40
3	22 – 34	0.75
4	34 – 50	0.97
5	50 – 82	1.00

(a) Determine the absolute frequencies for each category.
(b) What proportion of people live in a flat of at least 34 m^2?

Exercise 2.5 Consider a survey in which 100 people were asked to rate on a scale
from 1 to 10 how much they agree with the statement that "there is too much football
on television". The results are summarized below:

Score	0	1	2	3	4	5	6	7	8	9	10
Responses	0	1	3	8	8	27	30	11	6	4	2

(a) Calculate and draw the ECDF of the scores.
(b) Determine $F(3)$ and $F(9)$.

(c) Consider the situation where the data are summarized in the two categories "disagree" (score ≤ 5) and "agree" (score >5). What would the ECDF look like under the approach outlined in (2.11)? Determine $F(3)$ and $F(9)$ for the summarized data.

(d) Explain the differences between (b) and (c).

Exercise 2.6 It is possible to produce professional graphics in R. However, it is advantageous to go beyond the default options. To demonstrate this, consider Example 2.1.3 about the pizza delivery data which is described in Appendix A.4.1.

(a) Set the working directory in R (`setwd()`), read in the data (`read.csv()`), and attach the data. Draw a histogram of the variable "temperature". Type `?hist` and view the options. Adjust the histogram so that you are satisfied with (i) axes labelling, (ii) axes range and (iii) colour. Now use the `lines()` command to add a dashed vertical line at 65°C (which is the minimum temperature the pizza should have at the time of delivery).

(b) Consider a different approach, which constructs plots by means of multiple layers using `ggplot2`. You need an internet connection to install the package using the command `install.packages('ggplot2')`. Browse through the help pages on https://ggplot2.tidyverse.org/reference/. Look specifically at the examples for `ggplot`, `qplot`, `scale_histogram` and `scale_y_continuous`. Try to understand the roles of "aesthetics" and "geoms". Now, after loading the library via `library(ggplot2)`, create a ggplot object for the pizza data which declares "temperature" to be the x-variable. Now add a layer with `geom_histogram` to create a histogram with interval width of 2.5 and dark grey bars which are 50% transparent. Change the y-axis labelling by adding the relevant layer using `scale_y_continuous`. Plot the graph.

(c) Now create a normal bar chart for the variable "driver" in R. Type `?barplot` and `?par` to see the options one can pass on to `barchart()` to adjust the graph. Make the graph look good.

(d) Now create the same bar chart with `ggplot2`. Use `qplot` instead of `ggplot` to create the plot. Use an option which makes each bar to consist of segments relating to the day of delivery, so that one can see the number of deliveries by driver to highlight during which days the drivers delivered most often. Browse through "themes" and "scales" on the help page and add layers that make the background black and white and the bars on a grey scale.

$\rightarrow$ Solutions to all exercises in this chapter can be found in Appendix B

Measures of Central Tendency and Dispersion

3

A data set may contain many variables and observations. However, we are not always interested in each of the measured values but rather in a few summary measures of the data. Statistical functions fulfill the purpose of summarizing the data in a meaningful yet concise way.

Example 3.0.1 Suppose someone from Munich (Germany) plans a holiday in Bangkok (Thailand) during the month of December and would like to get information about the weather when preparing for the trip. Suppose last year's maximum temperatures during the day (in degrees Celsius) for December 1–31 are as follows:

$$22, 24, 21, 22, 25, 26, 25, 24, 23, 25, 25, 26, 27, 25, 26,$$
$$25, 26, 27, 27, 28, 29, 29, 29, 28, 30, 29, 30, 31, 30, 28, 29.$$

How do we draw conclusions from this data? Looking at the individual values gives us a feeling about the temperatures one can experience in Bangkok, but it does not provide us with a clear summary. It is evident that the average of these 31 values as "Sum of all values/Total number of observations" $(22 + 24 + \ldots + 28 + 29)/31 = 26.48$ is meaningful in the sense that we know what temperature to expect "on average". To choose the right clothing for the holidays, we may also be interested in knowing the temperature range to understand the variability in temperature, which is between 21 and 31 degrees Celsius. Summarizing 31 individual values with only three numbers (26.48, 21, 31) will provide sufficient information to plan the holidays.

© Springer Nature Switzerland AG 2022
C. Heumann et al., *Introduction to Statistics and Data Analysis*,
https://doi.org/10.1007/978-3-031-11833-3_3

In this chapter we focus on the most important statistical concepts to summarize data: these are measures of central tendency and variability. The applications of each measure depend on the scale of the variable of interest, see Appendix D.1 for a detailed summary.

3.1   Measures of Central Tendency

A natural human tendency is to make comparisons with the "average". For example, a student scoring 40% in an examination will be happy with the result if the average score of the class is 25%. If the average class score is 90%, then the student may not feel happy even if he got 70% right. Some other examples of the use of "average" values in common life are mean body height, mean temperature in July in some town, the most often selected study subject, the most popular TV show in 2015, average income, etc. Various statistical concepts refer to the "average" of the data, but the right choice depends upon the nature and scale of the data as well as the objective of the study. We call statistical functions which describe the average or centre of the data **location parameters** or **measures of central tendency**.

3.1.1   Arithmetic Mean

The **arithmetic mean** is one of the most intuitive measures of central tendency. Suppose a variable of size n consists of the values $x_1, x_2, \ldots, x_n$. The arithmetic mean of this data is defined as

$$\bar{x} = \frac{1}{n} \sum_{i=1}^{n} x_i \,.$$

(3.1)

In informal language, we often speak of "the average" or just "the mean" when using the formula (3.1).

To calculate the arithmetic mean for grouped data we need the following frequency table:

Class intervals a_j	$a_1 = e_0 - e_1$	$a_2 = e_1 - e_2$	...	$a_k = e_{k-1} - e_k$
mid-value m_j	$(e_0 + e_1)/2$	$(e_1 + e_2)/2$	...	$(e_{k-1} + e_k)/2$
Absolute freq. n_j	n_1	n_2	...	n_k
Relative freq. f_j	f_1	f_2	...	f_k

Note that $a_1, a_2, \ldots, a_k$ are the k class intervals and each interval $a_j (j = 1, 2, \ldots, k)$ contains n_j observations with $\sum_{j=1}^{k} n_j = n$. The relative frequency of the jth class is $f_j = n_j/n$ and $\sum_{j=1}^{k} f_j = 1$. The mid-value of the jth class interval is defined as $m_j = (e_{j-1} + e_j)/2$ which is the mean of the lower and upper limits of the interval. The **weighted arithmetic mean** for grouped data is defined as

$$\bar{x} = \frac{1}{n} \sum_{j=1}^{k} n_j m_j = \sum_{j=1}^{k} f_j m_j. \tag{3.2}$$

Example 3.1.1 Consider again Example 3.0.1 where we looked at the temperature in Bangkok during December. The measurements were

$$22, 24, 21, 22, 25, 26, 25, 24, 23, 25, 25, 26, 27, 25, 26,$$
$$25, 26, 27, 27, 28, 29, 29, 29, 28, 30, 29, 30, 31, 30, 28, 29.$$

The arithmetic mean is therefore

$$\bar{x} = \frac{22 + 24 + 21 + \ldots + 28 + 29}{31} = 26.48\,°C.$$

In R the arithmetic mean can be calculated using the mean command:

```
weather <- c(22,24,21,...,30,28,29)
mean(weather)
[1] 26.48387
```

Let us assume the data in Example 3.0.1 is summarized in categories as follows:

Class intervals	< 20	(20 − 25]	(25, 30]	(30, 35]	> 35
Absolute frequencies	$n_1 = 0$	$n_2 = 12$	$n_3 = 18$	$n_4 = 1$	$n_5 = 0$
Relative frequencies	$f_1 = 0$	$f_2 = \frac{12}{31}$	$f_3 = \frac{18}{31}$	$f_4 = \frac{1}{31}$	$f_5 = 0$
mid values	−	22.5	27.5	32.5	−

We can calculate the (weighted) arithmetic mean as:

$$\bar{x} = \sum_{j=1}^{k} f_j m_j = 0 + \frac{12}{31} \cdot 22.5 + \frac{18}{31} \cdot 27.5 + \frac{1}{31}32.5 + 0 \approx 25.7.$$

In R, we use the weighted.mean function to obtain the result. The function requires to specify the (hypothesized) means for each group, for example the middle values of the class intervals, as well as the weights.

```
weighted.mean(c(22.5,27.5,32.5),c(12/31,18/31,1/31))
```

Interestingly, the results of the mean and the weighted mean differ. This is because we use the middle of each class as an approximation of the mean within the class. The implication is that we assume that the values are uniformly distributed within each interval. This assumption is obviously not met. If we had knowledge about the mean in each class, like in this example, we would obtain the correct result as follows:

$$\bar{x} = \sum_{j=1}^{k} f_j \bar{x}_j = 0 + \frac{12}{31} \cdot 23.83333 + \frac{18}{31} \cdot 28 + \frac{1}{31} 32.5 + 0 = 26.48387 \,.$$

However, the weighted mean is meant to estimate the arithmetic mean in those situations where only grouped data is available. It is therefore typically used to obtain an approximation of the true mean.

3.1.1.1 Properties of the Arithmetic Mean

(i) The sum of the deviations of each variable around the arithmetic mean is zero:

$$\sum_{i=1}^{n}(x_i - \bar{x}) = \sum_{i=1}^{n} x_i - n\bar{x} = n\bar{x} - n\bar{x} = 0 \,. \tag{3.3}$$

(ii) If the data is linearly transformed as $y_i = a + bx_i$, where a and b are known constants, it holds that

$$\bar{y} = \frac{1}{n} \sum_{i=1}^{n} y_i = \frac{1}{n} \sum_{i=1}^{n}(a + bx_i) = \frac{1}{n} \sum_{i=1}^{n} a + \frac{b}{n} \sum_{i=1}^{n} x_i = a + b\bar{x} \,. \tag{3.4}$$

Example 3.1.2 Recall Examples 3.0.1 and 3.1.1 where we considered the temperatures in December in Bangkok. We measured them in degrees Celsius, but someone from the United States might prefer to know them in degrees Fahrenheit. With a linear transformation, we can create a new temperature variable as

$$\text{Temperature in } °F = 32 + 1.8 \cdot \text{Temperature in } °C \,.$$

Using $\bar{y} = a + b\bar{x}$ we get $\bar{y} = 32 + 1.8 \cdot 26.48 \approx 79.7°F$.

3.1.2   Median and Quantiles

The median is the value which divides the observations into two equal parts such that at least 50% of the values are greater than or equal to the median and at least 50% of the values are less than or equal to the median. The median is denoted by $\tilde{x}_{0.5}$; then, in terms of the empirical cumulative distribution function, the condition $F(\tilde{x}_{0.5}) = 0.5$ is satisfied. Consider the n observations $x_1, x_2, \ldots, x_n$ which can be ordered as $x_{(1)} \le x_{(2)} \le \ldots \le x_{(n)}$. The calculation of the median depends on whether the number of observations n is odd or even. When n is odd, then $\tilde{x}_{0.5}$ is the middle ordered value. When n is even, then $\tilde{x}_{0.5}$ is the arithmetic mean of the two middle ordered values:

$$\tilde{x}_{0.5} = \begin{cases} x_{((n+1)/2)} & \text{if } n \text{ is odd} \\ \frac{1}{2}(x_{(n/2)} + x_{(n/2+1)}) & \text{if } n \text{ is even.} \end{cases} \tag{3.5}$$

Example 3.1.3 Consider again Examples 3.0.1–3.1.2 where we evaluated the temperature in Bangkok in December. The ordered values $x_{(i)}$, $i = 1, 2, \ldots, 31$, are as follows:

°C	21	22	22	23	24	24	25	25	25	25	25	25	26	26	26	26
(i)	1	2	3	4	5	6	7	8	9	10	11	12	13	14	15	16
°C	27	27	27	28	28	28	29	29	29	29	29	30	30	30	31	
(i)	17	18	19	20	21	22	23	24	25	26	27	28	29	30	31	

We have $n = 31$, and therefore $\tilde{x}_{0.5} = x_{((n+1)/2)} = x_{((31+1)/2)} = x_{(16)} = 26$. Therefore at least 50% of the 31 observations are greater than or equal to 26 and at least 50% are less than or equal to 26. If one value was missing, let us say the last observation, then the median would be calculated as $\frac{1}{2}(x_{(30/2)} + x_{(30/2+1)}) = \frac{1}{2}(26 + 26) = 26$. In R we would have obtained the results using the median command:

```
median(weather)
```

If we deal with grouped data we can calculate the median under the assumption that the values within each class are equally distributed. Let $K_1, K_2, \ldots, K_k$ be k classes with observations of size $n_1, n_2, \ldots, n_k$ respectively. First we need to determine which class is the median class, i.e. the class that includes the median. We define the median class as the class K_m for which

$$\sum_{j=1}^{m-1} f_j < 0.5 \quad \text{and} \quad \sum_{j=1}^{m} f_j \geq 0.5 \tag{3.6}$$

hold. Then we can determine the median as

$$\tilde{x}_{0.5} = e_{m-1} + \frac{0.5 - \sum_{j=1}^{m-1} f_j}{f_m} d_m \tag{3.7}$$

where e_{m-1} denotes the lower limit of the interval K_m and d_m is the width of the interval K_m.

Example 3.1.4 Recall Example 3.1.1 where we looked at the grouped temperature data:

Class intervals	< 20	(20 − 25]	(25, 30]	(30, 35]	> 35
n_j	$n_1 = 0$	$n_2 = 12$	$n_3 = 18$	$n_4 = 1$	$n_5 = 0$
f_j	$f_1 = 0$	$f_2 = \frac{12}{31}$	$f_3 = \frac{18}{31}$	$f_4 = \frac{1}{31}$	$f_5 = 0$
$\sum_j f_j$	0	$\frac{12}{31}$	$\frac{30}{31}$	1	1

For the third class ($m = 3$), we have

$$\sum_{j=1}^{m-1} f_j = \frac{12}{31} < 0.5 \quad \text{and} \quad \sum_{j=1}^{m} f_j = \frac{30}{31} \geq 0.5 .$$

We can therefore calculate the median as

$$\tilde{x}_{0.5} = e_{m-1} + \frac{0.5 - \sum_{j=1}^{m-1} f_j}{f_m} d_m = 25 + \frac{0.5 - \frac{12}{31}}{\frac{18}{31}} \cdot 5 \approx 25.97.$$

3.1.2.1 Comparing the Mean with the Median

In the above examples, the mean and the median turn out to be quite similar to each other. This is because we looked at data which is symmetrically distributed around its centre, i.e. on average we can expect 26 degrees Celsius with deviations that are similar above and below the average temperature. A similar example is given in Fig. 3.1a: we see that the raw data is summarized by using ticks at the bottom of the graph and by using a kernel density estimator. The mean and the median are similar here because the distribution of the observations is symmetric around the centre. If we have skewed data (Fig. 3.1b), then the mean and the median may differ. If the data has more than one centre, such as in Fig. 3.1c, neither the median nor the mean have meaningful interpretations. If we have outliers (Fig. 3.1d), then it is wise to use the median because the mean is sensitive to outliers. These examples show that depending on the situation of interest either the mean, the median, both or neither of them can be useful.

3.1.2.2 Quantiles

Quantiles are a generalization of the idea of the median. The median is the value which splits the data into two equal parts. Similarly, a quantile partitions the data into other proportions. For example, a 25%-quantile splits the data into two parts such that at least 25% of the values are less than than or equal to the quantile and at least 75% of the values are greater than or equal to the quantile. In general, let α be a number between zero and one. The $(\alpha \times 100)$%-quantile, denoted as $\tilde{x}_\alpha$, is defined as the value which divides the data in proportions of $(\alpha \times 100)$% and $(1 - \alpha) \times 100$% such that at least $\alpha \times 100$% of the values are less than or equal to

Fig. 3.1 Arithmetic mean and median for different data

the quantile and at least $(1 - \alpha) \times 100\%$ of the values are greater than or equal to the quantile. In terms of the empirical cumulative distribution function we can write $F(\tilde{x}_\alpha) = \alpha$. It follows immediately that for n observations, at least $n\alpha$ values are less than or equal to $\tilde{x}_\alpha$ and at least $n(1 - \alpha)$ observations are greater than or equal to $\tilde{x}_\alpha$. The median is the 50%-quantile $\tilde{x}_{0.5}$. If α takes the values $0.1, 0.2, \ldots, 0.9$, the quantiles are called **deciles**. If $\alpha \cdot 100$ is an integer number (e.g. $\alpha \times 100 = 95$), the quantiles are called **percentiles**, i.e. the data is divided into 100 equal parts. If α takes the values $0.2, 0.4, 0.6, 0.8$, the quantiles are known as **quintiles** and they divide the data into five equal parts. If α takes the values $0.25, 0.5, 0.75$, the quantiles are called **quartiles**.

Consider n ordered observations $x_{(1)} \leq x_{(2)} \leq \ldots \leq x_{(n)}$. The $\alpha \cdot 100\%$-quantile $\tilde{x}_\alpha$ is calculated as

$$\tilde{x}_\alpha = \begin{cases} x_{(k)} & \text{if } n\alpha \text{ is not an integer number,} \\ & \text{choose } k \text{ as the smallest integer } > n\alpha, \\ \frac{1}{2}(x_{(n\alpha)} + x_{(n\alpha+1)}) & \text{if } n\alpha \text{ is an integer.} \end{cases} \qquad (3.8)$$

Example 3.1.5 Recall Examples 3.0.1–3.1.4 where we evaluated the temperature in Bangkok in December. The ordered values $x_{(i)}$, $i = 1, 2, \ldots, 31$ are as follows:

°C	21	22	22	23	24	24	25	25	25	25	25	25	26	26	26	26
(i)	1	2	3	4	5	6	7	8	9	10	11	12	13	14	15	16
°C	27	27	27	28	28	28	29	29	29	29	29	30	30	30	31	
(i)	17	18	19	20	21	22	23	24	25	26	27	28	29	30	31	

To determine the quartiles, i.e. the 25%, 50%, and 75% quantiles, we calculate $n\alpha$ as $31 \cdot 0.25 = 7.75$, $31 \cdot 0.5 = 15.5$ and $31 \cdot 0.75 = 23.25$. Using (3.8) it follows that

$$\tilde{x}_{0.25} = x_{(8)} = 25\,, \qquad \tilde{x}_{0.5} = x_{(16)} = 26\,,$$
$$\tilde{x}_{0.75} = x_{(24)} = 29\,.$$

In R we obtain the same results using the `quantile` function. The `probs` argument is used to specify α. By default, the quartiles are reported.

```
quantile(weather)
quantile(weather, probs=c(0,0.25,0.5,0.75,1))
```

However, please note that R offers nine different ways to obtain quantiles, each of which can be chosen by the `type` argument. See Hyndman and Fan (1996) for more details.

3.1.3   Quantile–Quantile Plots (QQ-Plots)

If we plot the quantiles of two variables against each other, we obtain a Quantile–Quantile plot (QQ-plot). This provides a simple summary of whether the distributions of the two variables are similar with respect to their location or not.

Example 3.1.6 Consider again the pizza data which is described in Appendix A.4.1. We may be interested in the delivery time for different drivers to see if their performance is the same. Figure 3.2a shows a QQ-plot for the delivery time of driver Luigi and the delivery time of driver Domenico. Each point refers to the $\alpha\%$ quantile of both drivers. If the point lies on the bisection line, then they are identical and we

(a) QQ-plot for Luigi and Domenico   (b) QQ-plot for Mario and Salvatore

Fig. 3.2 QQ-plots for the pizza delivery time for different drivers

conclude that the quantiles of the both drivers are the same. If the point is below the line, then the quantile is higher for Luigi, and if the point is above the line, then the quantile is lower for Luigi. So if all the points lie exactly on the line, we can conclude that the distributions of both the drivers are the same. We see that all the reported quantiles lie below the line which implies that all the quantiles of Luigi have higher values than those of Domenico. This means that not only on an average, but also in general, the delivery times are higher for Luigi. If we look at two other drivers, as displayed in Fig. 3.2b, the points lie very much on the bisection line. We can therefore conclude that the delivery times of these two drivers do not differ much.

In R we can generate QQ-plots by using the qqplot command:

```
qqplot()
```

As a summary, let us consider four important patterns:

(a) If all the pairs of quantiles lie (nearly) on a straight line at an angle of 45% from the x-axis, then the two samples have similar distributions (Fig. 3.3a).
(b) If the y-quantiles are lower than the x-quantiles, then the y-values have a tendency to be lower than the x-values (Fig. 3.3b).
(c) If the x-quantiles are lower than the y-quantiles, then the x-values have a tendency to be lower than the y-values (Fig. 3.3c).
(d) If the QQ-plot is like Fig. 3.3d, it indicates that there is a breakpoint up to which the y-quantiles are lower than the x-quantiles and after that point, the y-quantiles are higher than the x-quantiles.

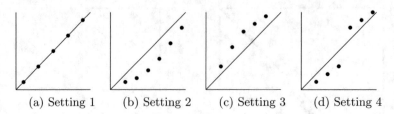

(a) Setting 1   (b) Setting 2   (c) Setting 3   (d) Setting 4

Fig. 3.3 Different patterns for a QQ-plot

3.1.4   Mode

Consider a situation in which an ice cream shop owner wants to know which flavour of ice cream is the most popular among his customers. Similarly, a footwear shop owner like to find out what design and size of shoes are in highest demand. To answer this type of questions, one can use the mode which is another measure of central tendency.

The mode $\bar{x}_M$ of n observations $x_1, x_2, \ldots, x_n$ is the value which occurs the most compared with all other values, i.e. the value which has maximum absolute frequency. It may happen that two or more values occur with the same frequency in which case the mode is not uniquely defined. A formal definition of the mode is

$$\bar{x}_M = a_j \Leftrightarrow n_j = \max\{n_1, n_2, \ldots, n_k\}. \tag{3.9}$$

The mode is typically applied to any type of variable for which the number of different values is not too large. If metric data is summarized in groups, then the mode can be used as well.

Example 3.1.7 Recall the pizza data set described in Appendix A.4.1. The pizza delivery service has three branches, in the East, West and Centre respectively. Suppose we want to know which branch delivers the most pizzas. We find that most of the deliveries have been made in the West, see Fig. 3.4a; therefore the mode is $\bar{x}_M =$ West. Similarly, suppose we also want to find the mode for the categorized pizza delivery time: if we group the delivery time in intervals of 5 min, then we see that the most frequent delivery time is the interval "30−35" min, see Fig. 3.4b. The mode is therefore $\bar{x}_M = [30, 35)$.

3.1.5   Geometric Mean

Consider n observations $x_1, x_2, \ldots, x_n$ which are all positive and collected on a quantitative variable. The geometric mean $\bar{x}_G$ of this data is defined as

$$\bar{x}_G = \sqrt[n]{\prod_{i=1}^{n} x_i} = \left(\prod_{i=1}^{n} x_i\right)^{\frac{1}{n}}. \tag{3.10}$$

(a) Bar chart for branch   (b) Histogram for grouped delivery time

Fig. 3.4 Results from the pizza data set

The geometric mean plays an important role in fields where we are interested in products of observations, such as when we look at percentage changes in quantities. We illustrate its interpretation and use by looking at the average growth of a quantity in the sense that we allow a starting value, such as a certain amount of money or a particular population, to change over time. Suppose we have a starting value at some baseline time point 0 (zero), which may be denoted as B_0. At time t, this value may have changed and we therefore denote it as $B_t, t = 1, 2, \ldots, T$. The ratio of B_t and B_{t-1},

$$x_t = \frac{B_t}{B_{t-1}},$$

is called the tth growth factor. The growth rate r_t is defined as

$$r_t = ((x_t - 1) \cdot 100)\,\%$$

and gives us an idea about the growth or decline of our value at time t. We can summarize these concepts in the following table:

Time	Inventory	Growth factor	Growth rate
t	B_t	x_t	r_t
0	B_0	–	–
1	B_1	$x_1 = B_1/B_0$	$((x_1 - 1) \cdot 100)\%$
2	B_2	$x_2 = B_2/B_1$	$((x_2 - 1) \cdot 100)\%$
$\vdots$	$\vdots$	$\vdots$	$\vdots$
T	B_T	$x_T = B_T/B_{T-1}$	$((x_T - 1) \cdot 100)\%$

We can calculate B_t $(t = 1, 2, \ldots, T)$ by using the growth factors:

$$B_t = B_0 \cdot x_1 \cdot x_2 \cdot \ldots \cdot x_t \,.$$

The average growth factor from B_0 to B_T is the geometric mean or geometric average of the growth factors:

$$\bar{x}_G = \sqrt[T]{x_1 \cdot x_2 \cdot \ldots \cdot x_T}$$

$$= \sqrt[T]{\frac{B_0 \cdot x_1 \cdot x_2 \cdot \ldots \cdot x_T}{B_0}}$$

$$= \sqrt[T]{\frac{B_T}{B_0}}. \tag{3.11}$$

Therefore, B_t at time t can be calculated as $B_t = B_0 \cdot \bar{x}_G^t$.

Example 3.1.8 Suppose someone wants to deposit money, say €1000, in a bank. The bank advisor proposes a 5 yrs savings plan with the following plan for interest rates: 1% in the first year, 1.5% in the second year, 2.5% in the third year and 3% in the last 2 yrs. Now he would like to calculate the average growth factor and average growth rate for the invested money. The concept of the geometric mean can be used as follows:

Year	Euro	Growth factor	Growth rate
0	1000	–	–
1	1010	1.01	1.0%
2	1025.15	1.015	1.5%
3	1050.78	1.025	2.5%
4	1082.30	1.03	3.0%
5	1114.77	1.03	3.0%

The geometric mean is calculated as

$$\bar{x}_G = (1.01 \cdot 1.015 \cdot 1.025 \cdot 1.03 \cdot 1.03)^{\frac{1}{5}} = 1.021968$$

which means that he will have on average about 2.2% growth per year. The savings after 5 yrs can be calculated as

$$€\ 1000 \cdot 1.021968^5 = €\ 1114.77.$$

It is easy to compare two different saving plans with different growth strategies using the geometric mean.

3.1.6 Harmonic Mean

The harmonic mean is typically used whenever different x_i contribute to the mean with a different weight w_i, i.e. when we implicitly assume that the weight of each x_i is not one. It can be calculated as:

$$\bar{x}_H = \frac{w_1 + w_2 + \ldots + w_k}{\frac{w_1}{x_1} + \frac{w_2}{x_2} + \ldots + \frac{w_k}{x_k}} = \frac{\sum_{i=1}^{k} w_i}{\sum_{i=1}^{k} \frac{w_i}{x_i}}. \tag{3.12}$$

For example, when calculating the average speed, each weight relates to the relative distance travelled, n_i/n, with speed x_i. Using $w_i = n_i/n$, $n = n_1 + \ldots + n_k$ and $\sum_i w_i = \sum_i n_i/n = 1$, the harmonic mean can be written as

$$\bar{x}_H = \frac{1}{\sum_{i=1}^{k} \frac{w_i}{x_i}}. \tag{3.13}$$

Example 3.1.9 Suppose an investor bought shares worth €1000 for two consecutive months. The price for a share was €50 in the first month and €200 in the second month. What is the average purchase price? The number of shares purchased in the first month is $1000/50 = 20$. The number of shares purchased in the second month is $1000/200 = 5$. The total number of shares purchased is thus $20 + 5 = 25$ and the total investment is €2000. It is evident that the average purchase price is $2000/25 =$ €80. This is in fact the harmonic mean calculated as

$$\bar{x}_H = \frac{1}{\frac{0.5}{50} + \frac{0.5}{200}} = 80$$

because the weight of each purchase is $n_i/n = 1000/2000 = 0.5$. If the investment was €1200 in the first month and €800 in the second month, then we could use the harmonic mean with weights $1200/2000 = 0.6$ and $800/2000 = 0.4$ respectively to obtain the results.

3.2 Measures of Dispersion

Measures of central tendency, as introduced earlier, give us an idea about the location where most of the data is concentrated. However, two different data sets may have the same value for the measure of central tendency, say the same arithmetic means, but they may have different concentrations around the mean. In this case the location measures may not be adequate enough to describe the distribution of the data. The concentration or dispersion of observations around any particular value is another property which characterizes the data and its distribution. We now introduce statistical methods which describe the **variability** or **dispersion** of data.

Example 3.2.1 Suppose three students Christine, Andreas and Sandro arrive at different times in the class to attend their lectures. Let us look at their arrival time in the class after or before the starting time of lecture, i.e. let us look how early or late they were (in minutes).

Week	1	2	3	4	5	6	7	8	9	10
Christine	0	0	0	0	0	0	0	0	0	0
Andreas	−10	+10	−10	+10	−10	+10	−10	+10	−10	+10
Sandro	3	5	6	2	4	6	8	4	5	7

We see that Christine always arrives on time (time difference of zero). Andreas arrives sometimes 10 min early and sometimes 10 min late. However, the arithmetic mean of both students is the same—on average they both arrive on time! This interpretation is obviously not meaningful. The difference between both students is the variability in arrival times that can't be measured with the mean or median. For this reason, we need to introduce measures of dispersion (variability). With the knowledge of both location and dispersion, we can give a much more nuanced comparison between the different arrival times. For example, consider the third student Sandro. He is always late; sometimes more, sometimes less. However, while on average he comes late, his behaviour is more predictable than that of Andreas. Both location and dispersion are needed to give a fair comparison.

Example 3.2.2 Consider another example in which a supplier for the car industry needs to deliver 10 car doors with an exact width of 1.00 m. He supplies 5 doors with a width of 1.05 m and the remaining 5 doors with a width of 0.95 m. The arithmetic mean of all the 10 doors is 1.00 m. Based on the arithmetic mean, one may conclude that all the doors are good but the fact is that none of the doors are usable as they will not fit into the car. This knowledge can be summarized by a measure of dispersion.

The above examples highlight that the distribution of a variable needs to be characterized by a measure of dispersion in addition to a measure of location (central tendency). Now we introduce various measures of dispersion.

3.2.1   Range and Interquartile Range

Consider a variable X with n observations $x_1, x_2, \ldots, x_n$. Order these n observations as $x_{(1)} \le x_{(2)} \le \ldots \le x_{(n)}$. The range is a measure of dispersion defined as the difference between the maximum and minimum value of the data as

$$R = x_{(n)} - x_{(1)}. \tag{3.14}$$

The **interquartile range** is defined as the difference between the 75th and 25th quartiles as

$$d_Q = \tilde{x}_{0.75} - \tilde{x}_{0.25}. \tag{3.15}$$

It covers the centre of the distribution and contains 50% of the observations.

Remark 3.2.1 Note that the interquartile range is defined as the interval $[\tilde{x}_{0.25}; \tilde{x}_{0.75}]$ in some literature . However, in line with most of the statistical literature, we define the interquartile range to be a measure of dispersion, i.e. the difference between $\tilde{x}_{0.75}$ and $\tilde{x}_{0.25}$.

Example 3.2.3 Recall Examples 3.0.1–3.1.5 where we looked at the temperature in Bangkok during December. The ordered values $x_{(i)}$, $i = 1, ..., 31$, are:

°C	21	22	22	23	24	24	25	25	25	25	25	25	26	26	26	26
(i)	1	2	3	4	5	6	7	8	9	10	11	12	13	14	15	16
°C	27	27	27	28	28	28	29	29	29	29	29	30	30	30	31	
(i)	17	18	19	20	21	22	23	24	25	26	27	28	29	30	31	

We obtained the quantiles in Example 3.1.5 as $\tilde{x}_{0.25} = 25$ and $\tilde{x}_{0.75} = 29$. The interquartile range is therefore $d_Q = 29 - 25 = 4$ which means that 50% of the data is centred between 25 and 29 degrees Celsius. The range is $R = 31 - 21 = 10$ degrees Celsius, meaning that the temperature is varying at most by 10 degrees Celsius. In R, there are several ways to obtain quartiles, minimum and maximum values, e.g. by using min, max, quantiles, range, among others. All numbers can be easily obtained by the summary command which we recommend using:

```
summary(weather)
```

3.2.2 Absolute Deviation, Variance and Standard Deviation

Another measure of dispersion is the variance. The variance is one of the most important measures in statistics and is needed throughout this book. We use the idea of "absolute deviation" to give some more background and motivation for understanding the variance as a measure of dispersion, followed by some examples.

Consider the deviations of n observations around a certain value "A" and combine them together, for instance via the arithmetic mean of all the deviations:

$$D = \frac{1}{n} \sum_{i=1}^{n} (x_i - A).$$ (3.16)

This measure has the drawback that the deviations $(x_i - A)$, $i = 1, 2, \ldots, n$, can be either positive or negative and, consequently, their sum can potentially be very small or even zero. Using D as a measure of variability is therefore not a good idea since D may be small even for a large variability in the data, see for example the student Andreas in Example 3.2.1.

Using absolute values of the deviations solves this problem and we introduce the following measure of dispersion:

$$D(A) = \frac{1}{n} \sum_{i=1}^{n} |x_i - A|.$$ (3.17)

It can be shown that the absolute deviation attains its minimum when A corresponds to the median of the data:

$$D(\tilde{x}_{0.5}) = \frac{1}{n} \sum_{i=1}^{n} |x_i - \tilde{x}_{0.5}| . \tag{3.18}$$

We call $D(\tilde{x}_{0.5})$ the **absolute median deviation**. When $A = \bar{x}$, we speak of the **absolute mean deviation** given by

$$D(\bar{x}) = \frac{1}{n} \sum_{i=1}^{n} |x_i - \bar{x}| . \tag{3.19}$$

Another solution to avoid the positive and negative signs of deviation in (3.16) is to consider the squares of deviations $x_i - A$, rather than using the absolute value. This provides another measure of dispersion as

$$s^2(A) = \frac{1}{n} \sum_{i=1}^{n} (x_i - A)^2 \tag{3.20}$$

which is known as the **mean squared error** (MSE) with respect to A. The MSE is another important measure in statistics, see Chap. 9, equation (9.4), for details. It can be shown that $s^2(A)$ attains its minimum value when $A = \bar{x}$. This is the (sample) **variance**

$$\tilde{s}^2 = \frac{1}{n} \sum_{i=1}^{n} (x_i - \bar{x})^2 . \tag{3.21}$$

After expanding $\tilde{s}^2$, we can write (3.21) as:

$$\tilde{s}^2 = \frac{1}{n} \sum_{i=1}^{n} x_i^2 - \bar{x}^2 . \tag{3.22}$$

The positive square root of the variance is called the (sample) **standard deviation**, defined as

$$\tilde{s} = \sqrt{\frac{1}{n} \sum_{i=1}^{n} (x_i - \bar{x})^2} . \tag{3.23}$$

The standard deviation has the same unit of measurement as the data whereas the unit of the variance is the square of the units of the observations. For example, if X is weight, measured in kg, then $\bar{x}$ and $\tilde{s}$ are also measured in kg, while $\tilde{s}^2$ is measured in kg^2 (which may be more difficult to interpret). The variance is a measure which we use in other chapters to obtain measures of association between variables and to draw conclusions from a sample about a population of interest; however, the standard deviation is typically preferred for a descriptive summary of the dispersion of data.

The standard deviation measures how much the observations vary or how they are dispersed around the arithmetic mean. A low value of the standard deviation indicates that the values are highly concentrated around the mean. A high value of the standard deviation indicates lower concentration of the observations around the mean, and some of the observed values may even be far away from the mean. If there are extreme values or outliers in the data, then the arithmetic mean is more sensitive to outliers than the median. In such a case, the absolute median deviation (3.18) may be preferred over the standard deviation.

Example 3.2.4 Consider again Example 3.2.1 where we evaluated the arrival times of Christine, Andreas and Sandro in their lecture. Using the arithmetic mean, we concluded that both Andreas and Christine arrive on average on time, whereas Sandro is always late; however, we saw that the variation of arrival times differs substantially among the three students. To describe and quantify this variability formally, we calculate the variance and absolute median deviation:

$$\tilde{s}_C^2 = \frac{1}{10} \sum_{i=1}^{10} (x_i - \bar{x})^2 = \frac{1}{10}((0 - 0)^2 + \dots + (0 - 0)^2) = 0$$

$$\tilde{s}_A^2 = \frac{1}{10} \sum_{i=1}^{10} (x_i - \bar{x})^2 = \frac{1}{10}((-10 - 0)^2 + \dots + (10 - 0)^2) \approx 111.1$$

$$\tilde{s}_S^2 = \frac{1}{10} \sum_{i=1}^{10} (x_i - \bar{x})^2 = \frac{1}{10}((3 - 5)^2 + \dots + (7 - 5)^2) \approx 3.3$$

$$D(\tilde{x}_{0.5,C}) = \frac{1}{10} \sum_{i=1}^{n} |x_i - \tilde{x}_{0.5}| = \frac{1}{10}(|0 - 0| + \dots + |0 - 0|) = 0$$

$$D(\tilde{x}_{0.5,A}) = \frac{1}{10} \sum_{i=1}^{n} |x_i - \tilde{x}_{0.5}| = \frac{1}{10}(|-10 - 0| + \dots + |10 - 0|) = 10$$

$$D(\tilde{x}_{0.5,S}) = \frac{1}{10} \sum_{i=1}^{n} |x_i - \tilde{x}_{0.5}| = \frac{1}{10}(|3 - 5| + \dots + |7 - 5|) = 1.4$$

We observe that the variation/dispersion/variability is the lowest for Christine and highest for Andreas. Both median absolute deviation and variance allow a comparison between the two students. If we take the square root of the variance, we obtain the standard deviation. For example, $\tilde{s}_S = \sqrt{3.3} \approx 1.8$ which means that the average difference of the observations from the arithmetic mean is 1.8.

In R, we can use the var command to calculate the variance. However, note that R uses $1/(n-1)$ instead of $1/n$ in calculating the variance. The idea behind the multiplication by $1/(n-1)$ in place of $1/n$ is discussed in Chap. 9, see also Theorem 9.2.1.

3.2.2.1 Variance for Grouped Data

The variance for grouped data can be calculated using

$$s_b^2 = \frac{1}{n} \sum_{j=1}^{k} n_j (a_j - \bar{x})^2 = \frac{1}{n} \left(\sum_{j=1}^{k} n_j a_j^2 - n\bar{x}^2 \right) = \frac{1}{n} \sum_{j=1}^{k} n_j a_j^2 - \bar{x}^2 \quad (3.24)$$

where a_j is the middle value of the jth interval. However, when the data is artificially grouped and the knowledge about the original un-grouped data is available, we can also use the arithmetic mean of the jth class:

$$s_b^2 = \frac{1}{n} \sum_{j=1}^{k} n_j (\bar{x}_j - \bar{x})^2 . \quad (3.25)$$

The two expressions (3.24) and (3.25) represent the **variance between the different classes**, i.e. they describe the variability of the class specific means $\bar{x}_j$, weighted by the size of each class n_j, around the overall mean $\bar{x}$. Note that $\bar{x}$ is either $\sum_{j=1}^{k} (n_j/n) \cdot a_j = n^{-1} \sum_{j=1}^{k} n_j a_j$ or $n^{-1} \sum_{j=1}^{k} n_j \bar{x}_j$. It is evident that the variance *within* each class is not taken into account in these formulae. The variability of measurements in each class, i.e. the variability of $\forall x_i \in K_j$, is another important component to determine the overall variance in the data. It is therefore not surprising that using only the between variance $\tilde{s}_b^2$ will underestimate the total variance and therefore

$$s_b^2 \le \tilde{s}^2 . \quad (3.26)$$

If the data within each class is known, we can use the Theorem of Variance Decomposition (see Sect. 7.5 for the theoretical background) to determine the variance. This allows us to represent the total variance as the sum of the **variance between the different classes** and the **variance within the different classes** as

$$\tilde{s}^2 = \underbrace{\frac{1}{n} \sum_{j=1}^{k} n_j (\bar{x}_j - \bar{x})^2}_{\text{between}} + \underbrace{\frac{1}{n} \sum_{j=1}^{k} n_j \tilde{s}_j^2}_{\text{within}} . \quad (3.27)$$

In (3.27), $\tilde{s}_j^2$ is the variance of the jth class:

$$\tilde{s}_j^2 = \frac{1}{n_j} \sum_{x_i \in K_j} (x_i - \bar{x}_j)^2 . \quad (3.28)$$

The proof of (3.27) is given in Appendix C.1.

Example 3.2.5 Recall the weather data used in Examples 3.0.1–3.2.3 and the grouped data specified as follows:

Class intervals	< 20	(20 − 25]	(25, 30]	(30, 35]	> 35
n_j	$n_1 = 0$	$n_2 = 12$	$n_3 = 18$	$n_4 = 1$	$n_5 = 0$
$\bar{x}_j$	–	23.83	28	31	–
$\tilde{s}_j^2$	–	1.972	2	0	–

We know that $\bar{x} = 26.48$ and $n = 31$. The first step is to calculate the mean and variances in each class using (3.28). We then obtain $\bar{x}_j$ and s_j^2 as listed above. The within and between variances are:

$$\frac{1}{n} \sum_{j=1}^{k} n_j \tilde{s}_j^2 = \frac{1}{31}(12 \cdot 1.972 + 18 \cdot 2 + 1 \cdot 0) \approx 1.925$$

$$\frac{1}{n} \sum_{j=1}^{k} n_j (\bar{x}_j - \bar{x})^2 = \frac{1}{31}(12 \cdot [23.83 - 26.48]^2 + 18 \cdot [28 - 26.48]^2$$

$$+1 \cdot [31 - 26.48]^2) \approx 4.71$$

The total variance is therefore $\tilde{s}^2 \approx 6.64$. Estimating the variance using all 31 observations would yield the same results. However, it becomes clear that without knowledge about the variance within each class, we cannot reliably estimate $\tilde{s}^2$. In the above example, the variance within the classes is 3 times lower than the total variance which is a serious underestimation.

3.2.2.2 Linear Transformations

Let us consider a linear transformation $y_i = a + bx_i$ $(b \neq 0)$ of the original data x_i, $(i = 1, 2, \ldots, n)$. We get the arithmetic mean of the transformed data as $\bar{y} = a + b\bar{x}$ and for the variance:

$$\tilde{s}_y^2 = \frac{1}{n} \sum_{i=1}^{n} (y_i - \bar{y})^2 = \frac{b^2}{n} \sum_{i=1}^{n} (x_i - \bar{x})^2$$

$$= b^2 \tilde{s}_x^2. \tag{3.29}$$

Example 3.2.6 Let x_i, $i = 1, 2, \ldots, n$, denote measurements on time. These data could have been recorded and analysed in hours, but we may be interested in a summary in minutes. We can make a linear transformation $y_i = 60 x_i$. Then $\bar{y} = 60\bar{x}$ and $\tilde{s}_y^2 = 60^2 \tilde{s}_x^2$. If the mean and variance of the x_i's have already been obtained, then the mean and variance of the y_i's can be obtained directly using these transformations.

3.2.2.3 Standardization

A variable is called standardized if its mean is zero and its variance is 1. Standardization can be achieved by using the following transformation:

$$y_i = \frac{x_i - \bar{x}}{\tilde{s}_x} = -\frac{\bar{x}}{\tilde{s}_x} + \frac{1}{\tilde{s}_x}x_i = a + bx_i . \tag{3.30}$$

It follows that $\bar{y} = \frac{1}{n}\sum_{i=1}^{n}(x_i - \bar{x})/\tilde{s}_x = 0$ and $\tilde{s}_y^2 = \frac{1}{n}\sum_{i=1}^{n}(x_i - \bar{x})^2/\tilde{s}_x^2 = 1$. There are many statistical methods which require standardization, see, for example, Sect. 10.3.1 for details in the context of statistical tests.

Example 3.2.7 Let X be a variable which measures air pollution by using the concentration of atmospheric particulate matter (in $\mu g/m^3$). Suppose we have the following 10 measurements:

$$30 \quad 25 \quad 12 \quad 45 \quad 50 \quad 52 \quad 38 \quad 39 \quad 45 \quad 33.$$

We calculate $\bar{x} = 36.9$, $\tilde{s}_x^2 = 136.09$ and $\tilde{s}_x = 11.67$. To get a standardized variable Y, we transform all the observations x_i's as

$$y_i = \frac{x_i - \bar{x}}{\tilde{s}_x} = -\frac{\bar{x}}{\tilde{s}_x} + \frac{1}{\tilde{s}_x}x_i = -\frac{36.9}{11.67} + \frac{1}{11.67}x_i = -3.16 + 0.086x_i .$$

Now $y_1 = -3.16 + 0.086 \cdot 30 = -0.58$, $y_2 = -3.16 + 0.086 \cdot 25 = -1.01$, ..., are the standardized observations. The scale command in R allows standardization and we can obtain the standardized observations corresponding to the 10 measurements as

```
air <- c(30,25,12,45,50,52,38,39,45,33)
scale(air)
```

Please note that the scale command uses $1/(n-1)$ for calculating the variance, as already outlined above. Thus the results provided by scale are not identical to those using (3.30).

3.2.3   Coefficient of Variation

Consider a situation where two different variables have arithmetic means $\bar{x}_1$ and $\bar{x}_2$ with standard deviations $\tilde{s}_1$ and $\tilde{s}_2$, respectively. Suppose we want to compare the variability of hotel prices in Munich (measured in euros) and London (measured in British pounds). How can we provide a fair comparison? Since the prices are measured in different units, and therefore likely have arithmetic means which differ substantially, it does not make much sense to compare the standard deviations directly. The coefficient of variation v is a measure of dispersion which uses both the standard deviation and mean and thus allows a fair comparison. It is properly defined only when all the values of a variable are measured on a ratio scale and are positive such that $\bar{x} > 0$ holds. It is defined as

$$v = \frac{\tilde{s}}{\bar{x}}. \tag{3.31}$$

The coefficient of variation is a unit-free measure of dispersion. It is often used when the measurements of two variables are different but can be put into relation by using a linear transformation $y_i = bx_i$. It is possible to show that if all values x_i of a variable X are transformed into a variable Y with values $y_i = b \cdot x_i$, $b > 0$, then v dose not change.

Example 3.2.8 If we want to compare the variability of hotel prices in two selected cities in Germany and England, we could calculate the mean prices, together with their standard deviations. Suppose a sample of prices of say 100 hotels in two selected cities in Germany and England is available and suppose we obtain the mean and standard deviations of the two cities as $x_1 = €130$, $x_2 = £230$, $s_1 = €99$ and $s_2 = £212$. Then $v_1 = 99/130 \approx 0.72$ and $v_2 = 212/230 = 0.92$. This indicates higher variability in hotel prices in England. However if the data distribution is skewed or bimodal, then it may be wise not to choose the arithmetic mean as a measure of central tendency and likewise the coefficient of variation.

3.3 Box Plots

So far we have described various measures of central tendency and dispersion. It can be tedious to list those measures in summary tables. A simple and powerful graph is the **box plot** which summarizes the distribution of a metric (or sometimes an ordinal) variable by using its median, quartiles, minimum, maximum and extreme values.

Figure 3.5a shows a typical box plot. The vertical length of the box is the interquartile range $d_Q = \tilde{x}_{0.75} - \tilde{x}_{0.25}$ which shows the region that contains 50% of the data. The bottom end of the box refers to the first quartile, the top end of the box refers to the third quartile. The thick line in the box is the median. It becomes immediately clear that the box indicates the symmetry of the data: if the median is in the middle of the box, the data should be symmetric, otherwise it is skewed. The *whiskers* at the end of the plot mark the minimum and maximum values of the data. Looking at the box plot as a whole tells us about the data distribution and the range and variability of observations. Sometimes it may be advisable to understand which values are extreme in the sense that they are "far away" from the centre of the distribution. In many software packages, including R, values are defined to be extreme if they are greater than 1.5 box lengths away from the first or third quartile. Sometimes they are called outliers. Outliers and extreme values are defined differently in some software packages and books.

The `boxplot` command in R draws a box plot. The `range` option controls whether extreme values should be plotted, and if yes, how one wants to define such values.

(a) Box plot without extreme values   (b) Box plot with extreme values

Fig. 3.5 Box plots (a) without extreme values (b) with extreme values

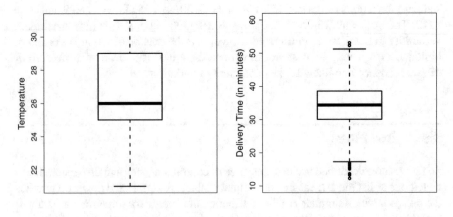

(a) Boxplot for weather data   (b) Boxplot for pizza data

Fig. 3.6 Box plots (a) for weather data (b) for pizza data

```
boxplot(variable, range=1.5)
```

Example 3.3.1 Recall Examples 3.0.1–3.2.5 where we looked at the temperature in Bangkok during December. We have already calculated the median (26°C) and the quartiles (25°C, 29°C). The minimum and maximum values are 21°C and 31°C. The box plot for this data is shown in Fig. 3.6a. One can see that the temperature distribution is slightly skewed with more variability for lower temperatures. The interquartile range is 4, and therefore any value $> 29 + 4 \times 1.5 = 35$ or $< 25 - 4 \times 1.5 = 19$ would be an extreme value. However, there are no extreme values in the data.

Example 3.3.2 Consider again the pizza data described in Appendix A.4.1. We use R to plot the box plot for the delivery time via `boxplot(time)` (Fig. 3.6b). We see a symmetric distribution with a median delivery time of about 35 min. Most of the deliveries took between 30 and 40 min. The extreme values indicate that there were some exceptionally short and long delivery times.

3.4 Measures of Concentration

A completely different concept used to describe a quantitative variable is the idea of concentration. For a variable X, it summarizes the proportion of each observation with respect to the sum of all observations $\sum_{i=1}^{n} x_i$. Let us look at a simple example to demonstrate its usefulness.

Example 3.4.1 Consider a village with 5 farms. Each farmer has a farm of a certain size. How can we evaluate the land distribution? Do all farmers have a similar amount of land or do one or two farmers have a big advantage because they have considerably more space?

Table 3.1 shows two different situations: in the table on the left, we see an equal distribution of land, i.e. each farmer owns 20 hectares of farmland. This means X is *not* concentrated, rather it is equally distributed. A statistical function describing the concentration could return a value of zero in such a case. Consider another extreme where one farmer owns all the farmland and the others do not own anything, as shown on the right side of Table 3.1. This is an extreme concentration of land: one person owns everything and thus we say the concentration is high. A statistical function describing the concentration could return a value of one in such a case.

3.4.1 Lorenz Curve

The **Lorenz curve** is a popular method to display concentrations graphically. Consider n observations $x_1, x_2, \ldots, x_n$ of a variable X. Assume that all the observations are positive. The sum of all the observations is $\sum_{i=1}^{n} x_i = n\bar{x}$ if the data is ungrouped. First, we need to order the data: $0 \leq x_{(1)} \leq x_{(2)} \leq \ldots \leq x_{(n)}$. To plot the Lorenz curve, we need

$$u_i = \frac{i}{n}, \quad i = 0, \ldots, n, \tag{3.32}$$

and

$$v_i = \frac{\sum_{j=1}^{i} x_{(j)}}{\sum_{j=1}^{n} x_{(j)}}, \quad i = 1, \ldots, n; \ v_0 := 0, \tag{3.33}$$

Table 3.1 Concentration of farm land: two different situations

Farmer (i)	x_i (Area, in hectare)
1	20
2	20
3	20
4	20
5	20
	$\sum_{i=1}^{5} x_i = 100$
Farmer (i)	x_i (Area, in hectare)
1	0
2	0
3	0
4	0
5	100
	$\sum_{i=1}^{5} x_i = 100$

where $\sum_{j=1}^{i} x_{(j)}$ is the cumulative total of observations up to the ith observation. The idea is that v_i describes the contribution of all values $\leq i$ in comparison to the sum of all values. Plotting u_i against v_i for all i shows how much the sum of all x_i, for all observations $\leq i$, contributes to the total sum. In other words, the point (u_i, v_i) says that $u_i \cdot 100\%$ of observations contain $v_i \cdot 100\%$ of the sum of all x_i less than or equal to i. Obviously, if all x_i are identical, the Lorenz curve will be a straight diagonal line, also known as the identity line or **line of equality**. If the x_i are of different sizes, then the Lorenz curve falls below the line of equality. This is illustrated in the following example.

Example 3.4.2 Recall Example 3.4.1 where we looked at the distribution of farm-land among 5 farmers. On the left side of Table 3.1, we observed an equal distribution of land among the farmers: $x_1 = 20, x_2 = 20, x_3 = 20, x_4 = 20, x_5 = 20$. We obtain $u_1 = 1/5, u_2 = 2/5, \ldots, u_5 = 1$ and $v_1 = 20/100, v_2 = 40/100, \ldots, v_5 = 1$. This yields a Lorenz curve as displayed on the left side of Fig. 3.7: there is no concentration. We can interpret each point. For example, $(u_2, v_2) = (0.4, 0.4)$ means that 40% of farmers own 40% of the land.

The right side of Table 3.1 describes the situation with strong concentration. For this table we obtain $u_1 = 1/5, u_2 = 2/5, \ldots, u_5 = 1$ and $v_1 = 0, v_2 = 0, \ldots, v_5 = 1$. Therefore, for example, 80% of farmers own 0% of the land which shows strong inequality. Most often we do not have such extreme situations. In this case, the Lorenz curve is bent towards the lower right corner of the plot, see the right side of Fig. 3.7.

We can plot the Lorenz curve in R using the Lc command in the library ineq. The Lorenz curve for the left table of Example 3.4.1 is plotted in R as follows:

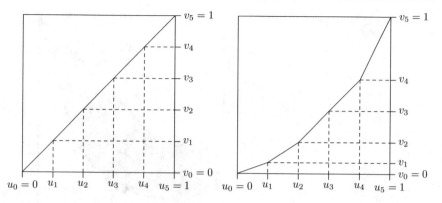

Fig. 3.7 Lorenz curves for no concentration (left) and some concentration (right)

```
library(ineq)
x <- c(20,20,20,20,20)
plot(Lc(x))
```

We can use the same approach as above to obtain the Lorenz curve when we have grouped data. We simply describe the contributions for each class rather than for each observation and approximate the values in each class by using its midpoint. More formally we can write:

$$\tilde{u}_i = \sum_{j=1}^{i} f_j, \quad i = 1, 2, \ldots, k; \ \tilde{u}_0 := 0 \tag{3.34}$$

and

$$\tilde{v}_i = \frac{\sum_{j=1}^{i} f_j a_j}{\sum_{j=1}^{k} f_j a_j} = \frac{\sum_{j=1}^{i} n_j a_j}{n\bar{x}}, \quad i = 1, 2, \ldots, k; \ \tilde{v}_0 := 0. \tag{3.35}$$

3.4.2   Gini Coefficient

We have seen in Sect. 3.4.1 that the Lorenz curve corresponds to the identity line, that is the diagonal line of equality, for no concentration. When there is some concentration, then the curve deviates from this line. The amount of deviation depends on the strength of concentration. Suppose we want to design a measure of concentration which is 0 for no concentration and 1 for perfect (i.e. extreme) concentration. We can simply measure the area between the Lorenz curve and the identity line and multiply it by 2. For no concentration, the area will be zero and hence the measure will be

Fig. 3.8 Lorenz Curve and
the Gini Coefficient

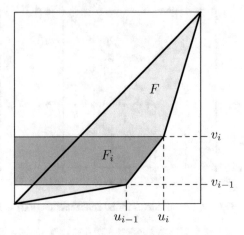

zero. If there is perfect concentration, then the curve will coincide with the axes, the
area will be close to 0.5 and twice the area will be close to one. The measure based
on such an approach is called the Gini coefficient:

$$G = 2 \cdot F . \tag{3.36}$$

Note that F is the area between the curve and the bisection or diagonal line.

The Gini coefficient can be estimated by adding up the areas of the trapeziums
F_i as displayed in Fig. 3.8:

$$F = \sum_{i=1}^{n} F_i - 0.5$$

where

$$F_i = \frac{u_{i-1} + u_i}{2} (v_i - v_{i-1}) .$$

It can be shown that this corresponds to

$$G = 1 - \frac{1}{n} \sum_{i=1}^{n} (v_{i-1} + v_i) , \tag{3.37}$$

but the proof is omitted. The same formula can be used for grouped data except that
$\tilde{v}$ is used instead of v. Since

$$0 \le G \le \frac{n-1}{n}, \tag{3.38}$$

one may prefer to use the standardized Gini coefficient

$$G^+ = \frac{n}{n-1} G \tag{3.39}$$

which takes a maximum value of 1.

Fig. 3.9 Lorenz curve for
Example 3.4.3

Example 3.4.3 We return to our farmland example. Suppose we have 7 farmers with
farms of different sizes:

Farmer	1	2	3	4	5	6	7
Farmland size x_i	20	14	59	9	36	23	3

Using the ordered values, we can calculate u_i and v_i using (3.32) and (3.33):

i	$x_{(i)}$	u_i	v_i
1	3	$\frac{1}{7} = 0.1429$	$\frac{3}{164} = 0.0183$
2	9	$\frac{2}{7} = 0.2857$	$\frac{12}{164} = 0.0732$
3	14	$\frac{3}{7} = 0.4286$	$\frac{26}{164} = 0.1585$
4	20	$\frac{4}{7} = 0.5714$	$\frac{46}{164} = 0.2805$
5	23	$\frac{5}{7} = 0.7143$	$\frac{69}{164} = 0.4207$
6	36	$\frac{6}{7} = 0.8571$	$\frac{105}{164} = 0.6402$
7	59	$\frac{7}{7} = 1.0000$	$\frac{164}{164} = 1.0000$

The Lorenz curve is displayed in Fig. 3.9. Using this information, it is easy to
calculate the Gini coefficient:

$$
G = 1 - \frac{1}{7}(0.0183 + [0.0183 + 0.0732] + [0.0732 + 0.1585] + [0.1585 + 0.2805]
$$
$$
+[0.2805 + 0.4207] + [0.4207 + 0.6402] + [0.6402 + 1]) = 0.402
$$

We know that $G = 0.4024 \leq \frac{6}{7} = \frac{n-1}{n}$. To standardize the coefficient, we therefore have to use (3.39):

$$G^+ = \frac{7}{6}G = \frac{7}{6} \cdot 0.4024 = 0.4695 \,.$$

In R, we can obtain the non-standardized Gini Coefficient using the `ineq` function in the library `ineq`.

```
library(ineq)
farm <- c(20,14,59,9,36,23,3)
ineq(farm)
```

3.5   Key Points and Further Issues

> **Note:**
>
> ✓ A summary on how to descriptively summarize data is given in Appendix D.1.
>
> ✓ The median is preferred over the arithmetic mean when the data distribution is skewed or there are extreme values.
>
> ✓ If data of a metric variable are grouped, and the original un-grouped data are not known, additional assumptions are needed to calculate measures of central tendency and dispersion. However, in some cases these assumptions may not be satisfied and the formulae provided may give imprecise results.
>
> ✓ QQ-plots are not only descriptive summaries but can also be used to test modelling assumptions, see Chap. 11.9 for more details.
>
> ✓ The distribution of a metric variable can be easily summarized using a box plot.

3.6   Exercises

Exercise 3.1 A hiking enthusiast has a new app for his smartphone which summarizes his hikes by using a GPS device. Let us look at the distance hiked (in km) and maximum altitude (in m) for the last 10 hikes:

Distance	12.5	29.9	14.8	18.7	7.6	16.2	16.5	27.4	12.1	17.5
Altitude	342	1245	502	555	398	670	796	912	238	466

(a) Calculate the arithmetic mean and median for both distance and altitude.

(b) Determine the first and third quartile for both the distance and the altitude variables. Discuss the shape of the distribution given the results of (a) and (b).

(c) Calculate the interquartile range, absolute median deviation and standard deviation for both variables. What is your conclusion about the variability of the data?

(d) One metre corresponds to approximately 3.28 feet. What is the average altitude when measured in feet rather than in metres?

(e) Draw and interpret the box plot for both distance and altitude.

(f) Assume distance is measured as only short (5–15 km), moderate (15–20 km), and long (20–30 km). Summarize the grouped data in a frequency table. Calculate the weighted arithmetic mean under the assumption that the raw data is not known. Determine the weighted median under the assumption that the values within each class are equally distributed.

(g) What is the variance for the grouped data when the raw data is known, i.e. when one has knowledge about the variance in each class? How does it compare with the variance one obtains when the raw data is unknown?

(h) Use R to reproduce the results of (a), (b), (c), (e), and (f).

Exercise 3.2 A gambler notes down his wins and losses (in €) from playing 10 games of roulette in a casino.

Round	1	2	3	4	5	6	7	8	9	10
Won/Lost	200	600	−200	−200	−200	−100	−100	−400	0	

(a) Assume $\bar{x} = -€90$ and $s = €294.7881$. What is the result of round 10?

(b) Determine the mode and the interquartile range.

(c) A different gambler plays 33 rounds of roulette. His results are $\bar{x} = €12$ and $s = €1000$. Is it meaningful to compare the variability of results of the two players by using the coefficient of variation? If yes, determine the coefficients of variation; if no, why is a comparison not possible?

Exercise 3.3 A fashion boutique has summarized its daily sales of designer socks in different groups: men's socks, women's socks, children's socks. Unfortunately the data for men's socks was lost. Determine the missing values.

	n	Arithmetic mean in €	Standard deviation in €
Women's wear	45	16	$\sqrt{6}$
Men's wear	?	?	?
Children's clothing	20	7.5	$\sqrt{3}$
Total	100	15	$\sqrt{19.55}$

Exercise 3.4 The number of members of a millionaires' club were as follows:

Year	2011	2012	2013	2014	2015	2016
Members	23	24	27	25	30	28

(a) What is the average growth rate of the membership?
(b) Based on the results of a), how many members would one expect in 2018?
(c) The president of the club is interested in the number of members in 2025, the year when his presidency ends. Would it make sense to predict the number of members for 2025?

In 2015 the members invested €250 million on the stock market. 10 members contributed 16% of the investment sum, 8 members contributed €60 million, 8 members contributed €70 million and another 4 members contributed the remaining amount.

(d) Draw the Lorenz curve for this data.
(e) Calculate and interpret the standardized Gini coefficient.

Exercise 3.5 Consider the monthly salaries Y (in Swiss francs) of a well-reputed software company, as well as the length of service (in months, X), and gender (Z). Figure 3.10 shows the QQ-plots for both Y and X given Z. Interpret both graphs.

Exercise 3.6 There is no built-in function in R to calculate the mode of a variable. Program such a function yourself. Hint: type ?table and ?names to recall the functionality of these functions. Combine them in an intelligent way.

Exercise 3.7 Consider a country in which 90 % of the wealth is owned by 20 % of the population, the so-called upper class. For simplicity, let us assume that the wealth is distributed equally within this class.

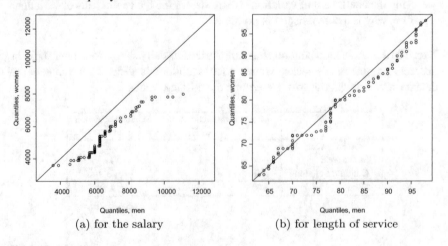

(a) for the salary   (b) for length of service

Fig. 3.10 QQ-plots

(a) Draw the Lorenz curve for this country.
(b) Now assume a revolution takes place in the country and all members of the upper class have to give away their wealth which is then distributed equally across the remaining population. Draw the Lorenz curve for this scenario.
(c) What would the curve from b) look like if the entire upper class left the country?

Exercise 3.8 A bus route in the mountainous regions of Romania has a length of 418 km. The manager of the bus company serving the route wants his buses to finish a trip within 8 hours. The bus travels the first 180 km with an average speed of 48 km/h, the next 117 km with an average speed of 37 km/h, and the last section with an average speed of 52 km/h.

(a) What is the average speed with which the bus travels?
(b) Will the bus finish the trip in time?

Exercise 3.9 Four friends have a start-up company which sells vegan ice cream. Their initial financial contributions are as follows:

Person	1	2	3	4
Contribution (in €)	800	10300	4700	2220

(a) Calculate and draw the Lorenz curve.
(b) Determine and interpret the standardized Gini coefficient.
(c) Does G^+ change if each of the friends contributes only half the amount of money? If yes, how much? If no, why not?
(d) Use R to draw the above Lorenz curve and to calculate the Gini coefficient.

Exercise 3.10 Recall the pizza delivery data which is described in Appendix A.4.1. Use R to read in and analyse the data.

(a) Calculate the mean, median, minimum, maximum, first quartile and third quartile for all quantitative variables.
(b) Determine and interpret the 99% quantile for delivery time and temperature.
(c) Write a function which calculates the absolute mean deviation. Use the function to calculate the absolute mean deviation of temperature.
(d) Scale the delivery time and calculate the mean and variance for this variable.
(e) Draw a box plot for delivery time and temperature. The box plots should not highlight extreme values.
(f) Use the `cut` command to create a new variable which summarizes delivery time in steps of 10 min. Calculate the arithmetic mean of this variable.
(g) Reproduce the QQ-plots shown in Example 3.1.6.

→ Solutions to all exercises in this chapter can be found in Appendix B

Association of Two Variables

4

In Chaps. 2 and 3 we discussed how to analyse a single variable using graphs and summary statistics. However, in many situations we may be interested in the interdependence of two or more variables. For example, suppose we want to know whether male and female students in a college have any preference between the subjects mathematics and biology, i.e. if there is any evidence that male students prefer mathematics over biology and female students prefer biology over mathematics or vice versa. Suppose we choose an equal number of male and female students and ask them about their preferred subject. We expect that if there is no association between the two variables "gender of student" (male or female) and "subject" (mathematics or biology), then an equal proportion of male and female students should choose the subjects biology and mathematics respectively. Any difference in the proportions may indicate a preference of males or females for a particular subject. Similarly, in another example, we may want to find out whether female employees of an organization are paid less than male employees or vice versa. Let us assume again that we choose an equal number of male and female employees and assume further that the salary is measured as a binary variable (low versus high salary group). We then expect that if there is no gender discrimination, the number of male and female employees in the lower and higher salary groups in the organization should be approximately equal. In both examples, the variables considered are binary and nominal (although the salary can also be seen as ordinal) and the data is summarized in terms of frequency measures. There may however be situations in which we are interested in associations between ordinal or metrical variables. Consider a data set in which height, weight and age of infants are given. Usually the height and weight of infants increase with age. Also, the height of infants increases with their weight and vice versa. Clearly, there is an interrelation or association among the three variables. In another example, two persons have to judge participants of a dance competition and rank them according to their performance. Now if we want to learn about the fairness

© Springer Nature Switzerland AG 2022
C. Heumann et al., *Introduction to Statistics and Data Analysis*,
https://doi.org/10.1007/978-3-031-11833-3_4

in the judgment, we expect that both the judges give similar ranks to each candidate, i.e. both judges give high ranks to good candidates and low ranks to not so good candidates. We are therefore interested in studying the association between the ranks given by the two judges. In all these examples, the intention lies in measuring the degree of association between two (or more) variables. We therefore need to study different ways of measuring the association level for different types of variables. In this chapter we present measures and graphical summaries for the association of two variables—dependent on their scale.

4.1   Summarizing the Distribution of Two Discrete Variables

When both variables are discrete, then it is possible to list all combinations of values of the two variables and to count how often these combinations occur in the data. Consider the salary example in the introduction to this chapter in which both the variables were binary. There are four possible combinations of variable categories (female and low salary group, female and high salary group, male and low salary group, male and high salary group). A complete description of the joint occurrence of these two variables can be given by counting, for each combination, the number of units for which this combination is observed. In the following we generalize this concept to two variables where each can have an arbitrary (but fixed) number of values or categories.

4.1.1   Contingency Tables for Discrete Data

Suppose we have data on two discrete variables. This data can be described in a two-dimensional **contingency table**.

Example 4.1.1 An airline conducts a customer satisfaction survey. The survey includes questions about travel class and satisfaction levels with respect to different categories such as seat comfort, in-flight service, meals, safety and other indicators. Consider the information on X, denoting the travel class (Economy = "E", Business = "B", First = "F"), and "Y", denoting the overall satisfaction with the flight on a scale from 1 to 4 as 1 (poor), 2 (fair), 3 (good) and 4 (very good). A possible response from 12 customers may look as follows:

	Passenger number											
i	1	2	3	4	5	6	7	8	9	10	11	12
Travel class	E	E	E	B	E	B	F	E	E	B	E	B
Satisfaction	2	4	1	3	1	2	4	3	2	4	3	3

Table 4.1 Contingency table for travel class and satisfaction

		Overall rating of flight quality				
		Poor	Fair	Good	Very good	Total (row)
Travel class	Economy	2	2	2	1	7
	Business	0	1	2	1	4
	First	0	0	0	1	1
	Total (column)	2	3	4	3	12

Table 4.2 $k \times l$ contingency table

		Y						Total (rows)
		y_1		y_j		y_l		
	x_1	n_{11}	$\cdots$	n_{1j}	$\cdots$	n_{1l}		n_{1+}
	x_2	n_{21}	$\cdots$	n_{2j}	$\cdots$	n_{2l}		n_{2+}
	$\vdots$	$\vdots$		$\vdots$		$\vdots$		$\vdots$
X	x_i	n_{i1}	$\cdots$	n_{ij}	$\cdots$	n_{il}		n_{i+}
	$\vdots$	$\vdots$		$\vdots$		$\vdots$		$\vdots$
	x_k	n_{k1}	$\cdots$	n_{kj}	$\cdots$	n_{kl}		n_{k+}
	Total (columns)	n_{+1}	$\cdots$	n_{+j}	$\cdots$	n_{+l}		n

We can calculate the absolute frequencies for each combination of observed values. For example, there are 2 passengers (passenger numbers 3 and 5) who were flying in economy class and rated the flight quality as poor, there were no passengers from both business class and first class who rated the flight quality as poor; there were 2 passengers who were flying in economy class and rated the quality as fair (2), and so on. Table 4.1 is a two-dimensional table summarizing this information.

Note that we not only summarize the joint frequency distribution of the two variables but also the distributions of the individual variables. Summing up the rows and columns of the table gives the respective frequency distributions. For example, the last column of the table demonstrates that 7 passengers were flying in economy class, 4 passengers were flying in business class and 1 passenger in first class.

Now we extend this example and discuss a general framework to summarize the absolute frequencies of two discrete variables in contingency tables. We use the following notations: Let $x_1, x_2, \ldots, x_k$ be the k classes of a variable X and let $y_1, y_2, \ldots, y_l$ be the l classes of another variable Y. We assume that both X and Y are discrete variables. It is possible to summarize the absolute frequencies n_{ij} related to $(x_i, y_j), i = 1, 2, \ldots, k, j = 1, 2, \ldots, l$, in a $k \times l$ **contingency table** as shown in Table 4.2.

We denote the sum of the ith row as $n_{i+} = \sum_{j=1}^{l} n_{ij}$ and the sum over the jth column as $n_{+j} = \sum_{i=1}^{k} n_{ij}$. The total number of observations is therefore

$$n = \sum_{i=1}^{k} n_{i+} = \sum_{j=1}^{l} n_{+j} = \sum_{i=1}^{k} \sum_{j=1}^{l} n_{ij} . \tag{4.1}$$

Remark 4.1.1 Note that it is also possible to use the relative frequencies $f_{ij} = n_{ij}/n$ instead of the absolute frequencies n_{ij} in Table 4.2, see Example 4.1.2.

4.1.2   Joint, Marginal, and Conditional Frequency Distributions

When the data on two variables are summarized in a contingency table, there are several concepts which can help us in studying the characteristics of the data. For example, how the values of both the variables behave jointly, how the values of one variable behave when another variable is kept fixed etc. These features can be studied using the concepts of joint frequency distribution, marginal frequency distribution and conditional frequency distribution. If relative frequency is used instead of absolute frequency, then we speak of the joint relative frequency distribution, marginal relative frequency distribution and conditional relative frequency distribution.

Definition 4.1.1 Using the notations of Table 4.2, we define the following:

- The frequencies n_{ij} represent the **joint frequencies** of X and Y, $i = 1, \ldots, k$, $j = 1, \ldots, l$.
- The frequencies n_{i+} represent the **marginal frequencies** of X, $i = 1, \ldots, k$.
- The frequencies n_{+j} represent the **marginal frequencies** of Y, $j = 1, \ldots, l$.
- We define $f_{i|j}^{X|Y} = n_{ij}/n_{+j}$ to be the **conditional frequencies** of X given $Y = y_j$, $i = 1, \ldots, k$, $j = 1, \ldots, l$.
- We define $f_{j|i}^{Y|X} = n_{ij}/n_{i+}$ to be the **conditional frequencies** of Y given $X = x_i$, $i = 1, \ldots, k$, $j = 1, \ldots, l$.
- The frequencies $f_{ij} = n_{ij}/n$ represent the **joint relative frequencies** of X and Y, $i = 1, \ldots, k$, $j = 1, \ldots, l$.
- The frequencies $f_{i+} = \sum_{j=1}^{l} f_{ij}$ represent the **marginal relative frequencies** of X, $i = 1, \ldots, k$.
- The frequencies $f_{+j} = \sum_{i=1}^{k} f_{ij}$ represent the **marginal relative frequencies** of Y, $j = 1, \ldots, l$.
- We define $f_{i|j}^{X|Y} = f_{ij}/f_{+j}$ to be the **conditional relative frequencies** of X given $Y = y_j$, $i = 1, \ldots, k$, $j = 1, \ldots, l$.
- We define $f_{j|i}^{Y|X} = f_{ij}/f_{i+}$ to be the **conditional relative frequencies** of Y given $X = x_i$, $i = 1, \ldots, k$, $j = 1, \ldots, l$.

- We denote the **joint frequency distribution** as $f_{XY} = f(X, Y)$, the **marginal distributions** as $f_X = f(X)$ and $f_Y = f(Y)$, respectively, and the **conditional distributions** as $f_{X|Y}$ and $f_{Y|X}$, respectively. The distributions are a collection of the respective relative (or absolute) frequencies. Overall, there are k conditional distributions for Y, i.e. $f_{Y|X_i} \forall i = 1, \ldots, k$ and l conditional distributions for X, i.e. $f_{X|Y_j} \forall j = 1, \ldots, l$.

Note that since $f_{ij} = n_{ij}/n$ and $f_{i+} = n_{i+}/n$, it holds that, for example, $f_{j|i}^{Y|X} = f_{ij}/f_{i+} = \frac{n_{ij}/n}{n_{i+}/n} = n_{ij}/n_{i+}$ is the same as the respective conditional frequency distribution. Note also that for a bivariate joint (relative) frequency distribution, there will only be two marginal (relative) frequency distributions but possibly more than two conditional (relative) frequency distributions.

Example 4.1.2 Recall the setup of Example 4.1.1. We now collect and evaluate the responses of 100 customers (instead of 12 passengers as in Example 4.1.1) regarding their choice of the travel class and their overall satisfaction with the flight quality.

The data is provided in Table 4.3 where each of the cell entries illustrates how many out of 100 passengers answered x_i *and* y_j: for example, the first entry "10" indicates that 10 passengers were flying in economy class *and* described the overall service quality as poor.

- The marginal frequency distributions are displayed in the last column and last row respectively. For example, the marginal distribution of X refers to the frequency table of "travel class" (X) and tells us that 62 passengers were flying in economy class, 25 in business class and 13 in first class. Similarly, the marginal distribution of "overall rating of flight quality" (Y) tells us that 10 passengers rated the quality as poor, 36 as fair, 40 as good and 14 as very good.
- The conditional frequency distributions give us an idea about the behaviour of one variable when the other one is kept fixed. For example, the conditional distribution of the "overall rating of flight quality" (Y) among passengers who were flying in economy class $(f_{Y|X=\text{Economy}})$ gives $f_{1|1}^{Y|X} = 10/62 \approx 16\%$ which means that approximately 16% of the customers in economy class are rating the quality as poor, $f_{2|1}^{Y|X} = 33/62 \approx 53\%$ of the customers in economy class are rating the

Table 4.3 Contingency table for travel class and satisfaction

		Overall rating of flight quality				
		Poor	Fair	Good	Very good	Total (rows)
Travel class	Economy	10	33	15	4	62
	Business	0	3	20	2	25
	First	0	0	5	8	13
	Total (columns)	10	36	40	14	100

quality as fair, $f_{3|1}^{Y|X} = 15/62 \approx 24\%$ of the customers in economy class are rating the quality as good and $f_{4|1}^{Y|X} = 4/62 \approx 7\%$ of the customers in economy class are rating the quality as very good. Similarly, $f_{3|2}^{Y|X} = 20/25 \approx 80\%$ which means that 80% of the customers in business class are rating the quality as good and so on.

- The conditional frequency distribution of the "travel class" (X) of passengers given the "overall rating of flight quality" (Y) is obtained by $f_{X|Y=\text{Satisfaction level}}$. For example, $f_{X|Y=\text{good}}$ gives $f_{1|3}^{X|Y} = 15/40 = 37.5\%$ which means that 37.5% of the passengers who rated the flight to be good travelled in economy class, $f_{2|3}^{X|Y} = 20/40 = 50\%$ of the passengers who rated the flight to be good travelled in business class and $f_{3|3}^{X|Y} = 5/40 = 12.5\%$ of the passengers who rated the flight to be good travelled in first class.

- In total, we have 100 customers and hence

$$n = \sum_{i=1}^{k} n_{i+} = 62 + 25 + 13 = \sum_{j=1}^{l} n_{+j} = 10 + 36 + 40 + 14$$

$$= \sum_{i=1}^{k} \sum_{j=1}^{l} n_{ij} = 10 + 33 + 15 + 4 + 3 + 20 + 2 + 5 + 8 = 100$$

- Alternatively, we can summarize X and Y using the relative frequencies as follows:

		Overall rating of flight quality				Total (rows)
		Poor	Fair	Good	Very good	
Travel	Economy	$\frac{10}{100}$	$\frac{33}{100}$	$\frac{15}{100}$	$\frac{4}{100}$	$\frac{62}{100}$
class	Business	0	$\frac{3}{100}$	$\frac{20}{100}$	$\frac{2}{100}$	$\frac{25}{100}$
	First	0	0	$\frac{5}{100}$	$\frac{8}{100}$	$\frac{13}{100}$
	Total (columns)	$\frac{10}{100}$	$\frac{36}{100}$	$\frac{40}{100}$	$\frac{14}{100}$	1

To produce the frequency table without the marginal distributions, we can use the R command `table(X,Y)`. To obtain the full contingency table including the marginal distributions in R, one can use the function `addmargins()`. For the relative frequencies, the function `prop.table()` can be used. In summary, a full contingency table is obtained by using

```
addmargins(table(X,Y))
addmargins(prop.table(table(X,Y)))
```

(a) Categories next to each other   (b) Categories stacked

Fig. 4.1 Bar charts for travel class and rating of satisfaction

4.1.3 Graphical Representation of Two Nominal or Ordinal Variables

Bar charts (see Sect. 2.3.1) can be used to graphically summarize the association between two nominal or two ordinal variables. The bar chart is drawn for X and the categories of Y are represented by separated bars or stacked bars for each category of X. In this way we summarize the joint distribution of the contingency table.

Example 4.1.3 Consider Example 4.1.2. There are 62 passengers flying in economy class. From these 62 passengers, 10 rated the quality of the flight as poor, 33 as fair, 15 as good, and 4 as very good. This means for $X = x_1 (= \text{Economy})$, we can either place 4 bars next to each other, as in Fig. 4.1a, or we can stack them on top of each other, as in Fig. 4.1b. The same can be done for the other categories of X, see Fig. 4.1. Stacked and stratified bar charts are prepared in R either simply by applying `barplot` to a table (i.e., `barplot(table(Class,Rating))`) or by loading the library `lattice` and using the function `barchart`. The latter option works as follows:

```
Class <- c(rep('1: Economy=',62),rep('2: Business',25),...)
Rating <- c(rep('1=poor',10),rep('2=fair',33),...)
library(lattice)
barchart(table(Class,Rating),horizontal=FALSE,stack=FALSE)
barchart(table(Class,Rating),horizontal=FALSE,stack=TRUE)
```

Remark 4.1.2 There are several other options in R to specify stratified bar charts. We refer the interested reader to Exercise 2.6 to explore how the R package ggplot2 can be used to make such graphics. Sometimes it can also be useful to visualize the difference of two variables and not stack or stratify the bars, see Exercise 2.1.

Independence and Expected Frequencies. An important statistical concept is **independence**. In this section we touch upon its descriptive aspects, see Chaps. 6 (Sect. 6.5) and 7 (Sect. 7.5) for more theoretical details. Two variables are considered to be independent if the observations on one variable do not influence the observations on the other variable. For example, suppose two different persons roll a die separately, then the outcomes of their rolls do not depend on each other. So we can say that the two observations are independent. In the context of contingency tables, two variables are independent of each other when the joint relative frequency equals the product of the marginal relative frequencies of the two variables, i.e. the following equation holds:

$$f_{ij} = f_{i+} f_{+j} . \tag{4.2}$$

The **expected absolute frequencies under independence** are obtained by

$$\tilde{n}_{ij} = n\, f_{ij} = n\frac{n_{i+}}{n}\frac{n_{+j}}{n} = \frac{n_{i+}n_{+j}}{n} . \tag{4.3}$$

Note that the absolute frequencies are always integers but the expected absolute frequencies may not always be integers.

Example 4.1.4 Recall Example 4.1.2. The expected absolute frequencies for the contingency table can be calculated using (4.3). For example,

$$\tilde{n}_{11} = \frac{62 \cdot 10}{100} = 6.2, \quad \tilde{n}_{12} = \frac{62 \cdot 36}{100} = 22.32 \quad \text{etc.}$$

Table 4.4 lists both the observed absolute frequency and expected absolute frequency (in brackets).

To calculate the expected absolute frequencies in R, we can access the "expected" object returned from a χ^2-test applied to the respective contingency table as follows:

```
chisq.test(table(Class,Rating))$expected
```

A detailed motivation and explanation of this command is given in Sect. 10.8.

Table 4.4 Observed and expected absolute frequencies for the airline survey

		Overall rating of flight quality				
		Poor	Fair	Good	Very good	Total
Travel class	Economy	10 (6.2)	33 (22.32)	15 (24.8)	4 (8.68)	62
	Business	0 (2.5)	3 (9.0)	20 (10.0)	2 (3.5)	25
	First	0 (1.3)	0 (4.68)	5 (5.2)	8 (1.82)	13
	Total	10	36	40	14	100

4.2 Measures of Association for Two Discrete Variables

When two variables are not independent, then they are associated. Their association can be weak or strong. Now we describe some popular measures of association. Measures of association describe the degree of association between two variables and can have a direction as well. Note that if variables are defined on a nominal scale then nothing can be said about the direction of association, only about the strength.

Let us first consider a 2×2 contingency table which is a special case of a $k \times l$ contingency table, see Table 4.5.

The variables X and Y are independent if

$$\frac{a}{a+c} = \frac{b}{b+d} = \frac{a+b}{n} \tag{4.4}$$

or equivalently if

$$a = \frac{(a+b)(a+c)}{n}. \tag{4.5}$$

Note that some other forms of the conditions (4.4)–(4.5) can also be derived in terms of a, b, c and d (for example, $ad = bc$).

Example 4.2.1 Suppose a vaccination against flu (influenza) is given to 200 persons. Some of the persons may get affected by flu despite the vaccination. The data is summarized in Table 4.6. Using the notations of Table 4.5, we have $a = 90, b = 10, c = 40, d = 60$ and thus $(a+b)(a+c)/n = 100 \cdot 130/200 = 65$ which is less than $a = 90$. Hence being affected by flu is not independent of the vaccination, i.e. whether one is vaccinated or not has an influence on getting affected by flu. In the vaccinated group only 10 of 100 persons are affected by flu while in the

Table 4.5 2×2 contingency table

		Y		
		y_1	y_2	Total (row)
X	x_1	a	b	$a+b$
	x_2	c	d	$c+d$
	Total (column)	$a+c$	$b+d$	n

Table 4.6 2×2 contingency table

		Persons		
		Not affected	Affected	Total (row)
Vaccination	Vaccinated	90	10	100
	Not vaccinated	40	60	100
	Total (column)	130	70	200

group not vaccinated 60 of 100 persons are affected. Another interpretation is that if independence holds then we would expect 65 persons to be not affected by flu in the vaccinated group but we observe 90 persons. This shows that vaccination has a protective effect.

To gain a better understanding about the strength of association between two variables, we need to develop the concept of dependence and independence further. The following three subsections illustrate this in more detail.

4.2.1   Pearson's χ^2 Statistic

We now introduce Pearson's χ^2 statistic which is used for measuring the association between variables in a contingency table and plays an important role in the construction of statistical tests, see Sect. 10.8. The χ^2 statistic or χ^2 coefficient for a $k \times l$ contingency table is given as

$$\chi^2 = \sum_{i=1}^{k}\sum_{j=1}^{l} \frac{\left(n_{ij} - \tilde{n}_{ij}\right)^2}{\tilde{n}_{ij}} = \sum_{i=1}^{k}\sum_{j=1}^{l} \frac{\left(n_{ij} - \frac{n_{i+}n_{+j}}{n}\right)^2}{\frac{n_{i+}n_{+j}}{n}}. \tag{4.6}$$

A simpler formula for 2×2 contingency tables is

$$\chi^2 = \frac{n(ad - bc)^2}{(a+b)(c+d)(a+c)(b+d)}. \tag{4.7}$$

The idea behind the χ^2 coefficient is that when the relationship between two variables is stronger, then the deviations between observed and expected frequencies are expected to be higher (because the expected frequencies are calculated assuming independence) and this indicates a stronger relationship between the two variables. If observed and expected frequencies are identical or similar, then this is an indication that the association between the two variables is weak and the variables may even be independent. The χ^2 statistic for a $k \times l$ contingency table sums up all the differences between the observed and expected frequencies, squares them and scales them with respect to the expected frequencies. The squaring of the differences makes the statistic independent of the positive and negative signs of the differences between observed and expected frequencies. The range of values for χ^2 is

$$0 \leq \chi^2 \leq n(\min(k,l) - 1). \tag{4.8}$$

Note that $\min(k,l)$ is the minimum function and simply returns the smaller of the two numbers k and l. For example, $\min(3,4)$ returns the value 3. Consequently the values of χ^2 obtained from (4.6) can be compared with the range from (4.8). A value of χ^2 close to zero indicates a weak association and a value of χ^2 close to $n(\min(k,l) - 1)$ indicates a strong association between the two variables. Note that the range of χ^2 depends on n, k and l, i.e. the sample size and the dimension of the contingency table.

The χ^2 statistic is a *symmetric* measure in the sense that its value does not depend on which variable is defined as X and which as Y.

Example 4.2.2 Consider Examples 4.1.2 and 4.1.4. Using the values from Table 4.4, we can calculate the χ^2 statistic as

$$\chi^2 = \frac{(10 - 6.2)^2}{6.2} + \frac{(33 - 22.32)^2}{22.32} + ... + \frac{(8 - 1.82)^2}{1.82} = 57.95064$$

The maximum possible value for the χ^2 statistic is $100(\min(4,3) - 1) = 200$. Thus $\chi^2 \approx 57$ indicates a moderate association between "travel class" and "overall rating of flight quality" of the passengers. In R, we obtain this result as follows:

```
chisq.test(table(Class,Rating))$statistic
```

4.2.2 Cramer's V Statistic

A problem with Pearson's χ^2 coefficient is that the range of its maximum value depends on the sample size and the size of the contingency table. These values may vary in different situations. To overcome this problem, the coefficient can be standardized to lie between 0 and 1 so that it is independent of the sample size as well as the dimension of the contingency table. Since $n(\min(k,l) - 1)$ was the maximal value of the χ^2 statistic, dividing χ^2 by this maximal value automatically leads to a scaled version with maximal value 1. This idea is used by Cramer's V statistic which for a $k \times l$ contingency table is given by

$$V = \sqrt{\frac{\chi^2}{n(\min(k,l) - 1)}}. \tag{4.9}$$

The closer the value of V gets to 1, the stronger the association between the two variables.

Example 4.2.3 Consider Example 4.2.2. The obtained χ^2 statistic is 57.95064. To obtain Cramer's V, we just need to calculate

$$V = \sqrt{\frac{\chi^2}{n(\min(k, l) - 1)}} = \sqrt{\frac{57.95064}{100(3 - 1)}} \approx 0.54. \tag{4.10}$$

This indicates a moderate association between "travel class" and "overall rating of flight quality" because 0.54 lies in the middle of 0 and 1. In R, there are two options to calculate V: (i) to calculate the χ^2 statistic and then adjust it as in (4.9), (ii) to use the functions assocstats and xtabs contained in the package vcd as follows:

```
library(vcd)
assocstats(xtabs(~Class+Rating))
```

4.2.3   Contingency Coefficient C

Another option to standardize χ^2 is given by a corrected version of Pearson's contingency coefficient C:

$$C_{\text{corr}} = \frac{C}{C_{\max}} = \sqrt{\frac{\min(k, l)}{\min(k, l) - 1}} \sqrt{\frac{\chi^2}{\chi^2 + n}}, \tag{4.11}$$

with

$$C = \sqrt{\frac{\chi^2}{\chi^2 + n}} \quad \text{and} \quad C_{\max} = \sqrt{\frac{\min(k, l) - 1}{\min(k, l)}}. \tag{4.12}$$

It always lies between 0 and 1. The closer the value of C is to 1, the stronger the association.

Example 4.2.4 We know from Example 4.2.2 that the χ^2 statistic for travel class and satisfaction level is 57.95064. To calculate C_{corr} we need the following calculations:

$$C = \sqrt{\frac{57.95064}{57.95064 + 100}} = 0.606 \, , \, C_{\max} = \sqrt{\frac{\min(4, 3) - 1}{\min(4, 3)}} = \sqrt{\frac{2}{3}} = 0.816,$$

$$C_{\text{corr}} = \frac{C}{C_{\max}} = \frac{0.606}{0.816} \approx 0.74 \, .$$

There is a moderate to strong association between "travel class" and "overall rating of flight quality" of the passengers. We can compute C in R using the vcd package as follows:

```
library(vcd)
Cmax = sqrt((min(c(3,4))-1)/min(c(3,4)))
assocstats(xtabs(~Class+Rating))$cont/Cmax
```

4.2.4 Relative Risks and Odds Ratios

We now introduce the concepts of odds ratios and relative risks. Consider a 2×2 contingency table as introduced in Table 4.5. Now suppose we have two variables X and Y with their conditional distributions $f_{i|j}^{X|Y}$ and $f_{j|i}^{Y|X}$. In the context of a 2×2 contingency table, $f_{1|1}^{X|Y} = n_{11}/n_{+1}$, $f_{1|2}^{X|Y} = n_{12}/n_{+2}$, $f_{2|2}^{X|Y} = n_{22}/n_{+2}$ and $f_{2|1}^{X|Y} = n_{21}/n_{+1}$. The relative risks are defined as the ratio of two conditional distributions, for example

$$\frac{f_{1|1}^{X|Y}}{f_{1|2}^{X|Y}} = \frac{n_{11}/n_{+1}}{n_{12}/n_{+2}} = \frac{a/(a+c)}{b/(b+d)} \quad \text{and} \quad \frac{f_{2|1}^{X|Y}}{f_{2|2}^{X|Y}} = \frac{n_{21}/n_{+1}}{n_{22}/n_{+2}} = \frac{c/(a+c)}{d/(b+d)} . \quad (4.13)$$

The odds ratio is defined as the ratio of these relative risks from (4.13) as

$$OR = \frac{f_{1|1}^{X|Y}/f_{1|2}^{X|Y}}{f_{2|1}^{X|Y}/f_{2|2}^{X|Y}} = \frac{f_{1|1}^{X|Y} f_{2|2}^{X|Y}}{f_{2|1}^{X|Y} f_{1|2}^{X|Y}} = \frac{a\,d}{b\,c} . \quad (4.14)$$

Alternatively, the odds ratio can be defined as the ratio of the chances for "disease", a/b (number of smokers with the disease divided by the number of non-smokers with the disease), and no disease, c/d (number of smokers with no disease divided by the number of non-smokers with no disease).

The relative risks compare proportions, while the odds ratio compares odds.

Example 4.2.5 A classical example refers to the possible association of smoking with a particular disease. Consider the following data on 240 individuals:

		Smoking		Total (row)
		Yes	No	
Disease	Yes	34	66	100
	No	22	118	140
	Total (column)	56	184	240

We calculate the following relative risks:

$$\frac{f_{1|1}^{X|Y}}{f_{1|2}^{X|Y}} = \frac{34/56}{66/184} \approx 1.69 \quad \text{and} \quad \frac{f_{2|1}^{X|Y}}{f_{2|2}^{X|Y}} = \frac{22/56}{118/184} \approx 0.61 . \quad (4.15)$$

Thus, the proportion of individuals with the disease is 1.69 times higher among smokers when compared with non-smokers. Similarly, the proportion of healthy individuals is 0.61 times smaller among smokers when compared with non-smokers.

The relative risks are calculated to compare the proportion of sick or healthy patients between smokers and non-smokers. Using these two relative risks, the odds ratio is obtained as

$$OR = \frac{34 \times 118}{66 \times 22} = 2.76.$$

We can interpret this outcome as follows: (i) the chances of smoking are 2.76 times higher for individuals with the disease compared with healthy individuals (follows from definition (4.14)). We can also say that (ii) the chances of having the particular disease is 2.76 times higher for smokers compared with non-smokers. If we interchange either one of the "Yes" and "No" columns or the "Yes" and "No" rows, we obtain $OR = 1/2.76 \approx 0.36$, giving us further interpretations: (iii) the chances of smoking are 0.36 times lower for individuals without disease compared with individuals with the disease and (iv) the chance of having the particular disease is 0.36 times lower for non-smokers compared with smokers. Note that all four interpretations are correct and one needs to choose the right interpretation in the light of the experimental situation and the question of interest.

4.3   Association Between Ordinal and Metrical Variables

4.3.1   Graphical Representation of Two Metrical Variables

A simple way to graphically summarize the association between two metrical variables is to plot the paired observations of the two variables in a two-dimensional coordinate system. If n paired observations for two metrical variables X and Y are available as $(x_i, y_i), i = 1, 2, \ldots, n$, then all such observations can be plotted in a single graph. This graph is called a **scatter plot**. Such a plot reveals possible relationships and trends between the two variables. For example, Figs. 4.2 and 4.3 show scatter plots with six different types of association.

- Figure 4.2a shows increasing values of Y for increasing values of X. We call this relationship positive association. The relationship between X and Y is nearly linear because all the points lie around a straight line.
- Figure 4.2b shows decreasing values of Y for increasing values of X. We call this relationship negative association.
- Figure 4.2c tells us the same as Fig. 4.2a, except that the positive association is weaker.
- Figure 4.2d tells us the same as Fig. 4.2b, except that the negative association is weaker.

(a) Strong positive linear relationship   (b) Strong negative linear relationship

(c) Moderate positive relationship   (d) Moderate negative relationship

Fig. 4.2 Scatter plots

- Figure 4.3a shows that as the X values increase, the values of Y neither increase nor decrease. This indicates that there is no clear relationship between X and Y and highlights the lack of association between X and Y.
- Figure 4.3b illustrates a nonlinear relationship between X and Y values.

Example 4.3.1 To explore the possible relationship between the overall number of tweets with the number of followers on Twitter, we take a sample of 10 influencers and obtain the following data:

$r_{Bravais-Pearson}$ =0.03,　$r_{Spearman}$ =−0.01　　　　　　　$r_{Bravais-Pearson}$ =0.64,　$r_{Spearman}$ =0.99

(a) No clear relationship　　　　　　　(b) Nonlinear relationship

Fig. 4.3 Continues Fig. 4.2—more scatter plots

	Tweets	Followers
Influencer 1	25	7194
Influencer 2	11,800	43,400,000
Influencer 3	99	324,000
Influencer 4	1934	2,330,000
Influencer 5	199	39,000
Influencer 6	2539	189,000
Influencer 7	4334	639,000
Influencer 8	952	688,000
Influencer 9	3245	2,690,000
Influencer 10	2468	110,000

The tweets are denoted by x_i and the followers are denoted by y_i, $i = 1, 2, \ldots, 10$. We plot paired observations (x_i, y_i) into a cartesian coordinate system. For example, we plot $(x_1; y_1) = (25; 7194)$ for the first influencer, $(x_2; y_2) = (11, 800; 43, 400, 000)$ for the second, and so on. Figure 4.4 shows the scatter plot for the number of tweets and the number of followers (on a log-scale).

One can see that there is a positive association between the number of tweets and the number of followers. This does however *not* imply a causal relationship: it is not necessarily *because* someone tweets more he/she has more followers or *because* someone has more followers he/she tweets more; the scatter plot just describes that those with more tweets have more followers. In R, we produce this scatter plot by the plot command:

```
tweets <- c(25,11800,99,...)
followers <- c(7194,43400000,...)
plot(tweets,followers)
```

Fig. 4.4 Scatter plot between tweets and followers

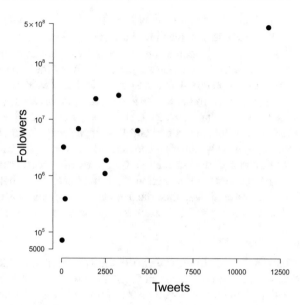

4.3.2   Correlation Coefficient

Suppose two variables X and Y are both measured on a metrical scale and are approximately linearly related like $Y \approx a + b X$ where a and b are constant values. The **correlation coefficient** $r(X, Y) = r$ measures the degree of *linear* relationship between X and Y using

$$
r = \frac{\sum\limits_{i=1}^{n}(x_i - \bar{x})(y_i - \bar{y})}{\sqrt{\sum\limits_{i=1}^{n}(x_i - \bar{x})^2 \cdot \sum\limits_{i=1}^{n}(y_i - \bar{y})^2}} = \frac{S_{xy}}{\sqrt{S_{xx}S_{yy}}}, \tag{4.16}
$$

with

$$
S_{xx} = \sum_{i=1}^{n}(x_i - \bar{x})^2 = n\tilde{s}_X^2, \quad S_{yy} = \sum_{i=1}^{n}(y_i - \bar{y})^2 = n\tilde{s}_Y^2, \tag{4.17}
$$

and

$$
S_{xy} = \sum_{i=1}^{n}(x_i - \bar{x})(y_i - \bar{y}) = \sum_{i=1}^{n} x_i y_i - n\bar{x}\bar{y}. \tag{4.18}
$$

Karl Pearson (1857–1936) presented the first rigorous treatment of correlation and acknowledged Auguste Bravais (1811–1863) for ascertaining the initial mathematical formulae for correlation. This is why the correlation coefficient is also known as the **Bravais–Pearson correlation coefficient**.

The correlation coefficient is independent of the units of measurement of X and Y. For example, if someone measures the height and weight in metres and kilograms respectively and another person measures them in centimetres and grams respectively, then the correlation coefficient between the two sets of data will be the same. The correlation coefficient is symmetric, i.e. $r(X, Y) = r(Y, X)$. The limits of r are $-1 \leq r \leq 1$. If all the points in a scatter plot lie exactly on a straight line, then the linear relationship between X and Y is perfect and $|r| = 1$, see also Exercise 4.7. If the relationship between X and Y is (i) perfectly linear and increasing, then $r = +1$ and (ii) perfectly linear and decreasing, then $r = -1$. The signs of r thus determine the direction of the association. If r is close to zero, then it indicates that the variables are independent or the relationship is not linear. Note that if the relationship between X and Y is nonlinear, then the degree of linear relationship may be low and r is then close to zero even if the variables are clearly not independent. Note that $r(X, X) = 1$ and $r(X, -X) = -1$.

Example 4.3.2 Look again at the scatter plots in Figs. 4.2 and 4.3. We observe strong positive linear correlation in Fig. 4.2a ($r = 0.91$), strong negative linear correlation in Fig. 4.2b ($r = -0.92$), moderate positive linear correlation in Fig. 4.2c ($r = 0.50$), moderate negative linear association in Fig. 4.2d ($r = -0.56$), no visible correlation in Fig. 4.3a ($r = 0.03$), and strong nonlinear (but not so strong linear) correlation in Fig. 4.3b ($r = 0.64$).

Example 4.3.3 In a decathlon competition, a group of athletes are competing with each other in 10 different track and field events. Suppose we are interested in how the results of the 100 m race relate to the results of the long jump competition. The correlation coefficient for the 100 m race (X, in seconds) and the long jump event (Y, in metres) for 5 athletes participating in the 2004 Olympic Games (see also Appendix A.4.2) are listed in Table 4.7.

To calculate the correlation coefficient, we need the following summary statistics:

$$\bar{x} = \frac{1}{5}(10.85 + 10.44 + 10.50 + 10.89 + 10.62) = 10.66$$

$$\bar{y} = \frac{1}{5}(7.84 + 7.96 + 7.81 + 7.47 + 7.74) = 7.764$$

$$S_{xx} = (10.85 - 10.66)^2 + (10.44 - 10.66)^2 + \ldots + (10.62 - 10.66)^2 = 0.1646$$

$$S_{yy} = (7.84 - 7.764)^2 + (7.96 - 7.764)^2 + \ldots + (7.74 - 7.764)^2 = 0.13332$$

$$S_{xy} = (10.85 - 10.66)(7.84 - 7.764) + \ldots + (10.62 - 10.66)(7.74 - 7.764)$$
$$= -0.1027$$

The correlation coefficient therefore is

$$r = \frac{S_{xy}}{\sqrt{S_{xx} S_{yy}}} = \frac{-0.1027}{\sqrt{0.1646 \times 0.13332}} \approx -0.69 \, .$$

Table 4.7 Results of 100 m race and long jump of 5 athletes

i	x_i	y_i
Roman Sebrle	10.85	7.84
Bryan Clay	10.44	7.96
Dmitriy Karpov	10.50	7.81
Dean Macey	10.89	7.47
Chiel Warners	10.62	7.74

Since -0.69 is negative, we can say that (i) there is a negative correlation between the 100 m race and the long jump event, i.e. shorter running times result in longer long jump results and (ii) this association is moderate to strong.

In R, we can obtain the results (after attaching the data) as follows:

```
cor(X.100m,X.Long.jump, method='pearson')
```

4.3.3 Spearman's Rank Correlation Coefficient

Consider a situation where n objects are ranked with respect to two variables X and Y. For instance, the variables could represent the opinion of two different judges in a talent competition who rank the participants with respect to their performance. This means that for each judge, the worst participant (with the lowest score x_i) is assigned rank 1, the second worst participant (with the second lowest score x_i) will receive rank 2, and so on. Thus, every participant has been given two ranks by two different judges. Suppose we want to measure the degree of association between the two different judgments, that is the two different sets of ranks. We expect that under perfect agreement, both the judges give the same judgment in the sense that they give the same ranks to each candidate. However, if they are not in perfect agreement then there may be some variation in the ranks assigned by them. To measure the degree of agreement, or, in general, the degree of association, one can use **Spearman's rank correlation coefficient**. As the name says, this correlation coefficient uses only the ranks of the values and not the values themselves. Thus, this measure is suitable for both ordinal and metrical variables. We introduce the following notations: let $R(x_i)$ denote the rank of the ith observation on X, i.e. the rank of x_i among the ordered values of X. Similarly, $R(y_i)$ denotes the rank of the ith observation of y. The difference between the two rank values is $d_i = R(x_i) - R(y_i)$. Spearman's rank correlation coefficient is defined as

$$R = 1 - \frac{6 \sum_{i=1}^{n} d_i^2}{n(n^2 - 1)}. \tag{4.19}$$

The values of R lie between -1 and $+1$ and measure the degree of correlation between the ranks of X and Y. Note that it doesn't matter whether we choose an ascending or descending order of the ranks, the value of R remains the same. When all the observations are assigned exactly the same ranks, then $R = 1$ and when all the observation are assigned exactly the opposite ranks, then $R = -1$.

Example 4.3.4 Look again at the scatter plots in Figs. 4.2 and 4.3. We observe strong positive correlation in Fig. 4.2a ($R = 0.87$), strong negative correlation in Fig. 4.2b ($R = -0.92$), moderate positive correlation in Fig. 4.2c ($R = 0.51$), moderate negative association in Fig. 4.2d ($R = -0.55$), no visible correlation in Fig. 4.3a ($R = -0.01$) and strong nonlinear correlation in Fig. 4.3b ($R = 0.99$).

Example 4.3.5 Let's follow Example 4.3.3 a bit further and calculate Spearman's rank correlation coefficient for the first five observations of the decathlon data. Again we list the results of the 100 m race (X) and the results of the long jump competition (Y). In addition, we assign ranks to both X and Y. For example, the shortest time receives rank 1, whereas the longest time receives rank 5. Similarly, the shortest long jump result receives rank 1, the longest long jump result receives rank 5.

i	x_i	$R(x_i)$	y_i	$R(y_i)$	d_i	d_i^2
Roman Sebrle	10.85	4	7.84	4	0	0
Bryan Clay	10.44	1	7.96	5	-4	16
Dmitriy Karpov	10.50	2	7.81	3	-1	1
Dean Macey	10.89	5	7.47	1	-4	16
Chiel Warners	10.62	3	7.74	2	-1	1
Total						34

Using (4.19), Spearman's rank correlation coefficient can be calculated as

$$R = 1 - \frac{6 \sum_{i=1}^{n} d_i^2}{n(n^2 - 1)} = 1 - \frac{6 \cdot 34}{5 \cdot 24} = -0.7.$$

We therefore have a moderate to strong negative association between the 100 m race and the long jump event. We now know that for the 5 athletes above longer running times relate to shorter jumping distances which in turn means that a good performance in one discipline implies a good performance in the other discipline. In R, we can obtain the same results by using the cor command:

```
cor(X.100m,X.Long.jump, method='spearman')
```

If two or more observations take the same values for x_i (or y_i), then there is a **tie**. In such situations the respective ranks can simply be averaged, though more complicated solutions also exist (one of which is implemented in the R function cor). For example if in Example 4.3.5 Bryan Clay's was 10.50 s instead of 10.44 s, then both Bryan Clay and Dmitriy Karpov had the same time. Instead of assigning the ranks 1 and 2 to them we assign the ranks 1.5 to each of them.

The differences between the correlation coefficient and the rank correlation coefficient are manifold: firstly, Pearson's correlation coefficient can be used for metrical variables only, but not for nominal or ordinal variables. The rank correlation coefficient can be used for either two metrical or two ordinal variables or a combination of an ordinal and a metric variable, but not for two nominal variables. Moreover, the rank correlation coefficient responds to any type of relationship whereas Pearson's correlation measures the degree of a linear relationship only—see also Fig. 4.3b. Another difference between the two correlation coefficients is that Pearson uses the entire information contained in the metric data in contrast to the rank correlation coefficient which uses only ordinal information contained in the ordered data.

4.3.4 Measures Using Discordant and Concordant Pairs

Another concept which uses ranks to measure the association between ordinal variables is based on **concordant** and **discordant** observation pairs. It is best illustrated by means of an example.

Example 4.3.6 Suppose an online book store conducts a survey on their customer's satisfaction with respect to both the timeliness of deliveries (X) and payment options (Y). Let us consider the following 2×3 contingency table with a summary of the responses of 100 customers. We assume that the categories for both variables can be ordered and ranks can be assigned to different categories, see the numbers in brackets in Table 4.8. There are 100 observation pairs (x_i, y_i) which summarize the response of the customers with respect to both X and Y. For example, there are 7 customers who were unsatisfied with the timeliness of the deliveries and complained that there are not enough payment options. If we compare two responses (x_{i_1}, y_{i_1}) and (x_{i_2}, y_{i_2}), it might be possible that one customer is more happy (or more unhappy) than the other customer with respect to both X and Y or that one customer is more happy with respect to X but more unhappy with respect to Y (or vice versa). If the former is the case, then this is a concordant observation pair; if the latter is true, then it is a discordant pair. For instance, a customer who replied "enough" and "satisfied" is more happy than a customer who replied "not enough" and "unsatisfied" because he is more happy with respect to both X and Y.

Table 4.8 Payment options and timeliness survey with 100 participating customers

		Timeliness			
		Unsatisfied (1)	Satisfied (2)	Very satisfied (3)	Total
Payment options	Not enough (1)	7	11	26	44
	Enough (2)	10	15	31	56
	Total	17	26	57	100

In general, a pair is

- **concordant** if $i_2 > i_1$ and $j_2 > j_1$ (or $i_2 < i_1$ and $j_2 < j_1$),
- **discordant** if $i_2 < i_1$ and $j_2 > j_1$ (or $i_2 > i_1$ and $j_2 < j_1$),
- **tied** if $i_1 = i_2$ (or $j_1 = j_2$).

Obviously, if we have only concordant observations, then there is a strong positive association because a higher value of X (in terms of the ranking) implies a higher value of Y. However, if we have only discordant observations then there is a clear negative association. The measures which are introduced below simply put the number of concordant and discordant pairs into relation. This idea is reflected in **Goodman and Kruskal's** γ which is defined as

$$\gamma = \frac{K}{K+D} - \frac{D}{K+D} = \frac{K-D}{K+D}, \qquad (4.20)$$

where

$$K = \sum_{i<m}\sum_{j<n} n_{ij} n_{mn}, \qquad D = \sum_{i<m}\sum_{j>n} n_{ij} n_{mn}$$

describe the number of concordant and discordant observation pairs respectively. An alternative measure is **Stuart's** τ_c given as

$$\tau_c = \frac{2\min(k,l)(K-D)}{n^2(\min(k,l)-1)}. \qquad (4.21)$$

Both measures are standardized to lie between -1 and 1, where larger values indicate a stronger association and the sign indicates the direction of the association.

Example 4.3.7 Consider Example 4.3.6. A customer who replied "enough" and "satisfied" is more happy than a customer who replied "not enough" and "unsatisfied" because the observation pairs, using ranks, are $(2, 2)$ and $(1, 1)$ and therefore $i_2 > i_1$

(a)	y_1	y_2	y_3
x_1		t	t
x_2	t	c	c

(b)	y_1	y_2	y_3
x_1	t		t
x_2	d	t	c

(c)	y_1	y_2	y_3
x_1	t	t	
x_2	d	d	t

(d)	y_1	y_2	y_3
x_1	t	d	d
x_2		t	t

(e)	y_1	y_2	y_3
x_1	c	t	d
x_2	t		t

(f)	y_1	y_2	y_3
x_1	c	c	t
x_2	t	t	

Fig. 4.5 Scheme to visualize concordant (c), discordant (d), and tied (t) pairs in a 2×3 contingency table

and $j_2 > j_1$. There are 7×15 such pairs. Similarly those who said "not enough" and "unsatisfied" are less happy than those who said "enough" and "very satisfied'" (7×31 pairs). Figure 4.5 summarizes the comparisons in detail.

Figure 4.5a shows that $(x_1, y_1) = $ (not enough, unsatisfied) is concordant to $(x_2, y_2) = $ (enough, satisfied) and $(x_2, y_3) = $ (enough, very satisfied) and tied to $(x_2, y_1) = $ (enough, unsatisfied), $(x_1, y_2) = $ (not enough, satisfied), and $(x_1, y_3) = $ (not enough, very satisfied). Thus for these comparisons, we have 0 discordant pairs, $(7 \times 15) + (7 \times 31)$ concordant pairs and $7 \times (10 + 11 + 26)$ tied pairs. Figure 4.5b–f show how the task can be completed. While tiresome, systematically working through the table (and making sure to not count pairs more than once) yields

$$K = 7 \times (15 + 31) + 11 \times 31 = 663$$
$$D = 10 \times (11 + 26) + 15 \times 26 = 760$$

As a visual rule of thumb, working from the top left to the bottom right yields the concordant pairs; and working from the bottom left to the top right yields the discordant pairs. It follows that $\gamma = (663 - 760)/(663 + 760) \approx -0.07$ which indicates no clear relationship between the two variables. A similar result is obtained using τ_c which is $4 \times (760 - 663)/100^2 \approx 0.039$. This rather lengthy task can be made much quicker by using the ord.gamma and ord.tau commands from the R library ryouready:

```
library(ryouready)
ex <- matrix(c(7,11,26,10,15,31),ncol=3,byrow=T)
ord.gamma(ex)
ord.tau(ex)
```

4.4 Visualization of Variables from Different Scales

If we want to jointly visualize the association between a variable X, which is either nominal or ordinal and another variable Y, which is metric, then we can use any graph which is suitable for the metric variable (see Chaps. 2 and 3) and produce it

(a) Using box plots   (b) Using ECDF's

Fig. 4.6 Distribution of pizza delivery time stratified by branch

for each category of the nominal/ordinal variable. We recommend using stratified box plots or stratified ECDF's, as they are easy to read when summarized in a single figure; however, it is also possible to place histograms next to each other or on top of each other, or overlay kernel density plots, but we do not illustrate this here in more detail.

Example 4.4.1 Consider again our pizza example (Sect. A.4.1). If we are interested in the delivery times by branch, we may plot the box plots and ECDF's of delivery time by branch. Figure 4.6 shows that the shortest delivery times can be observed in the branch in the East. Producing these graphs in *R* is straightforward: The `boxplot` command can be used for two variables by separating them with the ~ sign. For the ECDF we have to produce a plot for each branch and overlay them with the "add = TRUE" option.

```
boxplot(time~branch)
plot.ecdf(time[branch=='East'])
plot.ecdf(time[branch=='West'], add=TRUE)
plot.ecdf(time[branch=='Centre'], add=TRUE)
```

4.5 Key Points and Further Issues

> **Note:**
>
> ✓ How to use different measures of association:
>
> | | | |
> |---|---|---|
> | 2 nominal variables | → | Pearson's χ^2, relative risks, odds ratio, Cramer's V and C_{corr} |
> | 2 ordinal variables | → | Spearman's rank correlation coefficient, γ, τ_c |
> | 2 metric variables | → | Pearson's correlation coefficient, Spearman's correlation coefficient |
>
> ✓ For two variables which are measured on different scales, for example metric/ordinal or ordinal/nominal, one should use measures of association suitable for the less informative of the two scales.
>
> ✓ Another graphical representation of both a continuous and discrete variable is stratified confidence interval plots (error plots), see Chap. 9.

4.6 Exercises

Exercise 4.1 A newspaper asks two of its staff to review the coffee quality at different trendy cafés. The coffee can be rated on a scale from 1 (miserable) to 10 (excellent). The results of the two coffee enthusiasts X and Y are as follows:

Café i	x_i	y_i
1	3	6
2	8	7
3	7	10
4	9	8
5	5	4

(a) Calculate and interpret Spearman's rank correlation coefficient.
(b) Does Spearman's R differ depending on whether ranks are assigned in a decreasing or increasing order?
(c) Suppose the coffee can only be rated as either good (>5) or bad (≤ 5). Do the chances of a good rating differ between the two journalists?

Exercise 4.2 A total of 150 customers of a petrol station are asked about their satisfaction with their car and motorbike insurance. The results are summarized below:

	Satisfied	Unsatisfied	Total
Car	33	25	58
Car (diesel engine)	29	31	60
Motorbike	12	20	32
Total	74	76	150

(a) Determine and interpret Pearson's χ^2 statistic, Cramer's V, and C_{corr}.
(b) Combine the categories "car" and "car (diesel engine)" and produce the corresponding 2×2 table. Calculate χ^2 as efficiently as possible and give a meaningful interpretation of the odds ratio.
(c) Compare the results from (a) and (b).

Exercise 4.3 Recently there has been a big debate about the usefulness of speed limits on public roads. Consider the following table which lists the speed limits for country roads (in miles/h) and traffic deaths (per 100 million km) for different countries in 1986 when the debate was particularly serious:

Country	Speed limit	Traffic deaths
Denmark	55	4.1
Japan	55	4.7
Canada	60	4.3
Netherlands	60	5.1
Italy	75	6.1

(a) Draw the scatter plot for the two variables.
(b) Calculate the Bravais–Pearson and Spearman correlation coefficients.
(c) What are the correlation coefficients if the speed limit is given in km/h rather than miles/h (1 mile/h $\approx$ 1.61 km/h)?
(d) Consider one more observation: the speed limit for England was 70 miles/h and the death rate was 3.1.

  (i) Add this observation to the scatter plot.
  (ii) Calculate the Bravais–Pearson correlation coefficient given this additional observation.

Exercise 4.4 The famous passenger liner *Titanic* hit an iceberg in 1912 and sank. A total of 337 passengers travelled in first class, 285 in second class, and 721 in third class. In addition, there were 885 staff members on board. Not all passengers could be rescued. Only the following were rescued: 135 from the first class, 160 from the second class, 541 from the third class and 674 staff.

(a) Determine and interpret the contingency table for the variables "travel class" and "rescue status".

(b) Use a contingency table to summarize the conditional relative frequency distributions of rescue status given travel class. Could there be an association of the two variables?

(c) What would the contingency table from (a) look like under the independence assumption? Calculate Cramer's V statistic. Is there any association between travel class and rescue status?

(d) Combine the categories "first class" and "second class" as well as "third class" and "staff". Create a contingency table based on these new categories. Determine and interpret Cramer's V, the odds ratio and relative risks of your choice.

(e) Given the results from (a)–(d), what are your conclusions?

Exercise 4.5 To study the association of the monthly average temperature (in °C, X) and hotel occupation (in %, Y), we consider data from three cities: Polenca (Mallorca, Spain) as a summer holiday destination, Davos (Switzerland) as a winter skiing destination and Basel (Switzerland) as a business destination.

Month	Davos X	Davos Y	Polenca X	Polenca Y	Basel X	Basel Y
Jan	−6	91	10	13	1	23
Feb	−5	89	10	21	0	82
Mar	2	76	14	42	5	40
Apr	4	52	17	64	9	45
May	7	42	22	79	14	39
Jun	15	36	24	81	20	43
Jul	17	37	26	86	23	50
Aug	19	39	27	92	24	95
Sep	13	26	22	36	21	64
Oct	9	27	19	23	14	78
Nov	4	68	14	13	9	9
Dec	0	92	12	41	4	12

(a) Calculate the Bravais–Pearson correlation coefficient. The following summary statistics are available: $\sum_{i=1}^{36} x_i y_i = 22{,}776$, $\bar{x} = 12.22$, $\bar{y} = 51.28$, $\tilde{s}_x^2 = 76.95$ and $\tilde{s}_y^2 = 706.98$.

(b) Interpret the scatter plot in Fig. 4.7 which visualizes temperature and hotel occupancy for Davos (D), Polenca (P) and Basel (B).

(c) Use R to calculate the correlation coefficient separately for each city. Interpret the results and discuss the use of the correlation coefficient if more than two variables are available.

Exercise 4.6 Consider a neighbourhood survey on the use of a local park. Respondents were asked whether the park may be used for summer music concerts and whether dog owners should put their dogs on a lead. The results are summarized in the following contingency table:

Fig. 4.7 Temperature and hotel occupancy for the different cities

		Put dogs on a lead		
	Agree	No opinion	Disagree	Total
Use for concerts Agree	82	4	0	86
No opinion	8	43	9	60
Disagree	0	2	10	12
Total	90	49	19	158

(a) Calculate and interpret Goodman and Kruskal's γ.
(b) Now ignore the ordinal structure of the data and calculate Cramer's V.
(c) Create the contingency table which is obtained when the categories "no opinion" and "agree" are combined.
(d) What is the relative risk of disagreement with summer concerts depending on the opinion about using leads?
(e) Calculate the odds ratio and offer two interpretations of it.
(f) Determine γ for the table calculated in (c).
(g) What is your final interpretation and what may be the best measure to use in this example?

Exercise 4.7 Consider n observations for which $y_i = a + bx_i$, $b > 0$, holds. Show that $r = 1$.

Exercise 4.8 Make yourself familiar with the Olympic decathlon data described in Appendix A.4.2. Read in and attach the data in R.

(a) Use R to calculate and interpret the Bravais–Pearson correlation coefficient between the results of the discus and the shot put events.
(b) There are 10 metric variables. How many different correlation coefficients can you calculate? How would you summarize them?
(c) Apply the `cor` command to the whole data and interpret the output.
(d) Omit the two rows which contain missing data and interpret the output again.

Exercise 4.9 We are interested in the pizza delivery data which is described in Appendix A.4.1.

(a) Read in the data and create two new binary variables which describe whether a pizza was hot ($>65\,^\circ$C) and the delivery time was short (<30 min). Create a contingency table for the two new variables.
(b) Calculate and interpret the odds ratio for the contingency table from (a).
(c) Use Cramer's V, Stuart's τ_c, Goodman and Kruskal's γ and a stacked bar chart to explore the association between the categorical time and temperature variables.
(d) Draw a scatter plot for the continuous time and temperature variables. Determine both the Bravais–Pearson and Spearman correlation coefficients.
(e) Use methods of your choice to explore the relationship between temperature and driver, operator, number of ordered pizzas and bill. Is it clear which of the variables influence the pizza temperature?

$\rightarrow$ Solutions to all exercises in this chapter can be found in Appendix B

Part II
Probability Calculus

Combinatorics

5

5.1 Introduction

Combinatorics is a special branch of mathematics. It has many applications not only in several interesting fields such as enumerative combinatorics (the classical application) but also in other fields, for example in graph theory and optimization.

First we try to motivate and understand the role of combinatorics in statistics. Consider a simple example in which someone goes to a cafe. The person would like a hot beverage and a cake. Assume that one can choose among three different beverages, for example cappuccino, hot chocolate and green tea, and three different cakes, let's say carrot cake, chocolate cake and lemon tart. The person may consider different beverage and cake combinations when placing the order, for example carrot cake and cappuccino, carrot cake and tea, hot chocolate and lemon tart and many more. From a statistical perspective, the customer is evaluating the possible combinations before making a decision. Depending on their preferences, the order will be placed by choosing one of the combinations.

In this example, it is easy to calculate the number of possible combinations. There are three different beverages and three different cakes to choose from, leading to nine different (3 × 3) beverage and cake combinations. However, suppose there is a choice of 15 hot beverages and 8 different cakes. How many orders can be made? (Answer: 15 × 8) What if the person decides to order two cakes, how will it affect the number of possible combinations of choices? It will be a tedious task to count all the possibilities. So we need a systematic approach to count such possible combinations. Combinatorics deals with the counting of different possibilities in a systematic approach.

People often use the urn model to understand the system in the counting process. The urn model deals with the drawing of balls from an urn. The balls in the urn represent the units of a population, or the features of a population. The balls may

© Springer Nature Switzerland AG 2022
C. Heumann et al., *Introduction to Statistics and Data Analysis*,
https://doi.org/10.1007/978-3-031-11833-3_5

Fig. 5.1 a Representation of the urn model. Drawing from the urn model **b** with replacement and **c** without replacement. Compositions of three drawn balls: **d** all balls are distinguishable and **e** some balls are not distinguishable

vary in colour or size to represent specific properties of a unit or feature. We illustrate this concept in more detail in Fig. 5.1.

Suppose there are 5 balls of three different colours—two black, one grey, and two white (see Fig. 5.1a). This can be generalized to a situation in which there are n balls in the urn and we want to draw m balls. Suppose we want to know

- how many different possibilities exist to draw m out of n balls (thus determining the number of distinguishable **combinations**).

To deal with such a question, we first need to decide whether a ball will be put back into the urn after it is drawn or not. Figure 5.1b illustrates that a grey ball is drawn from the urn and then placed back (illustrated by the two-headed arrow). We say the ball is drawn *with replacement*. Figure 5.1c illustrates a different situation in which the grey ball is drawn from the urn and is *not* placed back into the urn (illustrated by the one-headed arrow). We say the ball is drawn *without replacement*.

Further, we may be interested in knowing the

- total number of ways in which the chosen set of balls can be arranged in a distinguishable order (which we will define as **permutations** later in this chapter).

To answer the question how many permutations exist, we first need to decide whether all the chosen balls are distinguishable from each other or not. For example, in Fig. 5.1d, the three chosen balls have different colours, therefore they are distinguishable. There are many options on how they can be arranged. In contrast, some of the chosen balls in Fig. 5.1e have the same colour, they are therefore not distinguishable. Consequently the number of combinations is much more limited. The concept of balls and urns just represents the features of observations from a sample. We illustrate this in more detail in the following example.

Example 5.1.1 Say a father promises his daughter three scoops of ice cream if she cleans up her room. For simplicity, let's assume the daughter has a choice of four flavours: chocolate—banana—cherry—lemon. How many different choices does the daughter have? If each scoop has to be a different flavour she obviously has much less choice than if the scoops can have the same flavour. In the urn model this is represented by the concept of "with/without replacement". The urn contains 4 balls of 4 different colours which represent the ice cream flavours. For each of the three scoops a ball is drawn to determine the flavour. If we draw with replacement, each flavour can be potentially chosen multiple times; however, if we draw without replacement each flavour can be chosen only once. Then the number of possible combinations is easy to calculate: it is 4, i.e. (chocolate, banana, cherry)—(chocolate, banana, lemon)—(chocolate, cherry, lemon)—(banana, cherry, lemon). But what if we have more choices? Or if we can draw flavours multiple times? We then need calculation rules which help us counting the number of options.

Now, let's assume that the daughter picked the flavours (chocolate [C], banana [B], lemon [L]). Like many other children she prefers to eat her most favourite flavour (chocolate) last, and her least favourite flavour (cherry) first. Therefore, the order in which the scoops are place on top of the cone are important! In how many different ways can the scoops be placed on top of the cone? This relates to the question of the number of distinguishable *permutations*. The answer is 6: (C, B, L)—(C, L, B)—(B, L, C)—(B, C, L)—(L, B, C)—(L, C, B). But what if the daughter did pick a flavour multiple times, e.g. (chocolate, chocolate, lemon)? Since the two chocolate scoops are non-distinguishable, there are fewer permutations: (chocolate, chocolate, lemon)—(chocolate, lemon, chocolate)—(lemon, chocolate, chocolate).

The bottom line of this example is that the number of combinations/options is determined by (i) whether we draw with or without replacement (i.e. allow flavours to be chosen more than once) and (ii) whether the arrangement in a particular order (= permutation) is of any specific interest.

Consider the urn example again. Suppose three balls of different colours, black, grey and white are drawn. Now there are two options: The first option is to take into account the order in which the balls are drawn. In such a situation two possible sets of balls like (black, grey, white) and (white, black, grey) constitute two different sets. Such a set is called an *ordered set*. In the second option, we do not take into account the order in which the balls are drawn. In such a situation, the two possible sets of balls like (black, grey, white) and (white, black, grey) are the same sets and constitute an *unordered set* of balls.

Definition 5.1.1 A group of elements is said to be **ordered** if the order in which these elements are drawn is of relevance. Otherwise it is called **unordered**.

Examples

- Ordered samples:
  - The first three places in an Olympic 100 m race are determined by the order in which the athletes arrive at the finishing line. If 8 athletes are competing with each other, the number of possible results for the first three places is of interest. In the urn language, we are taking draws without replacement (since every athlete can only have one distinct place).
  - In a raffle with two prizes, the first drawn raffle ticket gets the first prize and the second raffle ticket gets the second prize.
  - There exist various esoteric tarot card games which claim to foretell someone's fortune with respect to several aspects of life. The order in which the cards are shown on the table is important for the interpretation.
- Unordered samples:
  - The selected members for a national football team. The order in which the selected names are announced is irrelevant.
  - Out of 10 economists, 10 medical doctors, and 10 statisticians, an advisory committee consisting of 4 economists, 3 medical doctors, and 2 statisticians is elected.
  - Fishing 20 fish from a lake.
  - A bunch of 10 flowers made from 21 flowers of 4 different colours.

5.2   Permutations

Definition 5.2.1 The factorial function $n!$ is defined as

$$n! = \begin{cases} 1 & \text{for } n = 0 \\ 1 \cdot 2 \cdot 3 \cdots n & \text{for } n > 0 \,. \end{cases} \tag{5.1}$$

Example 5.2.1 It follows from the definition of the factorial function that

$$0! = 1, \quad 1! = 1 \quad 2! = 1 \cdot 2 = 2, \quad 3! = 1 \cdot 2 \cdot 3 = 6 \,.$$

This can be calculated in R as follows:

```
factorial(n)
```

Definition 5.2.2 Consider a set of n elements. Each ordered composition of these n elements is called a **permutation**.

We distinguish between two cases: If all the elements are distinguishable, then we speak of *permutation without replacement*. However, if some or all of the elements are not distinguishable, then we speak of *permutation with replacement*. Please note that the meaning of "replacement" here is just a convention and does not directly refer to the drawings, e.g. from the urn model considered in Fig. 5.1.

5.2.1 Permutations Without Replacement

If all the n elements are distinguishable, then there are

$$n! \tag{5.2}$$

different compositions of these elements.

Example 5.2.2 There were three candidate cities for hosting the 2020 Olympic Games: Tokyo (T), Istanbul (I), and Madrid (M). Before the election, there were $3! = 6$ possible outcomes regarding the final rankings of the cities:

$$(M, T, I), (M, I, T), (T, M, I), (T, I, M), (I, M, T), (I, T, M).$$

5.2.2 Permutations with Replacement

Assume that not all n elements are distinguishable. The elements are divided into groups and these groups are distinguishable. Suppose there are s groups of sizes $n_1, n_2, \ldots, n_s$ and $n = n_1 + \ldots + n_s$. The total number of different ways to arrange the n elements in s groups is:

$$\frac{n!}{n_1! \, n_2! \, n_3! \, \cdots n_s!} . \tag{5.3}$$

Example 5.2.3 Consider the data in Fig. 5.1e. There are two groups consisting of two black balls ($n_1 = 2$) and one white ball ($n_2 = 1$). So there are the following three possible combinations to arrange the balls: (black, black, white), (black, white, black) and (white, black, black). This can be determined by calculating

$$\frac{3!}{2! \, 1!} = \frac{3 \cdot 2 \cdot 1}{2 \cdot 1 \cdot 1} = 3 .$$

5.3 Combinations

Definition 5.3.1 The Binomial coefficient for any integers m and n with $n \geq m \geq 0$ is denoted and defined as

$$\binom{n}{m} = \frac{n!}{m!\,(n-m)!} \,. \tag{5.4}$$

It is read as "n choose m" and can be calculated in R using the following command:

```
choose(n,m)
```

There are several calculation rules for the binomial coefficient:

$$\binom{n}{0} = 1, \quad \binom{n}{1} = n, \quad \binom{n}{m} = \binom{n}{n-m}, \quad \binom{n}{m} = \prod_{i=1}^{m} \frac{n+1-i}{i}. \tag{5.5}$$

We now answer the question of how many different possibilities exist to draw m out of n elements, i.e. m out of n balls from an urn. It is necessary to distinguish between the following four cases:

(1) Combinations **without** replacement and **without** consideration of the order of the elements.
(2) Combinations **without** replacement and **with** consideration of the order of the elements.
(3) Combinations **with** replacement and **without** consideration of the order of the elements.
(4) Combinations **with** replacement and **with** consideration of the order of the elements.

5.3.1 Combinations Without Replacement and Without Consideration of the Order

When there is no replacement and the order of the elements is also not relevant, then the total number of distinguishable combinations in drawing m out of n elements is

$$\binom{n}{m}. \tag{5.6}$$

Example 5.3.1 Suppose a company elects a new board of directors. The board consists of 5 members and 15 people are eligible to be elected. How many combinations for the board of directors exist? Since a person can't be elected twice, we have a situation where there is no replacement. The order is also of no importance: either one is elected or not. We can thus apply (5.6) which yields

$$\binom{15}{5} = \frac{15!}{10!5!} = 3003$$

possible combinations. This result can be obtained in R by using the command choose(15,5).

5.3.2 Combinations Without Replacement and with Consideration of the Order

The total number of different combinations for the setting without replacement and with consideration of the order is

$$\frac{n!}{(n-m)!} = \binom{n}{m}m! . \tag{5.7}$$

Example 5.3.2 Consider a horse race with 12 horses. A possible bet is to forecast the winner of the race, the second horse of the race and the third horse of the race. The total number of different combinations for the horses in the first three places is

$$\frac{12!}{(12-3)!} = 12 \cdot 11 \cdot 10 = 1320 .$$

This result can be explained intuitively: for the first place, there is a choice of 12 different horses. For the second place, there is a choice of 11 different horses (12 horses minus the winner). For the third place, there is a choice of 10 different horses (12 horses minus the first and second horses). The total number of combinations is the product $12 \cdot 11 \cdot 10$. This can be calculated in R by typing 12*11*10.

5.3.3 Combinations with Replacement and Without Consideration of the Order

The total number of different combinations with replacement and without consideration of the order is

$$\binom{n+m-1}{m} = \frac{(n+m-1)!}{m!(n-1)!} = \binom{n+m-1}{n-1} . \tag{5.8}$$

Note, that these are the two representations which follow from the definition of the binomial coefficient but typically only the first representation is used in textbooks. We will motivate the second representation after Example 5.3.3.

Example 5.3.3 A farmer has 2 fields and aspires to cultivate one out of 4 different organic products per field. Then the total number of choices he has is

$$\binom{4+2-1}{2} = \binom{5}{2} = \frac{5!}{2!\,3!} = \frac{3!\cdot 4\cdot 5}{1\cdot 2\cdot 3!} = 10. \tag{5.9}$$

If 4 different organic products are denoted as a, b, c and d, then the following combinations are possible:

$$
\begin{array}{llll}
(a, a) & (a, b) & (a, c) & (a, d) \\
 & (b, b) & (b, c) & (b, d) \\
 & & (c, c) & (c, d) \\
 & & & (d, d)
\end{array}
$$

Please note that, for example, (a, b) is identical to (b, a) because the order in which the products a and b are cultivated on the first or second field is not important in this example.

We now try to give an intuitive explanation of formula (5.9) using Example 5.3.3. We have $n = 4$ products and $m = 2$ fields and apply the following technical "trick": we sort the combinations by the product symbols (a, b, c, or d). When we switch from one product to the next (e.g. from b to c) we make a note by adding a vertical line |. Whenever a product is skipped, we add a line too. For example, the combination (a, c) is denoted by $a||c|$, the combination (d, d) by $|||dd$, (c, c) by $||cc|$ and (a, a) by $aa|||$. Therefore, the number of characters equates to the 2 chosen symbols of the set (a, b, c, d) plus the 3 vertical lines, in summary $(4 + 2) - 1 = 5$ places where $3 = n - 1$ places are selected for the vertical line |. How many different line/letter combinations exist? There are 3 out of 5 possible positions for |, i.e. $\binom{5}{3} = 10$ possible combinations—and this is nothing but the right-hand side of (5.9).

5.3.4 Combinations with Replacement and with Consideration of the Order

The total number of different combinations for the integers m and n with replacement and when the order is of relevance is

$$n^m . \tag{5.10}$$

Example 5.3.4 Consider a credit card with a four-digit personal identification number (PIN) code. The total number of possible combinations for the PIN is

$$n^m = 10^4 = 10{,}000.$$

Note that every digit in the first, second, third and fourth places $(m = 4)$ can be chosen out of ten digits from 0 to 9 $(n = 10)$.

5.4 Key Points and Further Issues

> **Note:**
>
> ✓ The rules of combinatorics are as follows:
>
> | Combinations | without replacement | with replacement |
> |---|---|---|
> | without order | $\binom{n}{m}$ | $\binom{n+m-1}{m}$ |
> | with order | $\binom{n}{m}m!$ | n^m |
>
> ✓ Combinations with and without *replacement* are also often called combinations with and without *repetition*.
>
> ✓ The permutation rules are as follows:
>
> | | without replacement | with replacement |
> |---|---|---|
> | Permutations | $n!$ | $\dfrac{n!}{n_1! \cdots n_s!}$ |

5.5 Exercises

Exercise 5.1 At a party with 10 guests, every guest shakes hands with each other guest. How many handshakes can be counted in total?

Exercise 5.2 A language teacher is concerned about the vocabularies of his students. He thus tests 5 students in each lecture. What are the total number of possible combinations

(a) if a student is tested only once per lecture and
(b) if a student is tested more than once per lecture?

Use R to quantify numbers which you can't calculate manually.

Exercise 5.3 "Gobang" is a popular game in which two players set counters on a board with 381 knots. One needs to place 5 consecutive counters in a row to win the game. There are also rules on how to remove counters from the other player. Consider a match where 64 counters have already been placed on the board. How many possible combinations exist to place 64 counters on the board?

Exercise 5.4 A shop offers a special tray of beer: "Munich's favourites". Customers are allowed to fill the tray, which holds 20 bottles, with any combination of Munich's 6 most popular beers (from 6 different breweries).

(a) What are the number of possible combinations to fill the tray?
(b) A customer insists of having at least one beer from each brewery in his tray. How many options does he have to fill the tray?

Exercise 5.5 The football world cup usually consists of 32 teams. How many combinations for the top 3 teams exist when

(a) taking into account the order of these top 3 teams and
(b) *without* taking into account the order of these top 3 teams?

Exercise 5.6 An online book store assigns membership codes to each member. For administrative reasons, these codes consist of four letters between "A" and "L". A special discount period increased the total number of members from 18,200 to 20,500. Are there enough combinations of codes left to be assigned for the new membership codes?

Exercise 5.7 In the old scoring system of ice skating (valid until 2004), each member of a jury of 9 people judged the performance of the skaters on a scale between 0 and 6. It was a decimal scale and thus scores such as 5.1, 5.2, etc. were possible. Calculate the number of possible score combinations from the jury.

Fig. 5.2 Excerpt from Pascal's triangle (left) and its representation by means of binomial coefficients (right)

Exercise 5.8 It is possible in Pascal's triangle (Fig. 5.2, left) to view each entry as the sum of the two entries directly above it. For example, the 3 on the fourth line from the top is the sum of the 1 and 2 above the 3. Another interpretation refers to a geometric representation of the binomial coefficient, $\binom{n}{k}$ (Fig. 5.2, right) with $k = 0, 1, 2, \ldots$ being the column index and $n = 0, 1, 2, \ldots$ being the row index.

(a) Show that each entry in the bold third diagonal line can be represented via $\binom{n}{2}$.
(b) Now show that the sum of two consecutive entries in the bold third diagonal line always corresponds to quadratic numbers.

$\rightarrow$ Solutions to all exercises in this chapter can be found in Appendix B

Elements of Probability Theory

6

Let us first consider some simple examples to understand the need for probability theory. Often one needs to make a decision whether to carry an umbrella or not when leaving the house; a company might wonder whether to introduce a new advertisement to possibly increase sales or to continue with their current advertisement; or someone may want to choose a restaurant based on where he can get his favourite dish. In all these situations, randomness is involved. For example, the decision of whether to carry an umbrella or not is based on the possibility or chance of rain. The sales of the company may increase, decrease or remain unchanged with a new advertisement. The investment in a new advertising campaign may therefore only be useful if the probability of its success is higher than that of the current advertisement. Similarly, one may choose the restaurant where one is most confident of getting the food of one's choice. In all such cases, an event may be happening or not and depending on its likelihood, actions are taken. The purpose of this chapter is to learn how to calculate such likelihoods of events happening and not happening.

6.1   Basic Concepts and Set Theory

A simple (not rigorous) definition of a **random experiment** requires that the experiment can be repeated any number of times under the same set of conditions and its outcome is known only after the completion of the experiment. A simple and classical example of a random experiment is the tossing of a coin or the rolling of a die. If tossing a coin, it is unknown what the outcome will be, head or tail, until the coin is tossed. The experiment can be repeated and different outcomes may be observed in each repetition. Similarly, if rolling a die, it is unknown how many dots will appear on the upper surface until the die is rolled. Again, the die can be rolled repeatedly

© Springer Nature Switzerland AG 2022
C. Heumann et al., *Introduction to Statistics and Data Analysis*,
https://doi.org/10.1007/978-3-031-11833-3_6

and different numbers of dots are obtained in each trial. A possible outcome of a random experiment is called a **simple event** (or **elementary event**) and denoted by ω_i. The set of all possible outcomes, $\{\omega_1, \omega_2, \ldots, \omega_k\}$, is called the **sample space**, and is denoted as Ω, i.e. $\Omega = \{\omega_1, \omega_2, \ldots, \omega_k\}$. Subsets of Ω are called **events** and are denoted by capital letters such as A, B, C etc. The set of all simple events that are contained in the event A is denoted by Ω_A. The event $\bar{A}$ refers to the non-occurring of A and is called **a composite or complementary event**. Also Ω is an event. Since it contains all possible outcomes, we say that Ω will always occur and we call it a **sure event** or **certain event**. On the other hand, if we consider the null set $\emptyset = \{\}$ as an event, then this event can never occur and we call it an **impossible event**. The sure event therefore is the set of all elementary events and the impossible event is the set with no elementary events.

The above concepts of "events" form the basis of a definition of "probability". Once we understand the concept of probability, we can develop a framework to make conclusions about the population of interest, using a sample of data.

Example 6.1.1 (*Rolling a die*) If a die is rolled once, then the possible outcomes are the number of dots on the upper surface: $1, 2, \ldots, 6$. Therefore the sample space is the set of simple events $\omega_1 = $"1", $\omega_2 = $"2", $\ldots$, $\omega_6 = $"6" and $\Omega = \{\omega_1, \omega_2, \ldots, \omega_6\}$. Any subset of Ω can be used to define an event. For example, an event A may be "an even number of dots on the upper surface of the die". There are three possibilities that this event occurs: ω_2, ω_4 or ω_6. If an odd number shows up then the composite event $\bar{A}$ occurs instead of A. If an event is defined to observe only one particular number, say $\omega_1 = $"1", then it is an elementary event. An example of a sure event is "a number which is greater than or equal to 1" because any number between 1 and 6 is greater than or equal to 1. An impossible event is "the number is 7".

Example 6.1.2 (*Rolling two dice*) Suppose we throw two dice simultaneously and an event is defined as the "number of dots observed on the upper surface of both the dice". Then there are 36 simple events defined as (number of dots on first die, number of dots on second die), i.e. $\omega_1 = (1, 1)$, $\omega_2 = (1, 2)$, $\ldots$, $\omega_{36} = (6, 6)$. Therefore Ω is

$$\Omega = \begin{matrix} \{(1,1), & (1,2), & (1,3), & (1,4), & (1,5), & (1,6) \\ (2,1), & (2,2), & (2,3), & (2,4), & (2,5), & (2,6) \\ (3,1), & (3,2), & (3,3), & (3,4), & (3,5), & (3,6) \\ (4,1), & (4,2), & (4,3), & (4,4), & (4,5), & (4,6) \\ (5,1), & (5,2), & (5,3), & (5,4), & (5,5), & (5,6) \\ (6,1), & (6,2), & (6,3), & (6,4), & (6,5), & (6,6)\}. \end{matrix}$$

One can define different events and their corresponding sample spaces. For example, if an event A is defined as "upper faces of both the dice contain the same number of dots"; then the sample space is $\Omega_A = \{(1, 1), (2, 2), (3, 3), (4, 4), (5, 5), (6, 6)\}$. If another event B is defined as "the sum of numbers on the upper faces is 6" then the sample space is $\Omega_B = \{(1, 5), (2, 4), (3, 3), (4, 2), (5, 1)\}$. A sure event is "get

Fig. 6.1 $A \cup B$ and $A \cap B$

Fig. 6.2 $A \backslash B$ and $\bar{A} = \Omega \backslash A$

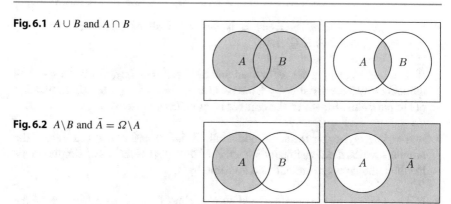

either an even number or an odd number"; an impossible event would be "the sum of the two dice is greater than 13".

It is possible to view events as sets of simple events. This helps to determine how different events relate to each other. A popular technique to visualize this approach is to use **Venn diagrams**. In Venn diagrams, two or more sets are visualized by circles. Overlapping circles imply that both events have one or more identical simple events. Separated circles mean that none of the simple events of event A are contained in the sample space of B. We use the following notations:

$A \cup B$ The union of events $A \cup B$ is the set of all simple events of A and B which occurs if at least one of the simple events of A *or* B occurs (Fig. 6.1, grey shaded area on left). Please note that we use the word "or" from a statistical perspective: "A or B" means that either a simple event from A occurs, or a simple event from B occurs, or a simple event which is part of both A and B occurs.

$A \cap B$ The intersection of events $A \cap B$ is the set of all simple events A and B which occur when a simple event occurs that belongs to A *and* B (Fig. 6.1, grey shaded area on right).

$A \backslash B$ The event $A \backslash B$ contains all simple events of A, which are not contained in B. The event "A but not B" or "A minus B" occurs, if A occurs but B does not occur. Also $A \backslash B = A \cap \bar{B}$ (Fig. 6.2, grey shaded area on left).

$\bar{A}$ The event $\bar{A}$ contains all simple events of Ω, which are not contained in A. The complementary event of A (which is "Not-A" or "$\bar{A}$" occurs whenever A does not occur (Fig. 6.2, grey shaded area on right).

$A \subseteq B$ A is a subset of B. This means that all simple events of A are also part of the sample space of B.

Example 6.1.3 Consider Example 6.1.1 where the sample space of rolling a die was determined as $\Omega = \{\omega_1, \omega_2, \ldots, \omega_6\}$ with $\omega_1 =$"1", $\omega_2 =$ "2", $\ldots$, $\omega_6 =$ "6".

- If $A = \{\omega_1, \omega_2, \omega_3, \omega_4, \omega_5\}$ and B is the set of all odd numbers, then $B = \{\omega_1, \omega_3, \omega_5\}$ and thus $B \subseteq A$.

- If $A = \{\omega_2, \omega_4, \omega_6\}$ is the set of even numbers and $B = \{\omega_3, \omega_6\}$ is the set of all numbers which are divisible by 3, then $A \cup B = \{\omega_2, \omega_3, \omega_4, \omega_6\}$ is the collection of simple events for which the number is either even or divisible by 3 or both.

- If $A = \{\omega_1, \omega_3, \omega_5\}$ is the set of odd numbers and $B = \{\omega_3, \omega_6\}$ is the set of the numbers which are divisible by 3, then $A \cap B = \{\omega_3\}$ is the set of simple events in which the numbers are odd and divisible by 3.

- If $A = \{\omega_1, \omega_3, \omega_5\}$ is the set of odd numbers and $B = \{\omega_3, \omega_6\}$ is the set of the numbers which are divisible by 3, then $A \backslash B = \{\omega_1, \omega_5\}$ is the set of simple events in which the numbers are odd but not divisible by 3.

- If $A = \{\omega_2, \omega_4, \omega_6\}$ is the set of even numbers, then $\bar{A} = \{\omega_1, \omega_3, \omega_5\}$ is the set of odd numbers.

Remark 6.1.1 Some textbooks also use the following notations:

$$
\begin{array}{lll}
A + B & \text{for} & A \cup B \\
AB & \text{for} & A \cap B \\
A - B & \text{for} & A \backslash B
\end{array}
$$

We can use these definitions and notations to derive the following properties of a particular event A:

$$
\begin{array}{ll}
A \cup A = A & A \cap A = A \\
A \cup \Omega = \Omega & A \cap \Omega = A \\
A \cup \emptyset = A & A \cap \emptyset = \emptyset \\
A \cup \bar{A} = \Omega & A \cap \bar{A} = \emptyset.
\end{array}
$$

Definition 6.1.1 Two events A and B are **disjoint** if $A \cap B = \emptyset$ holds, i.e. if both events cannot occur simultaneously.

Example 6.1.4 The events A and $\bar{A}$ are disjoint events.

Definition 6.1.2 The events $A_1, A_2, \ldots, A_m$ are said to be mutually or pairwise disjoint, if $A_i \cap A_j = \emptyset$ whenever $i \neq j = 1, 2, \ldots, m$.

Example 6.1.5 Recall Example 6.1.1. If $A = \{\omega_1, \omega_3, \omega_5\}$ and $B = \{\omega_2, \omega_4, \omega_6\}$ are the sets of odd and even numbers respectively, then the events A and B are disjoint.

Definition 6.1.3 The events $A_1, A_2, \ldots, A_m$ form a **complete decomposition** of Ω if and only if

$$A_1 \cup A_2 \cup \cdots \cup A_m = \Omega$$

and

$$A_i \cap A_j = \emptyset \quad (\text{for all } i \neq j).$$

Example 6.1.6 Consider Example 6.1.1. The elementary events $A_1 = \{\omega_1\}$, $A_2 = \{\omega_2\}, \ldots, A_6 = \{\omega_6\}$ form a complete decomposition. Other complete decompositions are, e.g.,

- $A_1 = \{\omega_1, \omega_3, \omega_5\}, \quad A_2 = \{\omega_2, \omega_4, \omega_6\}$
- $A_1 = \{\omega_1\}, \quad A_2 = \{\omega_2, \omega_3, \omega_4, \omega_5, \omega_6\}$
- $A_1 = \{\omega_1, \omega_2, \omega_3\}, \quad A_2 = \{\omega_4, \omega_5, \omega_6\}.$

6.2   Relative Frequency and Laplace Probability

There is a close connection between the relative frequency and probability of an event. A random experiment is described by its possible outcomes, for example getting a number between 1 and 6 when rolling a die. Suppose an experiment has m possible outcomes (events) $A_1, A_2, \ldots, A_m$ and the experiment is repeated n times. Now we can count how many times each of the possible outcomes has occurred. In other words, we can calculate the absolute frequency $n_i = n(A_i)$ which is equal to the number of times an event A_i, $i = 1, 2, \ldots, m$, occurs. The relative frequency $f_i = f(A_i)$ of a random event A_i, with n repetitions of the experiment, is calculated as

$$f_i = f(A_i) = \frac{n_i}{n}. \tag{6.1}$$

Example 6.2.1 Consider roulette, a game frequently played in casinos. The roulette table consists of 37 numbers from 0 to 36. Out of these 37 numbers, 18 numbers are red, 18 are black and one (zero) is green. Players can place their bets on either a single number or a range of numbers, the colours red or black, whether the number is odd or even, among many other choices. A casino employee spins a wheel (containing pockets representing the 37 numbers) in one direction and then spins a ball over the wheel in the opposite direction. The wheel and ball gradually slow down and the ball finally settles in a pocket. The pocket number in which the ball sits down when the wheel stops is the winning number. Consider three possible outcomes A_1: "red", A_2: "black" and A_3: "green (zero)". Suppose the roulette ball is spun $n = 500$ times.

All the outcomes are counted and recorded as follows: A_1 occurs 240 times, A_2 occurs 250 times and A_3 occurs 10 times. Then the absolute frequencies are given by $n_1 = n(A_1) = 240$, $n_2 = n(A_2) = 250$ and $n_3 = n(A_3) = 10$. We therefore get the relative frequencies as

$$f_1 = f(A_1) = \frac{240}{500} = 0.48, \quad f_2 = f(A_2) = \frac{250}{500} = 0.5,$$

$$f_3 = f(A_3) = \frac{10}{500} = 0.02.$$

If we assume that the experiment is repeated a large number of times (mathematically, this would mean that n tends to infinity) and the experimental conditions remain the same (at least approximately) over all the repetitions, then the relative frequency $f(A)$ converges to a limiting value for A. This limiting value is interpreted as the probability of A and denoted by $P(A)$, i.e.

$$P(A) = \lim_{n \to \infty} \frac{n(A)}{n}$$

where $n(A)$ denotes the number of times an event A occurs out of n times.

Example 6.2.2 Suppose a fair coin is tossed $n = 20$ times and we observe the number of heads $n(A_1) = 8$ times and number of tails $n(A_2) = 12$ times. The meaning of a fair coin in this case is that the probabilities of head and tail are equal (i.e. 0.5). Then the relative frequencies in the experiment are $f(A_1) = 8/20 = 0.4$ and $f(A_2) = 12/20 = 0.6$. When the coin is tossed a large number of times and n tends to infinity, then both $f(A_1)$ and $f(A_2)$ will have a limiting value 0.5 which is simply the probability of getting a head or tail in tossing a fair coin.

Example 6.2.3 In Example 6.2.1, the relative frequency of $f(\text{red}) = f(A_1)$ tends to 18/37 as n tends to infinity because 18 out of 37 numbers are red.

The reader will gain a more theoretical understanding of how repeated experiments relate to expected quantities in the following chapters after learning the Theorem of Large Numbers described in Appendix C.3.

A different definition of probability was given by Pierre-Simon Laplace (1749–1827). We call an experiment a **Laplace experiment** if the number of possible simple events is finite and all the outcomes are equally probable. The probability of an arbitrary event A is then defined as follows:

Definition 6.2.1 The proportion

$$P(A) = \frac{|A|}{|\Omega|} = \frac{\text{Number of "favourable simple events" for } A}{\text{Total number of possible simple events}} \tag{6.2}$$

is called the **Laplace probability** where $|A|$ is the cardinal number of A, i.e. the number of simple events contained in the set A, and $|\Omega|$ is the cardinal number of Ω, i.e. the number of simple events contained in the set Ω.

The cardinal numbers $|A|$ and $|\Omega|$ are often calculated using the combinatoric rules introduced in Chap. 5.

Example 6.2.4 (*Example* 6.1.2 *continued*) The sample space contains 36 simple events. All of these simple events have equal probability $1/36$. To calculate the probability of the event A, that the sum of the dots on the two dice is at least 4 and at most 6, we count the favourable simple events which fulfill this condition. The simple events are $(1, 3), (2, 2), (3, 1)$ (sum is 4), $(1, 4), (2, 3), (4, 1), (3, 2)$ (sum is 5) and $(1, 5), (2, 4), (3, 3), (4, 2), (5, 1)$ (sum is 6). In total, there are $(3 + 4 + 5) = 12$ favourable simple events, i.e.

$$A = \{(1, 3), (2, 2), (3, 1), (1, 4), (2, 3), (4, 1),$$
$$(3, 2), (1, 5), (2, 4), (3, 3), (4, 2), (5, 1)\} \ .$$

The probability of the event A is therefore $12/36 = 1/3$.

6.3 The Axiomatic Definition of Probability

An important foundation for modern probability theory was established by A.N. Kolmogorov in 1933 when he proposed the following **axioms of probability**.

Axiom 1 Every random event A has a probability in the (closed) interval $[0, 1]$, i.e.

$$0 \le P(A) \le 1.$$

Axiom 2 The sure event has probability 1, i.e.

$$P(\Omega) = 1.$$

Axiom 3 If A_1 and A_2 are disjoint events, then

$$P(A_1 \cup A_2) = P(A_1) + P(A_2).$$

holds.

Remark Axiom 3 also holds for three or more disjoint events and is called the **theorem of additivity of disjoint events**. For example, if A_1, A_2 and A_3 are disjoint events, then $P(A_1 \cup A_2 \cup A_3) = P(A_1) + P(A_2) + P(A_3)$.

Example 6.3.1 Suppose the two events in tossing a coin are A_1: "appearance of head" and A_2: "appearance of tail" which are disjoint. The event $A_1 \cup A_2$: "appearance of head or tail" has the probability

$$P(A_1 \cup A_2) = P(A_1) + P(A_2) = 1/2 + 1/2 = 1.$$

Example 6.3.2 Suppose an event is defined as the number of points observed on the upper surface of a die when rolling it. There are six events, i.e. the natural numbers $1, 2, 3, 4, 5, 6$. These events are disjoint and they have equal probability of occurring: $P(1) = P(2) = \cdots = P(6) = 1/6$. The probability of getting an even number is then

$$P(\text{"even number"}) = P(2) + P(4) + P(6) = 1/6 + 1/6 + 1/6 = 1/2.$$

6.3.1   Corollaries Following from Kolomogorov's Axioms

We already know that $A \cup \bar{A} = \Omega$ (sure event). Since A and $\bar{A}$ are disjoint, using Axiom 3 we have

$$P(A \cup \bar{A}) = P(A) + P(\bar{A}) = 1 .$$

Based on this, we have the following corollaries.

Corollary 1 *The probability of the complementary event of A, (i.e. $\bar{A}$) is*

$$P(\bar{A}) = 1 - P(A). \tag{6.3}$$

Example 6.3.3 Suppose a box of 30 chocolates contains chocolates of 6 different flavours with 5 chocolates of each flavour. Suppose an event A is defined as $A = \{\text{"marzipan flavour"}\}$. The probability of finding a marzipan chocolate (without looking into the box) is $P(\text{"marzipan"}) = 5/30$. Then the probability of the complementary event $\bar{A}$, i.e. the probability of not finding a marzipan chocolate is therefore

$$P(\text{"no marzipan flavour"}) = 1 - P(\text{"marzipan flavour"}) = 25/30.$$

Corollary 2 *The probability of occurrence of an impossible event $\emptyset$ is zero:*

$$P(\emptyset) = P(\bar{\Omega}) = 1 - P(\Omega) = 0 .$$

Corollary 3 *Let A_1 and A_2 be not necessarily disjoint events. The probability of occurrence of A_1 or A_2 is*

$$P(A_1 \cup A_2) = P(A_1) + P(A_2) - P(A_1 \cap A_2). \tag{6.4}$$

The rule in (6.4) is known as **the additive theorem of probability**. Again we use the word "or" in the statistical sense: either A_1 is occurring, or A_2 is occurring or both of them. This means we have to add the probabilities $P(A_1)$ and $P(A_2)$ but need to make sure that the simple events which are contained in both sets are not counted twice, thus we subtract $P(A_1 \cap A_2)$.

Example 6.3.4 There are 10 actors acting in a play. Two actors, one of whom is male, are portraying evil characters. In total, there are 6 female actors. Let an event A describe whether the actor is male and another event B describe whether the character is evil. Suppose we want to know the probability of a randomly chosen actor being male or evil. We can then calculate

$$P(\text{actor is male or evil}) =$$
$$= P(\text{actor is male}) + P(\text{actor is evil}) - P(\text{actor is male and evil})$$
$$= \frac{4}{10} + \frac{2}{10} - \frac{1}{10} = \frac{1}{2}.$$

Corollary 4 *If $A \subseteq B$ then $P(A) \leq P(B)$.*

Proof We use the representation $B = A \cup (\bar{A} \cap B)$ where A and $\bar{A} \cap B$ are the disjoint events. Then using Axioms 3 and 1 we get

$$P(B) = P(A) + P(\bar{A} \cap B) \geq P(A).$$

6.3.2   Calculation Rules for Probabilities

The introduced axioms and corollaries can be summarized as follows:

(1) $0 \leq P(A) \leq 1$

(2) $P(\Omega) = 1$

(3) $P(A_1 \cup A_2) = P(A_1) + P(A_2)$, if A_1 and A_2 are disjoint

(4) $P(\emptyset) = 0$

(5) $P(\bar{A}) = 1 - P(A)$

(6) $P(A_1 \cup A_2) = P(A_1) + P(A_2) - P(A_1 \cap A_2)$

(7) $P(A) \leq P(B)$, if $A \subseteq B$

6.4   Conditional Probability

Consider the following example to understand the concept of conditional probability: Suppose a new medical test is developed to diagnose a particular infection of the blood. The test is conducted on blood samples from 100 randomly selected patients and the outcomes of the tests are presented in Table 6.1.

There are the following four possible outcomes:

- The blood sample has an infection and the test diagnoses it, i.e. the test is correctly diagnosing the infection.
- The blood sample does not have an infection and the test does not diagnose it, i.e. the test is correctly diagnosing that there is no infection .
- The blood sample has an infection and the test does not diagnose it, i.e. the test is incorrect in stating that there is no infection.
- The blood sample does not have an infection but the test diagnoses it, i.e. the test is incorrect in stating that there is an infection.

Table 6.2 contains the relative frequencies of Table 6.1. In the following, we interpret the relative frequencies as probabilities, i.e. we assume that the values in Table 6.2 would be observed if the number n of patients was much larger than 100.

It can be seen that the probability that a test is positive is $P(T+) = 0.30 + 0.10 = 0.40$ and the probability that an infection is present is $P(IP) = 0.30 + 0.15 = 0.45$. If one already knows that the test is positive and wants to determine the probability that the infection is indeed present, then this can be achieved by the respective **conditional probability** $P(IP|T+)$ which is

$$P(IP|T+) = \frac{P(IP \cap T+)}{P(T+)} = \frac{0.3}{0.4} = 0.75 .$$

Table 6.1 Absolute frequencies of test results and infection status

		Infection		
		Present	Absent	Total (row)
Test	Positive (+)	30	10	40
	Negative (−)	15	45	60
	Total (column)	45	55	Total = 100

Table 6.2 Relative frequencies of patients and test

		Infection		
		Present (IP)	Absent (IA)	Total (row)
Test	Positive (+)	0.30	0.10	0.40
	Negative (−)	0.15	0.45	0.60
	Total (column)	0.45	0.55	Total = 1

Note that $I\,P \cap T+$ denotes the "relative frequency of blood samples in which the disease is present *and* the test is positive" which is 0.3.

More generally, recall Definition 4.1.1 from Chap. 4 where we defined conditional, joint, and marginal frequency distributions in contingency tables. The present example simply applies these rules to the contingency tables of relative frequencies and interprets the relative frequencies as an approximation to the probabilities of interest, as already explained.

We use the intersection operator $\cap$ to describe events which occur for $A = a$ *and* $B = b$. This relates to the joint relative frequencies. The marginal relative frequencies (i.e. probabilities $P(A = a)$) can be observed from the column and row sums respectively; and the conditional probabilities can be observed as the joint frequencies in relation to the marginal frequencies.

For simplicity, assume that all simple events in $\Omega = \{\omega_1, \omega_2, \ldots, \omega_k\}$ are equally probable, i.e. $P(\omega_j) = \frac{1}{k}$, $j = 1, 2, \ldots, k$. Let A and B be two events containing n_A and n_B numbers of simple events. Let further $A \cap B$ contain n_{AB} numbers of simple events. The Laplace probability using (6.2) is

$$P(A) = \frac{n_A}{k}, \quad P(B) = \frac{n_B}{k}, \quad P(A \cap B) = \frac{n_{AB}}{k}.$$

Assume that we have prior information that A has already occurred. Now we want to find out how the probability of B is to be calculated. Since A has already occurred, we know that the sample space is reduced by the number of simple events which are contained in A. There are n_A such simple events. Thus the total sample space Ω is reduced by the sample space of A. Therefore the simple events in $A \cap B$ are those simple events which are realized when B is realized. The Laplace probability for B under the prior information on A, or under the condition that A is known, is therefore

$$P(B|A) = \frac{n_{AB}/k}{n_A/k} = \frac{P(A \cap B)}{P(A)}. \tag{6.5}$$

This can be generalized to the case when the probabilities for simple events are unequal.

Definition 6.4.1 Let $P(A) > 0$. Then the **conditional probability** of event B occurring, given that event A has already occurred, is

$$P(B|A) = \frac{P(A \cap B)}{P(A)}. \tag{6.6}$$

The roles of A and B can be interchanged to define $P(A|B)$ as follows. Let $P(B) > 0$. The conditional probability of A given B is

$$P(A|B) = \frac{P(A \cap B)}{P(B)}. \tag{6.7}$$

We now introduce a few important theorems which are relevant to calculating conditional and other probabilities.

Theorem 6.4.1 (Multiplication Theorem of Probability) *For two arbitrary events A and B, the following holds:*

$$P(A \cap B) = P(A|B)P(B) = P(B|A)P(A). \tag{6.8}$$

This theorem follows directly from the two definitions (6.6) and (6.7) (but does not require that $P(A) > 0$ and $P(B) > 0$).

Theorem 6.4.2 (Law of Total Probability) *Assume that $A_1, A_2, \ldots, A_m$ are events such that $\cup_{i=1}^{m} A_i = \Omega$ and $A_i \cap A_j = \emptyset$ for all $i \neq j$, $P(A_i) > 0$ for all i, i.e. A_1, $A_2, \ldots, A_m$ form a complete decomposition of $\Omega = \cup_{i=1}^{m} A_i$ in pairwise disjoint events, then the probability of an event B can be calculated as*

$$P(B) = \sum_{i=1}^{m} P(B|A_i)P(A_i). \tag{6.9}$$

6.4.1   Bayes' Theorem

Bayes' Theorem gives a connection between $P(A|B)$ and $P(B|A)$. For events A and B with $P(A) > 0$ and $P(B) > 0$, using (6.6) and (6.7) or (6.8), we get

$$P(A|B) = \frac{P(A \cap B)}{P(B)} = \frac{P(A \cap B)}{P(A)} \frac{P(A)}{P(B)}$$
$$= \frac{P(B|A)P(A)}{P(B)}. \tag{6.10}$$

Let $A_1, A_2, \ldots, A_m$ be events such that $\cup_{i=1}^{m} A_i = \Omega$ and $A_i \cap A_j = \emptyset$ for all $i \neq j$, $P(A_i) > 0$ for all i, and B is another event than A, then using (6.9) and (6.10) we get

$$P(A_j|B) = \frac{P(B|A_j)P(A_j)}{\sum_i P(B|A_i)P(A_i)}. \tag{6.11}$$

The probabilities $P(A_i)$ are called **prior probabilities**, $P(B|A_i)$ are sometimes called **model probabilities** and $P(A_j|B)$ are called **posterior probabilities**.

Example 6.4.1 Suppose someone streams movies from two different online plat-forms. Sometimes it happens that the streaming does not work well because of technical problems. We consider the following events: A_i ($i = 1, 2$): "the movie is streamed from platform i". Further let B denote the event that the stream is func-tioning properly. Assume we know that $P(A_1) = 0.6$ and $P(A_2) = 0.4$ (note that $A_2 = \bar{A}_1$) and $P(B|A_1) = 0.95$, $P(B|A_2) = 0.75$ and we are interested in the prob-ability that a movie can be streamed well. We can then apply the Law of Total Probability and get

$$P(B) \overset{(6.9)}{=} P(B|A_1)P(A_1) + P(B|A_2)P(A_2)$$
$$= 0.6 \cdot 0.95 + 0.4 \cdot 0.75 = 0.87.$$

We may also be interested in the probability that the movie was streamed from platform 1 *and* is the stream is functioning well, which is

$$P(B \cap A_1) \overset{(6.8)}{=} P(B|A_1)P(A_1) = 0.95 \cdot 0.6 = 0.57.$$

Now suppose we have a properly functioning stream. What is the probability that it is streamed from platform 1? This is obtained as follows:

$$P(A_1|B) \overset{(6.7)}{=} \frac{P(A_1 \cap B)}{P(B)} = \frac{0.57}{0.87} = 0.6552.$$

Now assume we have a movie which can not be streamed well, i.e. $\bar{B}$ occurs. The probability that a movie stream is not working well given that it is from platform 1 is $P(\bar{B}|A_1) = 0.05$. Similarly, $P(\bar{B}|A_2) = 0.25$ for platform 2. We can now calculate the conditional probability that a movie is from platform 1 given that it the stream not working properly:

$$P(A_1|\bar{B}) \overset{(6.11)}{=} \frac{P(\bar{B}|A_1)P(A_1)}{P(\bar{B}|A_1)P(A_1) + P(\bar{B}|A_2)P(A_2)}$$
$$= \frac{0.05 \cdot 0.6}{0.05 \cdot 0.6 + 0.25 \cdot 0.4} = 0.2308.$$

The result about $P(\bar{B})$ used in the denominator can also be directly obtained by using $P(\bar{B}) = 1 - 0.87 = 0.13$.

6.5   Independence

Intuitively, two events are independent if the occurrence or non-occurrence of one event does not affect the occurrence or non-occurrence of the other event. In other words, two events A and B are independent if the probability of occurrence of B has no effect on the probability of occurrence of A. In such a situation, one expects that

$$P(A|B) = P(A) \quad \text{and} \quad P(A|\bar{B}) = P(A) .$$

Using this and (6.7), we can write

$$P(A|B) = \frac{P(A \cap B)}{P(B)}$$
$$= \frac{P(A \cap \bar{B})}{P(\bar{B})} = P(A|\bar{B}) . \tag{6.12}$$

This yields:

$$P(A \cap B)P(\bar{B}) = P(A \cap \bar{B})P(B)$$
$$P(A \cap B)(1 - P(B)) = P(A \cap \bar{B})P(B)$$
$$P(A \cap B) = (P(A \cap \bar{B}) + P(A \cap B))P(B)$$
$$P(A \cap B) = P(A)P(B) . \tag{6.13}$$

This leads to the following definition of stochastic independence.

Definition 6.5.1 Two random events A and B are called **(stochastically) independent** if

$$P(A \cap B) = P(A)P(B) , \tag{6.14}$$

i.e. if the probability of simultaneous occurrence of both events A and B is the product of the individual probabilities of occurrence of A and B.

This definition of independence can be extended to the case of more than two events as follows:

Definition 6.5.2 The n events $A_1, A_2, \ldots, A_n$ are stochastically mutually independent, if for any subset of m events $A_{i_1}, A_{i_2}, \ldots, A_{i_m}$ $(m \le n)$

$$P(A_{i_1} \cap A_{i_2} \cdots \cap A_{i_m}) = P(A_{i_1})P(A_{i_2}) \cdot \ldots \cdot P(A_{i_m}) \tag{6.15}$$

holds.

A weaker form of independence is pairwise independence. If condition (6.15) is fulfilled only for two arbitrary events, i.e. $m = 2$, then the events are called **pairwise independent**. The difference between pairwise independence and general stochastic independence is explained in the following example.

Example 6.5.1 Consider an urn with four balls. The following combinations of zeroes and ones are printed on the balls: $110, 101, 011, 000$. One ball is drawn from the urn. Define the following events:

$$A_1 : \text{The first digit on the ball is 1.}$$
$$A_2 : \text{The second digit on the ball is 1.}$$
$$A_3 : \text{The third digit on the ball is 1.}$$

Since there are two favourable simple events for each of the events A_1, A_2 and A_3 we get

$$P(A_1) = P(A_2) = P(A_3) = \frac{2}{4} = \frac{1}{2}.$$

The probability that all the three events simultaneously occur is zero because there is no ball with 111 printed on it. Therefore, A_1, A_2 and A_3 are not stochastically independent because

$$P(A_1)P(A_2)P(A_3) = \frac{1}{8} \neq 0 = P(A_1 \cap A_2 \cap A_3).$$

However,

$$P(A_1 \cap A_2) = \frac{1}{4} = P(A_1)P(A_2),$$
$$P(A_1 \cap A_3) = \frac{1}{4} = P(A_1)P(A_3),$$
$$P(A_2 \cap A_3) = \frac{1}{4} = P(A_2)P(A_3),$$

which means that the three events are pairwise independent.

6.6   Key Points and Further Issues

Note:

✓ We summarize some important theorems and laws:

- The Laplace probability is the ratio

$$P(A) = \frac{|A|}{|\Omega|} = \frac{\text{Number of "favourable simple events" for } A}{\text{Total number of possible simple events}}.$$

- The Law of Total Probability is

$$P(B) = \sum_{i=1}^{m} P(B|A_i) P(A_i).$$

- Bayes' Theorem is

$$P(A_j|B) = \frac{P(B|A_j) P(A_j)}{\sum_i P(B|A_i) P(A_i)}.$$

- n events $A_1, A_2, \ldots, A_n$ are (stochastically) independent, if

$$P(A_1 \cap A_2 \cdots \cap A_n) = P(A_1) P(A_2) \cdot \ldots \cdot P(A_n).$$

✓ In Sect. 10.8 we present the χ^2-independence test, which can test whether two discrete random variables (see Chap. 7) are independent or not.

6.7   Exercises

Exercise 6.1

(a) Suppose $\Omega = \{0, 1, \ldots, 15\}, A = \{0, 8\}, B = \{1, 2, 3, 5, 8, 10, 12\}, C = \{0, 4, 9, 15\}$. Determine $A \cap B, B \cap C, A \cup C, C \setminus A, \Omega \setminus (B \cup A \cup C)$.

(b) Now consider the three pairwise disjoint events E, F, G with $\Omega = E \cup F \cup G$ and $P(E) = 0.2$ and $P(F) = 0.5$. Calculate $P(\bar{F}), P(G), P(E \cap G), P(E \setminus E)$, and $P(E \cup F)$.

Exercise 6.2 A driving licence examination consists of two parts which are based on a theoretical and a practical examination. Suppose 25% of people fail the practical examination, 15% of people fail the theoretical examination and 10% of people fail both the examinations. If a person is randomly chosen, then what is the probability that this person

(a) fails at least one of the examinations?
(b) only fails the practical examination, but not the theoretical examination?
(c) successfully passes both the tests?
(d) fails any of the two examinations?

Exercise 6.3 A new board game uses a twelve-sided die. Suppose the die is rolled once, what is the probability of getting

(a) an even number?
(b) a number greater than 9?
(c) an even number greater than 9?
(d) an even number or a number greater than 9?

Exercise 6.4 The Smiths are a family of six. They are celebrating Christmas and there are 12 gifts, two for each family member. The name tags for each family member have been attached to the gifts. Unfortunately the name tags on the gifts are damaged by water. Suppose each family member draws two gifts at random. What is the probability that someone

(a) gets his/her two gifts, rather than getting the gifts for another family member?
(b) gets none of his/her gifts, but rather gets the gifts for other family members?

Exercise 6.5 A chef from a popular TV cookery show sometimes puts too much salt in his pumpkin soup and the probability of this happening is 0.2. If he is in love (which he is with probability 0.3) then the probability of using too much salt is 0.6.

(a) Create a contingency table for the probabilities of the two variables "in love" and "too much salt".
(b) Determine whether the two variables are stochastically independent or not.

Exercise 6.6 Dr. Obermeier asks his neighbour to take care of his basil plant while he is away on leave. He assumes that his neighbour does not take care of the basil with a probability of $\frac{1}{3}$. The basil dies with probability $\frac{1}{2}$ when someone takes care of it and with probability $\frac{3}{4}$ if no one takes care of it.

(a) Calculate the probability of the basil plant surviving after its owner's leave.
(b) It turns out that the basil eventually dies. What is the probability that Dr. Obermeier's neighbour didn't take care of the plant?

Exercise 6.7 A bank considers changing its credit card policy. Currently 5% of credit card owners are not able to pay their bills in any month, i.e. they never pay their bills. Among those who are generally able to pay their bills, there is still a 20% probability that the bill is paid too late in a particular month.

(a) What is the probability that someone is not paying his bill in a particular month?
(b) A credit card owner did not pay his bill in a particular month. What is the probability that he never pays back the money?
(c) Should the bank consider blocking the credit card if a customer does not pay his bill on time?

Exercise 6.8 There are epidemics which affect animals such as cows, pigs, and others. Suppose 200 cows are tested to see whether they are infected with a virus or not. Let event A describe whether a cow has been transported by a truck recently or not and let B denote the event that a cow has been tested positive with a virus. The data is summarized in the following table:

	B	$\bar{B}$	Total
A	40	60	100
$\bar{A}$	20	80	100
Total	60	140	200

(a) What is the probability that a cow is infected and has been transported by a truck recently?
(b) What is the probability of having an infected cow given that it has been transported by the truck?
(c) Determine and interpret $P(B)$.

Exercise 6.9 A football practice target is a portable wall with two holes (which are the target) in it for training shots. Suppose there are two players A and B. The probabilities of hitting the target by A and B are 0.4 and 0.5 respectively.

(a) What is the probability that at least one of the players succeeds with his shot?
(b) What is the probability that exactly one of the players hits the target?
(c) What is the probability that only B scores?

$\rightarrow$ Solutions to all exercises in this chapter can be found in Appendix B

Random Variables

7

In the first part of the book we highlighted how to *describe* data. Now we discuss
the concepts required to draw statistical conclusions from a sample of data about a
population of interest. For example, suppose we know the starting salary of a sample
of 100 students graduating in law. We can use this knowledge to draw conclusions
about the expected salary for the population of all students graduating in law. Similarly, if a newly developed drug is given to a sample of selected tuberculosis patients,
then some patients may show improvement and some patients may not; but we are
interested in the consequences for the entire population of patients. In the remainder
of this chapter we describe the theoretical concepts required for making such conclusions. They form the basis for statistical tests and inference which are introduced
in Chaps. 9–11.

7.1   Random Variables

Random variables help us to view the collected data as an outcome of a random
experiment. Consider the simple experiment of tossing a coin. If a coin is tossed, then
one can observe either "head" (H) or "tail" (T). The occurrence of "head" or "tail" is
random and the exact outcome will only be known after the coin is tossed. We can toss
the coin many times and obtain a sequence of outputs. For example, if a coin is tossed
seven times, then one of the outcomes may be H, H, T, H, T, T, T. This outcome
is the consequence of a random experiment and it may be helpful if we can distill the
sequence of outcomes in meaningful numbers. One option is to summarize them by
a variable X, which takes the values $x_1 = 1$ (denoting head) and $x_2 = 0$ (denoting
tail). We have learnt from Chap. 6 that this can be described in the framework of a
random experiment where $\Omega = \{\omega_1, \omega_2\}$ with the events $A_1 = \{\omega_1\} = 1 = $ head and

© Springer Nature Switzerland AG 2022
C. Heumann et al., *Introduction to Statistics and Data Analysis*,
https://doi.org/10.1007/978-3-031-11833-3_7

Table 7.1 Examples of random variables

X	Event	Realizations of X
Roll of a die	A_i: number i ($i = 1, 2, \ldots, 6$)	$x = i$
Lifetime of TV	A_i: survival time is i months ($i = 1, 2, \ldots$)	$x = i$
Roulette	A_1: red	$x_1 = 1$
	A_2: black	$x_2 = 2$
	A_3: green (zero)	$x_3 = 0$

$A_2 = \{\omega_2\} = 0 = $ tail. The random variable X is now mapped to real numbers and therefore it describes the possible outcome of *any* coin toss experiment. The observed outcomes H, H, T, H, T, T, T relate to a specific sample, a unique *realization* of this experiment. We can write $X(\omega_1) = 1$ and $X(\omega_2) = 0$ with $\omega_1, \omega_2 \in \Omega$ and $1, 0 \in \mathcal{R}$ where $\mathcal{R}$ is the set of real numbers. We know that in any coin tossing experiment, the probability of head being observed is $P(X(\omega_1) = 1) = 0.5$ and of tail being observed is $P(X(\omega_2) = 0) = 0.5$. We may therefore view X as a random variable which collects the possible outcomes of a random experiment and captures the uncertainty associated with them.

Definition 7.1.1 Let Ω represent the sample space of a random experiment and let $\mathcal{R}$ be the set of real numbers. A random variable is a function X which assigns to each element $\omega \in \Omega$ one and only one number $X(\omega) = x, x \in \mathcal{R}$, i.e.

$$X : \Omega \to \mathcal{R}. \tag{7.1}$$

Example 7.1.1 The features of a die roll experiment, a roulette game or the lifetime of a TV can all be described by a random variable, see Table 7.1. The events involve randomness and if we have knowledge about the random process, we can assign probabilities $P(X = x_i)$ to each event, e.g. when rolling a die, the probability of getting a "1" is $P(X = 1) = 1/6$, the probability of getting a "2" is $P(X = 2) = 1/6$, and so on.

Note that it is a convention to denote random variables by capital letters (e.g. X) and their values by small letters (e.g. x). It is evident from the coin tossing experiment that we need to know $P(X = x)$ to describe the respective random variable. We assume in this chapter that we have this knowledge. However, Chaps. 9–11 show how a sample of data can be used to estimate unknown probabilities and other quantities given a pre-specified uncertainty level. More generally, we can say that it is mandatory to know $P(X \in A)$ for all possible A which are subsets of $\mathcal{R}$. If we choose $A = (-\infty, x], x \in \mathcal{R}$, we have

$$P(X \in A) = P(X \in (-\infty, x]) = P(-\infty < X \leq x) = P(X \leq x).$$

This consideration gives rise to the definition of the cumulative distribution function. Recall that we developed the concept of the empirical cumulative distribution function (ECDF) in Chap. 2, Sect. 2.2, but the definition there was empirical. Now we develop it theoretically.

7.2 Cumulative Distribution Function (CDF)

Definition 7.2.1 The **cumulative distribution function (CDF)** of a random variable X is defined as

$$F(x) = P(X \leq x). \tag{7.2}$$

As in Chap. 2, we can see that the CDF is useful in obtaining the probabilities related to the occurrence of random events. Note that the empirical cumulative distribution function (ECDF, Sect. 2.2) and the cumulative distribution function are closely related and therefore have a similar definition and similar calculation rules. However in Chap. 2, we work with the cumulative distribution of *observed* values in a particular sample whereas in this chapter, we deal with random variables modelling the distribution of a general population.

The definition (7.2) implies the following properties of the cumulative distribution function:

- $F(x)$ is a monotonically nondecreasing function (if $x_1 \leq x_2$, it follows that $F(x_1) \leq F(x_2)$),
- $\lim_{x \to -\infty} F(x) = 0$ (the lower limit of F is 0),
- $\lim_{x \to +\infty} F(x) = 1$ (the upper limit of F is 1),
- $F(x)$ is continuous from the right,
- $0 \leq F(x) \leq 1$ for all $x \in \mathcal{R}$.

Another notation for $F(x) = P(X \leq x)$ is $F_X(x)$ but we use $F(x)$.

7.2.1 CDF of Continuous Random Variables

Before giving some examples about the meaning and interpretation of the CDF, we first need to consider some definitions and theorems.

Definition 7.2.2 A random variable X is said to be **continuous** if there is a function $f(x)$ such that for all $x \in \mathcal{R}$

$$F(x) = \int_{-\infty}^{x} f(t)dt \tag{7.3}$$

holds. $F(x)$ is the cumulative distribution function (CDF) of X, $f(x)$ is the probability density function (PDF) of x and $\frac{d}{dx}F(x) = f(x)$ for all x that are continuity points of f.

Theorem 7.2.1 *For a function $f(x)$ to be a **probability density function (PDF)** of X, it needs to satisfy the following conditions:*

(1) $f(x) \geq 0$ for all $x \in \mathcal{R}$,
(2) $\int_{-\infty}^{\infty} f(x)dx = 1$.

Theorem 7.2.2 *Let X be a continuous random variable with CDF $F(x)$. If $x_1 < x_2$, where x_1 and x_2 are known constants, $P(x_1 \leq X \leq x_2) = F(x_2) - F(x_1) = \int_{x_1}^{x_2} f(x)dx$.*

Theorem 7.2.3 *The probability of a continuous random variable taking a particular value x_0 is zero:*

$$P(X = x_0) = 0. \tag{7.4}$$

The proof is provided in Appendix C.2.

Example 7.2.1 Consider the continuous random variable "waiting time for the train". Suppose that a train arrives every 20 min and a person arrives randomly at the station. Therefore the waiting time of this person can be any time contained in the interval [0,20]. We can start describing the required probability density function as

$$f(x) = \begin{cases} k & \text{for } 0 \leq x \leq 20 \\ 0 & \text{otherwise} \end{cases}$$

where k is an unknown constant. Now using condition (2) of Theorem 7.2.1, we have

$$1 = \int_0^{20} f(x)dx = [kx]_0^{20} = 20k$$

which needs to be fulfilled. This yields $k = 1/20$ which is always greater than 0 and therefore condition (1) of Theorem 7.2.1 is also fulfilled. It follows that

$$f(x) = \begin{cases} \frac{1}{20} & \text{for } 0 \leq x \leq 20 \\ 0 & \text{otherwise} \end{cases}$$

is the probability density function describing the waiting time for the train. We can now use Definition 7.2.2 to determine the cumulative distribution function:

$$F(x) = \int_0^x f(t)dt = \int_0^x \frac{1}{20}dt = \frac{1}{20}[t]_0^x = \frac{1}{20}x.$$

(a) PDF   (b) CDF

Fig. 7.1 Probability density function (PDF) and cumulative distribution function (CDF) for waiting time in Example 7.2.1

Suppose we are interested in calculating the probability of a waiting time between 15 and 20 min. This can be calculated using Theorem 7.2.2:

$$P(15 \leq X \leq 20) = F(20) - F(15) = \frac{20}{20} - \frac{15}{20} = 0.25 \,.$$

We can obtain this probability from the graph of the CDF as well, see Fig. 7.1 where both the PDF and CDF of this example are illustrated.

Defining a function, for example the CDF, is simple in R: One can use the `function` command followed by specifying the variables the function evaluates in round brackets (e.g. x) and the function itself in braces (e.g. $x/20$). Functions can be plotted using the `curve` command:

```
cdf <- function(x){1/20*x}
curve(cdf,from=0,to=20)
```

Alternatively, the `plot` command can be used to plot vectors against each other; for example, after defining a function, we can define a sequence (`x<-seq(0,20,0.01)`), evaluate this sequence via the specified function (`cdf(x)`) and plot them against each other and connect the points from the sequence with a line (`plot(x,cdf(x),type='l')`).

This example illustrates how the cumulative distribution function can be used to obtain probabilities of interest. Most importantly, if we want to calculate the probability that the random variable X takes values in the interval $[x_1, x_2]$, we simply have to look at the difference of the respective CDF values at x_1 and x_2. Figure 7.2a highlights that the interval probability corresponds to the difference of the CDF values on the y-axis.

We can also use the probability density function to visualize $P(x_1 \leq X \leq x_2)$. We know from Theorem 7.2.1 that $\int_{-\infty}^{\infty} f(x)dx = 1$ and therefore the area under

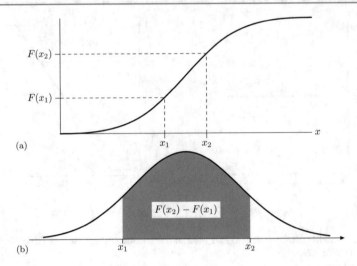

Fig. 7.2 Graphical representation of the probability $P(x_1 \leq X \leq x_2)$ **a** via the CDF and **b** via the PDF

the PDF equals 1. Thus we can interpret interval probabilities as the area under the PDF between x_1 and x_2. This is presented in Fig. 7.2b.

7.2.2 CDF of Discrete Random Variables

Definition 7.2.3 A random variable X is defined to be **discrete** if its probability space is either finite or countable, i.e. if it takes only a finite or countable number of values. Note that a set V is said to be **countable**, if its elements can be listed, i.e. there is a one-to-one correspondence between V and the positive integers.

Example 7.2.2 Consider the example of tossing of a coin where each trial results in either a head (H) or a tail (T), each occurring with the same probability 0.5. When the coin is tossed multiple times, we may observe sequences like $H, T, H, H, T, H, H, T, T, ...$ The sample space is $\Omega = \{H, T\}$. Let the random variable X denote the number of trials required to get the third head, then $X = 4$ for the given sequence above. Clearly, the space of X is the set $(3, 4, 5, \ldots)$. We can see that X is a discrete random variable because its space is finite and can be counted. We can also assign certain probabilities to each of these values, e.g. $P(X = 3) = p_1$, $P(X = 4) = p_2$, etc.

Definition 7.2.4 Let X be a discrete random variable which takes k different values. The **probability mass function** (PMF) of X is given by

$$f(X) = P(X = x_i) = p_i \quad \text{for each } i = 1, 2, \ldots, k . \tag{7.5}$$

It is required that the probabilities p_i satisfy the following conditions:

(1) $0 \leq p_i \leq 1$,   (2) $\sum_{i=1}^{k} p_i = 1$.

Definition 7.2.5 Given (7.5), we can write the CDF of a discrete random variable as

$$F(x) = \sum_{i=1}^{k} I_{\{x_i \leq x\}} p_i \, , \tag{7.6}$$

where I is an indicator function defined as

$$I_{\{x_i \leq x\}} = \begin{cases} 1 & \text{if} \ \ x_i \leq x \\ 0 & \text{otherwise} \, . \end{cases}$$

The CDF of a discrete variable is always a **step function**.

Working with the CDF for Discrete Random variables

We can easily calculate various types of probabilities for discrete random variables using the CDF. Let a and b be some known constants, then

$$P(X \leq a) = F(a) \, , \tag{7.7}$$
$$P(X < a) = P(X \leq a) - P(X = a) = F(a) - P(X = a) \, , \tag{7.8}$$
$$P(X > a) = 1 - P(X \leq a) = 1 - F(a) \, , \tag{7.9}$$
$$P(X \geq a) = 1 - P(X < a) = 1 - F(a) + P(X = a) \, , \tag{7.10}$$
$$P(a \leq X \leq b) = P(X \leq b) - P(X < a)$$
$$= F(b) - F(a) + P(X = a) \, , \tag{7.11}$$
$$P(a < X \leq b) = F(b) - F(a) \, , \tag{7.12}$$
$$P(a < X < b) = F(b) - F(a) - P(X = b) \, , \tag{7.13}$$
$$P(a \leq X < b) = F(b) - F(a) - P(X = b) + P(X = a) \, . \tag{7.14}$$

Remark 7.2.1 The Eqs. (7.7)–(7.14) can also be used for continuous variables, but in this case $P(X = a) = P(X = b) = 0$ (see Theorem 7.2.3), and therefore Eqs. (7.7)–(7.14) can be modified accordingly.

Example 7.2.3 Consider the experiment of rolling a die. There are 6 possible outcomes. If we define the random variable X as the number of dots observed on the upper surface of the die, then the six possible outcomes can be described as $x_1 = 1, x_2 = 2, \ldots, x_6 = 6$. The respective probabilities are $P(X = x_i) = 1/6$; $i = 1, 2, , \ldots, 6$. The PMF and CDF are therefore defined as follows:

Fig. 7.3 Probability density function and cumulative distribution function for rolling a die in Example 7.2.3. "•" relates to an included value, "∘" to an excluded value

$$f(x) = \begin{cases} 1/6 & \text{if} \quad x = 1 \\ 1/6 & \text{if} \quad x = 2 \\ 1/6 & \text{if} \quad x = 3 \\ 1/6 & \text{if} \quad x = 4 \\ 1/6 & \text{if} \quad x = 5 \\ 1/6 & \text{if} \quad x = 6 \\ 0 & \text{elsewhere.} \end{cases} \qquad F(x) = \begin{cases} 0 & \text{if} \quad -\infty < x < 1 \\ 1/6 & \text{if} \quad 1 \leq x < 2 \\ 2/6 & \text{if} \quad 2 \leq x < 3 \\ 3/6 & \text{if} \quad 3 \leq x < 4 \\ 4/6 & \text{if} \quad 4 \leq x < 5 \\ 5/6 & \text{if} \quad 5 \leq x < 6 \\ 1 & \text{if} \quad 6 \leq x < \infty. \end{cases}$$

Both the CDF and the PDF are displayed in Fig. 7.3.

We can use the CDF to calculate any desired probability, e.g. $P(X \leq 5) = F(5) = 5/6$. This can also be seen in Fig. 7.3b where for $X = 5$, we obtain $F(5) = 5/6$ when evaluating on the y-axis. Similarly, $P(3 < X \leq 5) = F(5) - F(3) = (5/6) - (3/6) = 2/6$ can be interpreted as the difference of $F(5)$ and $F(3)$ on the y-axis.

7.3   Expectation and Variance of a Random Variable

We have seen that both the probability density function (or probability mass function) and the cumulative distribution function are helpful in characterizing the features of a random variable. Some other features of random variables are characterized by the concepts of *expectation* and *variance*.

7.3.1 Expectation

Definition 7.3.1 The expectation of a continuous random variable X, having the probability density function $f(x)$ with $\int |x| f(x)dx < \infty$, is defined as

$$E(X) = \int_{-\infty}^{+\infty} x f(x)dx. \tag{7.15}$$

For a discrete random variable X, which takes the values $x_1, x_2, \ldots$ with respective probabilities $p_1, p_2, \ldots$, the **expectation** of X is defined as

$$E(X) = \sum_{i=1}^{k} x_i p_i = x_1 P(X = x_1) + x_2 P(X = x_2) + \ldots + x_k P(X = x_k).$$
$$\tag{7.16}$$

The *expectation* of X, i.e. $E(X)$, is usually denoted by $\mu = E(X)$ and relates to the arithmetic mean of the distribution of the population. It reflects the central tendency of the population.

Example 7.3.1 Consider again Example 7.2.1 where the waiting time for a train was described by the following probability density function:

$$f(x) = \begin{cases} \frac{1}{20} & \text{for } 0 \le x \le 20 \\ 0 & \text{otherwise.} \end{cases}$$

We can calculate the expectation as follows:

$$E(X) = \int_{-\infty}^{\infty} x f(x)\, dx = \int_{-\infty}^{0} x f(x)\, dx + \int_{0}^{20} x f(x)\, dx + \int_{20}^{\infty} x f(x)\, dx$$
$$= 0 + \int_{0}^{20} \frac{1}{20} x\, dx + 0 = \left[\frac{1}{40} x^2 \right]_{0}^{20} = \frac{400}{40} - 0 = 10.$$

The "average" waiting time for the train is therefore 10 min. This means that if a person has to wait for the train every day, then the person will experience waiting times varying randomly between 0 and 20 min and on average, has to wait for 10 min.

Example 7.3.2 Consider again the die roll experiment from Example 7.2.3. The probabilities for the occurrence of any x_i, $i = 1, 2, \ldots, 6$, are $P(X = x_i) = 1/6$.

The expectation can thus be calculated as

$$E(X) = \sum_{i=1}^{6} x_i\, p_i$$
$$= 1 \cdot P(X = 1) + 2 \cdot P(X = 2) + 3 \cdot P(X = 3) + 4 \cdot P(X = 4)$$
$$+ 5 \cdot P(X = 5) + 6 \cdot P(X = 6)$$
$$= (1 + 2 + 3 + 4 + 5 + 6)\frac{1}{6} = \frac{21}{6} = 3.5 \,.$$

7.3.2   Variance

The *variance* describes the variability of a random variable. It gives an idea about the concentration or dispersion of values around the arithmetic mean of the distribution.

Definition 7.3.2 The **variance** of a random variable X is defined as

$$\text{Var}(X) = E[X - E(X)]^2 \,. \tag{7.17}$$

The variance of a continuous random variable X is

$$\text{Var}(X) = \int_{-\infty}^{+\infty} (x - E(X))^2 f(x)dx \tag{7.18}$$

where $E(X) = \int_{-\infty}^{+\infty} x f(x)dx$. Similarly, the variance of a discrete random variable X is

$$\text{Var}(X) = \sum_{i=1} (x_i - E(X))^2 p_i \tag{7.19}$$

where $E(X) = \sum_i x_i\, p_i$. The variance is usually denoted by $\sigma^2 = \text{Var}(X)$.

Definition 7.3.3 The positive square root of the variance is called the **standard deviation**.

Example 7.3.3 Recall Examples 7.2.1 and 7.3.1. We can calculate the variance of the waiting time for a train using the probability density function

$$f(x) = \begin{cases} \frac{1}{20} & \text{for } 0 \le x \le 20 \\ 0 & \text{otherwise} \end{cases}$$

and $E(X) = 10$ (already calculated in Example 7.3.1). Using (7.18), we obtain:

$$
\begin{aligned}
\text{Var}(X) &= \int_{-\infty}^{\infty} (x - E(x))^2 f(x)\,dx = \int_{-\infty}^{\infty} (x - 10)^2 f(x)\,dx \\
&= \int_{-\infty}^{0} (x - 10)^2 f(x)\,dx + \int_{0}^{20} (x - 10)^2 f(x)\,dx + \int_{20}^{\infty} (x - 10)^2 f(x)\,dx \\
&= 0 + \int_{0}^{20} (x - 10)^2 \cdot \frac{1}{20}\,dx + 0 = \int_{0}^{20} \frac{1}{20}(x^2 - 20x + 100)\,dx \\
&= \left[\frac{1}{20}\left(\frac{1}{3}x^3 - 10x^2 + 100x \right) \right]_{0}^{20} = 33\frac{1}{3}\,.
\end{aligned}
$$

The standard deviation is $\sqrt{33\frac{1}{3}\ \text{min}^2} \approx 5.77\ \text{min}$.

Recall that in Chap. 3 we introduced the sample variance and the sample standard deviation. We already know that the standard deviation has the same unit of measurement as the variable whereas the unit of the variance is the square of the measurement unit. The standard deviation measures how the values of a random variable are dispersed around the population mean. A low value of the standard deviation indicates that the values are highly concentrated around the mean. A high value of the standard deviation indicates lower concentration of the data values around the mean and the observed values may be far away from the mean. These considerations are helpful in making connections between random variables and samples of data, see Chap. 9 for the construction of confidence intervals.

Example 7.3.4 Recall Example 7.3.2 where we calculated the expectation of a die roll experiment as $E(X) = 3.5$. With $x_i \in \{1, 2, 3, 4, 5, 6\}$ and $p_i = 1/6$ for all $i = 1, 2, 3, 4, 5, 6$, the variance for this example corresponds to

$$
\begin{aligned}
\text{Var}(X) &= \sum_{i=1} (x_i - E(X))^2 p_i = (1 - 3.5)^2 \cdot \frac{1}{6} + (2 - 3.5)^2 \cdot \frac{1}{6} + (3 - 3.5)^2 \cdot \frac{1}{6} \\
&+ (4 - 3.5)^2 \cdot \frac{1}{6} + (5 - 3.5)^2 \cdot \frac{1}{6} + (6 - 3.5)^2 \cdot \frac{1}{6} \approx 2.92\,.
\end{aligned}
$$

Theorem 7.3.1 *The variance of a random variable X can be expressed as*

$$
\text{Var}(X) = E(X^2) - [E(X)]^2\,. \tag{7.20}
$$

The proof is given in Appendix C.2.

Example 7.3.5 In Examples 7.2.1, 7.3.1, and 7.3.3 we evaluated the waiting time for a train using the PDF

$$
f(X) = \begin{cases} \frac{1}{20} & \text{for } 0 < X \le 20 \\ 0 & \text{otherwise.} \end{cases}
$$

We calculated the expectation and variance in Eqs. (7.15) and (7.17) as 10 min and $33\frac{1}{3}$ min^2, respectively. Theorem 7.3.1 tells us that we can calculate the variance in a different way as follows:

$$E(X^2) = \int_{-\infty}^{\infty} x^2 f(x)\, dx = \int_0^{20} \frac{1}{20} x^2\, dx$$

$$= \left[\frac{1}{60} x^3 \right]_0^{20} = 133\frac{1}{3}$$

$$\mathrm{Var}(X) = E(X^2) - [E(X)]^2 = 133\frac{1}{3} - 10^2 = 33\frac{1}{3}.$$

This yields the same result as Eq. (7.18) but is much faster.

7.3.3   Quantiles of a Distribution

We introduced the concept of quantiles in Chap. 3, Sect. 3.1.2. Now we define quantiles in terms of the distribution function.

Definition 7.3.4 The value x_p for which the cumulative distribution function is

$$F(x_p) = p \quad (0 < p < 1) \tag{7.21}$$

is called the **p-quantile**.

It follows from Definition 7.3.4 that x_p is the value which divides the cumulative distribution function into two parts: the probability of observing a value left of x_p is p whereas the probability of observing a value right of x_p is $1 - p$. For example, the 0.25-quantile $x_{0.25}$ describes the x-value for which the probability of observing $x_{0.25}$ or any smaller value is 0.25. Figure 7.4 shows the 0.25-quantile (first quartile), the 0.5-quantile (median) and the 0.75-quantile (third quartile) in a cumulative distribution function.

Fig. 7.4 First quartile, median and third quartile

Example 7.3.6 Recall Examples 7.2.1, 7.3.1, 7.3.5 and Fig. 7.1b where we described the waiting time for a train by using the following CDF:

$$F(x) = \frac{1}{20}x \,.$$

The first quartile $x_{0.25}$ is 5 because $F(5) = 5/20 = 0.25$. This means that the probability of waiting for the train for 5 min or less is 25%, and of waiting for longer than 5 min is 75%.

For continuous variables, there is a unique value which describes the p-quantile. However, for discrete variables this may not necessarily be true. In this case the p-quantile is chosen such that

$$F(x_p) \geq p \,,$$
$$F(x) < p \quad \text{for} \quad x < x_p$$

holds.

Example 7.3.7 The cumulative distribution function for rolling a die is described in Example 7.2.3 and Fig. 7.3b. The first quartile $x_{0.25}$ is 2 because $F(2) = 2/6 > 0.25$ and $F(x) < 0.25$ for $x < 2$.

7.3.4 Standardization

Standardization transforms a random variable in such a way that it has an expectation of zero and a variance of one. More details on the need for standardization is discussed in Chap. 10.

Definition 7.3.5 A random variable Y is called **standardized** when

$$E(Y) = 0 \quad \text{and} \quad \text{Var}(Y) = 1.$$

Theorem 7.3.2 *Suppose a random variable X has mean $E(X) = \mu$ and $\text{Var}(X) = \sigma^2$. Then it can be standardized as follows:*

$$Y = \frac{X - \mu}{\sigma} = \frac{X - E(X)}{\sqrt{\text{Var}(X)}} \,. \tag{7.22}$$

Example 7.3.8 In Examples 7.2.1, 7.3.1 and 7.3.5, we considered the waiting time X for a train. The random variable X can take values between 0 and 20 min and we calculated $E(X) = 10$ and $\text{Var}(X) = 33\frac{1}{3}$. The standardized variable of X is

$$Y = \frac{X - \mu}{\sigma} = \frac{X - 10}{\sqrt{33\frac{1}{3}}} \,.$$

One can show that $E(Y) = 0$ and $\text{Var}(Y) = 1$, see also Exercise 7.10 for more details.

7.4   Tschebyschev's Inequality

If we don't know the distribution of a random variable X, we can still make statements about the probability that X takes values in a certain interval (which has to be symmetric around the expectation μ) if the mean μ and the variance σ^2 of X are known.

Theorem 7.4.1 (Tschebyschev's inequality) *Let X be a random variable with $E(X) = \mu$ and $\text{Var}(X) = \sigma^2$. It holds that*

$$P(|X - \mu| \geq c) \leq \frac{\text{Var}(X)}{c^2}. \tag{7.23}$$

This is equivalent to

$$P(|X - \mu| < c) \geq 1 - \frac{\text{Var}(X)}{c^2}. \tag{7.24}$$

The proof is given in Appendix C.2.

Example 7.4.1 In Examples 7.2.1, 7.3.1 and 7.3.5 we have worked with a random variable which describes the waiting time for a train. We determined $E(X) = 10$ and $\text{Var}(X) = 33\frac{1}{3}$. We can calculate the probability of waiting between $10 - 7 = 3$ and $10 + 7 = 17$ min:

$$P(|X - \mu| < c) \geq 1 - \frac{\text{Var}(X)}{c^2}$$

$$P(|X - 10| < 7) \geq 1 - \frac{33\frac{1}{3}}{7^2} \approx 0.32.$$

The probability is therefore at least 0.32. However, if we apply our distributional knowledge that $F(x) = \frac{1}{20}x$ (for $0 \leq X \leq 20$) then we obtain a much more precise result which is

$$P(3 < X < 17) = F(17) - F(3) = \frac{17}{20} - \frac{3}{20} = 0.7.$$

We can clearly see that Tschebyschev's inequality gives us the correct answer, that is $P(3 < X < 17)$ is greater 0.32. Nevertheless the approximation to the exact probability, 0.7, is rather poor. One needs to keep in mind that only the lack of distributional knowledge makes the inequality useful.

7.5 Bivariate Random Variables

There are many situations in which we are interested in analysing more than one variable, say two variables. When we have more than one variable, then not only their individual distributions but also their joint probability density function can be of interest. For example, we know that driving a car after drinking alcohol is not necessarily safe. If we consider two variables, the blood alcohol content X and number of car accidents Y, then we may be interested in the probability of having a high blood alcohol content *and* a car accident at the same time. If we analyse (X, Y) jointly, then we are interested in their joint **bivariate** distribution $f_{XY}(x, y)$. This distribution can either be discrete or continuous.

Discrete Bivariate Random Variables. Suppose we have two categorical variables X and Y which can take the values $x_1, x_2, \ldots, x_I$ and $y_1, y_2, \ldots, y_J$ respectively. Their **joint probability distribution function** is characterized by

$$P(X = x_i, Y = y_j) = p_{ij} \quad (i = 1, 2, \ldots, I \,; j = 1, 2, \ldots, J)$$

with $\sum_{i=1}^{I} \sum_{j=1}^{J} p_{ij} = 1$. This means that the probability of observing x_i *and* y_j together is p_{ij}. We can summarize this information in a contingency table as follows:

		Y				
		1	2	$\ldots$	J	Total
	1	p_{11}	p_{12}	$\cdots$	p_{1J}	p_{1+}
	2	p_{21}	p_{22}	$\cdots$	p_{2J}	p_{2+}
X	$\vdots$	$\vdots$				$\vdots$
	I	p_{I1}	p_{I2}	$\cdots$	p_{IJ}	p_{I+}
	Total	p_{+1}	p_{+2}	$\cdots$	p_{+J}	1

Each cell contains a "piece" of the joint distribution. The entries p_{+1}, p_{+2}, $\ldots, p_{+J}$ in the bottom row of the table summarize the **marginal distribution** of Y, which is the distribution of Y without giving reference to X. The entries $p_{1+}, p_{2+}, \ldots, p_{I+}$ in the last column summarize the marginal distribution of X. The marginal distributions can therefore be expressed as

$$P(X = x_i) = \sum_{j=1}^{J} p_{ij} = p_{i+} \quad i = 1, 2, \ldots, I \,,$$

$$P(Y = y_j) = \sum_{i=1}^{I} p_{ij} = p_{+j} \quad j = 1, 2, \ldots, J \,.$$

The **conditional distributions** of X given $Y = y_j$ and Y given $X = x_j$ are given as follows:

$$P(X = x_i | Y = y_j) = p_{i|j} = \frac{p_{ij}}{p_{+j}} \quad i = 1, 2, \ldots, I,$$

$$P(Y = y_j | X = x_i) = p_{j|i} = \frac{p_{ij}}{p_{i+}} \quad j = 1, 2, \ldots, J.$$

They summarize the distribution of X for a given value of y_j (or the distribution of Y for a given value of x_i) and play a crucial role in the construction of regression models such as the linear regression model introduced in Chap. 11. Please also recall the definitions of Sect. 4.1 where we introduced conditional and marginal distributions for data samples rather than random variables.

Example 7.5.1 Suppose we have a contingency table on smoking behaviour X ($1 =$ never smoking, $2 =$ smoking sometimes and $3 =$ smoking regularly) and education level Y ($1 =$ primary education, $2 =$ secondary education, $3 =$ tertiary education):

		Y			
		1	2	3	Total
	1	0.10	0.20	0.30	0.60
X	2	0.10	0.10	0.10	0.30
	3	0.08	0.01	0.01	0.10
	Total	0.28	0.31	0.41	1

The cell entries represent the joint distribution of smoking behaviour and education level. We can interpret each entry as the probability of observing $X = x_i$ *and* $Y = y_j$ simultaneously. For example, $p_{23} = P$ ("smoking sometimes and tertiary education") $= 0.10$. The marginal distribution of X is contained in the last column of the table and lists the probabilities of smoking (unconditional on education level), e.g. the probability of being a non-smoker in this population is 60%. We can also interpret the conditional distributions: $P(X|Y = 3)$ represents the distribution of smoking behaviour among those who have tertiary education. If we are interested in the probability of smoking sometimes given tertiary education is completed, then we calculate $P(X = 2 | Y = 3) = p_{2|3} = \frac{0.10}{0.41} = 0.24$.

Continuous Bivariate Random Variables

Definition 7.5.1 A bivariate random variable (X, Y) is continuous if there is a function $f_{XY}(x, y)$ such that

$$F_{XY}(x, y) = P(X \le x, Y \le y) = \int_{-\infty}^{y} \int_{-\infty}^{x} f_{XY}(x, y) \, dx \, dy \tag{7.25}$$

holds.

Fig. 7.5 Area covering all points of (X, Y) with $(x_1 \leq X \leq x_2, y_1 \leq Y \leq y_2)$

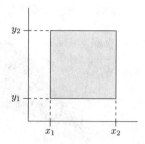

The function $F_{XY}(x, y)$ is the **joint cumulative distribution function** of X and Y; the joint distribution function is denoted by $f_{XY}(x, y)$ and $f_{XY}(x, y)$ has to fulfil the usual conditions of a density function. Necessary and sufficient conditions that a function $F_{XY}(x, y)$ is a bivariate cumulative distribution function are as follows:

$$\lim_{x \to -\infty} F_{XY}(x, y) = 0 \qquad \lim_{y \to -\infty} F_{XY}(x, y) = 0$$
$$\lim_{x \to \infty} F_{XY}(x, y) = 1 \qquad \lim_{y \to \infty} F_{XY}(x, y) = 1$$

and

$$F(x_2, y_2) - F(x_1, y_2) - F(x_2, y_1) + F(x_1, y_1) \geq 0 \text{ for all } x_1 < x_2, y_1 < y_2 .$$

The last condition is sometimes referred to as the *rectangle inequality*. As in the univariate case, we can use the cumulative distribution function to calculate interval probabilities; similarly we look at the rectangular area defined by (x_1, y_1), (x_1, y_2), (x_2, y_1), (x_2, y_2) in the bivariate case (instead of an interval $[a, b]$), see Fig. 7.5. We can calculate the desired probabilities as follows:

$$P(x_1 \leq X \leq x_2, y_1 \leq Y \leq y_2) = \int_{y_1}^{y_2} \int_{x_1}^{x_2} f_{XY}(x, y) \, dx \, dy .$$

The **marginal distributions** of X and Y are

$$f_X(x) = \int_{-\infty}^{\infty} f_{XY}(x, y) dy , \quad f_Y(y) = \int_{-\infty}^{\infty} f_{XY}(x, y) dx ,$$

respectively. Similarly to the discrete case, $f_X(x)$ and $f_Y(y)$ also describe the distribution of X unconditional on Y and the distribution of Y unconditional on X. The **cumulative marginal distributions** are

$$F_X(x) = \int_{-\infty}^{x} f_X(t) dt , \quad F_Y(y) = \int_{-\infty}^{y} f_Y(t) dt .$$

The **conditional distributions** can be obtained by the ratio of the joint and marginal distributions:

$$f_{X|Y}(x, y) = \frac{f(x, y)}{f(y)} , \quad f_{Y|X}(x, y) = \frac{f(x, y)}{f(x)} .$$

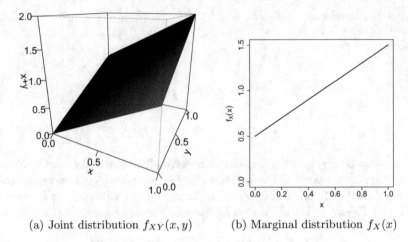

(a) Joint distribution $f_{XY}(x, y)$ (b) Marginal distribution $f_X(x)$

Fig. 7.6 Joint and marginal distribution for Example 7.5.2

Example 7.5.2 Consider the function

$$f_{XY}(x, y) = \begin{cases} x + y & \text{for } 0 \leq x \leq 1, \quad 0 \leq y \leq 1 \\ 0 & \text{elsewhere} \end{cases}$$

Suppose X and Y represent the concentrations of two drugs in the human body. Then $f_{XY}(x, y)$ may represent the sum of two drug concentrations in the human body. Since there are infinite possible realizations of both X and Y, we represent their joint distribution in a figure rather than a table, see Fig. 7.6a.

The marginal distributions for X and Y can be calculated as follows:

$$f_X(x) = \int_{-\infty}^{\infty} f_{XY}(x, y)dy = \int_0^1 (x + y) \, dy = \left[xy + \frac{1}{2}y^2 \right]_0^1 = x + \frac{1}{2},$$

$$f_Y(y) = \int_{-\infty}^{\infty} f_{XY}(x, y)dx = \int_0^1 (x + y) \, dx = \left[\frac{1}{2}x^2 + xy \right]_0^1 = y + \frac{1}{2}.$$

Figure 7.6b depicts the marginal distribution for X. The slope of the marginal distribution is essentially the slope of the surface of the joint distribution shown in Fig. 7.6a. It is easy to see in this simple example that the marginal distribution of X is nothing but a cut in the surface of the joint distribution. Note that the conditional distributions $f_{X|Y}(x, y)$ and $f_{Y|X}(x, y)$ can be easily calculated; for example $f_{X|Y}(x, y) = f(x, y)/f(y) = (x + y)/(y + 0.5)$.

Stochastic Independence

Definition 7.5.2 Two continuous random variables X and Y are said to be **stochastically independent** if

$$f_{XY}(x, y) = f_X(x)f_Y(y). \tag{7.26}$$

For discrete variables, this is equivalent to

$$P(X = x_i, Y = y_j) = P(X = x_i)P(Y = y_j) \qquad (7.27)$$

being valid for all (i, j).

Example 7.5.3 In Example 7.5.2, we considered the function

$$f_{XY}(x, y) = \begin{cases} x + y & \text{for } 0 \le x \le 1, \quad 0 \le y \le 1 \\ 0 & \text{elsewhere} \end{cases}$$

with the marginal distributions of X and Y as $f_X = x + 0.5$ and $f_Y = y + 0.5$ respectively. Since $f_X \cdot f_Y = (x + \frac{1}{2})(y + \frac{1}{2}) \ne f_{XY}$ it follows that X and Y are not independent. The interpretation is that the concentrations of the two drugs are not independent.

7.6   Calculation Rules for Expectation and Variance

Calculation rules for the expectation. For any constant values a and b, and any random variables X and Y, the following rules hold:

$$E(a) = a, \qquad (7.28)$$
$$E(bX) = bE(X), \qquad (7.29)$$
$$E(a + bX) = a + bE(X), \qquad (7.30)$$
$$E(X + Y) = E(X) + E(Y) \ (additivity). \qquad (7.31)$$

The proof of rule (7.29) is given in Appendix C.2.

Example 7.6.1 Consider again Example 7.2.3 where we illustrated how the outcome of a die roll experiment can be captured by a random variable. There were 6 events and X could take the values $x_1 = 1$, $x_2 = 2$, $\ldots$, $x_6 = 6$. The probability of the occurrence of any number was $P(X = x_i) = 1/6$ and the expectation was calculated as 3.5. Consider two different situations:

(i) Suppose the die takes the values $10, 20, 30, 40, 50, 60$ instead of the values $1, 2, 3, 4, 5, 6$. The random variable $Y = 10X$ describes this suitably and its expectation is

$$E(Y) = E(10X) = 10E(X) = 10 \cdot 3.5 = 35$$

which follows from (7.29).

(ii) If we are rolling two dice X_1 and X_2, then the expectation for the sum of the two outcomes is

$$E(X) = E(X_1 + X_2) = E(X_1) + E(X_2) = 3.5 + 3.5 = 7$$

due to (7.31).

Calculation rules for the variance. Let a and b be any known constants and X be a random variable (discrete or continuous). Then we have the following rules:

$$\mathrm{Var}(a) = 0, \tag{7.32}$$
$$\mathrm{Var}(bX) = b^2 \mathrm{Var}(X), \tag{7.33}$$
$$\mathrm{Var}(a + bX) = b^2 \mathrm{Var}(X). \tag{7.34}$$

The proof of rule (7.34) is given in Appendix C.2.

Example 7.6.2 In Examples 7.2.1, 7.3.1, 7.3.3 and 7.3.5, we evaluated a random variable describing the waiting time for a train. Now, suppose that a person has to also catch a bus after the train ride. If this bus arrives only every 60 min, then the PDF of the random variable Y denoting the waiting time for the bus is

$$f(Y) = \begin{cases} \frac{1}{60} & \text{for } 0 < x \le 60 \\ 0 & \text{otherwise}. \end{cases}$$

We can use Eqs. (7.15) and (7.17) to determine both the expectation and variance of Y. However, the waiting time for the bus is governed by the relation $Y = 3X$ where X is the waiting time for the train. Therefore we can calculate $E(Y) = E(3X) = 3E(X) = 3 \cdot 10 = 30\,\text{min}$ by using rule (7.29) and the variance as $\mathrm{Var}(Y) = \mathrm{Var}(3X) = 3^2 \mathrm{Var}(X) = 9 \cdot 33\frac{1}{3} = 300$ using rule (7.33). The total waiting time is the sum of the two waiting times.

7.6.1   Expectation and Variance of the Arithmetic Mean

Definition 7.6.1 We define the random variables $X_1, X_2, \ldots, X_n$ to be i.i.d. (independently identically distributed), if all X_i follow the same distribution and are stochastically independent of each other.

Let $X_1, X_2, \ldots, X_n$ be n i.i.d. random variables with $E(X_i) = \mu$ and $\mathrm{Var}(X_i) = \sigma^2$, $i = 1, 2, \ldots, n$. The arithmetic mean of these variables is given by

$$\bar{X} = \frac{1}{n} \sum_{i=1}^{n} X_i,$$

which is again a random variable that follows a distribution with certain expectation and variance. A function of random variables is called a **statistic**. By using (7.29), we obtain

$$E(\bar{X}) = \frac{1}{n} \sum_{i=1}^{n} E(X_i) = \mu. \tag{7.35}$$

If we apply (7.34) and recall that the variables are independent of each other, we can also calculate the variance as

$$Var(\bar{X}) = \frac{1}{n^2} \sum_{i=1}^{n} Var(X_i) = \frac{\sigma^2}{n}. \tag{7.36}$$

Example 7.6.3 If we toss a coin, we obtain either head or tail and therefore $P(\text{"head"}) = P(\text{"tail"}) = \frac{1}{2}$. If we toss the coin n times, we have for each toss

$$X_i = \begin{cases} 0 & \text{for "tail"} \\ 1 & \text{for "head"} \end{cases}, \qquad i = 1, \ldots, n.$$

It is straightforward to calculate the expectation and variance for each coin toss:

$$E(X_i) = 0 \cdot \frac{1}{2} + 1 \cdot \frac{1}{2} = \frac{1}{2},$$

$$Var(X_i) = (0 - \frac{1}{2})^2 \cdot \frac{1}{2} + (1 - \frac{1}{2})^2 \cdot \frac{1}{2} = \frac{1}{4} \cdot \frac{1}{2} + \frac{1}{4} \cdot \frac{1}{2} = \frac{1}{4}.$$

The arithmetic mean $\bar{X} = \frac{1}{n} \sum_{i=1}^{n} X_i$ describes the relative frequency of heads when the coin is tossed n times. We can now apply (7.35) and (7.36) to calculate

$$E(\bar{X}) = \frac{1}{n} \sum_{i=1}^{n} 1/2 = 1/2$$

and

$$Var(\bar{X}) = \frac{1}{n^2} \sum_{i=1}^{n} \frac{1}{4} = \frac{1}{4n}.$$

With this example the interpretation of formulae (7.35) and (7.36) becomes clearer: if the probability of head is 0.5 for a single toss, then it is also 0.5 for the mean of all tosses. If we toss a coin many times then the variance decreases when n increases. This means that a larger sample size yields a higher precision for the calculated arithmetic mean. This observation shows the basic conclusion of the next chapter: the higher the sample size, the more secure we are of our conclusions.

7.7   Covariance and Correlation

The variance measures the variability of a variable. Similarly, the covariance measures the co-variation or association between X and Y.

7.7.1   Covariance

Definition 7.7.1 The **covariance** between X and Y is defined as

$$\varrho = \text{Cov}(X, Y) = \text{E}[(X - \text{E}(X))(Y - \text{E}(Y))]. \qquad (7.37)$$

The covariance is positive if, on average, larger values of X correspond to larger values of Y; it is negative if, on average, greater values of X correspond to smaller values of Y.

The probability density function of any bivariate random variable (X, Y) is characterized by the expectation and variance of both X and Y,

$$\text{E}(X) = \mu_X, \quad \text{Var}(X) = \sigma_X^2,$$
$$\text{E}(Y) = \mu_Y, \quad \text{Var}(Y) = \sigma_Y^2,$$

as well as their **covariance**. We can summarize these features by using the expectation vector

$$\text{E}\begin{pmatrix} X \\ Y \end{pmatrix} = \begin{pmatrix} \text{E}(X) \\ \text{E}(Y) \end{pmatrix} = \begin{pmatrix} \mu_X \\ \mu_Y \end{pmatrix}$$

and the **covariance matrix**

$$\text{Cov}\begin{pmatrix} X \\ Y \end{pmatrix} = \begin{pmatrix} \text{Cov}(X, X) & \text{Cov}(X, Y) \\ \text{Cov}(Y, X) & \text{Cov}(Y, Y) \end{pmatrix} = \begin{pmatrix} \sigma_X^2 & \varrho \\ \varrho & \sigma_Y^2 \end{pmatrix}.$$

Important properties of covariance are

(i) $\text{Cov}(X, Y) = \text{Cov}(Y, X)$,
(ii) $\text{Cov}(X, X) = \text{Var}(X)$,
(iii) $\text{Cov}(aX + b, cY + d) = ac\,\text{Cov}(X, Y)$,
(iv) $\text{Cov}(X, Y) = \text{E}(XY) - \text{E}(X)\text{E}(Y)$ where $\text{E}(XY) = \int\int xyf(x, y)dxdy$ for continuous variables and $\text{E}(XY) = \sum_i \sum_j x_i y_j p_{ij}$ for discrete variables,
(v) If X and Y are independent it follows that $\text{E}(XY) = \text{E}(X)\text{E}(Y) = \mu_X \mu_Y$ and therefore $\text{Cov}(X, Y) = \mu_X \mu_Y - \mu_X \mu_Y = 0$.

Theorem 7.7.1 (Additivity Theorem) *The variance of the sum (subtraction) of X and Y is given by*

$$\text{Var}(X \pm Y) = \text{Var}(X) + \text{Var}(Y) \pm 2\text{Cov}(X, Y).$$

If X and Y are independent, it follows that $\text{Cov}(X, Y) = 0$ and therefore $\text{Var}(X \pm Y) = \text{Var}(X) + \text{Var}(Y)$. We omit the proof of this theorem.

Example 7.7.1 Recall Example 7.6.2 where we considered the waiting time Y for a bus and the waiting time X for a train. Suppose their joint bivariate probability density function can be written as

$$f_{XY}(x, y) = \begin{cases} \frac{1}{1200} & \text{for } 0 \le x \le 60, \quad 0 \le y \le 20 \\ 0 & \text{elsewhere.} \end{cases}$$

To calculate the covariance between X and Y, we need to calculate $\text{E}(XY)$:

$$\text{E}(XY) = \int_{-\infty}^{\infty} \int_{-\infty}^{\infty} xy f(x, y) dx\, dy = \int_{0}^{60} \int_{0}^{20} xy \frac{1}{1200} dx\, dy$$

$$= \int_{0}^{60} \left[\frac{x}{1200} \frac{y^2}{2} \right]_{0}^{20} dx = \int_{0}^{60} \frac{400x}{2400} dx = \left[\frac{1}{6} \frac{x^2}{2} \right]_{0}^{60} = \frac{3600}{12} = 300 .$$

We know from Example 7.6.2 that $\text{E}(X) = 10$, $\text{E}(Y) = 30$, $\text{Var}(X) = 33\frac{1}{3}$, and $\text{Var}(Y) = 300$. The covariance is thus

$$\text{Cov}(X, Y) = \text{E}(XY) - \text{E}(X)\text{E}(Y) = 300 - 30 \cdot 10 = 0 .$$

This makes sense as the waiting times for the train and the bus should be independent of each other. Using rule (7.31), we conclude that the total expected waiting time is

$$\text{E}(X + Y) = \text{E}(X) + \text{E}(Y) = 10 + 30 = 40$$

minutes. The respective variance is

$$\text{Var}(X + Y) = \text{Var}(X) + \text{Var}(Y) - 2\text{Cov}(X, Y) = 33\frac{1}{3} + 300 - 2 \cdot 0 = 333\frac{1}{3}$$

due to Theorem 7.7.1.

7.7.2   Correlation Coefficient

Definition 7.7.2 The **correlation coefficient** of X and Y is defined as

$$\rho(X, Y) = \frac{\text{Cov}(X, Y)}{\sqrt{\text{Var}(X)\text{Var}(Y)}} . \tag{7.38}$$

We already know from Chap. 4 that the correlation coefficient is a measure of the degree of linear relationship between X and Y. It can take values between -1

and 1, $-1 \le \rho(X, Y) \le 1$. However in Chap. 4, we considered the correlation of two samples, i.e. realizations of random variables; here we describe the correlation coefficient of the population. If $\rho(X, Y) = 0$, then X and Y are said to be uncorrelated. If there is a perfect linear relationship between X and Y then $\rho = 1$ for a positive relationship and $\rho = -1$ for a negative relationship, see Appendix C.2 for the proof.

Theorem 7.7.2 *If X and Y are independent, they are also uncorrelated. However, if they are uncorrelated then they are not necessarily independent.*

Example 7.7.2 In Example 7.6.2, we estimated the covariance between the waiting time for the bus and the waiting time for the train: $\text{Cov}(X, Y) = 0$. The correlation coefficient is therefore also 0 indicating no linear relationship between the waiting times for bus and train.

7.8   Key Points and Further Issues

Note:

✓ Note that there is a difference between the empirical cumulative distribution function introduced in Chap. 2 and the CDF introduced in this chapter. In Chap. 2, we work with the cumulative distribution of observed values in a particular sample whereas in this chapter, we deal with random variables modelling the distribution of a general population.

✓ The expectation and the variance of a random variable are defined as follows:

	Expectation	Variance
Discrete	$\sum_{i=1}^{n} x_i p_i$	$\sum_{i=1}^{n} (x_i - \text{E}(X))^2 p_i$
Continuous	$\int_{-\infty}^{+\infty} x f(x) dx$	$\int_{-\infty}^{+\infty} (x - \text{E}(X))^2 f(x) dx$

✓ Some important calculation rules are:

$$E(a + bX) = a + bE(X); \quad \text{Var}(a + bX) = b^2 \text{Var}(X);$$
$$E(X + Y) = E(X) + E(Y); \quad \text{Var}(X \pm Y) = \text{Var}(X) + \text{Var}(Y)$$
$$\pm 2\text{Cov}(X, Y)$$

✓ Bivariate random variables (X, Y) have a joint CDF $F_{XY}(x, y)$ which specifies the probability $P(X \le x; Y \le y)$. The conditional distribution of $X|Y$ $[Y|X]$ is the PDF of X $[Y]$ for a given value $Y = y$ $[X = x]$. The marginal distribution of X $[Y]$ is the distribution of X $[Y]$ without referring to the values of Y $[X]$.

7.9   Exercises

Exercise 7.1 Consider the following cumulative distribution function of a random variable X:

$$F(x) = \begin{cases} 0 & \text{if } x < 2 \\ -\frac{1}{4}x^2 + 2x - 3 & \text{if } 2 \le x \le 4 \\ 1 & \text{if } x > 4 \end{cases}$$

(a) What is the PDF of X?
(b) Calculate $P(X < 3)$ and $P(X = 4)$.
(c) Determine $E(X)$ and $\text{Var}(X)$.

Exercise 7.2 Joey manipulates a die to increase his chances of winning a board game against his friends. In each round a die is rolled and larger numbers are generally an advantage. Consider the random variable X denoting the outcome of the rolled die and the respective probabilities $P(X = 1 = 2 = 3 = 5) = 1/9$, $P(X = 4) = 2/9$, and $P(X = 6) = 3/9$.

(a) Calculate and interpret the expectation and variance of X.
(b) Imagine that the board game contains an action which makes the players use $1/X$ rather than X. What is the expectation of $Y = 1/X$? Is $E(Y) = E(1/X) = 1/E(X)$?

Exercise 7.3 An innovative wine maker experiments with new grapes and adds a new wine to his stock. The percentage sold by the end of the season depends on the weather and various other factors. It can be modelled using the random variable X with the CDF as

$$F(x) = \begin{cases} 0 & \text{if } x < 0 \\ 3x^2 - 2x^3 & \text{if } 0 \le x \le 1 \\ 1 & \text{if } x > 1 \end{cases}$$

(a) Plot the cumulative distribution function with R.
(b) Determine $f(x)$.
(c) What is the probability of selling at least one third of his wine, but not more than two thirds?
(d) Define the CDF in R and calculate the probability of (c) again.
(e) What is the variance of X?

Exercise 7.4 A quality index summarizes different features of a product by means of a score. Different experts may assign different quality scores depending on their experience with the product. Let X be the quality index for a tablet. Suppose the respective probability density function is given as follows:

$$f(x) = \begin{cases} cx(2 - x) & \text{if } 0 \le x \le 2 \\ 0 & \text{elsewhere} . \end{cases}$$

(a) Determine c such that $f(x)$ is a proper PDF.
(b) Determine the cumulative distribution function.
(c) Calculate the expectation and variance of X.
(d) Use Tschebyschev's inequality to determine the probability that X doesn't deviate more than 0.5 from its expectation.

Exercise 7.5 Consider the joint PDF for type of customer service X ($0 =$ telephonic hotline, $1 =$ Email) and of satisfaction score Y ($1 =$ unsatisfied, $2 =$ satisfied, $3 =$ very satisfied):

$X \backslash Y$	1	2	3
0	0	1/2	1/4
1	1/6	1/12	0

(a) Determine and interpret the marginal distributions of both X and Y.
(b) Calculate the 75% quantile for the marginal distribution of Y.
(c) Determine and interpret the conditional distribution of satisfaction level for $X = 1$.
(d) Are the two variables independent?
(e) Calculate and interpret the covariance of X and Y.

Exercise 7.6 Consider a continuous random variable X with expectation 15 and variance 4. Determine the smallest interval $[15 - c, 15 + c]$ which contains at least 90% of the values of X.

Exercise 7.7 Let X and Y be two random variables for which only 6 possible events—$A_1, A_2, A_3, A_4, A_5, A_6$—are defined:

i	1	2	3	4	5	6
$P(A_i)$	0.3	0.1	0.1	0.2	0.2	0.1
X_i	−1	2	2	−1	−1	2
Y_i	0	2	0	1	2	1

(a) What is the joint PDF of X and Y?
(b) Calculate the marginal distributions of X and Y.
(c) Are both variables independent?
(d) Determine the joint PDF for $U = X + Y$.
(e) Calculate $E(U)$ and $Var(U)$ and compare it with $E(X) + E(Y)$ and $Var(X) + Var(Y)$ respectively.

Exercise 7.8 Recall the urn model we introduced in Chap. 5. Consider an urn with 8 balls: 4 of them are white, 3 are black and 1 is red. Now two balls are drawn from the urn. The random variables X and Y are defined as follows:

$$X = \begin{cases} 1 & \text{black ball} \\ 2 & \text{red ball in the first draw} \\ 3 & \text{white ball} \end{cases}$$

$$Y = \begin{cases} 1 & \text{black ball} \\ 2 & \text{red ball in the second draw} \\ 3 & \text{white ball.} \end{cases}$$

(a) When are X and Y independent—when the two balls are drawn with replacement or without replacement?
(b) Assume the balls are drawn such that X and Y are dependent. Use the conditional distribution $P(Y|X)$ to determine the joint PDF of X and Y.
(c) Calculate $E(X)$, $E(Y)$ and $\rho(X, Y)$.

Exercise 7.9 If X is the amount of money spent on food and other expenses during a day (in €) and Y is the daily allowance of a businesswoman, the joint density of these two variables is given by

$$f_{XY}(x, y) = \begin{cases} c\left(\frac{100-x}{x}\right) & \text{if } 10 \le x \le 100, \quad 40 \le y \le 100 \\ 0 & \text{elsewhere.} \end{cases}$$

(a) Choose c such that $f_{XY}(x, y)$ is a probability density function.
(b) Find the marginal distribution of X.
(c) Calculate the probability that more than €75 are spent.
(d) Determine the conditional distribution of Y given X.

Exercise 7.10 Consider n i.i.d. random variables X_i with $E(X_i) = \mu$ and $Var(X_i) = \sigma^2$ and the standardized variable $Y = \frac{X-\mu}{\sigma}$. Show that $E(Y) = 0$ and $Var(Y) = 1$.

→ Solutions to all exercises in this chapter can be found in Appendix B

Probability Distributions

We introduced the concept of probability density and probability mass functions of random variables in the previous chapter. In this chapter, we are introducing some common standard discrete and continuous probability distributions which are widely used for either practical applications or for constructing statistical methods described later in this book. Suppose we are interested in determining the probability of a certain event. The determination of probabilities depends upon the nature of the study and various prevailing conditions which affect it. For example, the determination of the probability of a head when tossing a coin is different from the determination of the probability of rain in the afternoon. One can speculate that some mathematical functions can be defined which depict the behaviour of probabilities under different situations. Such functions have special properties and describe how probabilities are distributed under different conditions. We have already learned that they are called probability distribution functions. The form of such functions may be simple or complicated depending upon the nature and complexity of the phenomenon under consideration. Let us first recall and extend the definition of independent and identically distributed random variables:

Definition 8.0.1 The random variables $X_1, X_2, \ldots, X_n$ are called independent and identically distributed (i.i.d) if the $X_i \, (i = 1, 2, \ldots, n)$ have the same marginal cumulative distribution function (CDF) $F(x)$ and if they are mutually independent.

Example 8.0.1 Suppose a researcher plans a survey on the weight of newborn babies in a country. The researcher randomly contacts 10 hospitals with a maternity ward and asks them to randomly select 20 of the newborn babies (no twins) born in the last 6 months and records their weights. The sample therefore consists of $10 \times 20 = 200$ baby weights. Since the hospitals and the babies are randomly selected, the babies' weights are therefore not known beforehand. The 200 weights can be denoted by the

© Springer Nature Switzerland AG 2022
C. Heumann et al., *Introduction to Statistics and Data Analysis*,
https://doi.org/10.1007/978-3-031-11833-3_8

random variables $X_1, X_2, \ldots, X_{200}$. Note that the weights X_i are random variables because, depending on the size of the population, different samples consisting of 200 babies can be randomly selected. Also, the babies' weights can be seen as stochastically independent (an example of stochastically dependent weights would be the weights of twins if they are included in the sample). After collecting the weights of 200 babies, the researcher has a sample of 200 realized values (i.e. the weights in grams). The values are now known and denoted by $x_1, x_2, \ldots, x_{200}$.

8.1   Standard Discrete Distributions

First we discuss some standard distributions for discrete random variables.

8.1.1   Discrete Uniform Distribution

The discrete uniform distribution assumes that all possible outcomes have equal probability of occurrence. A more formal definition is given as follows:

Definition 8.1.1 A discrete random variable X with k possible outcomes $x_1, x_2, \ldots,$ x_k is said to follow a discrete **uniform** distribution if the probability mass function (PMF) of X is given by

$$P(X = x_i) = \frac{1}{k}, \quad \forall i = 1, 2, \ldots, k. \tag{8.1}$$

If the outcomes are the natural numbers $x_i = i$ $(i = 1, 2, \ldots, k)$, the mean and variance of X are obtained as

$$E(X) = \frac{k+1}{2}, \tag{8.2}$$

$$Var(X) = \frac{1}{12}(k^2 - 1). \tag{8.3}$$

Example 8.1.1 If we roll a fair die, the outcomes "1", "2", $\ldots$, "6" have equal probability of occurring and hence the random variable X "number of dots observed on the upper surface of the die" has a uniform discrete distribution with PMF

$$P(X = i) = \frac{1}{6}, \quad \text{for all } i = 1, 2, \ldots, 6.$$

The mean and variance of X are

$$E(X) = \frac{6+1}{2} = 3.5,$$

$$Var(X) = \frac{1}{12}(6^2 - 1) = 35/12.$$

Fig. 8.1 Frequency distribution of 1000 generated discrete uniform random numbers with possible outcomes (2, 5, 8, 10)

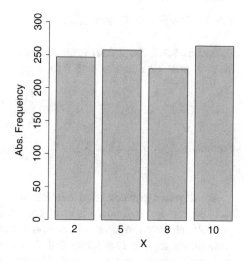

Using the function `sample()` in R, it is easy to generate random numbers from a discrete uniform distribution. The following command generates a random sample of size 1000 from a uniform distribution with the four possible outcomes 2, 5, 8, 10 and draws a bar chart of the observed numbers. The use of the `set.seed()` function allows to reproduce the generated random numbers at any time. It is necessary to use the option `replace=TRUE` to simulate draws with replacement, i.e. to guarantee that a value can occur more than once.

```
set.seed(123789)
x <- sample(x=c(2,5,8,10), size=1000, replace=T,
prob=c(1/4,1/4,1/4,1/4))
barchart(table(x), ylim=c(0,300))
```

A bar chart of the frequency distribution of the 1000 sampled numbers with the possible outcomes (2, 5, 8, 10) using the discrete uniform distribution is given in Fig. 8.1. We see that the 1000 generated random numbers are not exactly uniformly distributed, e.g. the numbers 5 and 10 occur more often than the numbers 2 and 8. In fact, they are only approximately uniform. We expect that the deviance from a perfect uniform distribution is getting smaller as we generate more and more random numbers but will probably never be zero for a finite number of draws. The random numbers reflect the practical situation that a sample distribution is only an approximation to the theoretical distribution from which the sample was drawn. More details on how to work with random variables in R are given in Appendix A.

8.1.2   Degenerate Distribution

Definition 8.1.2 A random variable X has a **degenerate distribution** at a, if a is the only possible outcome with $P(X = a) = 1$. The CDF in such a case is given by

$$F(x) = \begin{cases} 0 \text{ if } x < a \\ 1 \text{ if } x \geq a \end{cases}.$$

Further, $E(X) = a$ and $Var(X) = 0$.

The degenerate distribution indicates that there is only one possible fixed outcome and therefore no randomness is involved. It follows that we need at least two different possible outcomes to have randomness in the observations of a random variable or random experiment. The Bernoulli distribution is such a distribution where there are only two outcomes, e.g. success and failure or male and female. These outcomes are usually denoted by the values "0" and "1".

8.1.3   Bernoulli Distribution

Definition 8.1.3 A random variable X has a Bernoulli distribution if the PMF of X is given as

$$P(X = x) = \begin{cases} p & \text{if } x = 1 \\ 1 - p & \text{if } x = 0. \end{cases}$$

The cumulative distribution function (CDF) of X is

$$F(x) = \begin{cases} 0 & \text{if } x < 0 \\ 1 - p & \text{if } 0 \leq x < 1 \\ 1 & \text{if } x \geq 1. \end{cases}$$

The mean (expectation) and variance of a Bernoulli random variable are calculated as

$$E(X) = 1 \cdot p + 0 \cdot (1 - p) = p \tag{8.4}$$

and

$$Var(X) = (1 - p)^2 \cdot p + (0 - p)^2 \cdot (1 - p) = p(1 - p) \tag{8.5}$$

respectively.

A Bernoulli distribution is useful when there are only two possible outcomes and our interest lies in any of the two outcomes, e.g. whether a customer buys a certain product or not; or whether a hurricane hits an island or not etc. The outcome of an event A is usually coded as 1 which occurs with probability p. If the event of interest does not occur, i.e. the complementary event $\bar{A}$ occurs, the outcome is coded as 0 which occurs with probability $1 - p$. So p is the probability that the event of interest A occurs.

Example 8.1.2 A company organizes a raffle at an end-of-year function. There are 300 lottery tickets in total and 50 of them are marked as winning tickets. The event A of interest is "ticket wins" (coded as $X = 1$) and the probability p of having a winning ticket is a priori (i.e. before any lottery ticket has been drawn)

$$P(X = 1) = \tfrac{50}{300} = \tfrac{1}{6} = p \quad \text{and} \quad P(X = 0) = \tfrac{250}{300} = \tfrac{5}{6} = 1 - p.$$

According to (8.4) and (8.5), the mean (expectation) and variance of X are

$$E(X) = \tfrac{1}{6} \quad \text{and} \quad \text{Var}(X) = \tfrac{1}{6} \cdot \tfrac{5}{6} = \tfrac{5}{36} \text{ respectively.}$$

8.1.4 Binomial Distribution

Consider n independent trials or repetitions of a Bernoulli experiment. In each trial or repetition, we may observe either A or $\bar{A}$. At the end of the experiment, we have thus observed A between 0 and n times. Suppose we are interested in the probability of A occurring k times, then the Binomial distribution is useful.

Example 8.1.3 Consider a coin tossing experiment where a coin is tossed ten times and the event of interest is $A =$ "head". The random variable X "number of heads in 10 experiments" has the possible outcomes $k = 0, 1, \ldots, 10$. A question of interest may be: what is the probability that a head occurs in 7 out of 10 trials; or in 5 out of 10 trials? We assume that the order in which heads (and tails) appear is not of interest, only the total number of heads is of interest.

Questions of this kind are answered by the Binomial distribution. This distribution can either be motivated as a repetition of n Bernoulli experiments (as in the above coin tossing example) or by the urn model (see Chap. 5): assume there are M white and $N - M$ black balls in the urn. Suppose n balls are drawn randomly from the urn, the colour of the ball is recorded and the ball is placed back into the urn (sampling with replacement). Let A be the event of interest that a white ball is drawn from the urn. The probability of A is $p = M/N$ (the probability of drawing a black ball is $1 - p = (N - M)/N$). Since the balls are drawn with replacement these probabilities do not change from draw to draw. Further, let X be the random variable counting the number of white balls drawn from the urn in the n experiments. Since the order of the resulting colours of balls is not of interest in this case, there are $\binom{n}{k}$ combinations where k balls are white and $n - k$ balls are black. Since the balls are drawn with replacement, every outcome of the n experiments is independent of all others. The probability that $X = k, k = 0, 1, \ldots, n$, can therefore be calculated as

$$P(X = k) = \binom{n}{k} p^k (1 - p)^{n-k} \quad (k = 0, 1, \ldots, n). \tag{8.6}$$

Please note that we can use the product $p^k (1 - p)^{n-k}$ because the draws are independent. The binomial coefficient $\binom{n}{k}$ is necessary to count the number of possible orders of the black and white balls.

Definition 8.1.4 A discrete random variable X is said to follow a Binomial distribution with parameters n and p if its PMF is given by (8.6) We also write $X \sim B(n; p)$. The mean and variance of a binomial random variable X are given by

$$\mathrm{E}(X) = np, \tag{8.7}$$
$$\mathrm{Var}(X) = np(1 - p). \tag{8.8}$$

Remark 8.1.1 A Bernoulli random variable is therefore $B(1; p)$ distributed.

Example 8.1.4 Consider an unfair coin where the probability of observing a tail (T) is $p(T) = 0.6$. Let us denote tails by "1" and heads by "0". Suppose the coin is tossed 3 times. In total, there are the $2^3 = 8$ following possible outcomes:

Outcome	$X = x$
1 1 1	3
1 1 0	2
1 0 1	2
0 1 1	2
1 0 0	1
0 1 0	1
0 0 1	1
0 0 0	0

Note that the first outcome, viz. $(1, 1, 1)$ leads to $x = 3$, the next 3 outcomes, viz., $(1, 1, 0)$, $(1, 0, 1)$, $(0, 1, 1)$ obtained by $(= \binom{3}{2})$ lead to $x = 2$, the next 3 outcomes, viz., $(1, 0, 0)$, $(0, 1, 0)$, $(0, 0, 1)$ obtained by $(= \binom{3}{1})$ lead to $x = 1$, and the last outcome, viz. $(0, 0, 0)$ obtained by $(= \binom{3}{0})$ leads to $x = 0$. We can, for example, calculate

$$P(X = 2) = \binom{3}{2} 0.6^2 (1 - 0.6)^1 = 0.432 \quad (\text{or} \quad 43.2\%).$$

Further, the mean and variance of X are

$$\mathrm{E}(X) = np = 3 \cdot 0.6 = 1.8, \quad \text{and} \quad \mathrm{Var}(X) = np(1 - p) = 3 \cdot 0.6 \cdot 0.4 = 0.72.$$

Fig. 8.2 PMF of a
$B(50, 0.6)$ distribution

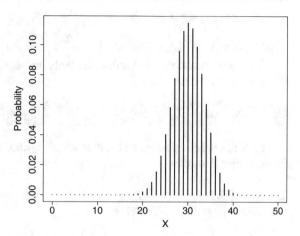

Functions for the Binomial distribution, as well as many other distributions, are implemented in *R*. For each of these distributions, we can easily determine the density function (PMF, PDF) for given values and parameters, determine the CDF, calculate quantiles and draw random numbers. Appendix A.3, gives more details. Nevertheless, we illustrate the concept of dealing with distributions in *R* in the following example.

Example 8.1.5 Suppose we toss an unfair coin 50 times with the probability of a tail $p_{tail} = 0.6$. We thus deal with a $B(50, 0.6)$ distribution which can be plotted using the dbinom command. The prefix d stands for "density".

```
n <- 50
p <- 0.6
k <- 0:n
pmf <- dbinom(k,n,p)
plot(k,pmf, type='h')
```

A plot of the PMF of a Binomial distribution with $n = 50$ and $p = 0.6$ (i.e. $B(50, 0.6)$) is given in Fig. 8.2.

Note that we can also calculate the CDF with *R*. We can use the pbinom(x,n,p) command, where the prefix p stands for probability, to calculate the CDF at any point. For example, suppose we are interested in $P(X \geq 30) = 1 - F(29)$, that is the probability of observing thirty or more tails; then we write

```
1-pbinom(29,50,0.6)
[1] 0.5610349
```

Similarly, we can determine quantiles. For instance, the 80% quantile q which describes that $P(X \leq q) \geq 0.8$ can be obtained by the qbinom(q,n,p) command as follows:

```
qbinom(0.8,50,0.6)
[1] 33
```

If we want to generate 100 random realizations from a $B(50, 0.6)$ distribution we can use the rbinom command.

```
rbinom(100,50,0.6)
```

The Binomial distribution has some nice properties. One of them is described in the following theorem:

Theorem 8.1.1 *Let $X \sim B(n; p)$ and $Y \sim B(m; p)$ and assume that X and Y are (stochastically) independent. Then*

$$X + Y \sim B(n + m; p) \,. \tag{8.9}$$

This is intuitively clear since we can interpret this theorem as describing the additive combination of two independent Binomial experiments with n and m trials, with equal probability p respectively. Since every Binomial experiment is a series of independent Bernoulli experiments, this is equivalent to a series of $n + m$ independent Bernoulli trials with constant success probability p which in turn is equivalent to a Binomial distribution with $n + m$ trials.

8.1.5 The Poisson Distribution

Consider a situation in which the number of events is very large and the probability of success is very small: for example, the number of alpha particles emitted by a radioactive substance entering a particular region in a given short time interval. Note that the number of emitted alpha particles is very high but only a few particles are transmitted through the region in a given short time interval. Some other examples where Poisson distributions are useful are the number of flu cases in a country within one year, the number of tropical storms within a given area in one year, or the number of bacteria found in a biological investigation.

Definition 8.1.5 A discrete random variable X is said to follow a *Poisson distribution* with parameter $\lambda > 0$ if its PMF is given by

$$P(X = x) = \frac{\lambda^x}{x!} \exp(-\lambda) \quad (x = 0, 1, 2, \ldots) \,. \tag{8.10}$$

We also write $X \sim Po(\lambda)$. The mean and variance of a Poisson random variable are identical:

$$E(X) = \text{Var}(X) = \lambda.$$

Example 8.1.6 Suppose a country experiences $X = 4$ tropical storms on average per year. Then the probability of suffering from only two tropical storms is obtained by using the Poisson distribution as

$$P(X = 2) = \frac{\lambda^x}{x!} \exp(-\lambda) = \frac{4^2}{2!} \exp(-4) = 0.146525 \,.$$

If we are interested in the probability that not more than 2 storms are experienced, then we can apply rules (7.7)–(7.13) from Chap. 7: $P(X \le 2) = P(X = 2) + P(X = 1) + P(X = 0) = F(2) = 0.2381033$. We can calculate $P(X = 1)$ and $P(X = 0)$ from (8.10) or using R. Similar to Example 8.1.5, we use the prefix d to obtain the PMF and the prefix p to work with the CDF, i.e we can use dpois(x, λ) and ppois(x, λ) to determine $P(X = x)$ and $P(X \le x)$ respectively.

```
dpois(2,4) + dpois(1,4) + dpois(0,4)
[1] 0.2381033
ppois(2,4)
[1] 0.2381033
```

8.1.6 The Multinomial Distribution

We now consider random experiments where k distinct or disjoint events A_1, $A_2, \ldots, A_k$ can occur with probabilities $p_1, p_2, \ldots, p_k$ respectively, with the restriction $\sum_{j=1}^{k} p_j = 1$. For example, if eight parties compete in a political election, we may be interested in the probability that a person votes for party A_j, $j = 1, 2, \ldots, 8$. Similarly one might be interested in the probability whether tuberculosis is detected in the lungs (A_1), in other organs (A_2), or both (A_3). Practically, we often use the multinomial distribution to model the distribution of categorical variables. This can be interpreted as a generalization of the Binomial distribution (where only two distinct events can occur) to the situation where more than two events or outcomes can occur. If the experiment is repeated n times independently, we are interested in the probability that

A_1 occurs n_1- times , A_2 occurs n_2- times , ..., A_k occurs n_k- times

with $\sum_{j=1}^{k} n_j = n$. Since several events can occur, the outcome of one (of the n) experiments is conveniently described by binary indicator variables. Let V_{ij}, $i = 1, \ldots, n$, $j = 1, \ldots, k$ denote the event "A_j is observed in experiment i", i.e.

$$V_{ij} = \begin{cases} 1 & \text{if } A_j \text{ occurs in experiment } i \\ 0 & \text{if } A_j \text{ does not occur in experiment } i \end{cases}$$

with probabilities $P(V_{ij} = 1) = p_j$, $j = 1, 2, \ldots, k$; then the outcome of one experiment is a vector of length k,

$$V_i = (V_{i1}, \ldots, V_{ij}, \ldots, V_{ik}) = (0, \ldots, 1, \ldots, 0) \,,$$

with "1" being present in only one position, i.e. in position j, if A_j occurs in experiment i. Now define (for each $j = 1, \ldots, k$) $X_j = \sum_{i=1}^{n} V_{ij}$. Then X_j is counting how often event A_j was observed in the n independent experiments (i.e. how often V_{ij} was 1 in the n experiments).

Definition 8.1.6 The random vector $\mathbf{X} = (X_1, X_2, \ldots, X_k)$ is said to follow a **multinomial distribution** if its PMF is given as

$$P(X_1 = n_1, X_2 = n_2, \ldots, X_k = n_k) = \frac{n!}{n_1! n_2! \cdots n_k!} \cdot p_1^{n_1} p_2^{n_2} \cdots p_k^{n_k} \quad (8.11)$$

with the restrictions $\sum_{j=1}^{k} n_j = n$ and $\sum_{j=1}^{k} p_j = 1$. We also write $\mathbf{X} \sim M(n; p_1, \ldots, p_k)$. The mean of $\mathbf{X}$ is the (component-wise) vector

$$E(\mathbf{X}) = (E(X_1), E(X_2), \ldots, E(X_k))$$
$$= (np_1, np_2, \ldots, np_k).$$

The $(i, j)^{th}$ element of the covariance matrix $V(\mathbf{X})$ is

$$\mathrm{Cov}(X_i, X_j) = \begin{cases} np_i(1 - p_i) & \text{if } i = j \\ -np_i p_j & \text{if } i \neq j. \end{cases}$$

Remark 8.1.2 Due to the restriction that $\sum_{j=1}^{k} n_j = \sum_{j=1}^{k} X_j = n$, $X_1, \ldots, X_k$ are not stochastically independent which is reflected by the negative covariance. This is also intuitively clear: if one X_j gets higher, another $X_{j'}$, $j \neq j'$ has to become lower to satisfy the restrictions.

We use the multinomial distribution to describe the randomness of categorical variables. Suppose we are interested in the variable "political party"; there might be eight political parties and we could thus summarize this variable by eight binary variables, each of them describing the event of party A_j, $j = 1, 2, \ldots, 8$, being voted for. In this sense, $\mathbf{X} = (X_1, X_2, \ldots, X_8)$ follows a multinomial distribution.

Example 8.1.7 Consider a simple example of the urn model. The urn contains 50 balls of three colours: 25 red balls, 15 white balls and 10 black balls. The balls are drawn from the urn with replacement. The balls are placed back into the urn after every draw, which means the draws are independent. Therefore the probability of drawing a red ball in every draw is $p_1 = \frac{25}{50} = 0.5$. Analogously, $p_2 = 0.3$ (for white balls) and $p_3 = 0.2$ (for black balls). Consider $n = 4$ draws. The probability of the random event of drawing "2 red balls, 1 white ball, 1 black ball" is:

$$P(X_1 = 2, X_2 = 1, X_3 = 1) = \frac{4!}{2!1!1!}(0.5)^2(0.3)^1(0.2)^1 = 0.18. \quad (8.12)$$

We would have obtained the same result in R using the dmultinom function:

```
dmultinom(c(2,1,1),prob=c(0.5,0.3,0.2))
```

This example demonstrates that the multinomial distribution relates to an experiment with replacement and without considering the order of the draws. Instead of the urn model, consider another example where we may interpret these three probabilities as probabilities of voting for candidate A_j, $j = 1, 2, 3$, in an election. Now, suppose we ask four voters about their choice, then the probability of candidate A_1 receiving 2 votes, candidate A_2 receiving 1 vote, and candidate A_3 receiving 1 vote is 18% as calculated in (8.12).

Remark 8.1.3 In contrast to most of the distributions, the CDF of the multinomial distribution, i.e. the function calculating $P(X_1 \leq x_1, X_2 \leq x_2, \ldots, X_k \leq x_k)$, is not contained in the base R distribution. Please note that for $k = 2$, the multinomial distribution reduces to the Binomial distribution.

8.1.7 The Geometric Distribution

Consider a situation in which we are interested in determining how many independent Bernoulli trials are needed until the event of interest occurs for the first time. For instance, we may be interested in how many tickets to buy in a raffle until we win for the first time; or how many different drugs to try to successfully tackle a severe migraine etc. The geometric distribution can be used to determine the probability that the event of interest happens at the kth trial for the first time.

Definition 8.1.7 A discrete random variable X is said to follow a *geometric distribution* with parameter p if its PMF is given by

$$P(X = k) = p(1 - p)^{k-1}, \qquad k = 1, 2, 3, \ldots \qquad (8.13)$$

The mean (expectation) and variance are given by $E(X) = 1/p$ and $\text{Var}(X) = 1/p(1/p - 1)$ respectively.

Example 8.1.8 Let's consider an experiment where a coin is tossed until "head" is obtained for the first time. The probability of getting a head is $p = 0.5$ for each toss. Using (8.13), we can determine the following probabilities:

$$P(X = 1) = 0.5$$
$$P(X = 2) = 0.5(1 - 0.5) = 0.25$$
$$P(X = 3) = 0.5(1 - 0.5)^2 = 0.125$$
$$P(X = 4) = 0.5(1 - 0.5)^3 = 0.0625$$

$$\ldots \qquad \ldots$$

Using the command structure for obtaining PMF's in R (Appendix A as well as Examples 8.1.5 and 8.1.6), we can determine the latter probability of $P(X = 4)$ as follows:

```
dgeom(3,0.5)
```

Note that the definition of X in R slightly differs from our definition. In R, k is the number of failures before the first success. This means we need to specify $k - 1$ in the dgeom function rather than k. The mean and variance for this setting are

$$E(X) = \frac{1}{0.5} = 2; \quad Var(X) = \frac{1}{0.5}\left(\frac{1}{0.5} - 1\right) = 2.$$

8.1.8 Hypergeometric Distribution

We can again use the urn model to motivate another distribution, the hypergeometric distribution.

M	white balls
$N - M$	black balls
N	total balls

Consider an urn with M white balls and $N - M$ black balls from which we randomly draw n balls without replacement, i.e. we do not place a ball back into the urn once it is drawn. The order in which the balls are drawn is assumed to be of no interest, only the number of drawn white balls is of relevance. We define the following random variable

$$X: \text{"number of white balls } (x)\text{among the } n \text{ drawn balls"}.$$

To be more precise, among the n drawn balls, x are white and $n - x$ are black. There are $\binom{M}{x}$ possibilities to choose x white balls from the total of M white balls and, analogously, there are $\binom{N-M}{n-x}$ possibilities to choose $(n - x)$ black balls from the total of $N - M$ black balls. In total, we draw n out of N balls. Recall the probability definition of Laplace as the number of simple favourable events divided by all possible events. The number of combinations for all possible events is $\binom{N}{n}$; the number of favourable events is $\binom{M}{x}\binom{N-M}{n-x}$ because we draw, independent of each other, x out of M balls and $n - x$ out of $N - M$ balls. Hence the PMF of the hypergeometric distribution is

$$P(X = x) = \frac{\binom{M}{x}\binom{N-M}{n-x}}{\binom{N}{n}} \tag{8.14}$$

for $x \in \{\max(0, n - (N - M)), \ldots, \min(n, M)\}$. The requirement that x is at least $\{\max(0, n - (N - M))$ is based on the insight that if there are less black balls in the urn $(N - M)$ than the size of the sample (n), then at least $n - (N - M)$ white balls have to be in the sample. Also, x can not be larger than $\min(n, M)$ because either $x = n$, or – if $M < n$ – not more than M white balls can be drawn.

Definition 8.1.8 A random variable X is said to follow a **hypergeometric** distribution with parameters n, M, N, i.e. $X \sim H(n, M, N)$, if its PMF is given by (8.14).

Example 8.1.9 The German national lottery draws 6 out of 49 balls from a rotating bowl. Each ball is associated with a number between 1 and 49. A simple bet is to choose 6 numbers between 1 and 49. If 3 or more chosen numbers correspond to the numbers drawn in the lottery, then one wins a certain amount of money. What is the probability of choosing 4 correct numbers? We can utilize the hypergeometric distribution with $x = 4$, $M = 6$, $N = 49$, and $n = 6$ to calculate such probabilities. The interpretation is that we "draw" (i.e. bet on) 4 out of the 6 winning balls and "draw" (i.e. bet on) another 2 out of the remaining 43 (49 - 6) losing balls. In total, we draw 6 out of 49 balls. Calculating the number of the favourable combinations and all possible combinations leads to the application of the hypergeometric distribution as follows:

$$P(X = 4) = \frac{\binom{M}{x}\binom{N-M}{n-x}}{\binom{N}{n}} = \frac{\binom{6}{4}\binom{43}{2}}{\binom{49}{6}} \approx 0.001 \text{ (or } 0.1\%).$$

We would have obtained the same results using the dhyper command. Its arguments are x, M, N, n and thus we specify

```
dhyper(4,6,43,6)
```

The $H(6, 43, 6)$ distribution is also visualized in Fig. 8.3. It is evident that the cumulative probability of choosing 2 or fewer correct numbers is approximately 98.1%, but it is very unlikely to have 3 or more numbers right. This may explain why the national lottery pays out money only for 3 or more correct numbers.

8.2   Standard Continuous Distributions

Now we discuss some standard probability distributions of (absolute) continuous random variables. Characteristics of continuous random variables are that the number of possible outcomes is uncountably infinite and that they have a continuous distribution function $F(x)$. It follows that the point probabilities are zero, i.e. $P(X = x) = 0$. Further, we assume a unique density function f exists, such that $F(x) = \int_{-\infty}^{x} f(t)dt$.

Fig. 8.3 The $H(6, 43, 6)$
distribution

8.2.1   Continuous Uniform Distribution

A continuous analogue to the discrete uniform distribution is the continuous uniform
distribution on a closed interval in $\mathbb{R}$.

Definition 8.2.1 A continuous random variable X is said to follow a (continuous)
uniform distribution in the interval $[a, b]$, i.e. $X \sim U(a, b)$, if its probability density
function (PDF) is given by

$$f(x) = \begin{cases} \frac{1}{b-a} & \text{if } a \leq x \leq b \quad (a < b) \\ 0 & \text{otherwise.} \end{cases}$$

The mean and variance of $X \sim U(a, b)$ are

$$E(X) = \frac{a+b}{2} \quad \text{and} \quad \text{Var}(X) = \frac{(b-a)^2}{12}.$$

respectively.

Example 8.2.1 Suppose a train arrives at a subway station regularly every 10 min.
If a passenger arrives at the station without knowing the timetable, then the waiting
time to catch the train is uniformly distributed with density

$$f(x) = \begin{cases} \frac{1}{10} & \text{if } 0 \leq x \leq 10 \\ 0 & \text{otherwise}. \end{cases}$$

The "average" waiting time is $E(X) = (10 + 0)/2 = 5\,\text{min}$. The probability of waiting for the train for less than 3 min is obviously 0.3 (or 30%) which can be calculated in R using the `punif(x,a,b)` command (see also Appendix A.3):

```
punif(3,0,10)
```

8.2.2 The Normal Distribution

The normal distribution is one of the most important distributions used in statistics. The name was given by Carl Friedrich Gauss (1777–1855), a German mathematician, astronomer, geodesist and physicist who observed that measurements in geodesy and astronomy randomly deviate in a symmetric way from their true values. The normal distribution is therefore also often called a Gaussian distribution.

Definition 8.2.2 A random variable X is said to follow a **normal distribution** with parameters μ and σ^2 if its PDF is given by

$$f(x) = \frac{1}{\sigma\sqrt{2\pi}} \exp\left(-\frac{(x - \mu)^2}{2\sigma^2}\right); \quad -\infty < x < \infty, -\infty < \mu < \infty, \sigma^2 > 0. \tag{8.15}$$

We write $X \sim N(\mu, \sigma^2)$. The mean and variance of X are

$$E(X) = \mu; \quad \text{and} \quad \text{Var}(X) = \sigma^2$$

respectively. If $\mu = 0$ and $\sigma^2 = 1$, then X is said to follow a *standard* normal distribution, $X \sim N(0, 1)$. The PDF of a standard normal distribution is given by

$$\phi(x) = \frac{1}{\sqrt{2\pi}} \exp(-\frac{x^2}{2}); \quad -\infty < x < \infty.$$

The density of a normal distribution has its maximum (see Fig. 8.4) at $x = \mu$. The density is also symmetric around μ. The inflexion points of the density are at $(\mu - \sigma)$ and $(\mu + \sigma)$ (Fig. 8.4). A lower σ indicates a higher concentration around the mean μ. A higher σ indicates a flatter density (Fig. 8.5).

The cumulative distribution function of $X \sim N(\mu, \sigma^2)$ is

$$F(x) = \int_{-\infty}^{x} \phi(t)dt \tag{8.16}$$

which is often denoted as $\Phi(x)$. The value of $\Phi(x)$ for various values of x can be obtained in R following the rules introduced in Appendix A.3. For example,

```
pnorm(1.96, mean = 0, sd = 1)
```

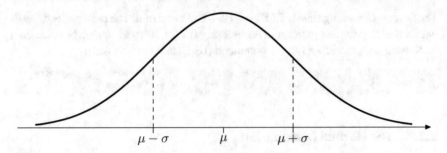

Fig. 8.4 PDF of a normal distribution

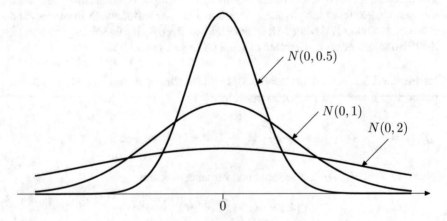

Fig. 8.5 PDF of $N(0, 2)$, $N(0, 1)$ and $N(0, 0.5)$ distributions

calculates $\Phi(1.96)$ as approximately 0.975. This means, for a standard normal distribution the probability $P(X \leq 1.96) \approx 0.975$.

Remark 8.2.1 There is no explicit formula to solve the integral in equation (8.16). It has to be solved by numerical (or computational) methods. This is the reason why CDF tables are presented in almost all statistical textbooks, see Table C.1 in the Appendix.

Example 8.2.2 An orange farmer sells his oranges in wooden boxes. The weights of the boxes vary and are assumed to be normally distributed with $\mu = 15$ kg and $\sigma^2 = \frac{9}{4}$ kg^2. The farmer wants to avoid customers being unsatisfied because the boxes are too low in weight. He therefore asks the following question: what is the probability that a box with a weight of less than 13 kg is sold? Using the pnorm(x, μ, σ) command in R, we get

```
pnorm(13,15,sqrt(9/4))
[1] 0.09121122
```

To calculate the probability in Example 8.2.2 manually, we first have to introduce some theoretical results.

8.2.2.1 Calculation Rules for Normal Random Variables

Let $X \sim N(\mu, \sigma^2)$. Using the transformation

$$Z = \frac{X - \mu}{\sigma} \sim N(0, 1), \tag{8.17}$$

every normally distributed random variable can be transformed into a *standard* normal random variable. We call this transformation the Z-transformation. We can use this transformation to derive convenient calculation rules. The probability for $X \leq b$ is

$$P(X \leq b) = P\left(\frac{X - \mu}{\sigma} \leq \frac{b - \mu}{\sigma}\right) = P\left(Z \leq \frac{b - \mu}{\sigma}\right) = \Phi\left(\frac{b - \mu}{\sigma}\right). \tag{8.18}$$

Consequently, the probability for $X > a$ is

$$P(X > a) = 1 - P(X \leq a) = 1 - \Phi\left(\frac{a - \mu}{\sigma}\right). \tag{8.19}$$

The probability that X realizes a value in the interval $[a, b]$ is

$$P(a \leq X \leq b) = P\left(\frac{a - \mu}{\sigma} \leq Z \leq \frac{b - \mu}{\sigma}\right) = \Phi\left(\frac{b - \mu}{\sigma}\right) - \Phi\left(\frac{a - \mu}{\sigma}\right). \tag{8.20}$$

Because of the symmetry of the probability density function $\phi(x)$ around its mean 0, the following equation holds for the distribution function $\Phi(x)$ of a standard normal random variable for any value a:

$$\Phi(-a) = 1 - \Phi(a). \tag{8.21}$$

It follows that $P(-a < Z < a) = 2 \cdot \Phi(a) - 1$, see also Fig. 8.6.

Example 8.2.3 Recall Example 8.2.2 where a farmer sold his oranges. He was interested in $P(X \leq 13)$ for $X \sim N(15, 9/4)$. Using (8.17), we get

$$P(X \leq 13) = \Phi\left(\frac{13 - 15}{\frac{3}{2}}\right)$$

$$= \Phi\left(-\frac{4}{3}\right) = 1 - \Phi\left(\frac{4}{3}\right) \approx 0.091 \text{ (or } 9.1\%).$$

To obtain $\Phi(4/3) \approx 90.9\%$, we could either use R (pnorm(4/3)) or use Table C.1.

Fig. 8.6 Distribution function of the standard normal distribution

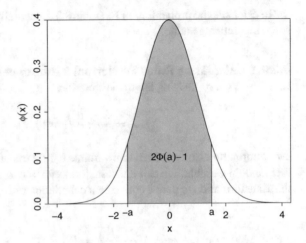

8.2.2.2 Distribution of the Arithmetic Mean

Assume that $X \sim N(\mu, \sigma^2)$. Consider a random sample $\mathbf{X} = (X_1, X_2, \ldots, X_n)$ of independent and identically distributed random variables X_i with $X_i \sim N(\mu, \sigma^2)$. Then the arithmetic mean $\bar{X} = \frac{1}{n} \sum_{i=1}^{n} X_i$ follows a normal distribution with mean

$$E(\bar{X}) = \frac{1}{n} \sum_{i=1}^{n} E(X_i) = \mu$$

and variance

$$\text{Var}(\bar{X}) = \frac{1}{n^2} \sum_{i=1}^{n} \text{Var}(X_i) = \frac{\sigma^2}{n} \qquad (8.22)$$

where $\text{Cov}(X_i, X_j) = 0$ for $i \neq j$. In summary, we get

$$\bar{X} \sim N\left(\mu, \frac{\sigma^2}{n}\right). \qquad (8.23)$$

Remark 8.2.2 In fact, in Eq. (8.22), we have used the fact that the sum of normal random variables also follows a normal distribution, i.e.

$$(X_1 + X_2) \sim N(\mu_1 + \mu_2, \sigma_1^2 + \sigma_2^2).$$

This result can be generalized to n (not necessarily identically distributed but independent) normal random variables. In fact, it holds that if $X_1, X_2, \ldots, X_n$ are independent normal variables with means $\mu_1, \mu_2, \ldots, \mu_n$ and variances $\sigma_1^2, \sigma_2^2, \ldots, \sigma_n^2$, then for any real numbers $a_1, a_2, \ldots, a_n$ it holds that

$$(a_1 X_1 + a_2 X_2 + \cdots + a_n X_n) \sim N(a_1 \mu_1 + a_2 \mu_2 + \cdots a_n \mu_n, a_1^2 \sigma_1^2 + a_2^2 \sigma_2^2 + \cdots a_n^2 \sigma_n^2) \,.$$

In general, it cannot be taken for granted that the sum of two random variables follows the same distribution as the two variables themselves. As an example, consider the sum of two independent uniform distributions with $X_1 \sim U[0, 10]$ and $X_2 \sim U[20, 30]$. It holds that $E(X_1 + X_2) = E(X_1) + E(X_2)$ and $Var(X_1 + X_2) = Var(X_1) + Var(X_2)$, but $X_1 + X_2$ is obviously not uniformly distributed.

8.2.3 The Exponential Distribution

The exponential distribution is useful in many situations, for example when one is interested in the waiting time, or life time, until an event of interest occurs. If we assume that the future life time is independent of the life time that has already taken place (i.e. no "ageing" process is working), the waiting times can be considered to be exponentially distributed.

Definition 8.2.3 A random variable X is said to follow an exponential distribution with parameter $\lambda > 0$ if its PDF is given by

$$f(x) = \begin{cases} \lambda \exp(-\lambda x) & \text{if } x \geq 0 \\ 0 & \text{otherwise.} \end{cases} \tag{8.24}$$

We write $X \sim Exp(\lambda)$. The mean and variance of an exponentially distributed random variable X are

$$E(X) = \frac{1}{\lambda} \quad \text{and} \quad Var(X) = \frac{1}{\lambda^2}$$

respectively. The CDF of the exponential distribution is given as

$$F(x) = \begin{cases} 1 - \exp(-\lambda x) & \text{if } x \geq 0 \\ 0 & \text{otherwise.} \end{cases} \tag{8.25}$$

Note, that $P(X > x) = 1 - F(x) = \exp(-\lambda x)$ $(x \geq 0)$. An interesting property of the exponential distribution is its **memorylessness**: if time t has already been reached, the probability of reaching a time greater than $t + \Delta$ does not depend on t. This can be written as

$$P(X > t + \Delta | X > t) = P(X > \Delta) \quad t, \Delta > 0 \,.$$

The result can be derived using basic probability rules as follows:

$$P(X > t + \Delta | X > t) = \frac{P(X > t + \Delta \text{ and } X > t)}{P(X > t)} = \frac{P(X > t + \Delta)}{P(X > t)}$$

$$= \frac{\exp[-\lambda(t + \Delta)]}{\exp[-\lambda t]} = \exp[-\lambda \Delta]$$

$$= 1 - F(\Delta) = P(X > \Delta).$$

For example, suppose someone stands in a supermarket queue for t minutes. Say the person forgot to buy milk, so she leaves the queue, gets the milk and stands in the queue again. If we use the exponential distribution to model the waiting time, we say that it doesn't matter what time it is: the random variable "waiting time from standing in the queue until paying the bill" is not influenced by how much time has elapsed already, it doesn't matter if we queued before or not. Please note that the memorylessness property is shared by the geometric and the exponential distributions.

There is also a relationship between the Poisson and the exponential distribution:

Theorem 8.2.1 *The number of events Y occurring within a continuum of time is Poisson distributed with parameter λ if and only if the time between two events is exponentially distributed with parameter λ.*

The continuum of time depends on the problem at hand. It may be a second, a minute, 3 months, a year or any other time period.

Example 8.2.4 Let Y be the random variable which counts the "number of accesses per second for a search engine". Assume that Y is Poisson distributed with parameter $\lambda = 10$ ($E(Y) = 10$, $Var(Y) = 10$). The random variable X, "waiting time until the next access" is then exponentially distributed with parameter $\lambda = 10$. We therefore get

$$E(X) = \frac{1}{10}, \quad Var(X) = \frac{1}{10^2}.$$

In this example, the continuum is one second. The expected number of accesses per second is therefore $E(Y) = 10$ and the expected waiting time between two accesses is $E(X) = 1/10$ s. The probability of experiencing a waiting time of less than 0.1 s is

$$F(0.1) = 1 - \exp(-\lambda x) = 1 - \exp(-10 \cdot 0.1) \approx 0.63.$$

In R, we can obtain the same result as

```
pexp(0.1,10)
[1] 0.6321206
```

8.3 Sampling Distributions

All the distributions introduced in this chapter up to now are motivated by practical applications. However there are theoretical distributions which play an important role in the construction and development of various statistical tools such as those introduced in Chaps. 9–11. We call these distributions "sampling distributions". Now we discuss the χ^2, t and F- distributions.

8.3.1 The χ^2-Distribution

Definition 8.3.1 Let $Z_1, Z_2 \ldots, Z_n$ be n independent and identically $N(0, 1)$-distributed random variables. The sum of their squares, $\sum_{i=1}^{n} Z_i^2$, is then χ^2-**distributed** with n degrees of freedom and is denoted as χ_n^2. The PDF of the χ^2-distribution is given in Eq. (C.7) in Appendix C.3.

The χ^2-distribution is not symmetric. A χ^2-distributed random variable can only realize values greater than or equal to zero. Figure 8.7a shows the χ_1^2, χ_2^2, and χ_5^2 distributions. It can be seen that the "degrees of freedom" specify the shape of the distribution. Their interpretation and meaning will nevertheless become clearer in the following chapters. The quantiles of the CDF of different χ^2-distributions can be obtained in R using the `qchisq(p,df)` command. They are also listed in Table C.3 for different values of n.

Theorem 8.3.1 *Consider two independent random variables which are χ_m^2 and χ_n^2 distributed respectively. The sum of these two random variables is χ_{n+m}^2-distributed.*

An important example of a χ^2-distributed random variable is the sample variance (S_X^2) of an i.i.d. sample of size n from a normally distributed population, i.e.

$$\frac{(n-1)S_X^2}{\sigma^2} \sim \chi_{n-1}^2 \,. \tag{8.26}$$

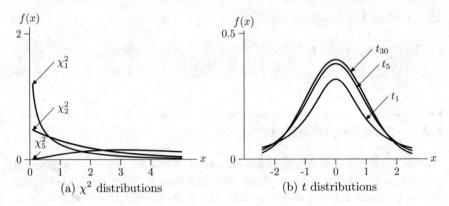

Fig. 8.7 Probability density function

8.3.2 The t-Distribution

Definition 8.3.2 Let X and Y be two independent random variables where $X \sim N(0, 1)$ and $Y \sim \chi_n^2$. The ratio

$$\frac{X}{\sqrt{Y/n}} \sim t_n$$

follows a t-**distribution** (Student's t-distribution) with n degrees of freedom. The PDF of the t-distribution is given in Eq. (C.8) in Appendix C.3.

Figure 8.7b visualizes the t_1, t_5, and t_{30} distributions. The quantiles of different t-distributions can be obtained in R using the `qt(p,df)` command. They are also listed in Table C.2 for different values of n.

   An application of the t-distribution is the following: if we draw a sample of size n from a normal population $N(\mu, \sigma^2)$ and calculate the arithmetic mean $\bar{X}$ and the sample variance S_X^2, then the following theorem holds:

Theorem 8.3.2 (Student's Theorem) *Let* $\mathbf{X} = (X_1, X_2, \ldots, X_n)$ *with* $X_i \overset{iid.}{\sim} N(\mu, \sigma^2)$. *The ratio*

$$\frac{(\bar{X} - \mu)\sqrt{n}}{S_X} = \frac{(\bar{X} - \mu)\sqrt{n}}{\sqrt{\frac{1}{n-1}\sum_i^n (X_i - \bar{X})^2}} \sim t_{n-1} \tag{8.27}$$

is then t-*distributed with* $n - 1$ *degrees of freedom.*

8.3.3 The F-Distribution

Definition 8.3.3 Let X and Y be independent χ_m^2 and χ_n^2-distributed random variables, then the distribution of the ratio

$$\frac{X/m}{Y/n} \sim F_{m,n} \tag{8.28}$$

follows the ***Fisher F-distribution*** with (m, n) degrees of freedom. The PDF of the F-distribution is given in Eq. (C.9) in Appendix C.3.

If X is a χ_1^2-distributed random variable, then the ratio (8.28) is $F_{1,n}$-distributed. The square root of this ratio is t_n-distributed since the square root of a χ_1^2-distributed random variable is $N(0, 1)$-distributed. If W is F-distributed, $F_{m,n}$, then $1/W$ is $F_{n,m}$-distributed. Figure 8.8 visualizes the $F_{5,5}$, $F_{5,10}$ and $F_{5,30}$ distributions. The quantiles of different F-distributions can be obtained in R using the `qf(p,df1,df2)` command.

One application of the F-distribution relates to the ratio of two sample variances of two independent samples of size m and n, where each sample is an i.i.d. sample from a normal population, i.e. $N(\mu_X, \sigma_X^2)$ and $N(\mu_Y, \sigma_Y^2)$. For the sample variances $S_X^2 = \frac{1}{m-1} \sum_{i=1}^{m} (X_i - \bar{X})^2$ and $S_Y^2 = \frac{1}{n-1} \sum_{i=1}^{n} (Y_i - \bar{Y})^2$ from the two populations, the ratio

$$\frac{S_X^2}{S_Y^2} \sim F_{m-1,n-1}$$

is F-distributed with $(m - 1)$ degrees of freedom in the numerator and $(n - 1)$ degrees of freedom in the denominator.

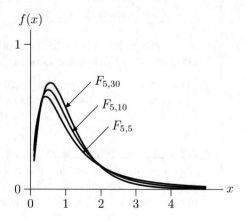

Fig. 8.8 Probability density functions for different F distributions

8.4   Key Points and Further Issues

Note:

✓ Examples of different distributions are:

Distribution	Example
Uniform	rolling a die (discrete)
	waiting for a train (continuous)
Bernoulli	any binary variable such as gender
Binomial	number of "heads" when tossing a coin n times
Poisson	number of particles emitted by a radioactive source
	entering a small area in a given time interval
Multinomial	categorical variables such as "party voted for"
Geometric	number of raffle tickets until first ticket wins
Hypergeometric	national lotteries; Fisher's test, see Appendix C.5
Normal	height or weight of women (men)
Exponential	survival time of a laptop
χ^2	sample variance; χ^2 tests, see Sect. 10.7 ff.
t	confidence interval for the mean, see Sect. 9.4.2
F	tests in the linear model, see Sect. 11.7

✓ One can use R to determine values of densities (PDF/PMF), cumulative probability distribution functions (CDF), quantiles of the CDF, and random numbers:

First letter	Function	Further letters	Example
d	density	`distribution name`	`dnorm`
p	probability	`distribution name`	`pnorm`
q	quantiles	`distribution name`	`qnorm`
r	random number	`distribution name`	`rnorm`

We encourage the use of R to obtain quantiles of sampling distributions, but Tables C.1–C.3 also list some of them.

✓ In this chapter, we assumed the parameters such as μ, σ, λ and others to be known. In Chap. 9 we will propose how to estimate these parameters from the data. In Chap. 10, we test statistical hypotheses about these parameters.

✓ For n i.i.d. random variables $X_1, X_2, \ldots, X_n$ the arithmetic mean $\bar{X}$ converges to a $N(\mu, \sigma^2/n)$ distribution as n tends to infinity. See Appendix C.3 as well as Exercise 8.11 for the Theorem of Large Numbers and the Central Limit Theorem respectively.

8.5 Exercises

Exercise 8.1 A company producing cereals offers a toy in every sixth cereal package in celebration of their 50th anniversary. A father immediately buys 20 packages.

(a) What is the probability of finding 4 toys in the 20 packages?
(b) What is the probability of finding no toy at all?
(c) The packages contain three toys. What is the probability that among the 5 packages that are given to the family's youngest daughter, she finds two toys?

Exercise 8.2 A study on breeding birds collects information such as the length of their eggs (in mm). Assume that the length is normally distributed with $\mu = 42.1$ mm and $\sigma^2 = 20.8^2$. What is the probability of

(a) finding an egg with a length greater than 50 mm?
(b) finding an egg between 30 and 40 mm in length?

Calculate the results both manually and by using R.

Exercise 8.3 A dodecahedron is a die with 12 sides. Suppose the numbers on the die are 1–12. Consider the random variable X which describes which number is shown after rolling the die once. What is the distribution of X? Determine $E(X)$ and $Var(X)$.

Exercise 8.4 Felix states that he is able to distinguish a freshly ground coffee blend from an ordinary supermarket coffee. One of his friends asks him to taste 10 cups of coffee and tell him which coffee he has tasted. Suppose that Felix has actually no clue about coffee and simply guesses the brand. What is the probability of at least 8 correct guesses?

Exercise 8.5 An advertising board is illuminated by several hundred bulbs. Some of the bulbs are fused or smashed regularly. If there are more than 5 fused bulbs on a day, the owner of the board replaces them, otherwise not. Consider the following data collected over a month which captures the number of days (n_i) on which i bulbs were broken:

Fused bulbs	0	1	2	3	4	5
n_i	6	8	8	5	2	1

(a) Suggest an appropriate distribution for X: "number of broken bulbs per day".
(b) What is the average number of broken bulbs per day? What is the variance?
(c) Determine the probabilities $P(X = x)$ using the distribution you chose in (a) and using the average number of broken bulbs you calculated in (b). Compare the probabilities with the proportions obtained from the data.

(d) Calculate the probability that at least 6 bulbs are fused, which means they need to be replaced.
(e) Consider the random variable Y: "time until next bulb breaks". What is the distribution of Y?
(f) Calculate and interpret $E(Y)$.

Exercise 8.6 Marco's company organizes a raffle at an end-of-year function. There are 4000 raffle tickets to be sold, of which 500 win a prize. The price of each ticket is €1.50. The value of the prizes, which are mostly electrical appliances produced by the company, varies between €80 and €250, with an average value of €142.

(a) Marco wants to have a 99% guarantee of receiving three prizes. How much money does he need to spend? Use R to solve the question.
(b) Use R to plot the function which describes the relationship between the number of tickets bought and the probability of winning at least three prizes.
(c) Given the value of the prizes and the costs of the tickets, is it worth taking part in the raffle?

Exercise 8.7 A country has a ratio between male and female births of 1.05 which means that 51.22% of babies born are male.

(a) What is the probability for a mother that the first girl is born during the first three births?
(b) What is the probability of getting 2 girls among 4 babies?

Exercise 8.8 A fishermen catches, on average, three fish in an hour. Let Y be a random variable denoting the number of fish caught in one hour and let X be the time interval between catching two fishes. We assume that X follows an exponential distribution.

(a) What is the distribution of Y?
(b) Determine $E(Y)$ and $E(X)$.
(c) Calculate $P(Y = 5)$ and $P(Y < 1)$.

Exercise 8.9 A restaurant sells three different types of dessert: chocolate brownies, yogurt with seasonal fruits, and lemon tart. Years of experience have shown that the probabilities with which the desserts are chosen are 0.2, 0.3 and 0.5 respectively.

(a) What is the probability that out of 5 guests, 2 guests choose brownies, 1 guest chooses yogurt and the remaining 2 guests choose lemon tart?
(b) Suppose two out of the five guests are known to always choose lemon tart. What is the probability of the others choosing lemon tart as well?
(c) Determine the expectation and variance assuming a group of 20 guests.

Exercise 8.10 A reinsurance company works on a premium policy for natural disasters. Based on experience, it is known that W = "number of natural disasters from October to March" (winter) is Poisson distributed with $\lambda_W = 4$. Similarly, the random variable S = "number of natural disasters from April to September" (summer) is Poisson distributed with $\lambda_S = 3$. Determine the probability that there is at least 1 disaster during both summer and winter based on the assumption that the two random variables are independent.

Exercise 8.11 Read Appendix C.3 to learn about the Theorem of Large Numbers and the Central Limit Theorem.

(a) Draw 1000 realizations from a standard normal distribution using R and calculate the arithmetic mean. Repeat this process 1000 times. Evaluate the distribution of the arithmetic mean by drawing a kernel density plot and by calculating the mean and variance of it.
(b) Repeat the procedure in (a) with an exponential distribution with $\lambda = 1$. Interpret your findings in light of the Central Limit Theorem.
(c) Repeat the procedure in (b) using 10,000 rather than 1000 realizations. How do the results change and why?

$\rightarrow$ Solutions to all exercises in this chapter can be found in Appendix B

Part III
Inductive Statistics

Inference

9

9.1 Introduction

The first four chapters of this book illustrated how one can summarize a data set both numerically and graphically. The validity of interpretations made from such a descriptive analysis are valid only for the data set under consideration and cannot necessarily be generalized to other data. However, it is desirable to make conclusions about the entire population of interest and not only about the sample data. In this chapter, we describe the framework of **statistical inference** which allows us to infer from the sample data about the population of interest—at a given, pre-specified uncertainty level—and knowledge about the random process generating the data.

Consider an example where the objective is to forecast an election outcome. This requires us to determine the proportion of votes that each of the k participating parties is going to receive, i.e. to calculate or estimate $p_1, p_2, \ldots, p_k$. If it is possible to ask every voter about their party preference, then one can simply calculate the proportions $p_1, p_2, \ldots, p_k$ for each party. However, it is logistically impossible to ask all eligible voters (which form the population in this case) about their preferred party. It seems more realistic to ask only a small fraction of voters and infer from their responses to the responses of the whole population. It is evident that there might be differences in responses between the sample and the population—but the more voters are asked, the closer we are to the population's preference, i.e. the higher the precision of our estimates for $p_1, p_2, \ldots, p_k$ (the meaning of "precision" will become clearer later in this chapter). Also, it is intuitively clear that the sample must be a representative sample of the voters' population to avoid any discrepancy or bias in the forecasting. When we speak of a representative sample, we mean that all the characteristics present in the population are contained in the sample too. There are many ways to get representative random samples. In fact, there is a branch of statistics, called sampling theory, which studies this subject, see e.g. Groves, Fowler, Couper, Lepkowski, Singer and Tourangeau (2009) for more details. A simple random sample

© Springer Nature Switzerland AG 2022
C. Heumann et al., *Introduction to Statistics and Data Analysis*,
https://doi.org/10.1007/978-3-031-11833-3_9

is one where each voter has an equal probability of being selected in the sample and each voter is independently chosen from the same population. In the following, we will assume that all samples are simple random samples. To further formalize the election forecast problem, assume that we are interested in the true proportions which each party receives on the election day. It is practically impossible to make a perfect prediction of these proportions because there are too many voters to interview and moreover, a voter may possibly make their final decision possibly only when casting the vote and not before. The voter may change his/her opinion at any moment and may differ from what he/she claimed earlier. In statistics, we call these true proportions *parameters of the population*. The task is then to estimate these parameters on the basis of a sample. In the election example, the intuitive estimates for the proportions in the population are the proportions in the sample and we call them *sample estimates*. How to find good and precise estimates are some of the challenges that are addressed by the concept of *statistical inference*. Now it is possible to describe the election forecast problem in a statistical and operational framework: estimate the parameters of a population by calculating the sample estimates. An important property of every good statistical inference procedure is that it provides not only estimates for the population parameters but also information about the precision of these estimates.

Consider another example in which we would like to study the distribution of weight of children in different age categories and get an understanding of the "normal" weight. Again, it is not possible to measure the weight of all the children of a specific age in the entire population of children in a particular country. Instead, we draw a random sample and use methods of statistical inference to estimate the weight of children in each age group. More specifically, we have several populations in this problem. We could consider all boys of a specific age and all girls of a specific age as two different populations. For example, all 3-year-old boys will form one possible population. Then, a random sample is drawn from this population. It is reasonable to assume that the distribution of the weight of k-year-old boys follows a normal distribution with some unknown parameters μ_{kb} and σ^2_{kb}. Similarly, another population of k-year-old girls is assumed to follow a normal distribution with some unknown parameters μ_{kg} and σ^2_{kg}. The indices kb and kg are used to emphasize that the parameters may vary by age and gender. The task is now to calculate the estimates of the unknown parameters (in the population) of the normal distributions from the samples. Using quantiles, a range of "normal" weights can then be specified, e.g. the interval from the 1%-quantile to the 99%-quantile of the estimated normal distribution; or, alternatively, all weights which are not more than twice the standard deviation away from the mean. Children with weights outside this interval may be categorized as underweight or overweight. Note that we make a specific assumption for the distribution class, i.e. we assume a normal distribution for the weights and estimate its parameters. We call this a **parametric** estimation problem because it is based on distributional assumptions. Otherwise, if no distributional assumptions are made, we speak of a **non-parametric** estimation problem.

9.2 Properties of Point Estimators

As we discussed in the introduction, the primary goal in statistical inference is to find a good estimate of (a) population parameter(s). The parameters are associated with the probability distribution which is believed to characterize the population, e.g. μ and σ^2 are the parameters in a normal distribution $N(\mu, \sigma^2)$. If these parameters are known, then one can characterize the entire population. In practice, these parameters are unknown, so the objective is to estimate them. One can attempt to obtain them based on a function of the sample values. But what does this function look like; and if there is more than one such function, then which is the best one? What is the best approach to estimate the population parameters on the basis of a given sample of data? The answer is given by various statistical concepts such as bias, variability, consistency, efficiency, sufficiency, and completeness of the estimates. We are going to introduce them now.

Assume $x = (x_1, x_2 \ldots, x_n)$ are the observations of a random sample from a population of interest. The random sample represents the realized values of a random variable X. It can be said that $x_1, x_2 \ldots, x_n$ are the n observations collected on the random variable X. Any function of random variables is called a **statistic**. For example, $\bar{X} = \frac{1}{n} \sum_{i=1}^{n} X_i$, $\max(X_1, X_2, \ldots, X_n)$ etc. are functions of $X_1, X_2, \ldots, X_n$, so they are a statistic. It follows that a statistic is also a random variable. Consider a statistic $T(X)$ which is used to estimate a population parameter θ (which may be either a scalar or a vector). We say $T(X)$ is an **estimator** of θ. To indicate that we estimate θ using $T(X)$, we use the "hat" ($\hat{\ }$) symbol, i.e. we write $\hat{\theta} = T(X)$. When T is calculated from the sample values $x_1, x_2 \ldots, x_n$, we write $T(x)$ and call it an **estimate** of θ. It becomes clear that $T(X)$ is a random variable but $T(x)$ is its observed value (dependent on the actual sample). For example, $T(X) = \bar{X} = \frac{1}{n} \sum_{i=1}^{n} X_i$ is an estimator and a statistic, but $T(x) = \bar{x} = \frac{1}{n} \sum_{i=1}^{n} x_i$ is its estimated value from the realized sample values $x_1, x_2 \ldots, x_n$. Since the sample values are realizations from a random variable, each sample leads to a different value of the estimate of the population parameter. The population parameter is assumed to be a fixed value. Parameters can also be assumed to be random, for example in Bayesian statistics, but this is beyond the scope of this book.

9.2.1 Unbiasedness and Efficiency

Definition 9.2.1 An estimator $T(X)$ is called an *unbiased* estimator of θ if

$$E_\theta(T(X)) = \theta . \tag{9.1}$$

The index θ denotes that the expectation is calculated with respect to the distribution whose parameter is θ.

The bias of an estimator $T(X)$ is defined as

$$\text{Bias}_\theta(T(X)) = E_\theta(T(X)) - \theta . \tag{9.2}$$

It follows that an estimator is said to be unbiased if its bias is zero.

Definition 9.2.2 The variance of $T(X)$ is defined as

$$\text{Var}_\theta(T(X)) = \text{E}\left\{[T(X) - \text{E}(T(X))]^2\right\}\,. \tag{9.3}$$

Both bias and variance are measures which characterize the properties of an estimator. In statistical theory we search for "good" estimators in the sense that the bias and the variance are as small as possible and therefore the accuracy is as high as possible. Readers interested in a practical example may consult Examples 9.2.1 and 9.2.2, or the explanations for Fig. 9.1.

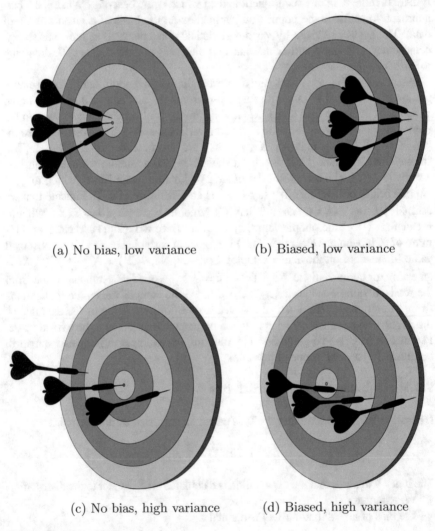

(a) No bias, low variance   (b) Biased, low variance

(c) No bias, high variance   (d) Biased, high variance

Fig. 9.1 Illustration of bias and variance

It turns out that we cannot minimize both measures simultaneously as there is always a so-called bias–variance tradeoff. A measure which combines bias and variance into one measure is the mean squared error.

Definition 9.2.3 The mean squared error (MSE) of $T(X)$ is defined as

$$\text{MSE}_\theta(T(X)) = \text{E}\left\{[T(X) - \theta]^2\right\} . \tag{9.4}$$

The expression (9.4) can be partitioned into two parts: the variance and the squared bias, i.e.

$$\text{MSE}_\theta(T(X)) = \text{Var}_\theta(T(X)) + [\text{Bias}_\theta(T(X))]^2 . \tag{9.5}$$

This can be proven as follows:

$$
\begin{aligned}
\text{MSE}_\theta(T(X)) &= \text{E}[T(X) - \theta]^2 \\
&= \text{E}[(T(X) - \text{E}_\theta(T(X)) + (\text{E}_\theta(T(X)) - \theta)]^2 \\
&= \text{E}[T(X) - \text{E}_\theta(T(X))]^2 + [\text{E}_\theta(T(X)) - \theta]^2 \\
&= \text{Var}_\theta(T(X)) + [\text{Bias}_\theta(T(X))]^2 .
\end{aligned}
$$

Note that the calculation is based on the result that the cross product term is zero. The mean squared error can be used to compare different biased estimators.

Definition 9.2.4 An estimator $T_1(X)$ is said to be MSE-better than another estimator $T_2(X)$ for estimating θ if

$$\text{MSE}_\theta(T_1(X)) < \text{MSE}_\theta(T_2(X)) ,$$

where $\theta \in \Theta$ and Θ is the parameter space, i.e. the set of all possible values of θ. Often, Θ is $\mathbb{R}$ or all positive real values $\mathbb{R}_+$. For example, for a normal distribution, $N(\mu, \sigma^2)$, μ can be any real value and σ^2 has to be a number greater than zero.

Unfortunately, we cannot find an MSE-optimal estimator in the sense that an estimator is MSE-better than all other possible estimators for all possible values of θ. This becomes clear if we define the constant estimator $T(x) = c$ (independent of the actual sample): if $\theta = c$, i.e. if the constant value equals the true population parameter we want to estimate, then the MSE of this constant estimator is zero (but it will be greater than zero for all other values of θ, and the bias increases more as we move c far away from the true θ). Usually, we can only find estimators which are locally best (in a certain subset of Θ). This is why classical statistical inference restricts the search for best estimators to the class of unbiased estimators. For unbiased estimators the MSE is equal to the variance of an estimator. In this context, the following definition is used for comparing two (unbiased) estimators.

Definition 9.2.5 An unbiased estimator $T_1(X)$ is said to be more efficient than another unbiased estimator $T_2(X)$ for estimating θ if

$$\text{Var}_\theta(T_1(X)) \leq \text{Var}_\theta(T_2(X)) , \ \forall \theta \in \Theta ,$$

and

$$\text{Var}_\theta(T_1(X)) < \text{Var}_\theta(T_2(X))$$

for at least one $\theta \in \Theta$. It turns out that restricting our search of best estimators to unbiased estimators is sometimes a successful strategy, i.e. for many problems, a best or most efficient estimate can be found. If such an estimator exists, it is said to be UMVU (uniformly minimum variance unbiased). Uniformly means that it has the lowest variance among all other unbiased estimators for estimating the population parameter(s) θ.

Consider the illustration in Fig. 9.1 to better understand the introduced concepts. Suppose we throw three darts at a target and the goal is to hit the centre of the target, i.e. the innermost circle of the dart board. The centre represents the population parameter θ. The three darts play the role of three estimates $\hat{\theta}_1, \hat{\theta}_2, \hat{\theta}_3$ (based on different realizations of the sample) of the population parameter θ. Four possible situations are illustrated in Fig. 9.1. For example, in Fig. 9.1b we illustrate the case of an estimator which is biased but has low variance: all three darts are "far" away from the centre of the target but they are "close" together. If we look at Fig. 9.1a and c, we see that all three darts are symmetrically grouped around the centre of the target, meaning that there is no bias; however in Fig. 9.1a there is much higher precision than in Fig. 9.1c. It is obvious that Fig. 9.1a presents an ideal situation: an estimator which is unbiased and has minimum variance.

Theorem 9.2.1 *Let $X = (X_1, X_2 \ldots, X_n)$ be an i.i.d. (random) sample of a random variable X with population mean $E(X_i) = \mu$ and population variance $Var(X_i) = \sigma^2$, for all $i = 1, 2, \ldots, n$. Then the arithmetic mean $\bar{X} = \sum_{i=1}^n X_i$ is an unbiased estimator of μ and the sample variance $S^2 = \frac{1}{n-1}\sum_{i=1}^n (X_i - \bar{X})^2$ is an unbiased estimator of σ^2.*

Note that the theorem holds, in general, for i.i.d. samples, irrespective of the choice of the distribution of the X_i's. Note again that we are looking at the situation *before* we have any observations on X. Therefore, we again use capital letters to denote that the X_i's are random variables which are not known beforehand (i.e. before we actually record the observations on our selected sampling units).

Remark 9.2.1 The empirical variance $\tilde{S}^2 = \frac{1}{n}\sum_{i=1}^n (X_i - \bar{X})^2$ is a biased estimate of σ^2 and its bias is $-\frac{1}{n}\sigma^2$.

Example 9.2.1 Let $X_1, X_2 \ldots, X_n$ be identically and independently distributed variables whose population mean is μ and population variance is σ^2. Then $\bar{X} = \frac{1}{n} \sum_{i=1}^{n} X_i$ is an unbiased estimator of μ. This can be shown as follows:

$$E(\bar{X}) = E\left(\frac{1}{n} \sum_{i=1}^{n} X_i\right) \overset{(7.29)}{=} \frac{1}{n} \sum_{i=1}^{n} E(X_i) = \frac{1}{n} \sum_{i=1}^{n} \mu = \mu.$$

The variance of $\bar{X}$ can be calculated as follows:

$$Var(\bar{X}) = \frac{1}{n^2} \sum_{i=1}^{n} Var(X_i) \ , \ \left[Cov(X_i, X_j) = 0 \ \text{ using independence of } \ X_i's\right]$$

$$= \frac{1}{n^2} \sum_{i=1}^{n} \sigma^2 = \frac{\sigma^2}{n}.$$

We conclude that $\bar{X}$ is an unbiased estimator of μ and its variance is $\frac{\sigma^2}{n}$ irrespective of the choice of the distribution of X. We have learned about the distribution of $\bar{X}$ already in Chap. 8, see also Appendix C.3 for the Theorem of Large Numbers and the Central Limit Theorem; however, we would like to highlight the property of "unbiasedness" in the current context.

Now we consider another example to illustrate that estimators may not always be unbiased but may have the same variance.

Example 9.2.2 Let $X_1, X_2 \ldots, X_n$ be identically and independently distributed variables whose population mean is μ and population variance is σ^2. Then $\tilde{X} = \bar{X} + 1 = \frac{1}{n} \sum_{i=1}^{n} (X_i + 1)$ is a biased estimator of μ. This can be shown as follows:

$$E(\tilde{X}) \overset{(7.31)}{=} E\left(\frac{1}{n} \sum_{i=1}^{n} X_i\right) + E\left(\frac{1}{n} \sum_{i=1}^{n} 1\right)$$

$$\overset{(7.29)}{=} \frac{1}{n} \sum_{i=1}^{n} E(X_i) + \frac{1}{n} \cdot n = \frac{1}{n} \sum_{i=1}^{n} \mu + 1$$

$$= \mu + 1 \neq \mu.$$

However, the variance of $\tilde{X}$ is

$$Var(\tilde{X}) = Var\left(\bar{X} + 1\right) \overset{(7.34)}{=} Var(\bar{X}) = \frac{\sigma^2}{n}.$$

If we compare the two estimators $\tilde{X} = \frac{1}{n} \sum_{i=1}^{n} (X_i + 1)$ and $\bar{X} = \frac{1}{n} \sum_{i=1}^{n} (X_i)$, we see that both have the same variance but the former ($\tilde{X}$) is biased. The efficiency of both estimators is thus the same. It further follows that the mean squared error of $\bar{X}$ is smaller than the mean squared error of $\tilde{X}$ because the MSE consists of the sum of the variance and the squared bias. Therefore $\bar{X}$ is MSE-better than $\tilde{X}$. The comparison of bias, variance and MSE tells us that we should prefer $\bar{X}$ over $\tilde{X}$ when estimating the population mean. This is intuitive, but the argument we make is a purely statistical one.

Theorem 9.2.1 contains the following special cases:

- The sample mean $\bar{X} = \frac{1}{n} \sum_{i=1}^{n} X_i$ based on an i.i.d. random sample $X_1, X_2, \ldots, X_n$ from a normally distributed population $N(\mu, \sigma^2)$, is an unbiased point estimator of μ.
- The sample variance $S^2 = \frac{1}{n-1} \sum_{i=1}^{n}(X_i - \bar{X})^2$ based on an i.i.d. random sample $X_1, X_2, \ldots, X_n$ from a normally distributed population $N(\mu, \sigma^2)$ is an unbiased point estimator of σ^2. The sample variance $\tilde{S}^2 = \frac{1}{n} \sum_{i=1}^{n}(X_i - \bar{X})^2$ is a biased estimator for σ^2, but it is asymptotically unbiased in the sense that its bias tends to zero as the sample size n tends to infinity.
- The sample mean $\bar{X} = \frac{1}{n} \sum_{i=1}^{n} X_i$ based on an i.i.d. random sample $X_1, X_2, \ldots, X_n$ from a Bernoulli distributed population $B(1, p)$ is an unbiased point estimator of the probability p.

For illustration we show the validity of the third statement. Let us consider an i.i.d. random sample $X_i, i = 1, 2 \ldots, n$, from a Bernoulli distribution, where $X_i = 1$ if an event occurs and $X_i = 0$ otherwise. Here, p is the probability of occurrence of an event in the population, i.e. $p = P(X_i = 1)$. Note that p is also the population mean: $E(X_i) = 1 \cdot p + 0 \cdot (1 - p) = p, i = 1, 2, \ldots, n$. The arithmetic mean (relative frequency) is an unbiased estimator of p because

$$E(\bar{X}) = \frac{1}{n} \sum_{i=1}^{n} E(X_i) = \frac{1}{n} \sum_{i=1}^{n} p = p,$$

and thus we can write the estimate of p as

$$\hat{p} = \frac{1}{n} \sum_{i=1}^{n} X_i. \tag{9.6}$$

Example 9.2.3 Suppose a random sample of size $n = 20$ of the weight of 10-year-old children in a particular city is drawn. Let us assume that the children's weight in the population follows a normal distribution $N(\mu, \sigma^2)$. The sample provides the following values of weights (in kg):

40.2, 32.8, 38.2, 43.5, 47.6, 36.6, 38.4, 45.5, 44.4, 40.3

34.6, 55.6, 50.9, 38.9, 37.8, 46.8, 43.6, 39.5, 49.9, 34.2

To obtain an estimate of the population mean μ, we calculate the arithmetic mean of the observations as

$$\hat{\mu} = \bar{x} = \frac{1}{n} \sum_{i=1}^{n} x_i = \frac{1}{20}(40.2 + 32.8 + \ldots + 34.2) = 41.97,$$

because it is an unbiased estimator of μ. Similarly we use S^2 to estimate σ^2 because it is unbiased in comparison to $\tilde{S}^2$. Using s_X^2 as an estimate for σ^2 for the given observations, we get

$$\hat{\sigma}^2 = s_x^2 = \frac{1}{n-1} \sum_{i=1}^{n}(x_i - \bar{x})^2$$

$$= \frac{1}{19}((40.2 - 41.97)^2 + \ldots + (34.2 - 41.97)^2) \approx 36.85.$$

The square root of 36.85 is approximately 6.07 which is the standard deviation. Note that the standard deviation based on the sample values divided by the square root of the sample size, i.e. $\hat{\sigma}/\sqrt{20}$, is called the **standard error** of the mean $\bar{X}$ (SEM). As already introduced in Chap. 3, we obtain these results in R using the mean and var commands.

Example 9.2.4 A library draws a random sample of size $n = 100$ members from the members' database to see how many members have to pay a penalty for returning books late, i.e. $x_i = 1$. It turns out that 39 members in the sample have to pay a penalty. Therefore an unbiased estimator of the population proportion of all members of the library who return books late is

$$\hat{p} = \bar{x} = \frac{1}{n} \sum_{i=1}^{n} x_i = \frac{1}{100} \cdot 39 = \frac{39}{100} = 0.39.$$

Remark 9.2.2 Unbiasedness and efficiency can also be defined asymptotically: we say, for example, that an estimator is asymptotically unbiased, if the bias approaches zero when the sample size tends to infinity. The concept of asymptotic efficiency involves some mathematical knowledge which is beyond the intended scope of this book. Loosely speaking, an asymptotic efficient estimator is an estimator which achieves the lowest possible (asymptotic) variance under given distributional assumptions. The estimators introduced in Sect. 9.3.1, which are based on the maximum-likelihood principle, have these properties (under certain mathematically defined regularity conditions).

Next we illustrate the properties of consistency and sufficiency of an estimator.

9.2.2 Consistency of Estimators

For a good estimator, as the sample size increases, the values of the estimator should get closer to the parameter being estimated. This property of estimators is referred to as consistency.

Definition 9.2.6 Let $T_1, T_2, \ldots, T_n, \ldots$ be a sequence of estimators for the parameter θ where $T_n = T_n(X_1, X_2, \ldots, X_n)$ is a function of $X_1, X_2, \ldots, X_n$. The sequence $\{T_n\}$ is a **consistent** sequence of estimators for θ if for every $\epsilon > 0$,

$$\lim_{n \to \infty} P\left[|T_n - \theta| < \epsilon\right] = 1$$

or equivalently

$$\lim_{n \to \infty} P\left[|T_n - \theta| \geq \epsilon\right] = 0 \,.$$

This definition says that as the sample size n increases, the probability that T_n is getting closer to θ is approaching 1. This means that the estimator T_n is getting closer to the parameter θ as n grows larger. Note that there is no information on how *fast* T_n is converging to θ in the sense of convergence defined above.

Example 9.2.5 Let $X_1, X_2 \ldots, X_n$ be identically and independently distributed variables with expectation μ and variance σ^2. Then for $\bar{X}_n = \frac{1}{n}\sum_{i=1}^{n} X_i$, we have $E(\bar{X}_n) = \mu$ and $\text{Var}(\bar{X}_n) = \sigma^2/n$. For any $\epsilon > 0$, we can write the following:

$$P\left[|\bar{X}_n - \mu| \geq \epsilon\right] = P\left[|\bar{X}_n - \mu| \geq \frac{c\sigma}{\sqrt{n}}\right]$$

where $\epsilon = c\sigma/\sqrt{n}$. Using Tschebyschev's inequality (Theorem 7.4.1) we get $\frac{1}{c^2} = \sigma^2/n\epsilon^2$, and therefore

$$P\left[|\bar{X}_n - \mu| \geq \frac{c\sigma}{\sqrt{n}}\right] \leq \frac{1}{c^2} = \frac{\sigma^2}{n\epsilon^2}$$

and

$$\lim_{n \to \infty} P\left[|\bar{X}_n - \mu| \geq \frac{c\sigma}{\sqrt{n}}\right] \leq \lim_{n \to \infty} \frac{\sigma^2}{n\epsilon^2} = 0 \,,$$

provided σ^2 is finite. Hence $\bar{X}_n, n = 1, 2, \ldots,$ converges to μ and therefore $\bar{X}_n$ is a consistent estimator of μ.

Remark 9.2.3 We call this type of consistency *weak* consistency. Another definition is *MSE*-consistency, which says that an estimator is MSE-consistent if $MSE \longrightarrow 0$ as $n \to \infty$. If the estimator is unbiased, it is sufficient that $\text{Var} \longrightarrow 0$ as $n \to \infty$. If $T_n(X)$ is MSE–consistent, it is also weakly consistent. Therefore, it follows that an unbiased estimator with its variance approaching zero as the sample size approaches infinity is both MSE-consistent and weakly consistent.

In Example 9.2.5, the variance of $T_n(X) = \bar{X}_n$ is σ^2/n which goes to zero as n goes to ∞ and therefore $\bar{X}_n$ is both weakly consistent and MSE-consistent.

9.2.3 Sufficiency of Estimators

Sufficiency is another criterion to judge the quality of an estimator. Before delving deeper into the subject matter, we first try to understand some basic concepts.

Consider two independent random variables X and Y, each following a $N(\mu, 1)$ distribution. We conclude that both X and Y contain information about μ. Consider two estimators of μ as $\hat{\mu}_1 = X + Y$ and $\hat{\mu}_2 = X - Y$. Suppose we want to know whether to use $\hat{\mu}_1$ or $\hat{\mu}_2$ to estimate μ. We notice that $E(\hat{\mu}_1) = E(X) + E(Y) = \mu + \mu = 2\mu$, $E(\hat{\mu}_2) = E(X) - E(Y) = \mu - \mu = 0$, $Var(\hat{\mu}_1) = Var(X) + Var(Y) = 1 + 1 = 2$ and $Var(\hat{\mu}_2) = Var(X) + Var(Y) = 1 + 1 = 2$. Using the additivity property of the normal distribution, which was introduced in Remark 8.2.2, we can say that $\hat{\mu}_1 \sim N(2\mu, 2)$ and $\hat{\mu}_2 \sim N(0, 2)$. So $\hat{\mu}_1$ contains information about μ whereas $\hat{\mu}_2$ does not contain any information about μ. In other words, $\hat{\mu}_2$ loses the information about μ. We call this property "loss of information".

If we want to make conclusions about μ using both X and Y, we need to acknowledge that the dimension of them is 2. On the other hand, if we use $\hat{\mu}_1$ or equivalently $\hat{\mu}_1/2 \sim N(\mu, \frac{1}{2})$, then we need to concentrate only on one variable and we say that it has dimension 1. It follows that $\hat{\mu}_1$ and $\hat{\mu}_1/2$ provide the same information about μ as provided by the entire sample on both X and Y. So we can say that either $\hat{\mu}_1$ or $\hat{\mu}_1/2$ is sufficient to provide the same information about μ that can be obtained on the basis of the entire sample. This is the idea behind the concept of sufficiency and it results in the reduction of dimension. In general, we can say that if all the information about μ contained in the sample of size n can be obtained, for example, through the sample mean then it is sufficient to use this one-dimensional summary statistic to make inference about μ.

Definition 9.2.7 Let $X_1, X_2, \ldots, X_n$ be a random sample from a probability density function (or probability mass function) $f(x, \theta)$. A statistic T is said to be sufficient for θ if the conditional distribution of $X_1, X_2, \ldots, X_n$ given $T = t$ is independent of θ.

The Neyman–Fisher Factorization Theorem provides a practical way to find sufficient statistics.

Theorem 9.2.2 (Neyman–Fisher Factorization Theorem (NFFT)) *Let* $X_1, X_2, \ldots,$ X_n *be a random sample from a probability density function (or probability mass function)* $f(x, \theta)$. *A statistic* $T = T(x_1, x_2, \ldots, x_n)$ *is said to be sufficient for* θ *if and only if the joint density of* $X_1, X_2, \ldots, X_n$ *can be factorized as*

$$f(x_1, x_2, \ldots, x_n; \theta) = g(t, \theta) \cdot h(x_1, x_2, \ldots, x_n)$$

where $h(x_1, x_2, \ldots, x_n)$ *is nonnegative and does not involve* θ; *and* $g(t, \theta)$ *is a nonnegative function of* θ *which depends on* $x_1, x_2, \ldots, x_n$ *only through* t, *which is a particular value of* T.

This theorem holds for discrete random variables too. Any one-to-one function of a sufficient statistic is also sufficient. A function f is called one-to-one if whenever $f(a) = f(b)$ then $a = b$.

Example 9.2.6 Let $X_1, X_2, \ldots, X_n$ be a random sample from $N(\mu, 1)$ where μ is unknown. We attempt to find a sufficient statistic for μ. Consider the following function as the joint distribution of $x_1, x_2, \ldots, x_n$ (whose interpretation will become clearer in the next section):

$$f(x_1, x_2, \ldots, x_n; \mu) = \left(\frac{1}{\sqrt{2\pi}}\right)^n \exp\left(-\frac{1}{2}\sum_{i=1}^{n}(x_i - \mu)^2\right)$$

$$= \left(\frac{1}{\sqrt{2\pi}}\right)^n \exp\left(-\frac{n\mu^2}{2} + \mu\sum_{i=1}^{n}x_i\right) \exp\left(-\frac{1}{2}\sum_{i=1}^{n}x_i^2\right).$$

Here

$$g(t, \mu) = \left(\frac{1}{\sqrt{2\pi}}\right)^n \exp\left(-\frac{n\mu^2}{2} + \mu\sum_{i=1}^{n}x_i\right),$$

$$h(x_1, x_2, \ldots, x_n) = \exp\left(-\frac{1}{2}\sum_{i=1}^{n}x_i^2\right),$$

$$t = t(x_1, x_2, \ldots, x_n) = \sum_{i=1}^{n}x_i.$$

Using the Neyman–Fisher Factorization Theorem, we conclude that $T = T(X_1, X_2, \ldots, X_n) = \sum_{i=1}^{n} X_i$ is a sufficient statistic for μ. Also, $T = T(X_1, X_2, \ldots, X_n) = \bar{X}$ is sufficient for μ as it is a one-to-one statistic of $\sum_{i=1}^{n} X_i$. On the other hand, $T = \bar{X}^2$ is not sufficient for μ as it is not a one-to-one function of $\sum_{i=1}^{n} X_i$. The important point here is that $\bar{X}$ is a function of the sufficient statistic and hence a good estimator for μ. It is thus summarizing the sample information about the parameter of interest in a complete yet parsimonious way. Another, multivariate, example of sufficiency is given in Appendix C.4.

9.3   Point Estimation

In the previous section we introduced and discussed various properties of estimators. In this section we want to show how one can find estimators with good properties. In the general case, properties such as unbiasedness and efficiency cannot be guaranteed for a finite sample. But often, the properties can be shown to hold asymptotically.

9.3.1   Maximum Likelihood Estimation

We have used several estimators throughout the book without stating explicitly that they are estimators. For example, we used the sample mean $(\bar{X})$ to estimate μ in a $N(\mu, \sigma^2)$–distribution; we also used the sample proportion (relative frequency) to estimate p in a $B(1, p)$–distribution, etc. The obvious question is how to obtain a good statistic to estimate an unknown parameter, for example how to determine that the sample mean can be used to estimate μ. We need a general framework for parameter estimation. The method of **maximum likelihood** provides such an approach. For the purpose of illustration we introduce the method of maximum likelihood estimation with an example using the Bernoulli distribution.

Example 9.3.1 Consider an i.i.d. random sample $X = (X_1, X_2, \ldots, X_n)$ from a Bernoulli population with $p = P(X_i = 1)$ and $(1 - p) = P(X_i = 0)$. The joint probability mass function for a given set of realizations $x_1, x_2, \ldots, x_n$ (i.e. the data) is

$$P(X_1 = x_1, X_2 = x_2, \ldots, X_n = x_n | p) = P(X_1 = x_1 | p) \cdot \ldots \cdot P(X_n = x_n | p)$$

$$= \prod_{i=1}^{n} p^{x_i} (1 - p)^{1-x_i} . \tag{9.7}$$

This is a function of $(x_1, x_2, \ldots, x_n)$ given the parameter p. The product results from the fact that the draws are independent and the fact that $p^{x_i} (1 - p)^{1-x_i} = p$ if $x_i = 1$ and $p^{x_i} (1 - p)^{1-x_i} = 1 - p$ if $x_i = 0$. That is, the term $p^{x_i} (1 - p)^{1-x_i}$ covers results from both possible outcomes. Now consider a random sample where the values $x = (x_1, x_2, \ldots, x_n)$ are known, for example $x = (0, 1, 0, 0, \ldots, 1)$. Then (9.7) can be seen as a function of p because $(x_1, x_2, \ldots, x_n)$ is known. In this case, after obtaining a sample of data, the function is called the likelihood function and can be written as

$$L(x_1, x_2, \ldots, x_n | p) = \prod_{i=1}^{n} p^{x_1} (1 - p)^{1-x_i} . \tag{9.8}$$

The joint density function of $X_1, X_2, \ldots, X_n$ is called the **likelihood function**.

For better understanding, consider a sample of size 5 with $x = (x_1 = 1, x_2 = 1, x_3 = 0, x_4 = 1, x_5 = 0)$. The likelihood (function) is

$$L(1, 1, 0, 1, 0|p) = p \cdot p \cdot (1 - p) \cdot p \cdot (1 - p) = p^3(1 - p)^2 . \tag{9.9}$$

The maximum likelihood estimation principle now says that the estimator $\hat{p}$ of p is the value of p which maximizes the likelihood (9.8) or (9.9). In other words, the maximum likelihood estimate is the value which maximizes the probability of observing the realized sample from the likelihood function. In general, i.e. for any sample, we have to maximize the likelihood function (9.9) with respect to p. We use the well-known principle of maxima-minima to maximize the likelihood function in this case. In principle, any other optimization procedure can also be used, for example numerical algorithms such as the Newton–Raphson algorithm. If the likelihood is differentiable, the first order condition for the maximum is that the first derivative with respect to p is zero. For maximization we can transform the likelihood by a strictly monotone increasing function. This guarantees that the potential maximum is taken at the same point as in the original likelihood. A good and highly common choice is the *natural logarithm* since it transforms products in sums and sums are easy to differentiate by differentiating each term in the sum. The log-likelihood in our example is therefore

$$l(1, 1, 0, 1, 0|p) = \ln L(1, 1, 0, 1, 0, |p) = \ln \left\{ p^3(1 - p)^2 \right\} \tag{9.10}$$
$$= 3 \ln(p) + 2 \ln(1 - p) \tag{9.11}$$

where ln denotes the natural logarithm function and we use the rules

$$\ln(a \cdot b) = \ln(a) + \ln(b) , a > 0, b > 0$$
$$\ln \left(\frac{a}{b} \right) = \ln(a) - \ln(b) , a > 0, b > 0$$
$$\ln \left(a^b \right) = b \ln(a) , a > 0 .$$

Taking the first derivative of (9.10) with respect to p results in

$$\frac{\partial l(1, 1, 0, 1, 0|p)}{\partial p} = \frac{3}{p} - \frac{2}{1 - p} . \tag{9.12}$$

Setting (9.12) to zero and solving for p leads to

$$\frac{3}{p} - \frac{2}{1 - p} = 0$$
$$\frac{3}{p} = \frac{2}{1 - p}$$
$$3(1 - p) = 2p$$
$$5p = 3$$
$$\hat{p}_{ML} = \frac{3}{5} = \frac{1}{5}(1 + 1 + 0 + 1 + 0) = \bar{x} .$$

(a) Likelihood   (b) Log-likelihood

Fig. 9.2 Illustration of the likelihood and log-likelihood function of a Binomial distribution

The value of the second order partial derivative of (9.9) with respect to p at $p = \hat{p}_{\mathrm{ML}}$ is negative which ensures that $\hat{p}_{\mathrm{ML}}$ maximizes the likelihood function. It follows from this example that the maximum likelihood estimate for p leads to the well-known arithmetic mean. Figure 9.2 shows the likelihood function and the log-likelihood function as functions of p, where $p \in [0, 1]$. The figures show that the likelihood function and the log-likelihood function have the same maxima at $p = 3/5 = 0.6$.

Maximum likelihood estimators have some important properties: they are usually consistent, asymptotically unbiased, asymptotically normally distributed, asymptotically efficient, and sufficient. Even if they are not, a function of a sufficient statistic can always be found which has such properties. This is the reason why maximum likelihood estimation is popular. By "asymptotically" we mean that the properties hold as n tends to infinity, i.e. as the sample size increases. There might be other good estimators in a particular context, for example estimators that are efficient and not only asymptotically efficient; however, in general the ML principle is a great choice in many circumstances. We are going to use it in the following sections and chapters, for instance for general point estimation and in the linear regression model (Chap. 11).

Remark 9.3.1 More examples of maximum likelihood estimators are given in Exercises 9.1–9.3.

9.3.2   Method of Moments

The **method of moments** is another well-known method to derive estimators for population parameters. Below, we outline this principle briefly by way of example.

The idea is that the population parameters of interest can be related to the moments (e.g. expectation, variance) of the distribution of the considered random variables.

A simple case is the estimator for the expected value $E(X) = \mu$ of a population using an i.i.d. random sample $X = (X_1, \ldots, X_n)$. In this case, $\hat{\mu} = \bar{X}$ is the natural moment estimator of μ. Further, since $E(X^2) = \sigma^2 + \mu^2$, an estimator of $\sigma^2 + \mu^2$ is $\frac{1}{n} \sum_{i=1}^{n} X_i^2$. Using $\bar{X}^2$ as an estimator for μ^2, this results in the biased, but asymptotically unbiased estimate

$$\hat{\sigma}^2 = \frac{1}{n} \sum_{i=1}^{n} X_i^2 - \left(\frac{1}{n} \sum_{i=1}^{n} X_i \right)^2 = \frac{1}{n} \sum_{i=1}^{n} (X_i - \bar{X})^2 \,.$$

An extension of this method is the *generalised method of moments* (GMM). GMM estimators have interesting properties: under relatively week conditions (not further discussed here), they are consistent and asymptotically normal, as well as efficient in the class of those estimators that do not use any additional information besides the information included in the moment conditions. Usually, they require a two-step estimating approach or an iterative estimating procedure.

The **least squares estimator** for a linear regression model with i.i.d. random errors, discussed in detail in Chap. 11, can be seen as a special case of a GMM estimator.

9.4 Interval Estimation

9.4.1 Introduction

Let us first consider an example to understand what we mean by interval estimation. Consider a situation in which a lady wants to know the time taken to travel from her home to the train station. Suppose she makes 20 trips and notes down the time taken. To get an estimate of the expected time, one can use the arithmetic mean. Let us say $\bar{x} = 25\,\text{min}$. This is the point estimate for the expected travelling time. It may not be appropriate to say that she will always take exactly 25 min to reach the train station. Rather the time may vary by a few minutes each time. To take this into account, the time can be estimated in the form of an interval: it may then be found that the time varies mostly between 20 and 30 min. Such a statement is more informative. Both expectation and variation of the data are taken into account. The interval (20 min, 30 min) provides a range in which most of the values are expected to lie. We call this concept interval estimation.

A point estimate on its own does not take into account the precision of the estimate. The deviation between the point estimate and the true parameter (e.g. $|\bar{x} - \mu|$) can be substantial, especially when the sample size is small. To incorporate the information about the precision of an estimate in the estimated value, a **confidence interval** can be constructed. It is a **random interval** with **lower and upper bounds**, $I_l(X)$ and $I_u(X)$, such that the unknown parameter θ is covered by a pre-specified probability of at least $1 - \alpha$:

$$P_\theta(I_l(\mathbf{X}) \leq \theta \leq I_u(\mathbf{X})) \geq 1 - \alpha. \tag{9.13}$$

The probability $1 - \alpha$ is called the **confidence level** or **confidence coefficient**, $I_l(\mathbf{X})$ is called the **lower confidence bound** or **lower confidence limit** and $I_u(\mathbf{X})$ is called the **upper confidence bound** or **upper confidence limit**. It is important to note that the bounds are random and the parameter is a fixed value. This is the reason why we say that the true parameter is covered by the interval with probability $1 - \alpha$ and **not** that the probability that the interval contains the parameter is $1 - \alpha$. Please note that some software packages use the term "error bar" when referring to confidence intervals.

Frequency interpretation of the confidence interval: Suppose N independent samples $\mathbf{X}^{(j)}$, $j = 1, 2, \ldots, N$, of size n are sampled from the same population and N confidence intervals of the form $[I_l(\mathbf{X}^{(j)}), I_u(\mathbf{X}^{(j)})]$ are calculated. If N is large enough, then on an average $N(1 - \alpha)$ of the intervals (9.13) cover the true parameter.

An alternative interpretation that refers to compatibility with the data is explained in Sect. 10.9.1.

Example 9.4.1 Let a random variable follow a normal distribution with $\mu = 10$ and $\sigma^2 = 1$. Suppose we draw a sample of $n = 10$ observations repeatedly. The sample will differ in each draw and hence the mean and the confidence interval will also differ. The data sets are realizations from random variables. Have a look at Fig. 9.3 which illustrates the mean and the 95% confidence intervals for 6 random samples. They vary with respect to the mean and the confidence interval width. Most of the means are close to $\mu = 10$, but not all. Similarly, most confidence intervals, but not all, include μ. This is the idea of the frequency interpretation of the confidence interval: different samples will yield different point and interval estimates. Most of the times the interval will cover μ, but not always. The coverage probability is specified by $1 - \alpha$, and the frequency interpretation means that we expect that (approximately) $(1 - \alpha) \cdot 100\%$ of the intervals to cover the true parameter μ. In that sense, the location of the interval will give us some idea about where the true but unknown population parameter μ lies, while the length of the interval reflects our uncertainty about μ: the wider the interval is, the higher is our uncertainty about the location of μ.

We now introduce the following confidence intervals:

- Confidence interval for the mean μ of a normal distribution.
- Confidence interval for the probability p of a Binomial random variable.
- Confidence interval for the odds ratio.

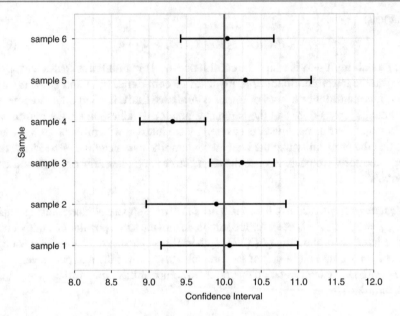

Fig. 9.3 Frequency interpretation of confidence intervals

9.4.2 Confidence Interval for the Mean of a Normal Distribution

9.4.2.1 Confidence Interval for μ When $\sigma^2 = \sigma_0^2$ is Known

Let $X_1, X_2, \ldots, X_n$ be an i.i.d. sample from a $N(\mu, \sigma_0^2)$ distribution where σ_0^2 is assumed to be known. We use the point estimate $\bar{X} = \frac{1}{n} \sum_{i=1}^{n} X_i$ to estimate μ and construct a confidence interval around the mean μ. Using the Central Limit Theorem (Appendix C.3), it follows that $\bar{X}$ follows a $N(\mu, \sigma_0^2/n)$ distribution. Therefore $\sqrt{n}(\bar{X} - \mu)/\sigma_0 \sim N(0, 1)$, and it follows that

$$P_\mu \left(\left| \frac{\sqrt{n}(\bar{X} - \mu)}{\sigma_0} \right| \leq z_{1-\frac{\alpha}{2}} \right) = 1 - \alpha \tag{9.14}$$

where $z_{1-\alpha/2}$ denotes the $(1 - \alpha/2)$ quantile of the standard normal distribution $N(0, 1)$. We solve this inequality for the unknown μ and get the desired confidence interval as follows

$$P_\mu \left[-z_{1-\frac{\alpha}{2}} \leq \left(\frac{\sqrt{n}(\bar{X} - \mu)}{\sigma_0} \right) \leq z_{1-\frac{\alpha}{2}} \right] = 1 - \alpha$$

or

$$P_\mu \left[\bar{X} - z_{1-\alpha/2} \frac{\sigma_0}{\sqrt{n}} \leq \mu \leq \bar{X} + z_{1-\alpha/2} \frac{\sigma_0}{\sqrt{n}} \right] = 1 - \alpha.$$

The confidence interval for μ is thus obtained as

$$[I_l(\mathbf{X}), I_u(\mathbf{X})] = \left[\bar{X} - z_{1-\alpha/2}\frac{\sigma_0}{\sqrt{n}}, \bar{X} + z_{1-\alpha/2}\frac{\sigma_0}{\sqrt{n}} \right]. \tag{9.15}$$

This is known as $(1 - \alpha)\%$ confidence interval for μ or the confidence interval for μ with confidence coefficient α.

We can use the R function qnorm or Table C.1 to obtain $z_{1-\frac{\alpha}{2}}$, see also Sects. 8.4, A.3, and C.9. For example, for $\alpha = 0.05$ and $\alpha = 0.01$ we get $z_{1-\frac{\alpha}{2}} = z_{0.975} = 1.96$ and $z_{1-\frac{\alpha}{2}} = z_{0.995} = 2.576$ using qnorm(0.975) and qnorm(0.995). This gives us the quantiles we need to determine a 95% and 99% confidence interval respectively.

Example 9.4.2 We consider again Example 9.2.3 where we evaluated the weight of 10-year-old children. Assume that the variance is known to be 36; then the upper and lower limits of a 95% confidence interval for the expected weight μ can be calculated as follows:

$$I_l(X) = \bar{X} - z_{1-\alpha/2}\frac{\sigma_0}{\sqrt{n}} = 41.97 - 1.96\frac{\sqrt{36}}{\sqrt{20}} \approx 39.34,$$

$$I_u(X) = \bar{X} + z_{1-\alpha/2}\frac{\sigma_0}{\sqrt{n}} = 41.97 + 1.96\frac{\sqrt{36}}{\sqrt{20}} \approx 44.59.$$

We get the confidence interval $[I_u(X), I_o(X)] = [39.34, 44.59]$. With 95% confidence, the true parameter μ is covered by the interval $[39.34, 44.59]$.

9.4.2.2 Confidence Interval for μ When σ^2 is Unknown

Let $X_1, X_2, \ldots, X_n$ be an i.i.d. sample from $N(\mu, \sigma^2)$ where σ^2 is assumed to be unknown and is being estimated by the sample variance S_X^2. We know from Sect. 8.3.1 that

$$\frac{(n-1)S_X^2}{\sigma^2} \sim \chi_{n-1}^2.$$

It can be shown that $\bar{X}$ and S_X^2 are stochastically independent. Thus we know that

$$\frac{\sqrt{n}(\bar{X} - \mu)}{S_X} \sim t_{n-1}$$

follows a t-distribution with $n - 1$ degrees of freedom. We can use this result to determine the confidence interval for μ as

$$P_\mu \left[-t_{1-\frac{\alpha}{2},n-1} \leq \left(\frac{\sqrt{n}(\bar{X} - \mu)}{S_X} \right) \leq t_{1-\frac{\alpha}{2},n-1} \right] = 1 - \alpha$$

or

$$P_\mu \left[\bar{X} - t_{1-\frac{\alpha}{2},n-1}\frac{S_X}{\sqrt{n}} \leq \mu \leq \bar{X} + t_{1-\frac{\alpha}{2},n-1}\frac{S_X}{\sqrt{n}} \right] = 1 - \alpha.$$

The confidence interval for μ is thus obtained as

$$[I_l(\mathbf{X}), I_u(\mathbf{X})] = \left[\bar{X} - t_{n-1;1-\alpha/2} \cdot \frac{S_X}{\sqrt{n}} \,, \; \bar{X} + t_{n-1;1-\alpha/2} \cdot \frac{S_X}{\sqrt{n}} \right] \qquad (9.16)$$

which is the $100(1 - \alpha)\%$ confidence interval for μ or the confidence interval for μ with confidence coefficient α.

The interval (9.16) is, in general, wider than the interval (9.15) for identical α and identical sample size n, since the unknown parameter σ^2 is estimated by S_X^2 which induces additional uncertainty. The quantiles for the t-distribution can be obtained using the R command qt or Table C.2.

Example 9.4.3 Consider Example 9.4.2 where we evaluated the weight of 10-year-old children. We have already calculated the point estimate of μ as $\bar{x} = 41.97$. With $t_{19;0.975} = 2.093$, obtained via qt(0.975,19) or Table C.2, the upper and lower limits of a 95% confidence interval for μ are obtained as

$$I_u(X) = \bar{x} - t_{19;0.975} \cdot \frac{s_X}{\sqrt{n}} = 41.97 - 2.093 \cdot \frac{6.07}{\sqrt{20}} \approx 39.12 \,,$$

$$I_o(X) = \bar{x} + t_{19;0.975} \cdot \frac{s_X}{\sqrt{n}} = 41.97 + 2.093 \cdot \frac{6.07}{\sqrt{20}} \approx 44.81 \,.$$

Therefore, the confidence interval is $[I_l(X), I_u(X)] = [39.13, 44.81]$. In R, we can use the conf.int value of the t.test command to get a confidence interval for the mean, see also Example 10.3.3 for more details on t.test. The default is a 95% confidence interval, but it can be changed easily if desired:

```
x <- c(40.2, 32.8, 38.2, 43.5, ..., 49.9, 34.2)
t.test(x,conf.level = 0.95)$conf.int
[1] 39.12384 44.80616
```

There is no unique best way to draw the calculated confidence intervals in R. Among many other options, one can simply work with the plot functionality or use geom_errorbar in conjunction with a ggplot object created with the library ggplot2, or use the plotCI command in the library plotrix.

9.4.3   Confidence Interval for a Binomial Probability

Let $X_1, X_2, \ldots, X_n$ be an i.i.d. sample from a Bernoulli distribution $B(1, p)$. Then $Y = \sum_{i=1}^{n} X_i$ has a Binomial distribution $B(n, p)$.

We have already introduced $\hat{p}$ as an estimator for p:

$$\hat{p} = \frac{1}{n} \sum_{i=1}^{n} X_i = \frac{1}{n} Y \,.$$

From (8.8), we know that $\mathrm{Var}(Y) = np(1 - p)$. Applying rule (7.33), the variance of the estimator $\hat{p}$ is

$$\mathrm{Var}(\hat{p}) = \frac{p(1 - p)}{n}$$

and it can be estimated by

$$S_{\hat{p}}^2 = \frac{\hat{p}(1 - \hat{p})}{n} \,.$$

Nowadays the exact confidence intervals of the binomial distribution function can be easily calculated using computer implementations. Nevertheless, (i) for a sufficiently large sample size n, (ii) if p is not extremely low or high, and (iii) if the condition $np(1 - p) \geq 9$ is fulfilled, we can use an approximation based on the normal distribution to calculate confidence intervals. To be more specific, one can show that

$$Z = \frac{\hat{p} - p}{\sqrt{\hat{p}(1 - \hat{p})/n}} \overset{approx.}{\sim} N(0, 1) \,. \tag{9.17}$$

This gives us

$$P\left[\hat{p} - z_{1-\alpha/2}\sqrt{\frac{\hat{p}(1 - \hat{p})}{n}} \leq p \leq \hat{p} + z_{1-\alpha/2}\sqrt{\frac{\hat{p}(1 - \hat{p})}{n}} \right] \approx 1 - \alpha, \tag{9.18}$$

and we get a confidence interval for p as

$$\left[\hat{p} - z_{1-\alpha/2}\sqrt{\frac{\hat{p}(1 - \hat{p})}{n}} \,, \; \hat{p} + z_{1-\alpha/2}\sqrt{\frac{\hat{p}(1 - \hat{p})}{n}} \right] \,. \tag{9.19}$$

Example 9.4.4 We look again at Example 9.2.4 where we evaluated the proportion of members who had to pay a penalty. Out of all borrowers, 39% brought back their books late and thus had to pay a fee. A 95% confidence interval for the probability p of bringing back a book late can be constructed using the normal approximation, since $n\hat{p}(1 - \hat{p}) = 100 \cdot 0.39 \cdot 0.61 = 23.79 > 9$. With $z_{1-\alpha/2} = z_{0.975} = 1.96$ and $\hat{p} = 0.39$ we get the 95% confidence interval as

$$\left[0.39 - 1.96\sqrt{\frac{0.39 \cdot 0.61}{100}} \,, \; 0.39 + 1.96\sqrt{\frac{0.39 \cdot 0.61}{100}} \right] = [0.294, 0.486] \,.$$

In R, an exact confidence interval can be found using the function `binom.test`:

```
binom.test(x=39,n=100)$conf.int
[1] 0.2940104 0.4926855
```

One can see that the exact and approximate confidence limits differ slightly due to the normal approximation which approximates the exact binomial probabilities.

9.4.4 Confidence Interval for the Odds Ratio

In Chap. 4, we introduced the odds ratio to determine the strength of association between two binary variables. One may be interested in the dispersion of the odds ratio and hence calculate a confidence interval for it. Recall the notation for 2×2 contingency tables:

		Y		
		y_1	y_2	Total (row)
X	x_1	a	b	$a+b$
	x_2	c	d	$c+d$
	Total (column)	$a+c$	$b+d$	n

In the spirit of the preceding sections we can interpret the entries in this contingency table as population parameters. For example, a describes the absolute frequency of observations in the population for which $Y = y_1$ and $X = x_1$. If we have a sample then we can estimate a by the number of *observed* observations n_{11} for which $Y = y_1$ and $X = x_1$. We can thus view n_{11} to be an estimator for a, n_{12} to be an estimator for b, n_{21} to be an estimator for c, and n_{22} to be an estimator for d. It follows that

$$\widehat{OR} = \frac{n_{11}n_{22}}{n_{12}n_{21}} \tag{9.20}$$

serves as the point estimate for the population odds ratio $OR = ad/bc$. To construct a confidence interval for the odds ratio, we need to work on a log-scale. The log odds ratio,

$$\theta_0 = \ln OR = \ln a - \ln b - \ln c + \ln d\,, \tag{9.21}$$

takes the natural logarithm of the odds ratio. It is evident that it can be estimated using the observed absolute frequencies of the joint frequency distribution of X and Y:

$$\hat{\theta}_0 = \ln \widehat{OR} = \ln \frac{n_{11}n_{22}}{n_{12}n_{21}}\,. \tag{9.22}$$

It can be shown that $\hat{\theta}_0$ follows approximately a normal distribution with expectation θ_0 and standard deviation

$$\hat{\sigma}_{\hat{\theta}_0} = \left(\frac{1}{n_{11}} + \frac{1}{n_{22}} + \frac{1}{n_{12}} + \frac{1}{n_{21}} \right)^{\frac{1}{2}} . \qquad (9.23)$$

Following the reasoning explained in the earlier section on confidence intervals for binomial probabilities, we can calculate the $100(1 - \alpha)\%$ confidence interval for θ_0 under a normal approximation as follows:

$$\left[\hat{\theta}_0 - z_{1-\frac{\alpha}{2}} \hat{\sigma}_{\hat{\theta}_0}, \hat{\theta}_0 + z_{1-\frac{\alpha}{2}} \hat{\sigma}_{\hat{\theta}_0} \right] = [I_u, I_o] . \qquad (9.24)$$

Since we are interested in the confidence interval of the odds ratio, and not the log odds ratio, we need to transform back the lower and upper bound of the confidence interval as

$$\left[\exp(I_u), \exp(I_o) \right] . \qquad (9.25)$$

Example 9.4.5 Recall Example 4.2.5 from Chap. 4 where we were interested in the association of smoking with a particular disease. The data are summarized in the following 2×2 contingency table:

		Smoking Yes	No	Total (row)
Disease	Yes	34	66	100
	No	22	118	140
	Total (column)	56	184	240

The odds ratio was estimated to be 2.76 and we therefore concluded that the chances of having the particular disease is 2.76 times higher for smokers compared with non-smokers. To calculate a 95% confidence intervals, we need $\hat{\theta}_0 = \ln(2.76)$, $z_{1-\frac{\alpha}{2}} \approx 1.96$ and

$$\hat{\sigma}_{\hat{\theta}_0} = \left(\frac{1}{n_{11}} + \frac{1}{n_{22}} + \frac{1}{n_{12}} + \frac{1}{n_{21}} \right)^{\frac{1}{2}}$$

$$= \left(\frac{1}{34} + \frac{1}{118} + \frac{1}{66} + \frac{1}{22} \right)^{\frac{1}{2}} \approx 0.314 .$$

The confidence interval for the log odds ratio is

$$[\ln(2.76) - 1.96 \cdot 0.314, \ln(2.76) + 1.96 \cdot 0.314] \approx [0.40, 1.63] .$$

Exponentiation of the confidence interval bounds yields the 95% confidence interval for the odds ratio as

$$[1.49, 5.11] .$$

There are many ways to obtain the same results in *R*. One option is to use the oddsratio function of the library epitools, another is to use methods introduced in Chap. 12 (see also Exercise 12.1). Note that we need to specify "wald" under the methods option to get confidence intervals which use a normal approximation (similar to what we did).

```
library(epitools)
smd <- matrix(c(34,22,66,118),ncol=2,nrow=2)    #data
oddsratio(smd,method='wald')
```

9.5   Sample Size Determinations

Confidence intervals help us estimating the precision of point estimates. What if we are required to adhere to a pre-specified precision level? We know that the variance decreases as the sample size increases. In turn, confidence intervals become narrower. On the other hand, increasing the sample size has its own consequences. For example, the cost and time involved in setting up experiments, or conducting a survey, increases. In these situations it is important to find a balance between the variability of the estimates and the sample size. We cannot control the variability in the data in most of the situations, but it is possible to control the sample size and therefore the precision of our estimates. For example, we can control the number of people to be interviewed in a survey—given the resources which are available. We discuss how to determine the number of observations needed to get a particular precision (length) of the confidence interval. We find the answers to such questions using the formulae for confidence intervals.

9.5.1   Sample Size Calculation for μ

Let us consider the situation where we are interested in estimating the population mean μ. The length of the confidence interval (9.15) for the point estimate $\bar{X}$ is

$$2z_{1-\alpha/2}\frac{\sigma_0}{\sqrt{n}} . \tag{9.26}$$

We would now like to fix the width of the confidence interval and come up with a sample size which is required to achieve this width. Let us fix the length of the confidence interval as

$$\Delta = 2z_{1-\alpha/2}\frac{\sigma_0}{\sqrt{n}}. \tag{9.27}$$

Assume we have knowledge of σ_0. The knowledge about σ_0 can be obtained, for example, through a pilot study or past experience with the experiment. We are interested in obtaining the value of n for which a confidence interval has a fixed confidence width of Δ or less. Re-arranging (9.27) gives us

$$n \geq \left[2\frac{z_{1-\alpha/2}\sigma_0}{\Delta}\right]^2. \tag{9.28}$$

This means a minimum or optimum sample size is

$$n_{opt} = \left[2\frac{z_{1-\alpha/2}\sigma_0}{\Delta}\right]^2. \tag{9.29}$$

The sample size n_{opt} ensures that the $1 - \alpha$ confidence interval for μ has at most length Δ. But note that we have assumed that σ_0 is known. If we do not know σ_0 (which is more likely in practice), we have to make an assumption about it, e.g. by using an estimate from a former study, a pilot study, or other external information. Practically, (9.28) is used in the case of known and unknown σ_0^2.

Example 9.5.1 A call centre is interested in determining the expected length of a telephone call as precisely as possible. The requirements are that the 95% confidence interval for μ should have a width of 1 min. Suppose that the call centre has developed a pilot study in which σ_0 was estimated to be 5 min. The sample size n that is needed to estimate the expected length of the phone calls with the desired precision is:

$$n \geq \left[\frac{2z_{1-\alpha/2}\sigma_0}{\Delta}\right]^2 = \left[\frac{2 \times 1.96 \times 5}{1}\right]^2 \approx 384.$$

This means that at least 384 calls are required to get the desired confidence interval width.

9.5.2 Sample Size Calculation for p

We can follow the earlier reasoning and determine the optimum sample size for a specific confidence interval width using the confidence interval definition (9.19). Since the width of the confidence interval is

$$\Delta = 2z_{1-\alpha/2}\sqrt{\frac{\hat{p}(1-\hat{p})}{n}},$$

we get

$$n \geq \left[2\frac{z_{1-\alpha/2}}{\Delta}\right]^2 \hat{p}(1-\hat{p}). \tag{9.30}$$

Example 9.5.2 A factory may be interested in the probability of an error in an operating process. The length of the confidence interval should be $\pm 2\%$, i.e. $\Delta = 0.04$. Suppose it is speculated that the error probability is 10%; we may then use $\hat{p} = 0.1$ as our prior judgment for the true value of p. This yields

$$n \geq \left[2\frac{z_{1-\alpha/2}}{\Delta}\right]^2 \hat{p}(1-\hat{p}) = \left[2 \times \frac{1.96}{0.04}\right]^2 0.1 \cdot (1-0.1) \approx 865. \tag{9.31}$$

This means we need a sample size of at least 865 to obtain the desired width of the confidence interval for p.

The above examples for both μ and p have shown us that without external knowledge about the research question of interest it is difficult to come up with an appropriate sample size. Results may vary considerably depending on what type of information is assumed to be known. With limited knowledge it can be useful to report results for different widths of confidence intervals and hypothesized values of p or σ_0.

Sample size calculations can be highly complex in many practical situations and may not remain as simple as in the examples considered here. For example, Chap. 10 uses additional concepts in the context of hypothesis testing, such as the power, which can be taken into consideration when estimating sample sizes. However, in this case, calculations and interpretations become more difficult and complex. A detailed overview of sample size calculations can be found in Chow, Wang and Shao (2007).

9.6   Key Points and Further Issues

> **Note:**
>
> ✓ We have introduced important point estimates for the parameters of a normal and a Binomial distribution:
>
> $$\bar{x} \text{ for } \mu, \qquad S^2 \text{ for } \sigma^2, \qquad \bar{x} \text{ for } p.$$
>
> In general, the choice of these point estimates is not arbitrary but follows some principles of statistical inference like maximum likelihood estimation, or least squares estimation (introduced in Chap. 11).
>
> ✓ The maximum likelihood estimator is usually consistent, asymptotically unbiased, asymptotically normally distributed, and asymptotically efficient.
>
> ✓ The validity of all results in this chapter depend on the assumption that the data are complete and have no missing values. Incomplete data may yield different conclusions.
>
> ✓ A confidence interval is defined in terms of upper and lower confidence limits and covers the true target parameter with probability $1 - \alpha$. Confidence intervals are often constructed as follows:
>
> $$\text{point estimate} \pm \text{quantile} \cdot \underbrace{\sqrt{\text{variance of point estimate}}}_{\text{standard error}}.$$
>
> ✓ More detailed introductions to inference are presented in Casella and Berger (2002) and Young and Smith (2005).

9.7   Exercises

Exercise 9.1 Consider an i.i.d. sample of size n from a $Po(\lambda)$ distributed random variable X.

(a) Determine the maximum likelihood estimate for λ.

(b) What does the log-likelihood function look like for the following realizations: $x_1 = 4, x_2 = 3, x_3 = 8, x_4 = 6, x_5 = 6$? Plot the function using R. Hint: The curve command can be used to plot functions.

(c) Use the Neyman–Fisher Factorization Theorem to argue that the maximum likelihood estimate obtained in a) is a sufficient statistic for λ.

Exercise 9.2 Consider an i.i.d. sample of size n from a $N(\mu, \sigma^2)$ distributed random variable X.

(a) Determine the maximum likelihood estimator for μ under the assumption that $\sigma^2 = 1$.
(b) Now determine the maximum likelihood estimator for μ for an arbitrary σ^2.
(c) What is the maximum likelihood estimate for σ^2?

Exercise 9.3 Let $X_1, X_2, \ldots, X_n$ be n i.i.d. random variables which follow a uniform distribution, $U(0, \theta)$. Write down the likelihood function and argue, without differentiating the function, what the maximum likelihood estimate of θ is.

Exercise 9.4 Let $X_1, X_2, \ldots, X_n$ be n i.i.d. random variables which follow an exponential distribution. An intelligent statistician proposes to use the following two estimators to estimate $\mu = 1/\lambda$:

(i) $T_n(X) = nX_{\min}$ with $X_{\min} = \min(X_1, \ldots, X_n)$ and $X_{\min} \sim \text{Exp}(n\lambda)$,
(ii) $V_n(X) = n^{-1} \sum_{i=1}^n X_i$.

(a) Are both $T_n(X)$ and $V_n(X)$ (asymptotically) unbiased for μ?
(b) Calculate the mean squared error of both estimators. Which estimator is more efficient?
(c) Is $V_n(X)$ MSE consistent, weakly consistent, both, or not consistent at all?

Exercise 9.5 A national park in Namibia determines the weight (in kg) of a sample of common eland antelopes:

$$450\ 730\ 700\ 600\ 620\ 660\ 850\ 520\ 490\ 670\ 700\ 820$$
$$910\ 770\ 760\ 620\ 550\ 520\ 590\ 490\ 620\ 660\ 940\ 790$$

Calculate

(a) the point estimate of μ and σ^2 and
(b) the confidence interval for μ ($\alpha = 0.05$)

under the assumption that the weight is normally distributed.

(c) Use R to reproduce the results from (b).

Exercise 9.6 We are interested in the heights of the players of the two basketball teams "Brose Baskets Bamberg" and "Bayer Giants Leverkusen" as well as the football team "SV Werder Bremen". The following summary statistics are given:

	N	Minimum	Maximum	Mean	Std. dev.
Bamberg	16	185	211	199.06	7.047
Leverkusen	14	175	210	196.00	9.782
Bremen	23	178	195	187.52	5.239

Calculate a 95% confidence interval for μ for all three teams and interpret the results.

Exercise 9.7 A married couple tosses a coin after each dinner to determine who has to wash the dishes. If the coin shows "head" then the husband has to wash the dishes, if the coin shows "tails" then the wife has to wash the dishes. After 98 dinners the wife notes that the coin has shown head 59 times.

(a) Estimate the probability that the wife has to wash the dishes.
(b) Calculate and interpret the 95% confidence interval for p.
(c) How many dinners are needed to estimate the true probability for the coin showing "head" with a precision of $\pm 0.5\%$ under the assumption that the coin is fair?

Exercise 9.8 Suppose 93 out of 104 pupils have passed the final examination at a certain school.

(a) Calculate a 95% confidence interval for the probability of failing the exam both by manual calculations and by using R, and compare the results.
(b) At county level 3.2% of pupils failed the examination. Are the school's pupils worse than those in the whole county?

Exercise 9.9 To estimate the audience rate for several TV stations 3000 households are asked to allow a device, which records which TV station is watched, to be installed on their TVs. 2500 agreed to participate. Assume it is of interest to estimate the probability of someone switching on the TV and watching the show "Germany's next top model".

(a) What is the precision with which the probability can be estimated?
(b) What source of bias could potentially influence the estimates?

Exercise 9.10 An Olympic decathlon athlete is interested in his performance compared with the performance of other athletes. He is a good runner and interested in his 100 m results compared with those of other athletes.

(a) He uses the decathlon data from this book (Appendix A.2) to come up with $\hat{\sigma} = s = 0.233$. What sample size does he need to calculate a 95% confidence interval for the mean running time which is precise to ± 0.1 s?

(b) Calculate a 95% confidence interval for the mean running time ($\bar{x} = 10.93$) of the 30 athletes captured in the data set in Appendix A.2. Interpret the width of this interval compared with the width determined in (a).

(c) The runner's own best time is 10.86 s. He wants to be among the best 10% of all athletes. Calculate an appropriate confidence interval to compare his time with the 10% best times.

Exercise 9.11 Consider the pizza delivery data described in Appendix A.4.1. We distinguish between pizzas delivered on time (i.e. in less than 30 min) and not delivered on time (i.e. in more than 30 min). The contingency table for delivery time and operator looks as follows:

	Operator		
	Laura	Melissa	Total
< 30 min	163	151	314
≥ 30 min	475	477	952
Total	638	628	1266

(a) Calculate and interpret the odds ratio and its 95% confidence interval.

(b) Reproduce the results from a) using R.

→ Solutions to all exercises in this chapter can be found in Appendix B

Hypothesis Testing

10

10.1   Introduction

We introduced point and interval estimation of parameters in the previous chapter. Sometimes, the research question is less ambitious in the sense that we are not interested in precise estimates of a parameter but we only want to examine whether a statement about a parameter of interest or the research hypothesis, is true or not (although we will see later in this chapter that there is a connection between confidence intervals and statistical tests, called *duality*). Another related issue is that once an analyst estimates the parameters on the basis of a random sample, (s)he would like to infer something about the value of the parameter in the population. Statistical hypothesis tests facilitate the comparison of estimated values with hypothetical values.

Example 10.1.1 As a simple example, consider the case where we want to find out whether the proportion of votes for a party P in an election will exceed 30% or not. Typically, before the election, we will try to get representative data about the election proportions for different parties (e.g. by telephone interviews) and then make a statement like "yes", we expect that P will get more than 30% of the votes or "no", we do not have enough evidence that P will get more than 30% of the votes. In such a case, we will only know after the election whether our statement was right or wrong. Note that the term representative data only means that the sample is similar to the population with respect to the distributions of some key variables, e.g. age, gender and education. Since we use one sample to compare it with a fixed value (30%), we call it a **one-sample problem**.

Example 10.1.2 Consider another example in which a clinical study is conducted to compare the effectiveness of a new drug (B) to an established standard drug (A)

© Springer Nature Switzerland AG 2022
C. Heumann et al., *Introduction to Statistics and Data Analysis*,
https://doi.org/10.1007/978-3-031-11833-3_10

for a specific disease, for example too high blood pressure. Assume that, as a first step, we want to find out whether the new drug causes a higher reduction in blood pressure than the already established older drug. A frequently used study design for this question is a randomized (i.e. patients are randomly allocated to one of the two treatments) controlled clinical trial (double blinded, i.e. neither the patient nor the doctor know which of the drugs a patient is receiving during the trial), conducted in a fixed time interval, say 3 months. A possible hypothesis is that the average change in the blood pressure in group B is higher than in group A, i.e. $\delta_B > \delta_A$ where $\delta_j = \mu_{j0} - \mu_{j3}$, $j = A, B$ and μ_{j0} is the average blood pressure at baseline before measuring the blood pressure again after 3 months (μ_{j3}). Note that we expect both the differences δ_A and δ_B to be positive, since otherwise we would have some doubt that either drug is effective at all. As a second step (after statistically proving our hypothesis), we are interested in whether the improvement of B compared to A is relevant in a medical or biological sense and is valid for the entire population or not. This will lead us again to the estimation problems of the previous chapter, i.e. quantifying an effect using point and interval estimation. Since we are comparing two drugs, we need to have two samples from each of the drugs, hence we have a **two-sample problem**. Since the patients receiving A are different from those receiving B in this example, we refer to it as a "two *independent* samples problem".

Example 10.1.3 In another example, we consider an experiment in which a group of students receives extra mathematical tuition. Their ability to solve mathematical problems is evaluated before and after the extra tuition. We are interested in knowing whether the ability to solve mathematical problems increases after the tuition, or not. Since the same group of students is used in a pre-post experiment, this is called a "two *dependent* samples problem" or a "paired data problem".

10.2   Basic Definitions

10.2.1 One- and Two- Sample Problems

In one-sample problems, the data are usually assumed to arise as *one* sample from a defined population. In two-sample problems, the data originate in the form of *two samples* possibly from two different populations. The heterogeneity is often modelled by assuming that the two populations only differ in some parameters or key quantities like expectation (i.e. mean), median, or variance. As in our introductory example, the samples can either be independent (as in the drug Example 10.1.2) or dependent (as in the evaluation Example 10.1.3).

10.2.2   Hypotheses

A researcher may have a research question for which the truth about the population of interest is unknown. Suppose data can be obtained using a survey, observation, or an experiment: if, given a pre-specified uncertainty level, a statistical test based on the data supports the hypothesis about the population, we say that this hypothesis is statistically proven. Note that the research question has to be operationalized before it can be tested by a statistical test. Consider the drug Example 10.1.2: we want to examine whether the new drug B has a greater blood-pressure lowering effect than the standard drug A. We have several options to operationalize this research question into a statistical setup. One is to test whether the *average* reduction (from baseline to 3 months) of the blood pressure is higher (and positive) for drug B than drug A. We then state our hypotheses in terms of expected values (i.e. μ). Why do we have to use the expected values μ and not simply compare the arithmetic means $\bar{x}$? The reason is that the superiority of B shown in the sample will only be valid for this sample and not necessarily for another sample. We need to show the superiority of B in the entire population, and hence our hypothesis needs to reflect this. Another option would be, for example, to use median changes in blood pressure values instead of mean changes in blood pressure values. An important point is that the research hypothesis which we want to prove has to be formulated as the statistical alternative hypothesis, often denoted by H_1. The reason for this will become clearer later in this chapter. The opposite of the research hypothesis has to be formulated as the statistical null hypothesis, denoted by H_0. In the drug example, the alternative and null hypotheses are

$$H_1 : \delta_B > \delta_A$$

and

$$H_0 : \delta_B \leq \delta_A$$

respectively. We note that the two hypotheses are disjoint and the union of them covers all possible differences of δ_B and δ_A. There is a boundary value ($\delta_B = \delta_A$) which separates the two hypotheses. Since we want to show the superiority of B, the hypothesis was formulated as a one-sided hypothesis. Note that there are different ways to formulate two-sample hypotheses; for example $H_1 : \delta_B > \delta_A$ is equivalent to $H_1 : \delta_B - \delta_A > 0$. In fact, it is very common to formulate two-sample hypotheses as differences, which we will see later in this chapter.

10.2.3   One- and Two-Sided Tests

We distinguish between one-sided and two-sided hypotheses and tests. In the previous section we gave an example of a one-sided test.

For an unknown population parameter θ (e.g. μ) and a fixed value θ_0 (e.g. 5), the following three cases have to be distinguished:

Case	Null hypothesis	Alternative hypothesis	
(a)	$\theta = \theta_0$	$\theta \neq \theta_0$	two-sided test problem
(b)	$\theta \geq \theta_0$	$\theta < \theta_0$	one-sided test problem
(c)	$\theta \leq \theta_0$	$\theta > \theta_0$	one-sided test problem

Example 10.2.1 One-sample problems often test whether a target value is achieved or not. For example, consider the null hypothesis as

- H_0 : average filling weight of packages of flour = 1 kg
- H_0 : average body height (men) = 178 cm.

The alternative hypothesis H_1 is formulated as deviation from the target value. If deviations in both directions are interesting, then H_1 is formulated as a two-sided hypothesis,

- H_1 : average body height (men) $\neq$ 178 cm.

If deviations in a specific direction are the subject of interest, then H_1 is formulated as a one-sided hypothesis, e.g.,

- H_1 : average filling weight of flour packages is lower than 1 kg.
- H_1 : average filling weight of flour packages is greater than 1 kg.

Two-sample problems often examine differences of two samples. Suppose the null hypothesis H_0 is related to the average weight of flour packages filled by two machines, say 1 and 2. Then the null hypothesis is

- H_0 : average weight of flour packages filled by machine 1 = average weight of flour packages filled by machine 2.

Then H_1 can be formulated as a one-sided or two-sided hypothesis. If we want to prove that machine 1 and machine 2 have different filling weights, then H_1 would be formulated as a two-sided hypothesis

- H_1 : average filling weight of machine 1 $\neq$ average filling weight of machine 2.

If we want to prove that machine 1 has lower average filling weight than machine 2, H_1 would be formulated as a one-sided hypothesis

- H_1 : average filling weight of machine 1 < average filling weight of machine 2.

If we want to prove that machine 2 has lower filling weight than machine 1, H_1 would be formulated as a one sided hypothesis

- H_1 : average filling weight of machine 1 > average filling weight of machine 2.

Remark 10.2.1 Note that we have *not* considered the following situation: $H_0 : \theta \neq \theta_0, H_1 : \theta = \theta_0$. In general, with the tests described in this chapter, we cannot prove the equality of a parameter to a pre-defined value and neither can we prove the equality of two parameters, as in $H_0 : \theta_1 \neq \theta_2, H_1 : \theta_1 = \theta_2$. We can, for example, not prove (statistically) that machines 1 and 2 in the previous example provide equal filling weight. This would lead to the more complex class of equivalence tests, which is a topic beyond the scope of this book.

10.2.4 Type I and Type II Error

If we undertake a statistical test, two types of error can occur.

- The hypothesis H_0 is true but is rejected; this error is called **type I error**.
- The hypothesis H_0 is not rejected although it is wrong; this is called **type II error**.

When a hypothesis is tested, then the following four situations are possible:

	H_0 is true	H_0 is not true
H_0 is not rejected	Correct decision	Type II error
H_0 is rejected	Type I error	Correct decision

The significance level is the probability of type I error, $P(H_1|H_0) = \alpha$, which is the probability of rejecting H_0 (accepting H_1) if H_0 is true. If we construct a test, the significance level α is pre-specified, e.g. $\alpha = 0.05$. A significance test is constructed such that the probability of a type I error does not exceed α while the probability of a type II error depends on the true but unknown parameter values in the population(s) and the sample size. Therefore, the two errors are not symmetrically treated in a significance test. In fact, the type II error β, $P(H_0|H_1) = \beta$ is not controlled by the construction of the test and can become very high, sometimes up to $1 - \alpha$. This is the reason why a test not rejecting H_0 is not a (statistical) proof of H_0. In mathematical statistics one searches for the best test which maintains α and minimizes β. Minimization of both α and β simultaneously is not possible. The reason is that when α increases then β decreases and vice versa. So one of the errors needs to be fixed and the other error is minimized. Consequently, the error which is considered more serious is fixed and then the other error is minimized. The tests discussed in the below sections are obtained based on the assumption that the type I error is more serious than the type II error. So the test statistics are obtained by fixing

α and then minimizing β. In fact, the null hypothesis is framed in such a way that it implies that the type I error is more serious than the type II error. The probability $1 - \beta = P(H_1|H_1)$ is called the **power** of the test. It is the probability of making a decision in favour of the research hypothesis H_1, if it is true, i.e. the probability of detecting a correct research hypothesis.

10.2.5 How to Conduct a Statistical Test

In general, we can follow the steps described below to test a hypothesis about a population parameter based on a sample of data.

(1) Define the distributional assumptions for the random variables of interest, and specify them in terms of population parameters (e.g. θ or μ and σ). This is necessary for parametric tests. There are other types of tests, so-called nonparametric tests, where the assumptions can be relaxed in the sense that we do not have to specify a particular distribution, see Sect. 10.6ff. Moreover, for some tests the distributional assumptions can be relaxed if the sample size is large.

(2) Formulate the null hypothesis and the alternative hypothesis as described in Sects. 10.2.2 and 10.2.3.

(3) Fix a significance value (often called type I error) α, for example $\alpha = 0.05$, see also Sect. 10.2.4.

(4) Construct a test statistic $T(\mathbf{X}) = T(X_1, X_2, \ldots, X_n)$. The distribution of T has to be known under the null hypothesis H_0. We note again that $(X_1, X_2, \ldots, X_n)$ refers to the random variables before drawing the actual sample and $x_1, x_2, \ldots, x_n$ are the realized values (observations) in the sample.

(5) Construct a critical region K for the statistic T, i.e. a region where—if T falls in this region—H_0 is rejected, such that

$$P_{H_0}(T(\mathbf{X}) \in K) \le \alpha .$$

The notation $P_{H_0}(\cdot)$ means that this inequality must hold for all parameter values θ that belong to the null hypothesis H_0. Since we assume that we know the distribution of $T(\mathbf{X})$ under H_0, the critical region is defined by those values of $T(\mathbf{X})$ which are unlikely (i.e. with probability of less than α) to be observed under the null hypothesis. Note that although $T(\mathbf{X})$ is a random variable, K is a well-defined region, see Fig. 10.1 for an example.

(6) Calculate $t(x) = T(x_1, x_2, \ldots, x_n)$ based on the realized sample values $X_1 = x_1, X_2 = x_2, \ldots, X_n = x_n$.

(7) Decision rule: if $t(x)$ falls into the critical region K, the null hypothesis H_0 is rejected. The alternative hypothesis is then statistically proven. If $t(x)$ falls outside the critical region, H_0 is not rejected.

$$t(x) \in K : H_0 \text{ rejected} \Rightarrow H_1 \text{ is statistically significant,}$$
$$t(x) \notin K : H_0 \text{ not rejected and therefore accepted.}$$

The next two paragraphs show how to arrive at the test decisions from step 7 in a different way. Readers interested in an example of a statistical test may jump to Sect. 10.3.1 and possibly also Example 10.3.1.

10.2.6 Test Decisions Using the p-Value

Statistical software usually does not show us all the steps of hypothesis testing as outlined in Sect. 10.2.5. It is common that instead of calculating and reporting the critical values, the test statistic is printed together with the so-called p-value. It is possible to use the p-value instead of critical regions for making test decisions. The p-value of the test statistic $T(\mathbf{X})$ is defined as follows:

$$\text{two-sided case:}\quad P_{H_0}(|T| \geq t(x)) = p\text{-value}$$
$$\text{one-sided case:}\quad P_{H_0}(T \geq t(x)) = p\text{-value}$$
$$P_{H_0}(T \leq t(x)) = p\text{-value}$$

It can be interpreted as the probability that the test statistic is equal to, or more extreme than its observed value if the null hypothesis was true. Additionally, every assumption used to compute the test statistic and the p-value needs to be correct too, see Sect. 10.9 for details. Then a common decision rule is

H_0 is rejected if the p-value is smaller than the pre-specified significance level α. Otherwise, H_0 cannot be rejected.

Example 10.2.2 Assume that we are dealing with a two-sided test and assume further that the test statistic $T(x)$ is $N(0, 1)$-distributed under H_0. The significance level is set as $\alpha = 0.05$. If we observe, for example, $t = 3$, then the p-value is $P_{H_0}(|T| \geq 3)$. This can be calculated in R as

```
2*(1-pnorm(3))
```

because `pnorm()` is used to calculate $P(X \leq x)$, and therefore `1-pnorm()` can be used to calculate $P(X > x)$. We have to multiply with two because we are dealing with a two-sided hypothesis. The result is $p = 0.002699796$. Therefore H_0 is rejected if we opt for a binary decision. The one-sided p-value is half of the two-sided p-value, i.e. $P(T \geq 3) = P(T \leq -3) = 0.001349898$ and is not necessarily reported by R. It is therefore important to look carefully at the R output when dealing with one-sided hypotheses.

The p-value is sometimes also called the *significance*, although we prefer the term p-value. We use the term *significance* only in the context of a test result: a test is (statistically) significant if (and only if) H_0 can be rejected.

Unfortunately, the p-value is often over-interpreted: interpreting the p-value as the probability that the null hypothesis is true is wrong. It is also incorrect to say that the p-value is the probability of making an error during the test decision. In our (frequentist) context, hypotheses are true or false and no probability is assigned to them. It is also important to highlight that statistical significance does not imply that a substantive important relationship has been detected: significance is not necessarily equal to relevance. Finally, note that the p-value itself is a random variable: under the null hypothesis it follows a uniform distribution, i.e. $p \sim U(0, 1)$.

It is often the case that no simple yes/no decision is required and acceptance or rejection of one particular hypothesis is not needed. Then a small p-value simply indicates the data as being unusual if the test hypothesis (and every assumption used to compute the p-value) is correct. Conversely, a large p-value suggests that the data are not very unusual if the test hypothesis (and every assumption used to compute the p-value) is correct. We now continue to illustrate how tests and p-values can be used to make binary test decisions and explain viable alternatives in Sect. 10.9.

10.2.7 Test Decisions Using Confidence Intervals

There is an interesting and useful relationship between confidence intervals and hypothesis tests. If the null hypothesis H_0 is rejected at the significance level α then there exists a $100(1 - \alpha)\%$ confidence interval which yields the same conclusion as the test: if the appropriate confidence interval does not contain the value θ_0 targeted in the hypothesis, then H_0 is rejected. We call this **duality**. For example, recall Example 10.1.2 where we were interested in whether the average change in blood pressure for drug B is higher than for drug A, i.e. $H_1 : \delta_B > \delta_A$. This hypothesis is equivalent to $H_1 : \delta_B - \delta_A > \delta_0 = 0$. In the following section, we develop tests to decide whether H_1 is statistically significant or not. Alternatively we could construct a $100(1 - \alpha)\%$ confidence interval for the difference $\delta_B - \delta_A$ and evaluate whether the interval contains $\delta_0 = 0$ or not; if yes, we accept H_0, otherwise we reject it. For some of the tests introduced in following section, we refer to the confidence intervals which lead to the same results as the corresponding test.

10.3 Parametric Tests for Location Parameters

10.3.1 Test for the Mean When the Variance is Known (One-Sample Gauss-Test)

We develop a hypothesis test to test whether the unknown mean (expectation) μ of a $N(\mu, \sigma^2)$-distributed random variable X either differs from a specific value $\mu = \mu_0$ or is smaller (or greater) than μ_0. We assume that the variance $\sigma^2 = \sigma_0^2$ is known. We apply the scheme of Sect. 10.2.5 step by step to develop the test procedure and then give an illustrative example.

1. Distributional assumption: The random variable X follows a $N(\mu, \sigma_0^2)$- distribution with known variance σ_0^2. We assume that an i.i.d. random sample is drawn from $X_1, X_2, \ldots, X_n$ where the X_i's follow the same distribution as $X, i = 1, 2, \ldots, n$.

2. Define any of the following sets of hypotheses H_0 and H_1:

$$H_0 : \mu = \mu_0 \quad \text{versus} \quad H_1 : \mu \neq \mu_0, \qquad \text{(two-sided test)}$$
$$H_0 : \mu \leq \mu_0 \quad \text{versus} \quad H_1 : \mu > \mu_0, \qquad \text{(one-sided test)}$$
$$H_0 : \mu \geq \mu_0 \quad \text{versus} \quad H_1 : \mu < \mu_0, \qquad \text{(one-sided test)}.$$

3. Specify the probability of a type I error α: Often $\alpha = 0.05 = 5\%$ is chosen.

4. Construct a test statistic: The unknown mean, i.e. the expectation μ, is usually estimated by the sample mean $\bar{x}$. We already know that if the X_i's are i.i.d. then the sample mean is normally distributed. Under the assumption that H_0 is true,

$$\bar{X} = \frac{1}{n} \sum_{i=1}^{n} X_i \overset{H_0}{\sim} N(\mu_0, \sigma_0^2/n),$$

where $\overset{H_0}{\sim}$ means the "distribution under H_0". If we standardize the mean under H_0, we get a $N(0, 1)$-distributed test statistic

$$T(\mathbf{X}) = \frac{\bar{X} - \mu_0}{\sigma_0} \sqrt{n} \overset{H_0}{\sim} N(0, 1),$$

see also Theorem 7.3.2. Note that $T(\mathbf{X})$ follows a normal distribution even if the X_i's are *not* normally distributed and if n is large enough which follows from the Central Limit Theorem (Appendix C.3.2). One can conclude that the distributional assumption from step 1 is thus particularly important for small samples, but not necessarily important for large samples. As a rule of thumb, $n \geq 30$ is considered to be a large sample. This rule is based on the knowledge that a t-distribution with more than 30 degrees of freedom gets very close to a $N(0, 1)$-distribution.

5. Critical region: Since the test statistic $T(\mathbf{X})$ is $N(0, 1)$-distributed, we get the following critical regions, depending on the hypothesis:

Case	H_0	H_1	Critical region K
(a)	$\mu = \mu_0$	$\mu \neq \mu_0$	$K = (-\infty, -z_{1-\alpha/2}) \cup (z_{1-\alpha/2}, \infty)$
(b)	$\mu \leq \mu_0$	$\mu > \mu_0$	$K = (z_{1-\alpha}, \infty)$
(c)	$\mu \geq \mu_0$	$\mu < \mu_0$	$K = (-\infty, z_\alpha = -z_{1-\alpha})$

For case (a) with H_0: $\mu = \mu_0$ and H_1: $\mu \neq \mu_0$, we are interested in extreme values of the test statistic on both tails: very small values and very large values of the test statistic give us evidence that H_0 is wrong (because the statistic is mainly driven by the difference of the sample mean and the test value μ_0 for a fixed variance), see Fig. 10.1. In such a two-sided test, when the distribution of the test statistic is

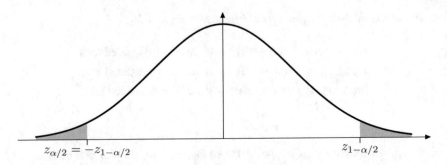

Fig. 10.1 Critical region of a two-sided one-sample Gauss- test H_0: $\mu = \mu_0$ versus H_1: $\mu \neq \mu_0$. The critical region $K = (-\infty, -z_{1-\alpha/2}) \cup (z_{1-\alpha/2}, \infty)$ has probability mass α if H_0 is true

symmetric, we divide the critical region into two equal parts and assign each region of size $\alpha/2$ to the left and right tails of the distribution. For $\alpha = 0.05$, 2.5% of the most extreme values towards the right end of the distribution and 2.5% of the most extreme values towards the left end of the distribution give us enough evidence that H_0 is wrong and can be rejected and that H_1 is accepted. It is also clear why α is the probability of a type I error: the most extreme values in the two tails together have 5% probability and are just the probability that the test statistic falls into the critical region although H_0 is true. Also, these areas are those which have the least probability of occurring if H_0 is true. For $\alpha = 0.05$ we get $z_{1-\frac{\alpha}{2}} = 1.96$.

For case (b), only one direction is of interest. The critical region lies on the right tail of the distribution of the test statistic. A very large value of the test statistic has a low probability of occurrence if H_0 is true. An illustration is given in Fig. 10.2: for $\alpha = 0.05$ we get $z_{1-\alpha} = 1.64$ and any values greater than 1.64 are unlikely to be observed under H_0. Analogously, the critical region for case (c) is constructed. Here, the shaded area (critical region) is on the left-hand side. In this case, for $\alpha = 0.05$, we get $z_\alpha = -z_{1-\alpha} = -1.64$.

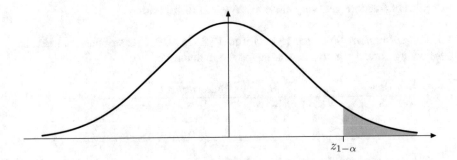

Fig. 10.2 Critical region of a one-sided one-sample Gauss- test H_0: $\mu \leq \mu_0$ vs. H_1: $\mu > \mu_0$. The critical region $K = (z_{1-\alpha}, \infty)$ has probability mass α if H_0 is true

Table 10.1 Rules to make test decisions for the one-sample Gauss test (and the two-sample Gauss test, the one-sample approximate Binomial test, and the two-sample approximate Binomial test—which are all discussed later in this chapter)

Case	H_0	H_1	Reject H_0 if
(a)	$\mu = \mu_0$	$\mu \neq \mu_0$	$|t(x)| > z_{1-\alpha/2}$
(b)	$\mu \geq \mu_0$	$\mu < \mu_0$	$t(x) < z_\alpha$
(c)	$\mu \leq \mu_0$	$\mu > \mu_0$	$t(x) > z_{1-\alpha}$

6. *Realization of the test statistic:* For an observed sample $x_1, x_2, \ldots, x_n$ the arithmetic mean

$$\bar{x} = \frac{1}{n} \sum_{i=1}^{n} x_i$$

is used to calculate the realized (observed) test statistic $t(x) = T(x_1, x_2, \ldots, x_n)$ as

$$t(x) = \frac{\bar{x} - \mu_0}{\sigma_0} \sqrt{n}.$$

7. *Test decision:* If the realized test statistic from step 6 falls into the critical region, H_0 is rejected (and therefore H_1 is statistically proven). Table 10.1 summarizes the test decisions depending on $t(x)$ and the quantiles defining the appropriate critical regions.

Example 10.3.1 A bakery supplies loaves of bread to super-markets. The stated selling weight (and therefore the required minimum expected weight) is $\mu = 2$ kg. However, not every package weighs exactly 2 kg because there is variability in the weights. It is therefore important to find out if the average weight of the loaves is significantly smaller than 2 kg. The weight X (measured in kg) of the loaves is assumed to be normally distributed. We assume that the variance $\sigma_0^2 = 0.1^2$ is known from experience. A supermarket draws a sample of $n = 20$ loaves and weighs them. The average weight is calculated as $\bar{x} = 1.97$ kg. Since the supermarket wants to be sure that the weights are, on average, not lower than 2 kg, a one-sided hypothesis is appropriate and is formulated as $H_0: \mu \geq \mu_0 = 2$ kg versus $H_1: \mu < \mu_0 = 2$ kg. The significance level is specified as $\alpha = 0.05$ and therefore $z_{1-\alpha} = 1.64$. The test statistic is calculated as

$$t(x) = \frac{\bar{x} - \mu_0}{\sigma_0} \sqrt{n} = \frac{1.97 - 2}{0.1} \sqrt{20} = -1.34.$$

The null hypothesis is not rejected, since $t(x) = -1.34 > -1.64 = -z_{1-0.05} = z_{0.05}$.

Interpretation: The sample average $\bar{x} = 1.97$ kg is below the target value of $\mu = 2$ kg. But there is not enough evidence to reject the hypothesis that the sample comes from a $N(2, 0.1^2)$-distributed population. The probability to observe a sample of size $n = 20$ with an average of at most 1.97 in a $N(2, 0.1^2)$-distributed population is greater than $\alpha = 0.05 = 5\%$. The difference between $\bar{x} = 1.97$ kg and the target value $\mu = 2$ kg is not statistically significant.

Remark 10.3.1 The Gauss test assumes the variance to be known, which is often not the case in practice. The t-test (Sect. 10.3.2) assumes that the variance needs to be estimated. The t-test is therefore commonly employed when testing hypotheses about the mean. Its usage is outlined below. In R, the command Gauss.test from the library compositions offers an implementation of the Gauss test.

10.3.2 Test for the Mean When the Variance is Unknown (One-Sample t-Test)

If the variance σ^2 is unknown, hypotheses about the mean μ of a normal random variable $X \sim N(\mu, \sigma^2)$ can be tested in a similar way to the one-sample Gauss-test. The difference is that the unknown variance is estimated from the sample. An unbiased estimator of σ^2 is the sample variance

$$S_X^2 = \frac{1}{n-1} \sum_{i=1}^{n} (X_i - \bar{X})^2 .$$

The test statistic is therefore

$$T(\mathbf{X}) = \frac{\bar{X} - \mu_0}{S_X} \sqrt{n} ,$$

which follows a t-distribution with $n-1$ degrees of freedom if H_0 is true, as we know from Theorem 8.3.2.

Critical Regions and Test Decisions

Since $T(\mathbf{X})$ follows a t-distribution under H_0, the critical regions refer to the regions of the t-distribution which are unlikely to be observed under H_0:

Case	H_0	H_1	Critical region K
(a)	$\mu = \mu_0$	$\mu \neq \mu_0$	$K = (-\infty, -t_{n-1;1-\alpha/2}) \cup (t_{n-1;1-\alpha/2}, \infty)$
(b)	$\mu \geq \mu_0$	$\mu < \mu_0$	$K = (-\infty, -t_{n-1;1-\alpha})$
(c)	$\mu \leq \mu_0$	$\mu > \mu_0$	$K = (t_{n-1;1-\alpha}, \infty)$

The hypothesis H_0 is rejected if the realized test statistic, i.e.

$$t(x) = \frac{\bar{x} - \mu_0}{s_X} \sqrt{n} ,$$

falls into the critical region. The critical regions are based on the appropriate quantiles of the t-distribution with $(n-1)$ degrees of freedom, as outlined in Table 10.2.

Table 10.2 Rules to make test decisions for the one-sample t-test (and the two-sample t-test, and the paired t-test, both explained below)

Case	H_0	H_1	Reject H_0, if
(a)	$\mu = \mu_0$	$\mu \neq \mu_0$	$|t(x)| > t_{n-1;1-\alpha/2}$
(b)	$\mu \geq \mu_0$	$\mu < \mu_0$	$t(x) < -t_{n-1;1-\alpha}$
(c)	$\mu \leq \mu_0$	$\mu > \mu_0$	$t(x) > t_{n-1;1-\alpha}$

Example 10.3.2 We again consider Example 10.3.1. Now we assume that the variance of the loaves is unknown. Suppose a random sample of size $n = 20$ has an arithmetic mean of $\bar{x} = 1.9668$ and a sample variance of $s^2 = 0.0927^2$. We want to test whether this result contradicts the two-sided hypothesis $H_0: \mu = 2$, that is case (a). The significance level is fixed at $\alpha = 0.05$. For the realized test statistic $t(x)$, we calculate

$$t(x) = \frac{\bar{x} - \mu_0}{s_X}\sqrt{n} = \frac{1.9668 - 2}{0.0927}\sqrt{20} = -1.60 .$$

H_0 is not rejected since $|t| = 1.60 < 2.09 = t_{19;0.975}$, where the quantiles ± 2.09 are defining the critical region (see Table C.2 or use R: qt(0.975,19)). The same results can be obtained in R using the t.test() function, see Example 10.3.3 for more details. Or, we can directly calculate the (two-sided) p-value as

```
2*(1-pt( abs(1.6), df=19))
```

This yields a p-value of 0.1260951 which is not smaller than α and therefore H_0 is not rejected.

10.3.3 Comparing the Means of Two Independent Samples

In a two-sample problem, we may be interested in comparing the means of two *independent* samples. Assume that we have two samples of two normally distributed variables $X \sim N(\mu_X, \sigma_X^2)$ and $Y \sim N(\mu_Y, \sigma_Y^2)$ of size n_1 and n_2, i.e. $X_1, X_2, \ldots, X_{n_1}$ are i.i.d. with the same distribution as X and $Y_1, Y_2, \ldots, Y_{n_2}$ are i.i.d. with the same distribution as Y. We can specify the following hypotheses:

Case	Null hypothesis	Alternative hypothesis	
(a)	$\mu_X = \mu_Y$	$\mu_X \neq \mu_Y$	two-sided test problem
(b)	$\mu_X \geq \mu_Y$	$\mu_X < \mu_Y$	one-sided test problem
(c)	$\mu_X \leq \mu_Y$	$\mu_X > \mu_Y$	one-sided test problem

We distinguish another three cases:

1. σ_X^2 and σ_Y^2 are known.
2. σ_X^2 and σ_Y^2 are unknown, but they are assumed to be equal, i.e. $\sigma_X^2 = \sigma_Y^2$.

3. Both σ_X^2 and σ_Y^2 are unknown and unequal ($\sigma_X^2 \neq \sigma_Y^2$).

Case 1: The Variances are Known (Two-Sample Gauss-Test)

If the null hypothesis H_0: $\mu_X = \mu_Y$ is true, then, using the usual rules for the normal distribution and the independence of the samples,

$$\bar{X} \sim N\left(\mu_X, \frac{\sigma_X^2}{n_1}\right),$$

$$\bar{Y} \sim N\left(\mu_Y, \frac{\sigma_Y^2}{n_2}\right),$$

and

$$(\bar{X} - \bar{Y}) \sim N\left(\mu_Y - \mu_Y, \frac{\sigma_X^2}{n_1} + \frac{\sigma_Y^2}{n_2}\right).$$

It follows that the test statistic

$$T(\mathbf{X}, \mathbf{Y}) = \frac{\bar{X} - \bar{Y}}{\sqrt{\frac{\sigma_X^2}{n_1} + \frac{\sigma_Y^2}{n_2}}} \tag{10.1}$$

follows a standard normal distribution, $T(\mathbf{X}, \mathbf{Y}) \sim N(0, 1)$. The realized test statistic is

$$t(x, y) = \frac{\bar{x} - \bar{y}}{\sqrt{\frac{\sigma_X^2}{n_1} + \frac{\sigma_Y^2}{n_2}}}. \tag{10.2}$$

The test procedure is identical to the procedure of the one-sample Gauss test introduced in Sect. 10.3.1, i.e. the test decision is based on Table 10.1.

Case 2: The Variances are Unknown, But Equal (Two-Sample t-Test)

We denote the unknown variance of both distributions as σ^2 (i.e. both the populations are assumed to have variance σ^2). We estimate σ^2 by using the pooled sample variance where each sample is assigned weights relative to the sample size:

$$S^2 = \frac{(n_1 - 1)S_X^2 + (n_2 - 1)S_Y^2}{n_1 + n_2 - 2}. \tag{10.3}$$

The test statistic

$$T(\mathbf{X}, \mathbf{Y}) = \frac{\bar{X} - \bar{Y}}{S}\sqrt{\frac{n_1 \cdot n_2}{n_1 + n_2}} \tag{10.4}$$

with S as in (10.3) follows a t-distribution with $n_1 + n_2 - 2$ degrees of freedom if H_0 is true. The realized test statistic is

$$t(x, y) = \frac{\bar{x} - \bar{y}}{s} \sqrt{\frac{n_1 \cdot n_2}{n_1 + n_2}}. \tag{10.5}$$

The test procedure is identical to the procedure of the one-sample t-test, i.e. the test decision is based on Table 10.2.

Case 3: The Variances are Unknown and Unequal (Welch-Test)

We test $H_0: \mu_X = \mu_Y$ versus $H_1: \mu_X \neq \mu_Y$ given $\sigma_X^2 \neq \sigma_Y^2$ and both σ_X^2 and σ_Y^2 are unknown. This problem is also known as the Behrens–Fisher problem and is the most frequently used test when comparing two means in practice. The test statistic can be written as

$$T(\mathbf{X}, \mathbf{Y}) = \frac{|\bar{X} - \bar{Y}|}{\sqrt{\frac{S_X^2}{n_1} + \frac{S_Y^2}{n_2}}}, \tag{10.6}$$

which is approximately t-distributed with v degrees of freedom:

$$v = \left(\frac{s_x^2}{n_1} + \frac{s_y^2}{n_2} \right)^2 \Bigg/ \left(\frac{(s_x^2/n_1)^2}{n_1 - 1} + \frac{(s_y^2/n_2)^2}{n_2 - 1} \right) \tag{10.7}$$

where s_x^2 and s_y^2 are the estimated values of $S_X^2 = \frac{1}{n-1} \sum_{i=1}^{n}(X_i - \bar{X})^2$ and $S_Y^2 = \frac{1}{n-1} \sum_{i=1}^{n}(Y_i - \bar{Y})^2$ respectively. The test procedure, using the observed test statistic

$$t(x, y) = \frac{|\bar{x} - \bar{y}|}{\sqrt{\frac{s_X^2}{n_1} + \frac{s_Y^2}{n_2}}}, \tag{10.8}$$

is identical to the procedure of the one-sample t-test, i.e. the test decision is based on Table 10.2 except that the degrees of freedom are not $n - 1$ but v. If v is not an integer, it can be rounded off to an integer value.

Example 10.3.3 A small bakery sells cookies in packages of 500 g. The cookies are hand made and the packaging is either done by the baker himself or his wife. Some customers conjecture that the wife is more generous than the baker. One customer does an experiment: he buys packages of cookies packed by the baker and his wife on 16 different days and weighs the packages. He gets the following two samples (one for the baker, one for his wife).

Weight (wife) (X)	512	530	498	540	521	528	505	523
Weight (baker) (Y)	499	500	510	495	515	503	490	511

We want to test whether the complaint of the customers is justified. Let us start with the following simple hypotheses:

$$H_0 : \mu_x = \mu_y \quad \text{versus} \quad H_1 : \mu_x \neq \mu_y \,,$$

i.e. we only want to test whether the weights are different, not that the wife is making heavier cookie packages. Since the variances are unknown, we assume that case 3 is the right choice. We calculate and obtain $\bar{x} = 519.625$, $\bar{y} = 502.875$, $s_X^2 = 192.268$ and $s_Y^2 = 73.554$. The test statistic is:

$$t(x, y) = \frac{|\bar{x} - \bar{y}|}{\sqrt{\frac{s_X^2}{n_1} + \frac{s_Y^2}{n_2}}} = \frac{|519.625 - 502.875|}{\sqrt{\frac{192.268}{8} + \frac{73.554}{8}}} \approx 2.91 \,.$$

The degrees of freedom are:

$$v = \left(\frac{192.268}{8} + \frac{73.554}{8} \right)^2 / \left(\frac{(192.268/8)^2}{7} + \frac{(73.554/8)^2}{7} \right) \approx 11.67 \approx 12 \,.$$

Since $|t(x)| = 2.91 > 2.18 = t_{12;0.975}$, it follows that H_0 is rejected. Therefore, H_1 is statistically significant. This means that the mean weight of the wife's packages is different from the mean weight of the baker's packages. Let's refine the hypothesis and try to find out whether the wife's packages have a higher mean weight. The hypotheses are now:

$$H_0 : \mu_x \leq \mu_y \quad \text{versus} \quad H_1 : \mu_x > \mu_y \,.$$

The test statistic remains the same but the critical region and the degrees of freedom change. Thus, H_0 is rejected if $t(x, y) > t_{v;1-\alpha}$. Using $t_{v;1-\alpha} = t_{12;0.95} \approx 1.78$ and $t(x, y) = 2.91$, it follows that the null hypothesis can be rejected. The mean weight of the wife's packages is greater than the mean weight of the baker's packages.

In R we would have obtained the same result using the t.test command:

```
x <- c(512,530,498,540,521,528,505,523)
y <- c(499,500,510,495,515,503,490,511)
t.test(x,y,alternative='greater')
```

```
        Welch Two Sample t-test

data:   x and y
t = 2.9058, df = 11.672, p-value = 0.006762
alternative hypothesis: true difference in means is greater
than 0...
```

Note that we have to specify the *alternative* hypothesis under the option `alternative`. The output shows us the test statistic (2.9058), the degrees of freedom (11.672), the alternative hypothesis—but not the decision rule. We know that H_0 is rejected if $t(x, y) > t_{v;1-\alpha}$, so the decision is easy in this case: we simply have to calculate $t_{12;0.95}$ using `qt(0.95,12)` in *R*. A simpler way to arrive at the same decision is to use the *p*-value. We know that H_0 is rejected if $p < \alpha$ which is the case in this example. It is also worthwhile mentioning that *R* displays the hypotheses slightly differently from ours: our alternative hypothesis is $\mu_x > \mu_y$ which is identical to the statement $\mu_x - \mu_y > 0$, as shown by *R*, see also Sect. 10.2.2.

If we specify `two.sided` as an alternative (which is the default), a confidence interval for the mean *difference* is also part of the output:

```
t.test(x,y,alternative='two.sided')
```

```
...
95 percent confidence interval:
 4.151321 29.348679
```

It can be seen that the confidence interval of the difference does not cover the "0". Therefore, the null hypothesis is rejected. This is the duality property referred to earlier in this chapter: the test decision is the same, no matter whether one evaluates (i) the confidence interval, (ii) the test statistic, or (iii) the *p*-value.

Any kind of *t*-test can be calculated with the `t.test` command: for example, the two-sample *t*-test requires to specify the option `var.equal=TRUE` while the Welch test is calculated when the (default) option `var.equal=FALSE` is set. We can also conduct a one-sample *t*-test. Suppose we are interested in whether the mean weight of the wife's packages of cookies is greater than 500 grams; then we could test the hypotheses:

$$H_0 : \mu_x \leq 500 \quad \text{versus} \quad H_1 : \mu_x > 500 .$$

In *R*, we simply have to specify μ_0:

```
t.test(x,mu=500,alternative='greater')
```

which gives us

```
        One Sample t-test

data:  x
t = 4.0031, df = 7, p-value = 0.002585
alternative hypothesis: true mean is greater than 500
...
```

10.3.4   Test for Comparing the Means of Two Dependent Samples (Paired t-Test)

Suppose there are two dependent continuous random variables X and Y with $E(X) = \mu_X$ and $E(Y) = \mu_Y$. They could be dependent because we measure the same variable twice on the same subjects at different times. Typically, this is the case in pre-post experiments, for example when we measure the weight of a person before starting a special diet and after finishing the diet; or when evaluating household expenditures on electronic appliances in two consecutive years. We then say that the samples are *paired*, or dependent. Since the same variable is measured twice on the same subject, it makes sense to calculate a difference between the two respective values. Let $D = X - Y$ denote the random variable "difference of X and Y". If $H_0\colon \mu_X = \mu_Y$ is true, then the expected difference is zero, and we get $E(D) = \mu_D = 0$. This means testing $H_0 : \mu_X = \mu_Y$ is identical to testing $\mu_X - \mu_Y = \mu_D = 0$. We further assume that D is normally distributed if $H_0\colon \mu_X = \mu_Y$ is true (or equivalently if $H_0\colon \mu_D = 0$ is true), i.e., $D \sim N(0, \sigma_D^2)$. For a random sample $(D_1, D_2, \ldots, D_n)$ of the differences, the test statistic

$$T(\mathbf{X}, \mathbf{Y}) = T(\mathbf{D}) = \frac{\bar{D}}{S_D}\sqrt{n} \qquad (10.9)$$

is t-distributed with $n - 1$ degrees of freedom. The sample mean is $\bar{D} = \sum_{i=1}^{n} D_i/n$ and the sample variance is

$$S_D^2 = \frac{\sum_{i=1}^{n}(D_i - \bar{D})^2}{n - 1}$$

which is an estimator of σ_D^2. The realized test statistic is thus

$$t(d) = \frac{\bar{d}}{s_d}\sqrt{n} \qquad (10.10)$$

where $\bar{d} = \sum_{i=1}^{n} d_i/n$ and $s_d^2 = \sum_{i=1}^{n}(d_i - \bar{d})^2/(n - 1)$.

The two-sided test $H_0\colon \mu_D = 0$ versus $H_1\colon \mu_D \neq 0$ and the one-sided tests $H_0\colon \mu_D \leq 0$ versus $H_1\colon \mu_D > 0$ or $H_0\colon \mu_D \geq 0$ versus $H_1\colon \mu_D < 0$ can be derived as in Sect. 10.3.2, i.e. the test decision is based on Table 10.2. In fact, the paired t-test is a one-sample t-test on the differences of X and Y.

Example 10.3.4 In an experiment, $n = 10$ students have to solve different tasks before and after drinking a cup of coffee. Let Y and X denote the random variables "number of points before/after drinking a cup of coffee". Assume that a higher number of points means that the student is performing better. Since the test is repeated on the same students, we have a paired sample. The data are given in the following table:

i	y_i (before)	x_i (after)	$d_i = x_i - y_i$	$(d_i - \bar{d})^2$
1	4	5	1	0
2	3	4	1	0
3	5	6	1	0
4	6	7	1	0
5	7	8	1	0
6	6	7	1	0
7	4	5	1	0
8	7	8	1	0
9	6	5	−1	4
10	2	5	3	4
Total			10	8

We calculate

$$\bar{d} = 1 \quad \text{and} \quad s_d^2 = \frac{8}{9} = 0.943^2,$$

respectively. For the realized test statistic $t(d)$, using $\alpha = 0.05$, we get

$$t(d) = \frac{1}{0.943}\sqrt{10} = 3.35 > t_{9;0.95} = 1.83,$$

such that H_0: $\mu_X \leq \mu_Y$ is rejected and H_1: $\mu_X > \mu_Y$ is accepted. We can conclude (for this example) that drinking coffee significantly increased the problem solving capacity of the students.

In R, we would have obtained the same results using the t.test function and specifying the option paired=TRUE:

```
yp <- c(4,3,5,6,7,6,4,7,6,2)
xp <- c(5,4,6,7,8,7,5,8,5,5)
t.test(xp,yp,paired=TRUE)
```

```
        Paired t-test

data:  xp and yp
t = 3.3541, df = 9, p-value = 0.008468
alternative hypothesis: true difference in means != 0
95 percent confidence interval:
 0.325555 1.674445
sample estimates:
mean of the differences
```

We can make the test decision using the R output in three different ways:

(i) We compare the test statistic ($t = 3.35$) with the critical value (1.83, obtained via qt(0.95,9)).
(ii) We evaluate whether the p-value (0.008468) is smaller than the significance level $\alpha = 0.05$.
(iii) We evaluate whether the confidence interval for the mean difference covers "0" or not.

10.4   Parametric Tests for Probabilities

10.4.1 One-Sample Binomial Test for the Probability p

Test Construction and Hypotheses

Let X be a Bernoulli $B(1; p)$ random variable with the two possible outcomes 1 and 0, which indicate occurrence and non-occurrence of an event of interest A. The probability for A in the population is p. From the sample $\mathbf{X} = (X_1, X_2, \ldots, X_n)$ of independent $B(1; p)$-distributed random variables, we calculate the mean (relative frequency) as $\hat{p} = \frac{1}{n} \sum_{i=1}^{n} X_i$ which is an unbiased estimate of p. The following hypotheses may thus be of interest:

Case	Null hypothesis	Alternative hypothesis	
(a)	$p = p_0$	$p \neq p_0$	two-sided problem
(b)	$p \geq p_0$	$p < p_0$	one-sided problem
(c)	$p \leq p_0$	$p > p_0$	one sided problem

In the following, we describe two possible solutions, one exact approach and an approximate solution. The approximate solution is based on the approximation of the Binomial distribution by the normal distribution, which is appropriate if n is sufficiently large and the condition $np(1 - p) \geq 9$ holds (i.e. p is neither to small nor too large). First, we present the approximate solution; then the exact one.

Test Statistic and Test Decisions

(a) **Approximate Binomial test**. We define the standardized test statistic as

$$T(\mathbf{X}) = \frac{\hat{p} - p_0}{\sqrt{p_0(1 - p_0)}} \sqrt{n}. \qquad (10.11)$$

It holds approximately that $T(\mathbf{X}) \sim N(0, 1)$, given that the conditions that (i) n is sufficiently large and (ii) $np(1 - p) \geq 9$ are satisfied. The test can then be conducted along the lines of the Gauss test in Sect. 10.3.1, i.e the test decision is based on Table 10.1.

Example 10.4.1 We return to Example 10.1.1. Let us assume that a representative sample of size $n = 2000$ has been drawn from the population of eligible voters, from which 700 (35%) have voted for the party of interest P. The research hypothesis (which has to be stated as H_1) is that more than 30% (i.e. $p_0 = 0.3$) of the eligible voters cast their votes for party P. The sample is in favour of H_1 because $\hat{p} = 35\%$, but to draw conclusions for the proportion of voters of party P in the population, we have to conduct a Binomial test. Since n is large and $np(1 - p) = 2000 \cdot 0.35 \cdot 0.65 = 455 \geq 9$, the assumptions for the use of the test statistic (10.11) are satisfied. We can write down the realized test statistic as

$$t(x) = \frac{\hat{p} - p_0}{\sqrt{p_0(1 - p_0)}} \sqrt{n} = \frac{0.35 - 0.3}{\sqrt{0.3(1 - 0.3)}} \sqrt{2000} = 4.8795 \,.$$

Using $\alpha = 0.05$, it follows that $T(X) = 4.8795 > z_{1-\alpha} = 1.64$ and thus the null hypothesis $H_0 : p \leq 0.3$ can be rejected. Therefore $H_1 : p > 0.3$ is statistically significant, i.e. the proportion of votes for party P is greater than 30%.

(b) The **exact Binomial test** can be constructed using the knowledge that under H_0, $Y = \sum_{i=1}^{n} X_i$ (i.e. the number of successes) follows a Binomial distribution. In fact, we can use Y directly as the test statistic:

$$T(\mathbf{X}) = Y \sim B(n, p_0) \,.$$

The observed test statistic is $t(x) = \sum_i x_i$. For the two-sided case (a), the two critical numbers c_l and c_r ($c_l < c_r$) which define the critical region, have to be found such that

$$P_{H_0}(Y \leq c_l) \leq \frac{\alpha}{2} \quad \text{and} \quad P_{H_0}(Y \geq c_r) \leq \frac{\alpha}{2} \,.$$

The null hypothesis is rejected if the test statistic, i.e. Y, is greater than or equal to c_r or less than or equal to c_l. For the one-sided case, a critical number c has to be found such that

$$P_{H_0}(Y \leq c) \leq \alpha$$

for hypotheses of type (b) and

$$P_{H_0}(Y \geq c) \leq \alpha$$

for hypotheses of type (c). If Y is less than the critical value c (for case (b)) or greater than the critical value (for case (c)) the null hypothesis is rejected.

Example 10.4.2 We consider again Example 10.1.1 where we looked at the population of eligible voters, from which 700 (35%) have voted for the party of interest P. The observed test statistic is $t(x) = \sum_i x_i = 700$ and the alternative hypothesis is $H_1 : p > 0.3$, as in case (c). There are at least two ways in which we can obtain the results:

(i) *Long way:* We can calculate the test statistic and compare it to the critical region. To get the critical region we search c such that

$$P_{p=0.3}(Y \geq c) \leq 0.05 ,$$

which equates to

$$P_{p=0.3}(Y < c) \geq 0.95$$

and can be calculated in R as:

```
qbinom(p=0.95, prob=0.3, size=2000)
[1]  634
```

Since $Y = 700 > c = 634$ we reject the null hypothesis. As in Example 10.4.1, we conclude that there is enough evidence that the proportion of votes for party P is greater than 30%.

(ii) *Short way:* The above result can be easily obtained in R using the binom.test() command. We need to specify the number of "successes" (here: 700), the number of "failures" ($2000 - 700 = 1300$), and the alternative hypothesis:

```
binom.test(c(700,1300),p=0.3,alternative='greater')
```

```
data:   c(700, 1300)
number of successes = 700, number of trials = 2000,
p-value = 8.395e-07
alternative hypothesis: true probability of success
is greater than 0.3
95 percent confidence interval:
 0.332378 1.000000
probability of success
              0.35
```

Both the p-value (which is smaller than $\alpha = 0.05$) and the confidence interval (for which we do not show the calculation) confirm the rejection of the null hypothesis.
Note that

```
binom.test(x=700,n=2000,p=0.3, alternative='greater')
```

returns the same result.

10.4.2 Two-Sample Binomial Test

Test Construction and Hypotheses

We consider now the case of two independent i.i.d. samples from Bernoulli distributions with parameters p_1 and p_2.

$$\mathbf{X} = (X_1, X_2, \ldots, X_{n_1}), \quad X_i \sim B(1; p_1)$$
$$\mathbf{Y} = (Y_1, Y_2, \ldots, Y_{n_2}), \quad Y_i \sim B(1; p_2).$$

The sums

$$X = \sum_{i=1}^{n_1} X_i \sim B(n_1; p_1), \quad Y = \sum_{i=1}^{n_2} Y_i \sim B(n_2; p_2).$$

follow Binomial distributions. One of the following hypotheses may be of interest:

Case	Null hypothesis	Alternative hypothesis	
(a)	$p_1 = p_2$	$p_1 \neq p_2$	two-sided problem
(b)	$p_1 \geq p_2$	$p_1 < p_2$	one-sided problem
(c)	$p_1 \leq p_2$	$p_1 > p_2$	one-sided problem

Similar to the one-sample case, both exact and approximate tests exist. Here we only present the approximate test. The **Exact Test of Fisher** is presented in Appendix C.5. Let n_1 and n_2 denote the sample sizes. Then X/n_1 and Y/n_2 are approximately normally distributed:

$$\frac{X}{n_1} \overset{approx.}{\sim} N\left(p_1, \frac{p_1(1 - p_1)}{n_1}\right),$$
$$\frac{Y}{n_2} \overset{approx.}{\sim} N\left(p_2, \frac{p_2(1 - p_2)}{n_2}\right).$$

Their difference D,

$$D \overset{approx.}{\sim} N\left(0, p(1 - p)\left(\frac{1}{n_1} + \frac{1}{n_2}\right)\right)$$

is normally distributed too under H_0 (given $p = p_1 = p_2$ holds). Since the probabilities p_1 and p_2 are identical under H_0, we can pool the two samples and estimate p by

$$\hat{p} = \frac{X + Y}{n_1 + n_2}. \tag{10.12}$$

Test Statistic and Test Decision

The test statistic

$$T(\mathbf{X}, \mathbf{Y}) = \frac{D}{\sqrt{\hat{p}(1 - \hat{p}) \left(\frac{1}{n_1} + \frac{1}{n_2} \right)}}, \tag{10.13}$$

follows a $N(0, 1)$-distribution if n_1 and n_2 are sufficiently large and p is not near the boundaries 0 and 1 (one could use, for example, again the condition $np(1 - p) > 9$ with $n = n_1 + n_2$). The realized test statistic can be calculated using the observed difference $\hat{d} = \hat{p}_1 - \hat{p}_2$. The test can be conducted for the one-sided and the two-sided case as the Gauss test introduced in Sect. 10.3.1, i.e. the decision rules from Table 10.1 can be applied.

Example 10.4.3 Two competing lotteries claim that every fourth lottery ticket wins. Suppose we want to test whether the probabilities of winning are different for the two lotteries, i.e. $H_0 : p_1 = p_2$ and $H_1 : p_1 \neq p_2$. We have the following data

	n	Winning	Not winning
Lottery A	63	14	49
Lottery B	45	13	32

We can estimate the probabilities of a winning ticket for each lottery, as well as the respective difference, as

$$\hat{p}_A = \frac{14}{63}, \quad \hat{p}_B = \frac{13}{45}, \quad \hat{d} = \hat{p}_A - \hat{p}_B = -\frac{1}{15}.$$

Under H_0, an estimate for p following (10.12) is

$$\hat{p} = \frac{14 + 13}{63 + 45} = \frac{27}{108} = 0.25.$$

The test statistic can be calculated as

$$t(x, y) = \frac{-\frac{1}{15}}{\sqrt{0.25(1 - 0.25) \left(\frac{1}{63} + \frac{1}{45} \right)}} = -0.79.$$

H_0 is not rejected since $|t(x, y)| = 0.79 < 1.96 = z_{1-0.05/2}$. Thus, there is no statistical evidence for different winning probabilities for the two lotteries. These hypotheses can be tested in R using the Test of Fisher, see Appendix C.5, for more details.

10.5 Tests for Scale Parameters

There are various tests available to test hypotheses about scale parameters. Such tests are useful when one is interested in the dispersion of a variable, for example in quality control where the variability of a process may be of interest. One-sample tests of hypotheses for the variance of a normal distribution, e.g. hypotheses such as $H_0 : \sigma^2 = \sigma_0^2$, can be tested by the χ^2-test for the variance, see Appendix C.5. Two-sample problems can be addressed by the F-test (which is explained in Appendix C.5); or by other tests such as the Levene test or Bartlett's test, which are also available in R (`leveneTest` in the package `car`, `bartlett` in the base distribution of R).

10.6 Wilcoxon–Mann–Whitney (WMW) U-Test

Test Construction and Hypotheses

The WMW U-test is often proposed as an alternative to the t-test because it also focuses on location but not on the expected value μ. It is a *nonparametric* test and useful in situations where skewed distributions are compared with each other. We consider two independent random samples $\mathbf{X} = (X_1, X_2, \ldots, X_{n_1})$ and $\mathbf{Y} = (Y_1, Y_2, , \ldots, Y_{n_2})$ from two populations with observed values $(x_1, x_2, \ldots, x_{n_1})$ and $(y_1, y_2, \ldots, y_{n_2})$ respectively. In this case, the null hypothesis H_0 considering the location can be formulated as

$$H_0 : P(X > Y) = P(Y > X) = \frac{1}{2} \, .$$

The null hypothesis can be interpreted in the following way: The probability that a randomly drawn observation from the first population has a value x that is greater (or lower) than the value y of a randomly drawn subject from the second population, is $\frac{1}{2}$. The alternative hypothesis H_1 is then

$$H_1 : P(X > Y) \neq P(Y > X) \, .$$

This means we are comparing the entire distribution of two variables. If there is a location shift in the sense that one distribution is shifted left (or right) compared with the other distribution, the null hypothesis will be rejected because this shift can be seen as part of the alternative hypothesis $P(X > Y) \neq P(Y > X)$. In fact, under some assumptions, the hypothesis can even be interpreted as comparing two medians, and this is what is often done in practice.

Observed Test Statistic

To construct the test statistic, it is necessary to merge $(x_1, x_2, \ldots, x_{n_1})$ and $(y_1, y_2, \ldots, y_{n_2})$ into one sorted sample, usually in ascending order, while keeping the information which value belongs to which sample. For now we assume that all values of the two samples are distinct, i.e. no ties are present. Then each observation has a rank between 1 and $(n_1 + n_2)$. Let R_{1+} be the sum of ranks of the x-sample and let R_{2+} be the sum of ranks of the y-sample. The test statistic is defined as U, where U is the minimum of the two values $U_1, U_2, U = \min(U_1, U_2)$ with

$$U_1 = n_1 \cdot n_2 + \frac{n_1(n_1 + 1)}{2} - R_{1+}, \tag{10.14}$$

$$U_2 = n_1 \cdot n_2 + \frac{n_2(n_2 + 1)}{2} - R_{2+}. \tag{10.15}$$

Test Decision

H_0 is rejected if $U < u_{n_1,n_2;\alpha}$. Here $u_{n_1,n_2;\alpha}$ is the critical value derived from the distribution of U under the null hypothesis. The exact (complex) distribution can, for example, be derived computationally (in R). We are presenting an approximate solution together with its implementation in R.

Since $U_1 + U_2 = n_1 \cdot n_2$, it is sufficient to compute only R_{i+} and $U = \min\{U_i, n_1 n_2 - U_i\}$ ($i = 1$ or $i = 2$ are chosen such that R_{i+} is calculated for the sample with the lower sample size). For $n_1, n_2 \geq 8$, one can use the approximation

$$T(\mathbf{X}, \mathbf{Y}) = \frac{U - \frac{n_1 \cdot n_2}{2}}{\sqrt{\frac{n_1 \cdot n_2 \cdot (n_1 + n_2 + 1)}{12}}} \overset{approx.}{\sim} N(0, 1) \tag{10.16}$$

as the test statistic. For two-sided hypotheses, H_0 is rejected if $|t(x, y)| > z_{1-\alpha/2}$; for one-sided hypotheses H_0 is rejected if $|t(x, y)| > z_{1-\alpha}$. In the case of ties, the denominator of the test statistic in (10.16) can be modified as

$$T(\mathbf{X}, \mathbf{Y}) = \frac{U - \frac{n_1 \cdot n_2}{2}}{\sqrt{\left[\frac{n_1 \cdot n_2}{n(n-1)}\right]\left[\frac{n^3 - n}{12} - \sum_{j=1}^{G} \frac{t_j^3 - t_j}{12}\right]}} \overset{approx.}{\sim} N(0, 1),$$

where G is the number of different (groups of) ties and t_j denotes the number of tied ranks in tie group j.

Example 10.6.1 In a study, the reaction times (in seconds) to a stimulus were measured for two groups. One group drank a strong coffee before the stimulus and the other group drank only the same amount of water. There were 9 study participants in the coffee group and 10 participants in the water group. The following reaction times were recorded:

Reaction time	1	2	3	4	5	6	7	8	9	10
Coffee group (C)	3.7	4.9	5.2	6.3	7.4	4.4	5.3	1.7	2.9	
Water group (W)	4.5	5.1	6.2	7.3	8.7	4.2	3.3	8.9	2.6	4.8

We test with the U-test whether there is a location difference between the two groups. First, the ranks of the combined sample are calculated as:

	1	2	3	4	5	6	7	8	9	10	Total
Value (C)	3.7	4.9	5.2	6.3	7.4	4.4	5.3	1.7	2.9		
Rank (C)	5	10	12	15	17	7	13	1	3		83
Value (W)	4.5	5.1	6.2	7.3	8.7	4.2	3.3	8.9	2.6	4.8	
Rank (W)	8	11	14	16	18	6	4	19	2	9	107

With $R_{C+} = 83$ and $R_{W+} = 107$, we get

$$U_1 = n_1 \cdot n_2 + \frac{n_1(n_1 + 1)}{2} - R_{C+} = 9 \cdot 10 + \frac{9 \cdot 10}{2} - 83 = 52 \,,$$

$$U_2 = n_1 \cdot n_2 + \frac{n_2(n_2 + 1)}{2} - R_{W+} = 9 \cdot 10 + \frac{10 \cdot 11}{2} - 107 = 38 \,.$$

With $n_1, n_2 \geq 8$ and $U = U_2 = 38$,

$$t(x, y) = \frac{U - \frac{n_1 \cdot n_2}{2}}{\sqrt{\dfrac{n_1 \cdot n_2 \cdot (n_1 + n_2 + 1)}{12}}} = \frac{38 - \frac{9 \cdot 10}{2}}{\sqrt{\dfrac{9 \cdot 10 \cdot (9 + 10 + 1)}{12}}} \approx -0.572 \,.$$

Since $|t(x, y)| = 0.572 < z_{1-\alpha/2} = 1.96$ the null hypothesis cannot be rejected, i.e. there is no statistical evidence that the two groups have different reaction times.

In R, one can use the `wilcox.test` command to obtain the results:

```
coffee <- c(3.7, 4.9, 5.2, 6.3, ..., 2.9)
water <- c(4.5, 5.1, 6.2, ..., 4.8)
wilcox.test(coffee, water)
```

The output is

```
    Wilcoxon rank sum test

data:   coffee.sample and water.sample
W = 38, p-value = 0.6038
alternative hypothesis: true location shift is not equal to 0
```

We can see that the null hypothesis is not rejected because $p = 0.6038 > \alpha = 0.05$. The displayed test statistic is W which equates to our statistic U_2. The alternative hypothesis in R is framed as location shift, an interpretation which has already been given earlier in the chapter. Note that the test also suggests that the medians of the two samples are not statistically different.

10.7 χ^2-Goodness of Fit Test

Test Construction

The χ^2-goodness of fit test is one of the most popular tests for testing the goodness of fit of the observed data to a distribution. The construction principle is very general and can be used for variables of any scale. The test statistic is derived such that the *observed* absolute frequencies are compared with the *expected* absolute frequencies *under the null hypothesis H_0*.

Example 10.7.1 Consider an experiment where a die is rolled $n = 60$ times. Under the null hypothesis H_0, we assume that the die is fair, i.e. $p_i = \frac{1}{6}, i = 1, 2, \ldots, 6$, where $p_i = P(X = i)$. We could have also said that H_0 is the hypothesis that the rolls are following a discrete uniform distribution. Thus the expected absolute frequencies under H_0 are $np_i = 60 \cdot \frac{1}{6} = 10$, while the observed frequencies in the sample are $N_i, i = 1, 2, \ldots, 6$. The N_i generally deviate from np_i. The χ^2-statistic is based on the squared differences, $\sum_{i=1}^{6}(N_i - np_i)^2$, and becomes large as the differences between the observed and the expected frequencies become larger. The χ^2-test statistic is a modification of this sum by scaling each squared difference by the expected frequencies, np_i, and is explained below.

With a nominal variable, we can proceed as in Example 10.7.1. If the scale of the variable is ordinal or metric, the number of different values can be large. Note that in the most extreme case, we can have as many different values as observations (n), leading to $N_i = 1$ for all $i = 1, 2, \ldots, n$. Then it is necessary to group the data into k intervals before applying the χ^2-test. The reason is that the general theory of the χ^2-test assumes that the number k (which was 6 in Example 10.7.1 above) is fixed and does not grow with the number of observations n, i.e., the theory says that the χ^2-test only works properly if k is fixed and n is large. For this reason, we group the sample $\mathbf{X} = (X_1, X_2, \ldots, X_n)$ into k classes as shown in Sect. 2.1.

Class	1	2	$\cdots$	k	Total
Number of observations	n_1	n_2	$\cdots$	n_k	n

The choice of the class intervals is somewhat arbitrary. As a rule of thumb $np_i > 5$ should hold for most class intervals. The general hypotheses can be formulated in the form of distribution functions:

$$H_0\colon F(x) = F_0(x) \text{ versus } H_1\colon F(x) \neq F_0(x).$$

Test Statistic

The test statistic is defined as

$$T(\mathbf{X}) = t(x) = \chi^2 = \sum_{i=1}^{k} \frac{(N_i - np_i)^2}{np_i}. \tag{10.17}$$

Here,

- N_i ($i = 1, 2, \ldots, k$) are the absolute frequencies of observations of the sample **X** in class i, N_i is a random variable with realization n_i in the observed sample;
- p_i ($i = 1, 2, \ldots, k$) are calculated from the distribution under H_0, $F_0(x)$, and are the (hypothetical) probabilities that an observation of X falls in class i;
- np_i are the expected absolute frequencies in class i under H_0.

Test Decision

For a significance level α, H_0 is rejected if $t(x)$ is greater than the $(1 - \alpha)$ quantile of the χ^2-distribution with $k - 1 - r$ degrees of freedom, i.e. if

$$t(x) = \chi^2 > c_{k-1-r,1-\alpha} .$$

Note that r is the number of parameters of $F_0(x)$, if these parameters are estimated from the sample. The χ^2-test statistic is only asymptotically χ^2-distributed under H_0.

Example 10.7.2 Let $F_0(x)$ be the distribution function of the test distribution. If one specifies a normal distribution such as $F_0(x) = N(3, 10)$, or a discrete uniform distribution with $p_i = 0.25$ ($i = 1, 2, 3, 4$), then $r = 0$, since no parameters have to be estimated from the data. Otherwise, if we simply want to test whether the data is generated from a normal distribution $N(\mu, \sigma^2)$ or the data follows a normal distribution $N(\mu, \sigma^2)$, then μ and σ^2 may be estimated from the sample by $\bar{x}$ and s^2. Then $r = 2$ and the number of degrees of freedom is reduced.

Example 10.7.3 Gregor Mendel (1822–1884) conducted crossing experiments with pea plants of different shape and colour. Let us look at the outcome of a pea crossing experiment with the following results:

Crossing result	Round Yellow	Round Green	Edged Yellow	Edged Green
Observations	315	108	101	32

Mendel had the hypothesis that the four different types occur in proportions of 9:3:3:1, that is

$$p_1 = \frac{9}{16}, \ p_2 = \frac{3}{16}, \ p_3 = \frac{3}{16}, \ p_4 = \frac{1}{16} .$$

The hypotheses are

$$H_0 : P(X = i) = p_i \quad \text{versus} \quad H_1 : P(X = i) \neq p_i, \quad i = 1, 2, 3, 4 .$$

With $n = 556$ observations, the test statistic can be calculated from the following observed and expected frequencies:

i	N_i	p_i	np_i
1	315	$\frac{9}{16}$	312.75
2	108	$\frac{3}{16}$	104.25
3	101	$\frac{3}{16}$	104.25
4	32	$\frac{1}{16}$	34.75

The χ^2-test statistic is calculated as

$$t(x) = \chi^2 = \frac{(315 - 312.75)^2}{312.75} + \dots + \frac{(32 - 34.75)^2}{34.75} = 0.47 .$$

Since $\chi^2 = 0.47 < 7.815 = \chi^2_{0.95,3} = c_{0.95,3}$, the null hypothesis is not rejected. Statistically, there is no evidence that Mendel was wrong with his 9:3:3:1 assumption. In R, the test can be conducted by applying the chisq.test command:

```
chisq.test(c(315, 108, 101, 32),
p=c(9/16,3/16,3/16,1/16))
qchisq(df=3, p=0.95)
```

which leads to the following output

```
    Chi-squared test for given probabilities

data:   c(315, 108, 101, 32)
X-squared = 0.47, df = 3, p-value = 0.9254
```

and the critical value is

```
[1] 7.814728
```

Remark 10.7.1 In this example, the data was already summarized in a frequency table. For raw data, the table command can be used to preprocess the data, i.e we can use chisq.test(table(var1,var2)).

Another popular goodness-of-fit test is the test of Kolmogorov–Smirnov. There are two different versions of this test, one for the one-sample scenario and one for the two-sample scenario. The null hypothesis for the latter is that the two independent samples come from the same distribution. In R, the command ks.test() can be used to perform Kolmogorov–Smirnov tests.

10.8 χ^2-Independence Test and Other χ^2-Tests

In Chap. 4, we introduced different methods to describe the association between two variables. Several association measures are possibly suitable if the variables are categorical, for example Cramer's V, Goodman's and Kruskal's γ, Spearman's rank correlation coefficient, the odds ratio, and others. If we are not interested in the strength of association but rather in finding out whether there is an association at all, one can use the χ^2-independence test.

Test Construction

In the following we assume that we observe a sample from a bivariate discrete distribution of two variables X and Y which can be summarized in a contingency table with absolute frequencies n_{ij}, $(i = 1, 2, \ldots, I; j = 1, 2 \ldots, J)$:

		Y				
		1	2	$\cdots$	J	
	1	n_{11}	n_{12}	$\cdots$	n_{1J}	n_{1+}
	2	n_{21}	n_{22}	$\cdots$	n_{2J}	n_{2+}
X	$\vdots$	$\vdots$	$\vdots$		$\vdots$	$\vdots$
	I	n_{I1}	n_{I2}	$\cdots$	n_{IJ}	n_{I+}
		n_{+1}	n_{+2}	$\cdots$	n_{+J}	n

Remember that

n_{i+}	is the ith row sum,
n_{+j}	is the jth column sum, and
n	is the total number of observations.

The hypotheses are H_0: X and Y are independent versus H_1 : X and Y are not independent. If X and Y are independent, then the expected frequencies m_{ij} are

$$\hat{m}_{ij} = n\hat{\pi}_{ij} = \frac{n_{i+}n_{+j}}{n} . \tag{10.18}$$

Test Statistic

Pearson's χ^2-test statistic was introduced in Chap. 4, Eq. (4.6). It is

$$T(\mathbf{X}, \mathbf{Y}) = \chi^2 = \sum_{i=1}^{I} \sum_{j=1}^{J} \frac{(n_{ij} - m_{ij})^2}{m_{ij}} ,$$

where $m_{ij} = n\pi_{ij} = n\pi_{i+}\pi_{+j}$ (expected absolute cell frequencies under H_0). Strictly speaking, m_{ij} are the true, unknown expected frequencies under H_0 and are estimated by $\hat{m}_{ij} = n\pi_{i+}\pi_{+j}$, such that the realized test statistic equates to

$$t(x, y) = \chi^2 = \sum_{i=1}^{I} \sum_{j=1}^{J} \frac{(n_{ij} - \hat{m}_{ij})^2}{\hat{m}_{ij}} . \tag{10.19}$$

Test Decision

The number of degrees of freedom under H_0 is $(I-1)(J-1)$, where $I-1$ are the parameters which have to be estimated for the marginal distribution of X, and $J-1$ are the number of parameters for the marginal distribution of Y. The test decision is:

$$\text{Reject } H_0, \text{ if } t(x,y) = \chi^2 > c_{(I-1)(J-1);1-\alpha}.$$

Note that the alternative hypothesis H_1 is very general. If H_0 is rejected, nothing can be said about the structure of the dependence of X and Y from the χ^2-value itself.

Example 10.8.1 Consider the following contingency table. Here X describes the educational level (1: primary, 2: secondary, 3: tertiary) and Y the preference for a specific political party (1: Party A, 2: Party B, 3: Party C). Our null hypothesis is that the two variables are independent, and we want to show the alternative hypothesis which says that there is a relationship between them.

		Y			
		1	2	3	Total
X	1	100	200	300	600
	2	100	100	100	300
	3	80	10	10	100
	Total	280	310	410	1000

For the (estimated) expected frequencies $\hat{m}_{ij} = \frac{n_{i+}n_{+j}}{n}$, we get

		Y		
		1	2	3
X	1	168	186	246
	2	84	93	123
	3	28	31	41

For example: $\hat{m}_{11} = 600 \cdot 280/1000 = 168$. The test statistic is

$$t(x,y) = \sum_{i=1}^{I} \sum_{j=1}^{J} \frac{(n_{ij} - \hat{m}_{ij})^2}{\hat{m}_{ij}}$$

$$= \frac{(100 - 168)^2}{168} + \dots + \frac{(10 - 41)^2}{41} \approx 182.54.$$

Since $\chi^2_{4;0.95} = 9.49 < t(x,y) = 182.54$, H_0 is rejected.

In R, either the summarized data (as shown below) can be used to calculate the test statistic or the raw data (summarized in a contingency table via table(var1,var2)):

```
ct <- matrix(nrow=3,ncol=3,byrow=T,
data=c(100,200,300,100,100,100,80,10,10))
chisq.test(ct)
qchisq(df=(3-1)*(3-1), p=0.95)
```

The output is

```
    Pearson's Chi-squared test

data:  contingency.table
X-squared = 182.5428, df = 4, p-value < 2.2e-16
```

with the critical value

```
[1] 9.487729
```

which confirms our earlier manual calculations. The p-value is smaller than $\alpha = 0.05$ which further confirms that the null hypothesis has to be rejected.

For a binary outcome, the χ^2-test of independence can be formulated as a test for the null hypothesis that the proportions of the binary variable are equal in several (≥ 2) groups, i.e. for a $K \times 2$ (or $2 \times K$) table. This test is called the χ^2-**test of homogeneity**.

Example 10.8.2 Consider two variables X and Y, where X is describing the rating of a coffee brand with the categories "bad taste" and "good taste" and Y denotes three age subgroups, e.g. "18–25", "25–35" and "35–45". The observed data are

		Y			
		18–25	25–35	35–45	Total
	Bad	10	30	65	105
X	Good	90	70	35	195
	Total	100	100	100	300

Assume H_0 is the hypothesis that the probabilities $P(X = \text{'good'}|Y = \text{'18–25'})$, $P(X = \text{'good'}|Y = \text{'25–35'})$, and $P(X = \text{'good'}|Y = \text{'35–45'})$ are all equal. Then we can use either the function prop.test or chisq.test in R to test this hypothesis:

```
prop.test(x=cbind(c(10,30,65), c(90,70,35) ))
chisq.test(x=cbind(c(10,30,65), c(90,70,35) ))
```

This produces the following outputs:

```
3-sample test for equality of proportions

data:   cbind(c(10, 30, 65), c(90, 70, 35))
X-squared = 68.1319, df = 2, p-value = 1.605e-15
alternative hypothesis: two.sided
sample estimates:
prop 1 prop 2 prop 3
  0.10   0.30   0.65
```

and

```
Pearson's Chi-squared test

data:   cbind(c(10, 30, 65), c(90, 70, 35))
X-squared = 68.1319, df = 2, p-value = 1.605e-15
```

The results (test statistic, p-value) are identical and H_0 is rejected. Note that prop.test strictly expects a $K \times 2$ table (i.e. exactly 2 columns).

Remark 10.8.1 For 2×2- tables with small sample sizes and therefore small cell frequencies, it is recommended to use the Exact Test of Fisher as described in Appendix C.5.

Remark 10.8.2 The test described in Example 10.8.2 is a special case (since one variable is binary) of the general χ^2-test of homogeneity. The χ^2-test of homogeneity is valid for any $K \times C$ table, where K is the number of subgroups of a variable Y and C is the number of values of the outcome X of interest. The null hypothesis H_0 assumes that the conditional distributions of X given Y are identical in all subgroups, i.e.

$$P(X = x_c | Y = y_k) = P(X = x_c | Y = y_{k'})$$

$\forall c = 1, 2, \ldots, C$ for all $k, k' = 1, 2, \ldots, K, k \neq k'$. Again, the usual χ^2-test statistic can be used.

10.9   Beyond Dichotomies

So far, we have emphasized how to arrive at binary decisions regarding hypotheses using statistical tests. This approach may not be ideal in some situations since it may falsely reject a hypothesis because the assumptions used in computing the test statistic may not be satisfied. This can happen, for example, if the observations are not i.i.d. or if a parametric distributional assumption about a random variable of interest is not met. Similarly, a null hypothesis may be incorrect but still accepted because of a small sample size. In either case, we have only focused on a single null hypothesis whereas there may be many scientifically interesting hypotheses and for each of those there may be more or less support. A nuanced look at different hypotheses could thus provide a basis to evaluate gradations of evidence, rather than dichotomous decisions. Before explaining these issues in more detail, we first introduce the concept of compatibility.

10.9.1 Compatibility

We defined the p-value as the probability that the test statistic is equal to, or more extreme than its observed value if both the null hypothesis and the assumptions used in its calculation were true. This p-value is a random variable following a uniform distribution under H_0. Now consider the *observed* p-value: it is a (continuous) measure of compatibility between the observed data and a target hypothesis (e.g., the null hypothesis)—given a set of assumptions in the background which are used along with the hypothesis to compute the p-value from the data. We can similarly interpret the confidence interval in terms of a compatibility interval (CoI): a 95% CoI summarizes the results of varying the test hypothesis over those parameter values for which $p > 0.05$. Thus, the interval contains the set of parameter values that are more compatible with the data than those values lying outside the interval (given the background assumptions).

The advantage of speaking of "compatibility" rather than "confidence" is that it may avoid incorrect interpretations such as "the results show that one can be 95% *confident* that the true parameter value is contained in the calculated interval". Moreover, the interpretation of a compatibility interval makes it clear that possibly more than one test hypothesis may be of interest. This consideration highlights the set of parameter values that are most compatible with the data rather than drawing immediate attention to a dichotomization at the 5% level into significant and non-significant zones. Further advantages will become clear in the following subsections; however, we first introduce an example.

Example 10.9.1 Suppose we are interested in knowing the effect of a new training program consisting of a set of physical exercises conducted daily over a period of 1 month on weight. The table below shows the recorded weight loss during the 1-month period for both a control group (who did not participate in the new training program; group 1) and a group of participants who followed the new program (group 2).

	Weight Difference (in kg)									
	1	2	3	4	5	6	7	8	9	10
Group 1	−1.0	0.7	−0.3	1.1	−0.1	−1.6	−1.2	−0.3	2	−1.7
Group 2	−1.4	−4.6	−3.4	−0.9	1.2	−3.6	−1.2	0.3	−3.2	0.6

We can use R to conduct a 2-sample t-test (Welch-test) to compare the reduction in the mean weight between groups 1 and 2, i.e. test $H_0 : \mu_{\text{diff},1} = \mu_{\text{diff},2}$:

```
weight_diff <- c(-1.0,0.7,...,-3.2,0.6)
program <- c(rep(1,10),rep(2,10))
t.test(weight_diff[program==1],weight_diff[program==2])
```

```
t = 1.8706, df = 14.847, p-value = 0.08124
alternative hypothesis: true difference in means
is not equal to 0
95 percent confidence interval:
 -0.193814  2.953814
sample estimates:
mean of x mean of y
    -0.24      -1.62
```

The results show a mean weight loss of of 0.24 kg in the control group and a mean weight loss of 1.62 kg in the group following the new training program. This equates to a mean difference in weight loss of $-0.24 - (-1.62) = 1.38$ kg. Making a binary decision using the t-test's result ($p = 0.08124$) and $\alpha = 0.05$ leads to not rejecting the null hypothesis of equal mean weight losses in both the groups. Saying that "the new training program therefore does not lead to a weight reduction" is incorrect: firstly as we can never "prove" a null hypothesis. Secondly, the sample size is small and with a larger sample size the CoI may be shorter and the p-value smaller. Thirdly, interpreting the confidence interval as a compatibility interval shows that any hypothesized value between a difference in weight loss of about -0.19 kg and a difference in weight loss of about 2.95 kg is more compatible with the data than hypothesized values outside this interval. Thus, there is considerable support to indicate some weight loss with the new training program.

10.9.2 The S-Value

In Sect. 8.1.4 we introduced the Binomial distribution which was defined as follows:

$$P(X = k) = \binom{n}{k} p^k (1 - p)^{n-k} \qquad (k = 0, 1, \ldots, n).$$

We considered an example of a coin tossing experiment in which we used a random variable X to denote "number of heads" to calculate the probability of obtaining k heads during n trials. Now suppose a coin is tossed 5 times and every toss results in a head, i.e. 5 heads are obtained. This is surprising for a fair coin (i.e., for $p = 0.5$) and we calculate the probability of this outcome as

$$P(X = 5) = \binom{5}{5} 0.5^5 (1 - 0.5)^{5-5} = 0.5^5 = 3.125\%.$$

In general, the probability of receiving k heads in k tosses is 0.5^k. Is it not surprising to observe 3 heads in a row? A bit maybe, but the probability of experiencing this outcome is still 12.5%. Observing 7 heads in a row seems to be unusual though (with a probability of $0.5^7 = 0.78125\%$); and 15 heads almost impossible ($\approx 0.003\%$). This outcome of the experiment is in obvious contrast to our hypothesis that $H_0 : p = 0.5$. Applying an exact binomial test (with `binom.test(c(15,0),p=0.5)`) yields a p-value of 0.00006104, which suggests rejection of the null hypothesis. Because the number of "successes" is used as the test statistic in the exact binomial test (see Sect. 10.4.1), and the statistic follows a Binomial distribution, the two-sided p-value of 0.00006104 can also be obtained by evaluating the CDF of the Binomial distribution through `2*pbinom(0,15,prob=0.5)`; and the one-sided p-value as `pbinom(0,15,prob=0.5)`, which yields approximately 0.003% ($=0.5^{15}$!). This example illustrates an intuitive relationship between the p-value and coin tossing experiments: the probability of observing k heads in k tosses is identical to the p-value of an exact binomial test for the one-sided hypothesis that the probability of heads is no greater than 0.5, that is 0.5^k. Taking this thought a step further, we can find what number k of heads in a coin tossing experiment is needed to come closest to p: we get the answer by solving $p = \frac{1}{2}^k$, which leads to $k = -\log_2(p)$. Renaming k as S gives us the definition of the binary Shannon-information, surprisal or simply S-**value** as

$$S = -\log_2(p). \tag{10.20}$$

The S-value rescales the p-value and tells us how much information can be used as evidence against the null hypothesis (for a given test with a given set of assumptions in the background). For example, a p-value of 0.05 results in an S-value of 4.3 bits of information against the null hypothesis, which is no more surprising than getting only heads in 4 fair coin tosses. We use the unit "bit" (binary digit) because a coin toss produces a binary outcome that can be coded in only two numbers or symbols.

Example 10.9.2 Consider again Example 10.9.1 where we looked at the effect of a new training program on weight loss. The p-value from the t-test comparing the mean weight loss between the training and control group was 0.08124. This equates to an S-value of $-\log_2(0.08124) = 3.62$ bits of information embedded in the test statistic, which can be implied as evidence against the null hypothesis. Rounding to the nearest integer we conclude that the data are no more surprising than getting 4 heads in 4 coin tosses.

Figure 10.3 and Table 10.3 show the relationship between p- and S-values.

We see that the difference in evidence against the null hypothesis between $p = 0.9$ and 0.99 is very small: both have much less than 1 bit of information against the test hypothesis. However, the difference between $p = 0.01$ and $p = 0.001$ is large: The S-values are 6.64 and 13.29 respectively, which equates to observing 5 additional heads in a row (given that we round off to the nearest integer). This illustrates

Fig. 10.3 The S-value as a function of the p-value for $p \geq 0.001$

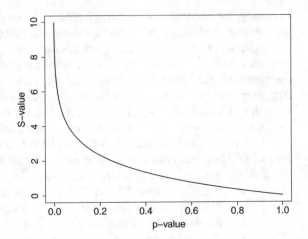

Table 10.3 Examples of S-values for particular p-values

p	0.99	0.9	0.5	0.25	0.1	0.05	0.025	0.01	0.001	1×10^{-7}
S	0.01	0.15	1	2	3.32	4.32	5.32	6.64	13.29	23.25

the benefit of paying attention to the actual information provided against the null hypothesis rather than solely looking at a binary cutoff such as $\alpha = 0.05$.

The S-value tells us that hypotheses relating to the values contained in a 95% CoI have at most 4.3 bits of information against them. This is because all parameter values contained in a 95% CoI have p-values greater than 0.05. As with the p-value, this interpretation is conditional on the background assumptions of the test being correct. Before explaining more about these background assumptions, we first introduce the concept of p- and S-value functions.

10.9.3 Graphs of p- and S-Values

As we have highlighted in the above sections, there might be multiple meaningful hypotheses of interest. In fact, in most examples there might be many meaningful null hypotheses that cannot be rejected. We could thus plot the p- or S-value against those hypotheses, to see which ones are most compatible with the data given the background assumptions. This approach is often described as plotting the p- or S-**value function** and is best illustrated by an example.

Example 10.9.3 Consider again Examples 10.9.1 and 10.9.2, where we looked at the effect of a new physical training program on weight loss. We calculated p- and S-values for $H_0 : \mu_{\text{diff},1} = \mu_{\text{diff},2}$, which is identical to testing $H_0 : \mu_{\text{diff},1} - \mu_{\text{diff},2} = 0$. Suppose we now test $H_0 : \mu_{\text{diff},1} - \mu_{\text{diff},2} = \theta_0 \ \forall \theta_0 \in [-c_1, c_2]$, where c_1 and c_2 are appropriate constants to define a meaningful set of hypotheses. We can then

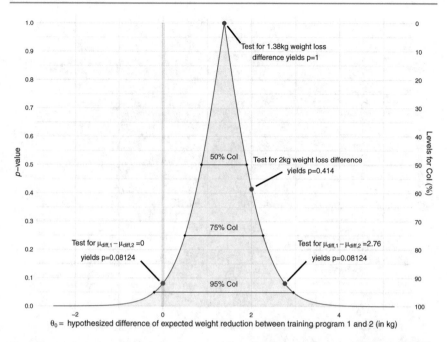

Fig. 10.4 p-values for a range of θ_0's, i.e. hypothesized differences of expected weight reductions over the training program period. This compatibility graph shows how p-values vary across different null hypotheses. Compatibility intervals (CoI) in percents can be read moving from the right-hand axis to the bottom axis

use the R-package `concurve` to plot p- and S-values against different θ_0: we use `curve_mean` to specify that we want to employ a t-test, use the command `ggcurve` to draw a curve, and use the option `type` to determine whether to plot S-values (`type='s'`) or p-values (`type='c'`) on the y-axis:

```
library(concurve)
training <- data.frame(cbind(weight_diff,program))
pc <- curve_mean(weight_diff[program==1],
weight_diff[program==2],data=training)
ggcurve(pc[[1]],type='c')    # p-value function
ggcurve(pc[[1]],type='s')    # surprisal function
```

Figure 10.4 shows the p-value function and Fig. 10.5 the corresponding S-value function.

We can read off the graph that $p = 0.08124$ for $H_0 : \mu_{\text{diff},1} - \mu_{\text{diff},2} = 0$, as calculated in Example 10.9.1. If we test $H_0 : \mu_{\text{diff},1} - \mu_{\text{diff},2} = 2.76\,\text{kg}$, we get the same p-value, but the respective hypothesis implies a considerable weight loss with the new training program. All hypotheses of a weight loss between 0 and 2.76 kg are more compatible with the data than hypotheses outside this interval; and we get the highest compatibility for $H_0 : \mu_{\text{diff},1} - \mu_{\text{diff},2} = 1.38$ kg., i.e. the estimated differ-

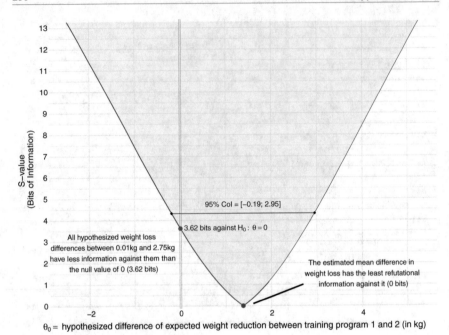

Fig. 10.5 S-values (surprisals) for a range of hypothesized differences of expected weight reductions over the training program period

ence in mean weight loss between the two groups. Different compatibility intervals can be read off the right y-axis.

Both graphs show that it may often not be ideal to frame the discussion in terms of a binary decision based on the cutoff $p < 0.05$: all hypothesized weight loss differences between 0.01 kg and 2.75 kg have less information against them than the null value of 0 (i.e. < 3.62 bits). This gives considerable support to the claim of at least moderate weight loss with the new training program, even though $p > 0.05$.

10.9.4   Unconditional Interpretations

We have already outlined in the above sections that the interpretation of the results of a statistical test, such as a test statistic or a p- or S-value, are based on the premise that the assumptions used to compute these results are correct. It is thus accurate to define the actual **test model** as the combination of the null hypothesis *and* the collection of background assumptions. This view makes it clear that a CoI shows the values of the parameter, which when combined with the background assumptions, produce a test model that is highly compatible with the data in the sense of having less than 4.32 bits of information against it.

But what are those background assumptions and when will they be satisfied? We have assumptions relating to the construction of the test, such as that the units are independent and identically distributed (i.i.d) or are drawn from a specified distribution; we further assume that there are no systematic errors and technical faults during data collection and data entry. If we want to make causal claims, we need additional assumptions (see Chap. 14), and to draw inferences with respect to some population of interest, we may have to make certain assumptions about the process of selection into the study (see also Chap. 13). For regression modeling, which is described in Chaps. 11 and 12, assumptions relating to model specification and model selection have to be made. Applications that face problems such as measurement error or missing data may require further assumptions: however, we do not explain those in this book.

Many physical-science *experiments* may be constructed in a way that allow for controlling of all relevant conditions, which would justify a conditional interpretation. Sometimes there are exceptions to this: for example, in a physics experiment, neutrinos were found to be faster than light because of a large test statistic and corresponding small *p*-value, but this was due to a faulty element of the experiment's fiber-optic timing system (see Rafi and Greenland, 2020).

In social sciences, economics and other areas, it is however common that not all relevant conditions can be controlled. Many analyses are based on surveys or observational data, with limited or no control over many conditions. Thus, the application of statistical tests (or other statistical methods) relies on the background assumptions, such as those explained above. Violating any of these assumptions may lead to a small *p*-value and therefore to a false conclusion with respect to the hypothesis made.

When we relate the results of a statistical test to the entire test model, which consists of both the hypothesis and the background assumptions, we reflect our knowledge that those results may have been observed not only by chance, but potentially also due to violations of assumptions.

Example 10.9.4 Recall Examples 10.9.1, 10.9.2 and 10.9.3, where we evaluated the effect of a new training program on weight loss. We found that hypothesized weight loss differences between 0.01 and 2.75 kg have less information against them than the null value of 0, hence supporting the hypothesis of at least moderate weight loss with the new training program. Unconditionally (and regardless of a frequency interpretation of the interval) the 95% CoI of $[-0.19; 2.95]$ shows the values of the parameter (i.e. expected difference in weight loss) which, when combined with the background assumptions, produce a test model which is highly compatible with the data in the sense of having less than 4.32 bits of information against it. The results were produced with a t-test, which assumes that the data come from a normal distribution. If the sample size is small then the test statistic may not be approximately normally distributed (see Sect. 10.3.3). In this case a larger sample may be needed to confirm the findings.

We now summarize some of the key points of this chapter, before presenting the section of exercises. The first few exercises focus on the choice of the most appropriate test for a given problem, its respective implementation and a traditional interpretation to arrive at a binary decision. Later exercises illustrate the difference between a mere dichotomic use of tests and the application of the concepts presented in this section.

10.10 Key Points and Further Issues

Note:

✓ A graphical summary on when to use the tests introduced in this chapter is given in Appendices D.2 and D.3.

✓ To arrive at a binary test decision, i.e. accept H_0 or reject it, it does not matter whether one compares the test statistic to the critical region, one uses the p-value obtained from statistical software, or one evaluates the appropriate confidence interval. However, it is important not to misinterpret the p-value (see Sect. 10.2.6) and to choose the correct confidence interval.

✓ There is a difference between relevance and significance. A test might be significant, but the point estimate of the quantity of interest may not be relevant from a substantive point of view. Similarly, a test might not be significant but the point and interval estimates may still yield relevant conclusions.

✓ The test statistic of the t-test (one-sample, two-sample, paired) is *asymptotically* normally distributed. This means that for relatively large n (as a rule of thumb > 30 per group) the sample does not need to come from a normal distribution. However, the application of the t-test makes sense only when the expectation μ can be interpreted meaningfully; this may not be the case for skewed distributions or distributions with outliers.

✓ As an alternative to arriving at binary test decisions, one can display the compatibility of the observed data with various hypotheses, for example using S-value functions. The respective interpretations target the entire test model, which consists of both the hypothesis and the background assumptions, see also Rafi and Greenland (2020) and the references therein.

10.11 Exercises

Exercise 10.1 Two people, A and B, are suspects for having committed a crime together. Both of them are interrogated in separate rooms. The jail sentence depends on who confesses to have committed the crime, and who does not:

	B does not confess	B does confess
A does not confess	each serves 1 year	A: 3 years; B: goes free
A does confess	A: goes free; B: 3 years	each serves 2 years

A has two hypotheses:

$$H_0 : \text{B does not confess} \quad \text{versus} \quad H_1 : \text{B does confess.}$$

Given the possible sentences he decides to not confess if H_0 is true, and to confess otherwise. Explain the concepts of type I error and type II error for this situation. Comment on the consequences if these errors are made.

Exercise 10.2 A producer of chocolate bars hypothesizes that his production does not adhere to the weight standard of 100 g. As a measure of quality control, he weighs 15 bars and obtains the following results in grams:

96.40, 97.64, 98.48, 97.67, 100.11, 95.29, 99.80, 98.80, 100.53, 99.41, 97.64, 101.11, 93.43, 96.99, 97.92

It is assumed that the production process is standardized in the sense that the variation is controlled to be $\sigma = 2$ and all background assumptions are met.

(a) What are the hypotheses regarding the expected weight μ for a two-sided test?
(b) Which test should be used to test these hypotheses?
(c) Conduct the test that was suggested to be used in (b). Use $\alpha = 0.05$.
(d) The producer wants to show that the expected weight is smaller than 100 g. What are the appropriate hypotheses to use?
(e) Conduct the test for the hypothesis in (d). Again use $\alpha = 0.05$.

Exercise 10.3 Christian decides to purchase a vinyl (together with a download link) for his collection. His first thought is to buy it online, via an online auction. He discovers that he can also buy the vinyl immediately, without bidding at an auction, from the same online store. He also looks at the price at an internet book store. He notes down the following prices (in €):

Internet book store: 16.95
Online store, no auction: 18.19, 16.98, 19.97, 16.98, 18.19, 15.99, 13.79, 15.90, 15.90, 15.90, 15.90, 15.90, 19.97, 17.72

Online store, auction: 10.50, 12.00, 9.54, 10.55, 11.99, 9.30, 10.59, 10.50, 10.01, 11.89, 11.03, 9.52, 15.49, 11.02

(a) Calculate and interpret the arithmetic mean, variance, standard deviation, and coefficient of variation for the online store, both for the auction and non-auction offers.
(b) Test the hypothesis that the mean price at the online store (no auction) is unequal to €16.95 ($\alpha = 0.05$). Assume, for simplicity, that every assumption used to calculate the test statistic is correct—here, and also in (d) to (h).
(c) Calculate a confidence interval for the mean price at the online store (no auction) and interpret your findings in the light of the hypothesis in (b).
(d) Test the hypothesis that the mean price at the online store (auction) is less than €16.95 ($\alpha = 0.05$).
(e) Test the hypothesis that the mean non-auction price is higher than the mean auction price. Assume that i) the variances are equal in both samples and ii) the variances are unequal ($\alpha = 0.05$).
(f) Test the hypothesis that the variance of the non-auction price is unequal to the variance of the auction price ($\alpha = 0.05$).
(g) Use the U-test to compare the location of the auction and non-auction prices. Compare the results with those of (e).
(h) Calculate the results of (a)–(g) with R.

Exercise 10.4 Ten of Leonard's best friends try a new diet: the "Banting" diet. Each of them weighs him/herself before and after the diet. The data are as follows:

Person (i)	1	2	3	4	5	6	7	8	9	10
Before diet(x_i)	80	95	70	82	71	70	120	105	111	90
After diet (y_i)	78	94	69	83	65	69	118	103	112	88

Choose a test and a confidence interval to test the hypothesis whether there is a difference between the mean weight before and after the diet ($\alpha = 0.05$).

Exercise 10.5 A company producing clothing often finds deficient T-shirts among its production.

(a) The company's controlling department decides that the production is no longer profitable when there are more than 10% deficient shirts. A sample of 230 shirts yields 30 shirts which contain deficiencies. Use the approximate Binomial test to decide whether the T-shirt production is profitable or not ($\alpha = 0.05$).
(b) Test the same hypothesis as in (a) using the exact Binomial test. You can use R to determine the quantiles needed for the calculation. Compare the results of (a) and (b) and interpret your findings.
(c) The company is offered a new cutting machine. To test whether the change of machine helps to improve the production quality, 115 sample shirts are evaluated,

7 of which have deficiencies. Use the two-sample Binomial test to decide whether the new machine yields improvement or not ($\alpha = 0.05$).

(d) Test the same hypothesis as in (c) using the test of Fisher in R.

Exercise 10.6 Two friends play a computer game and each of them repeats the same level 10 times. The scores obtained are:

	1	2	3	4	5	6	7	8	9	10
Player 1	91	101	112	99	108	88	99	105	111	104
Player 2	261	47	40	29	64	6	87	47	98	351

(a) Player 2 insists that she is the better player and suggests to compare their mean performance. Use an appropriate test ($\alpha = 0.05$) to test this hypothesis.

(b) Player 1 insists that she is the better player. She proposes to not focus on the mean and to use the U-test for comparison. What are the results ($\alpha = 0.05$)? What are the advantages and disadvantages of using this test compared with (a)?

Exercise 10.7 Otto loves gummy bears and buys 10 packets at a factory store. He opens all packets and sorts them by their colour. He counts 222 white gummy bears, 279 red gummy bears, 251 orange gummy bears, 232 yellow gummy bears, and 266 green ones. He is disappointed since white (pineapple flavour) is his favourite flavour. He hypothesizes that the producer of the bears does not uniformly distribute the bears into the packets. Choose an appropriate test to find out whether Otto's speculation could be true.

Exercise 10.8 We consider Exercise 4.4 where we evaluated which of the passengers from the *Titanic* were rescued. The data were summarized as follows:

	1. Class	2. Class	3. Class	Staff	Total
Rescued	202	125	180	211	718
Not rescued	135	160	541	674	1510

(a) The hypothesis derived from the descriptive analysis was that travel class and rescue status are not independent. Test this hypothesis.

(b) Interpret the following R output:

```
4-sample test for equality of proportions
data:  titanic
X-squared = 182.06, df = 3, p-value < 2.2e-16
alternative hypothesis: two.sided
sample estimates:
  prop 1    prop 2    prop 3    prop 4
0.5994065 0.4385965 0.2496533 0.2384181
```

(c) Summarize the data in a 2×2 table: passengers from the first and second class should be grouped together, and third class passengers and staff should be grouped together as well. Is the probability of being rescued higher in the first and second class? Provide an answer using the following three tests: exact test of Fisher, χ^2-independence test, χ^2-homogeneity test. You can use R to conduct the test of Fisher.

Exercise 10.9 We are interested in understanding how well the t-test can detect differences with respect to the mean. We use R to draw 3 samples each of 20 observations from three different normal distributions: $X \sim N(5, 2^2)$, $Y_1 \sim N(4, 2^2)$, $Y_2 \sim N(3.5, 2^2)$. The summary statistics of this experiment are as follows:

- $\bar{x} = 4.97, s_x^2 = 2.94,$
- $\bar{y}_1 = 4.55, s_{y_1}^2 = 2.46,$
- $\bar{y}_2 = 3.27, s_{y_2}^2 = 3.44.$

(a) Use the t-test to compare the means of X and Y_1.
(b) Use the t-test to compare the means of X and Y_2.
(c) Interpret the results from (a) and (b). What is the challenge and how could it be addressed?

Exercise 10.10 Access the theatre data described in Appendix A.4.3. The data summarize a survey conducted on visitors of a local Swiss theatre in terms of age, sex, annual income, general expenditure on cultural activities, expenditure on theatre visits, and the estimated expenditure on theatre visits in the year before the survey was done.

(a) Compare the mean expenditure on cultural activities for men and women using the Welch test ($\alpha = 0.05$).
(b) Would the conclusions change if the two-sample t-test or the U-test were used for comparison?
(c) Test the hypothesis that women spend on average more money on theatre visits than men ($\alpha = 0.05$).
(d) Compare the mean expenditure on theatre visits in the year of the survey and the preceding year using an appropriate test($\alpha = 0.05$).
(e) Investigate the hypothesis from (d) further: calculate and interpret both the S-value and an appropriate 95% compatibility interval. Then draw a S-value function and interpret it.

Exercise 10.11 Use R to read in and analyse the pizza data described in Appendix A.1 (assume $\alpha = 0.05$).

(a) The manager's aim is to deliver pizzas in less than 30 min and with a temperature of greater than 65 °C. Use an appropriate test to evaluate whether these aims have been reached on average.

(b) If it takes longer than 40 min to deliver the pizza, then the customers are promised a free bottle of wine. This offer is only profitable if less than 15% of deliveries are too late. Test the hypothesis $p < 0.15$.

(c) The manager wonders whether there is any relationship between the operator taking the phone call and the pizza temperature. Assume that a hot pizza is defined to be one with a temperature greater 65 °C. Use the Test of Fisher, the χ^2-independence test, and the χ^2-test of homogeneity to test his hypothesis.

(d) Each branch employs the same number of staff. It would thus be desirable if each branch receives the same number of orders. Use an appropriate test to investigate this hypothesis.

(e) Is the proportion of calls taken by each operator the same in each branch?

(f) Test whether there is a relationship between drivers and branches.

Exercise 10.12 The authors of this book went to visit historical sites in India. None of them has a particularly strong interest in photography and they speculated that each of them would take about the same number of pictures on their trip. After returning home they counted 110, 118, and 105 pictures respectively. Use an appropriate test to find out whether their speculation was correct ($\alpha = 0.01$).

$\rightarrow$ Solutions to all exercises in this chapter can be found in Appendix B

Linear Regression

11

We learnt about various measures of association in Chap. 4. Such measures are used to understand the degree of *relationship* or *association* between two variables. The correlation coefficient of Bravais–Pearson, for example, can help us to quantify the degree of a linear relationship, but it does not tell us about the functional form of a relationship. Moreover, any association of interest may depend on more than one variable and we would therefore like to explore *multivariate* relationships. For example, consider a situation where we measured the body weights and the long jump results (distances) of school children taking part in a competition. The correlation coefficient of Bravais–Pearson between the body weight and distance jumped can tell us how strong or weak the association between the two variables is. It can also tell us whether the distance jumped increases or decreases with the body weight in the sense that a negative correlation coefficient indicates shorter jumps for higher body weights whereas a positive coefficient implies longer jumps for higher body weights. Suppose we want to know how far a child with known body weight, say 50 kg, will jump. Such questions cannot be answered from the value and sign of the correlation coefficient. Another question of interest may be to explore whether the relationship between the body weight and distance jumped is linear or not. One may also be interested in the joint effect of age and body weight on the distance jumped. Could it be that older children, who have on average a higher weight than younger children, perform better? What would be the association of weight and the long jump results of children of the same age? Such questions are addressed by linear regression analysis which we introduce in this chapter.

In many situations, the outcome does not depend on one variable but on several variables. For example, the recovery time of a patient after an operation depends on several factors such as weight, hemoglobin level, blood pressure, body temperature, diet control, rehabilitation, and others. Similarly, the weather depends on many factors like temperature, pressure, humidity, speed of winds, and others. Multiple linear

© Springer Nature Switzerland AG 2022
C. Heumann et al., *Introduction to Statistics and Data Analysis*,
https://doi.org/10.1007/978-3-031-11833-3_11

regression, introduced in Sect. 11.6, addresses the issue where the outcome depends on more than one variable.

To begin with, we consider only two quantitative variables X and Y in which the outcome Y depends on X and we explore the quantification of their relationship.

Examples 11.0.1 Examples of associations in which we might be interested in are:

- body height (X) and body weight (Y) of persons,
- speed (X) and braking distance (Y) measured on cars,
- money invested (in €) in the marketing of a product (X) and sales figures for this product (in €) (Y) measured in various branches,
- amount of fertilizer used (X) and yield of rice (Y) measured on different acres,
- temperature (X) and hotel occupation (Y) measured in cities.

11.1 The Linear Model

Consider the scatter plots from Fig. 4.2 on page 83. Plotting X-values against Y-values enables us to visualize the relationship between two variables. Figure 11.1a reviews what a positive (linear) association between X and Y looks like: the higher the X values, the higher the Y values (and vice versa). The plot indicates that there may be a linear relationship between X and Y. On the other hand, Fig. 11.1b displays a scatter plot which shows no clear relationship between X and Y. The R^2 measure, shown in the two figures and explained in more detail later, equates to the squared correlation coefficient of Bravais–Pearson.

(a) Positive linear relationship   (b) No clear relationship

Fig. 11.1 Scatter plots

To summarize the observed association between X and Y, we can postulate the following linear relationship between them:

$$Y = \alpha + \beta X. \tag{11.1}$$

This Eq. (11.1) represents a straight line where α is the **intercept** and β represents the **slope** of the line. The slope indicates the change in the Y-value when the X value changes by one unit. If the sign of β is positive, it indicates that the value of Y increases as the value of X increases. If the sign of β is negative, it indicates that the value of Y decreases as the value of X increases. When $X = 0$, then $Y = \alpha$. If $\alpha = 0$ then $Y = \beta X$ represents a line passing through the origin. Suppose the height and body weights in the example of school children are represented in Fig. 11.1a. This has the following interpretation: when the height of a child increases by one centimetre, then the body weight increases by β kilograms. The slope β in Fig. 11.1a would certainly be positive because we have a positive linear relationship between X and Y. Note that in this example the intercept term has no particular meaning because when the height $X = 0$, the body weight $Y = \alpha = 0$. The scatter diagram in Fig. 11.1b does not exhibit any clear relationship between X and Y. Still, the slope of a possible line would likely be somewhere around 0.

It is obvious from Fig. 11.1a that by assuming a linear relationship between X and Y, any straight line will not exactly match the data points in the sense that it cannot pass through all the observations: the observations will lie above and below the line. The line represents a *model* to describe the process generating the data.

In our context, a model is a mathematical representation of the relationship between two or more variables. A model has two components— variables (e.g. X, Y) and parameters (e.g. α, β). A model is said to be *linear* if it is linear in its parameters. A model is said to be *nonlinear* if it is nonlinear in its parameters (e.g. β^2 instead of β). Now assume that each observation potentially deviates by e_i from the line in Fig. 11.1a. The **linear model** in Eq. (11.1) can then be written as follows to take this into account:

$$Y = \alpha + \beta X + e. \tag{11.2}$$

Suppose we have n observations $(x_1, y_1), (x_2, y_2), \ldots, (x_n, y_n)$, then each observation satisfies

$$y_i = \alpha + \beta x_i + e_i. \tag{11.3}$$

Each deviation e_i is called an *error*. It represents the deviation of the data points (x_i, y_i) from the **regression line**. The line in Fig. 11.1a is the fitted (regression) line which we will discuss in detail later. We assume that the errors e_i are identically and independently distributed with mean 0 and constant variance σ^2, i.e. $E(e_i) = 0$, $Var(e_i) = \sigma^2$ for all $i = 1, 2, \ldots, n$. We will discuss these assumptions in more detail in Sect. 11.7.

In the model (11.2), Y is called the **response, response variable, dependent variable** or **outcome**; X is called the **covariate, regressor** or **independent variable**. The scalars α and β are the parameters of the model and are described as **regression**

coefficients or **parameters** of the linear model. In particular, α is called the **intercept term** and β is called the **slope parameter**.

It may be noted that if the regression parameters α and β are known, then the linear model is completely known. An important objective in identifying the model is to determine the values of the regression parameters using the available observations for X and Y. It can be understood from Fig. 11.1a that an ideal situation would be when all the data points lie exactly on the line or, in other words, the error e_i is zero for each observation. It seems meaningful to determine the values of the parameters in such a way that the errors are minimized.

There are several methods to estimate the values of the regression parameters. In the remainder of this chapter, we describe the methods of least squares and maximum likelihood estimation.

11.2   Method of Least Squares

Suppose n sets of observations $P_i = (x_i, y_i)$, $i = 1, 2, \ldots, n$, are obtained on two variables $P = (X, Y)$ and are plotted in a scatter plot. An example of four observations, (x_1, y_1), (x_2, y_2), (x_3, y_3), (x_4, y_4), is given in Fig. 11.2.

The **method of least squares** says that a line can be fitted to the given data set such that the errors are minimized. This implies that one can determine α and β such that the sum of the squared distances between the data points and the line $Y = \alpha + \beta X$ is minimized. For example, in Fig. 11.2, the first data point (x_1, y_1) does not lie on the plotted line and the deviation is $e_1 = y_1 - (\alpha + \beta x_1)$. Similarly we obtain the difference of the other three data points from the line: e_2, e_3, e_4. The error is zero if the point lies exactly on the line. The problem we would like to solve in this example

Fig. 11.2 Regression line, observations and errors e_i

is to minimize the sum of squares of e_1, e_2, e_3 and e_4, i.e.

$$\min_{\alpha,\beta} \sum_{i=1}^{4} (y_i - \alpha - \beta x_i)^2. \tag{11.4}$$

We want the line to fit the data well. This can generally be achieved by choosing α and β such that the squared errors of all the n observations are minimized:

$$\min_{\alpha,\beta} \sum_{i=1}^{n} e_i^2 = \min_{\alpha,\beta} \sum_{i=1}^{n} (y_i - \alpha - \beta x_i)^2. \tag{11.5}$$

If we solve this optimization problem by the principle of maxima and minima, we obtain estimates of α and β as

$$\left.\begin{array}{l} \hat{\beta} = \dfrac{S_{xy}}{S_{xx}} = \dfrac{\sum(x_i-\bar{x})(y_i-\bar{y})}{\sum(x_i-\bar{x})^2} = \dfrac{\sum_{i=1}^{n} x_i \, y_i - n\bar{x}\bar{y}}{\sum_{i=1}^{n} x_i^2 - n\bar{x}^2} \\ \hat{\alpha} = \bar{y} - \hat{b}\bar{x} \end{array}\right\}, \tag{11.6}$$

see Appendix C.6 for a detailed derivation. Here $\hat{\alpha}$ and $\hat{\beta}$ represent the estimates of the parameters α and β respectively and are called the **least squares estimator** of α and β, respectively. This gives us the model $y = \hat{\alpha} + \hat{\beta}x$ which is called the *fitted model* or the *fitted regression line*. The literal meaning of "regression" is to move back. Since we are acquiring the data and then moving back to find the parameters of the model using the data, it is called a *regression model*. The fitted regression line $y = \hat{\alpha} + \hat{\beta}x$ describes the postulated relationship between Y and X. The sign of β determines whether the relationship between X and Y is positive or negative. If the sign of β is positive, it indicates that if X increases, then Y increases too. On the other hand, if the sign of β is negative, it indicates that if X increases, then Y decreases. For any given value of X, say x_i, the *predicted* value $\hat{y}_i$ is calculated by

$$\hat{y}_i = \hat{\alpha} + \hat{\beta}x_i$$

and is called the ith *fitted value*.

If we compare the observed data point (x_i, y_i) with the point suggested (predicted, fitted) by the regression line, $(x_i, \hat{y}_i)$, the difference between y_i and $\hat{y}_i$ is called the **residual** and is given as

$$\hat{e}_i = y_i - \hat{y}_i = y_i - (\hat{\alpha} + \hat{\beta}x_i). \tag{11.7}$$

This can be viewed as an estimator of the error e_i. Note that it is not really an estimator in the true statistical sense because e_i is random and so it cannot be estimated. However, since the e_i are unknown and $\hat{e}_i$ measures the same difference between the estimated and true values of the y's, see for example Fig. 11.2, it can be treated as estimating the error e_i.

Example 11.2.1 A physiotherapist advises 12 of his patients, all of whom had the same knee surgery done, to regularly perform a set of exercises. He asks them to record how long they practise. He then summarizes the average time they practised (X, time in minutes) and how long it takes them to regain their full range of motion again (Y, time in days). The results are as follows:

i	1	2	3	4	5	6	7	8	9	10	11	12
x_i	24	35	64	20	33	27	42	41	22	50	36	31
y_i	90	65	30	60	60	80	45	45	80	35	50	45

To estimate the linear regression line $y = \hat{\alpha} + \hat{\beta}x$, we first calculate $\bar{x} = 35.41$ and $\bar{y} = 57.08$. To obtain S_{xy} and S_{xx} we need the following table:

i	x_i	y_i	$(x_i - \bar{x})$	$(y_i - \bar{y})$	$(x_i - \bar{x})(y_i - \bar{y})$	$(x_i - \bar{x})^2$
1	24	90	−11.41	32.92	−375.61	130.19
2	35	65	−0.41	7.92	−3.25	0.17
3	64	30	28.59	−27.08	−774.22	817.39
4	20	60	−15.41	2.92	−45.00	237.47
5	33	60	−2.41	2.92	−7.27	5.81
6	27	80	−8.41	22.92	−192.75	70.73
7	42	45	6.59	−12.08	−79.61	43.43
8	41	45	5.59	−12.08	−67.53	31.25
9	22	80	−13.41	22.92	−307.36	179.83
10	50	35	14.59	−22.08	−322.14	212.87
11	36	50	0.59	−7.08	−4.18	0.35
12	31	45	−4.41	−12.08	53.27	19.45
Total					−2125.65	1748.94

Using (11.6), we can now easily find the least squares estimates $\hat{\alpha}$ and $\hat{\beta}$ as

$$\hat{\beta} = \frac{S_{xy}}{S_{xx}} = \frac{\sum(x_i - \bar{x})(y_i - \bar{y})}{\sum(x_i - \bar{x})^2} = \frac{-2125.65}{1748.94} \approx -1.22,$$

$$\hat{\alpha} = \bar{y} - \hat{\beta}\bar{x} = 57.08 - (-1.215) \cdot 35.41 = 100.28.$$

The fitted regression line is therefore

$$y = 100.28 - 1.22 \cdot x.$$

We can interpret the results as follows:

- For an increase of 1 minute in exercising, the recovery time decreases by 1.22 days because $\hat{\beta} = -1.22$. The negative sign of $\hat{\beta}$ indicates that the relationship between exercising time and recovery time is negative, i.e. as exercise time increases, the recovery time decreases.
- When comparing two patients with a difference in exercising time of 10 min, the linear model estimates a difference in recovery time of 12.2 days because $10 \cdot 1.22 = 12.2$.

Fig. 11.3 Scatter plot and
regression line for Example
11.2.1

- The model predicts an average recovery time of $100.28 - 1.22 \cdot 38 = 53.92$ days
  for a patient who practises the set of exercises for 38 min.
- If a patient didn't exercise at all, the recovery time would be predicted as $\hat{a} =$
  100.28 days by the model. However, see item (i) in Sect. 11.2.1 below about
  interpreting a regression line outside of the observed data range.

We can also obtain these results by using R. The command $\mathrm{lm}(Y{\sim}X)$ fits a linear
model and provides the estimates of $\hat{\alpha}$ and $\hat{\beta}$.

```
lm(Y~X)
```

We can draw the regression line onto a scatter plot using the command `abline`,
see also Fig. 11.3.

```
plot(X,Y)
abline(a=100.28,b=-1.22)
```

11.2.1 Properties of the Linear Regression Line

There are a couple of interesting results related to the regression line and the least
square estimates.

(i) As a rule of thumb, one should interpret the regression line $\hat{y}_i = \hat{\alpha} + \hat{\beta}x_i$ only in
     the interval $[x_{(1)}, x_{(n)}]$. For example, if X denotes "Year", with observed values
     from 1999 to 2015, and Y denotes the "annual volume of sales of a particular
     company", then a prediction for the year 2030 may not be meaningful or valid

because a linear relationship discovered in the past may not continue to hold in the future.

(ii) For the points $\hat{P}_i = (x_i, \hat{y}_i)$, forming the regression line, we can write

$$\hat{y}_i = \hat{\alpha} + \hat{\beta} x_i = \bar{y} + \hat{\beta}(x_i - \bar{x}). \tag{11.8}$$

(iii) It follows for $x_i = \bar{x}$ that $\hat{y}_i = \bar{y}$, i.e. the point $(\bar{x}, \bar{y})$ always lies on the regression line. The linear regression line therefore always passes through $(\bar{x}, \bar{y})$.

(iv) The sum of the residuals is zero. The ith residual is defined as

$$\hat{e}_i = y_i - \hat{y}_i = y_i - (\hat{\alpha} + \hat{\beta} x_i) = y_i - [\bar{y} + \hat{\beta}(x_i - \bar{x})].$$

The sum is therefore

$$\sum_{i=1}^{n} \hat{e}_i = \sum_{i=1}^{n} y_i - \sum_{i=1}^{n} \bar{y} - \hat{\beta} \sum_{i=1}^{n} (x_i - \bar{x})$$
$$= n\bar{y} - n\bar{y} - \hat{\beta}(n\bar{x} - n\bar{x}) = 0. \tag{11.9}$$

(v) The arithmetic mean of $\hat{y}$ is the same as the arithmetic mean of y:

$$\bar{\hat{y}} = \frac{1}{n} \sum_{i=1}^{n} \hat{y}_i = \frac{1}{n}(n\bar{y} + \hat{\beta}(n\bar{x} - n\bar{x})) = \bar{y}. \tag{11.10}$$

(vi) The least squares estimate $\hat{\beta}$ has a direct relationship with the correlation coefficient of Bravais–Pearson:

$$\hat{\beta} = \frac{S_{xy}}{S_{xx}} = \frac{S_{xy}}{\sqrt{S_{xx}}\sqrt{S_{yy}}} \cdot \sqrt{\frac{S_{yy}}{S_{xx}}} = r\sqrt{\frac{S_{yy}}{S_{xx}}}. \tag{11.11}$$

The slope of the regression line is therefore proportional to the correlation coefficient r: a positive correlation coefficient implies a positive estimate of β and vice versa. However, a stronger correlation does not necessarily imply a steeper slope in the regression analysis because the slope depends upon $\sqrt{S_{yy}/S_{xx}}$ as well.

11.3   Goodness of Fit

While one can easily fit a linear regression model this doesn't mean that the model is necessarily good. Consider again Fig. 11.1: In Fig. 11.1a, the regression line is clearly capturing the linear trend seen in the scatter plot. The fit of the model to the data seems good. Figure 11.1b shows however that the data points vary considerably around the line. The quality of the model doesn't seem to be very good. If we'd use the regression line to predict the data, we would likely obtain poor results. It is obvious from Fig. 11.1 that the model provides a good fit to the data when the observations are close to the line and capture a linear trend.

A look at the scatter plot provides a visual and qualitative approach to judging the quality of the fitted model. Consider Fig. 11.4a, b which both show a linear trend in the data, but the observations in Figure 11.4a are closer to the regression line than those in 11.4b. Therefore the goodness of fit is worse in the latter figure and any quantitative measure should capture this.

A quantitative measure for the goodness of fit can be derived by means of **variance decomposition** of the data. The total variation of y is partitioned into two components—sum of squares due to the fitted model and sum of squares due to random errors in the data:

$$\underbrace{\sum_{i=1}^{n}(y_i - \bar{y})^2}_{SQ_{\text{Total}}} = \underbrace{\sum_{i=1}^{n}(\hat{y}_i - \bar{y})^2}_{SQ_{\text{Regression}}} + \underbrace{\sum_{i=1}^{n}(y_i - \hat{y}_i)^2}_{SQ_{\text{Error}}} . \tag{11.12}$$

The proof of the above equation is given in Appendix C.6.

(a) Strong linear relationship   (b) Weak linear relationship

Fig. 11.4 Different goodness of fit for different data

The left-hand side of (11.12) represents the total variability in y with respect to $\bar{y}$. It is proportional to the sample variance (3.21) and is called SQ_{Total} (total sum of squares). The first term on the right-hand side of the equation measures the variability which is explained by the fitted linear regression model ($SQ_{\text{Regression}}$, sum of squares due to regression); the second term relates to the error sum of squares (SQ_{Error}) and reflects the variation due to random error involved in the fitted model. Larger values of SQ_{Error} indicate that the deviations of the observations from the regression line are large. This implies that the model provides a bad fit to the data and that the goodness of fit is low.

Obviously one would like to have a fit in which the error sum of squares is as small as possible. To judge the goodness of fit, one can therefore look at the error sum of squares in relation to the total sum of squares: in an ideal situation, if the error sum of squares equals zero, the total sum of squares is equal to the sum of squares due to regression and the goodness of fit is optimal. On the other hand, if the sum of squares due to error is large, it will make the sum of squares due to regression smaller and the goodness of fit should be bad. If the sum of squares due to regression is zero, it is evident that the model fit is the worst possible. These thoughts are reflected in the criterion for the goodness of fit, also known as R^2:

$$R^2 = \frac{SQ_{\text{Regression}}}{SQ_{\text{Total}}} = 1 - \frac{SQ_{\text{Error}}}{SQ_{\text{Total}}}. \tag{11.13}$$

It follows from the above definition that $0 \leq R^2 \leq 1$. The closer R^2 is to 1, the better the fit because SQ_{Error} will then be small. The closer R^2 is to 0, the worse the fit, because SQ_{Error} will then be large. If R^2 takes any other value, say $R^2 = 0.7$, it means that only 70% of the variation in data is explained by the fitted model and hence, in simple language, the model is 70% good. An important point to remember is that R^2 is defined only when there is an intercept term in the model (an assumption we make throughout this chapter). So it is not used to measure the goodness of fit in models without an intercept term.

Example 11.3.1 Consider again Fig. 11.1: In Fig. 11.1a the line and data points fit well together. As a consequence R^2 is high, $R^2 = 0.82$. Figure 11.1b shows data points with large deviations from the regression line; therefore R^2 is small, here $R^2 = 0.002$. Similarly, in Fig. 11.4a an R^2 of 0.97 relates to an almost perfect model fit whereas in Fig. 11.4b the model describes the data only moderately well ($R^2 = 0.57$).

Example 11.3.2 Consider again Example 11.2.1 where we analysed the relationship between exercise intensity and recovery time for patients convalescing from knee surgery. To calculate R^2 we need the following table:

i	y_i	$\hat{y}_i$	$(y_i - \bar{y})$	$(y_i - \bar{y})^2$	$(\hat{y}_i - \bar{y})$	$(\hat{y}_i - \bar{y})^2$
1	90	70.84	32.92	1083.73	13.76	189.34
2	65	57.42	7.92	62.73	0.34	0.12
3	30	22.04	−27.08	733.33	-35.04	1227.80
4	60	75.72	2.92	8.53	18.64	347.45
5	60	59.86	2.92	8.53	2.78	7.73
6	80	67.18	22.92	525.33	10.10	102.01
7	45	48.88	−12.08	145.93	−8.2	67.24
8	45	50.10	−12.08	145.93	−6.83	48.72
9	80	73.28	22.92	525.33	16.20	262.44
10	35	39.12	−22.08	487.53	−17.96	322.56
11	50	56.20	−7.08	50.13	-0.88	0.72
12	45	62.30	−12.08	145.93	5.22	27.25
Total				3922.96		2603.43

We calculate R^2 with these results as

$$R^2 = \frac{SQ_{\text{Regression}}}{SQ_{\text{Total}}} = \frac{\sum_{i=1}^{n}(\hat{y}_i - \bar{y})^2}{\sum_{i=1}^{n}(y_i - \bar{y})^2} = \frac{2603.43}{3922.96} = 0.66.$$

We conclude that the regression model provides a reasonable but not perfect fit to the data because 0.66 is not close to 0, but also not very close to 1. About 66% of the variability in the data can be explained by the fitted linear regression model. The rest is random error: for example, individual variation in the recovery time of patients due to genetic and environmental factors, surgeon performance, and others.

We can also obtain this result in R by looking at the summary of the linear model:

```
summary(lm(Y~X))
```

Please note that we give a detailed explanation of the model summary in Sect. 11.7.

There is a direct relationship between R^2 and the correlation coefficient of Bravais–Pearson r:

$$R^2 = r^2 = \left(\frac{S_{xy}}{\sqrt{S_{xx}S_{yy}}} \right)^2. \tag{11.14}$$

The proof is given in Appendix C.6.

Example 11.3.3 Consider again Examples 11.3.1 and 11.3.2 where we analysed the association of exercising time and time to regain full range of motion after knee surgery. We calculated $R^2 = 0.66$. We therefore know that the correlation coefficient is $r = \sqrt{R^2} = \sqrt{0.66} \approx 0.81$.

11.4 Linear Regression with a Binary Covariate

Until now we have assumed that the covariate X is metric. It is however also straightforward to fit a linear regression model when X is binary, i.e. if X has two categories. In this case, the values of X in the first category are usually coded as 0 and the values of X in the second category are coded as 1. For example, if the binary variable is "gender", we replace the word "male" with the number 0, and the word "female" with 1. We can then fit a linear regression model using these numbers, but the interpretation differs from the interpretations in case of a metric variable. Recall the definition of the linear model, $Y = \alpha + \beta X + e$ with $E(e) = 0$; if $X = 0$ (male) then $E(Y|X = 0) = \alpha$ and if $X = 1$ (female), then $E(Y|X = 1) = \alpha + \beta \cdot 1 = \alpha + \beta$. Thus α is the average value of Y for males, i.e. $E(Y|X = 0)$, and $\beta = E(Y|X = 1) - E(Y|X = 0)$. It follows that those subjects with $X = 1$ (e.g. females) have on average Y values which are β units higher than subjects with $X = 0$ (e.g. males).

Example 11.4.1 Recall Examples 11.2.1, 11.3.2, and 11.3.3 where we analysed the association of exercising time and recovery time after knee surgery. We keep the values of Y (recovery time, in days) and replace values of X (exercising time, in minutes) with 0 for patients exercising for less than 40 min and with 1 for patients exercising for 40 min or more. We have therefore a new variable X indicating whether a patient is exercising a lot ($X = 1$) or not ($X = 0$). To estimate the linear regression line $\hat{y} = \hat{\alpha} + \hat{\beta}x$, we first calculate $\bar{x} = 4/12$ and $\bar{y} = 57.08$. To obtain S_{xy} and S_{xx} we need the following table:

i	x_i	y_i	$(x_i - \bar{x})$	$(y_i - \bar{y})$	$(x_i - \bar{x})(y_i - \bar{y})$	$(x_i - \bar{x})^2$
1	0	90	$-\frac{4}{12}$	32.92	-10.97	0.11
2	0	65	$-\frac{4}{12}$	7.92	-2.64	0.11
3	1	30	$\frac{8}{12}$	-27.08	-18.05	0.44
4	0	60	$-\frac{4}{12}$	2.92	-0.97	0.11
5	0	60	$-\frac{4}{12}$	2.92	-0.97	0.11
6	0	80	$-\frac{4}{12}$	22.92	-7.64	0.11
7	1	45	$\frac{8}{12}$	-12.08	-8.05	0.44
8	1	45	$\frac{8}{12}$	-12.08	-8.05	0.44
9	0	80	$-\frac{4}{12}$	22.92	-7.64	0.11
10	1	35	$\frac{8}{12}$	-22.08	-14.72	0.44
11	0	50	$-\frac{4}{12}$	-7.08	2.36	0.11
12	0	45	$-\frac{4}{12}$	-12.08	4.03	0.11
Total					-72.34	2.64

We can now calculate the least squares estimates of α and β using (11.6) as

$$\hat{\beta} = \frac{S_{xy}}{S_{xx}} = \frac{\sum (x_i - \bar{x})(y_i - \bar{y})}{\sum (x_i - \bar{x})^2} = \frac{-72.34}{2.64} \approx -27.4 \, ,$$

$$\hat{\alpha} = \bar{y} - \hat{\beta}\bar{x} = 57.08 - (-27.4) \cdot \frac{4}{12} = 66.2.$$

Fig. 11.5 Scatter plot and
regression lines for
Examples 11.4.1 and 11.5.1

The fitted regression line is

$$y = 66.2 - 27.4 \cdot x .$$

The interpretation is as follows:

- The average recovery time for patients doing little exercise is $66.2 - 27.4 \cdot 0 = 66.2$ days.
- Patients exercising heavily ($x = 1$) have, on average, a 27.4 days shorter recovery period ($\beta = -27.4$) than patients with a short exercise time ($x = 0$).
- The average recovery time for patients exercising heavily is 38.8 days ($66.2 - 27.4 \cdot 1$ for $x = 1$).
- These interpretations are visualized in Fig. 11.5. Because "exercise time" (X) is considered to be a nominal variable (two categories representing the intervals $[0; 40)$ and $(40; \infty)$) we can plot the regression line on a scatter plot as two parallel lines.

11.5   Linear Regression with a Transformed Covariate

Recall that a model is said to be linear when it is linear in its parameters. The definition of a linear model is not at all related to the linearity in the covariate. For example,

$$Y = \alpha + \beta^2 X + e$$

is *not* a linear model because the right-hand side of the equation is not a linear function in β. However,

$$Y = \alpha + \beta X^2 + e$$

is a linear model. This model can be fitted as for any other linear model: we obtain $\hat{\alpha}$ and $\hat{\beta}$ as usual and simply use the squared values of X instead of X. This can be justified by considering $X^* = X^2$ and then the model is written as $Y = \alpha + \beta X^* + e$ which is again a linear model. Only the interpretation changes: for each unit increase in the squared value of X, i.e. X^*, Y increases by β units. Such an interpretation is often not even needed. One could simply plot the regression line of Y on X^* to visualize the functional form of the effect. This is also highlighted in the following example.

Example 11.5.1 Recall Examples 11.2.1, 11.3.2, 11.3.3, and 11.4.1 where we analysed the association of exercising time and recovery time after knee surgery. We estimated $\hat{\beta}$ as -1.22 by using X as it is, as a metric variable, see also Fig. 11.3. When using a binary X, based on a cutoff of 40 min, we obtained $\hat{\beta} = -27.4$, see also Fig. 11.5. If we now use $\sqrt{X}$ rather than X we obtain $\hat{\beta} = -15.1$. This means for an increase of 1 unit of the square root of exercising time, the recovery time decreases by 15.1 days. Such an interpretation will be difficult to understand for many people. It is better to plot the linear regression line $y = 145.8 - 15.1 \cdot \sqrt{x}$, see Fig. 11.5. We can see that the new nonlinear line (obtained from a *linear* model) fits the data nicely and it may even be preferable to a straight regression line. Moreover, the value of α substantially increased with this modelling approach, highlighting that no exercising at all may severely delay recovery from the surgery (which is biologically more meaningful). In R, we obtain these results by either creating a new variable $\sqrt{X}$ or by using the I() command which allows specifying transformations in regression models.

```
newX <- sqrt(X)
lm(Y~newX)          #option 1
lm(Y~I(sqrt(X)))    #option 2
```

Generally the covariate X can be replaced by any transformation of it, such as using $\log(X)$, $\sqrt{X}$, $\sin(X)$, or X^2. The details are discussed in Sect. 11.6.3. It is also possible to compare the quality and goodness of fit of models with different transformations, see Sect. 11.8.

11.6   Linear Regression with Multiple Covariates

Up to now, we have only considered two variables—Y and X. We therefore assume that Y is affected by only one covariate X. However, in many situations, there might be more than one covariate which affects Y and possibly all of them are relevant to the analysis (see also Sect. 11.10 on the difference between association and causation). For example, the yield of a crop depends on several factors like quantity of fertilizer, irrigation, temperature, among others. In another example, in the pizza data set (described in Appendix A.4.1), many variables could potentially be associated with delivery time—driver, amount of food ordered, operator handling the order, whether it is a weekend, and so on. A reasonable question to ask is: do different drivers have (on average) a different delivery time—given they deal with the same operator, the amount of food ordered is the same, the day is the same, etc. These kind of questions can be answered with **multiple linear regression**. The model contains more than one, say p, covariates $X_1, X_2, \ldots, X_p$. The linear model (11.2) can be extended as

$$Y = \beta_0 + \beta_1 X_1 + \cdots + \beta_p X_p + e .\tag{11.15}$$

Note that the intercept term is denoted here by β_0. In comparison with (11.2), $\alpha = \beta_0$ and $\beta = \beta_1$.

Example 11.6.1 For the pizza delivery data a particular linear model with multiple covariates could be specified as follows:

$$\text{Delivery Time} = \beta_0 + \beta_1 \text{Number pizzas ordered} + \beta_2 \text{Weekend(0/1)}$$
$$+ \beta_3 \text{Operator (0/1)} + e .$$

11.6.1 Matrix Notation

If we have a data set of n observations then every set of observations satisfies the model (11.15) and we can write

$$y_1 = \beta_0 + \beta_1 x_{11} + \beta_2 x_{12} + \cdots + \beta_p x_{1p} + e_1$$
$$y_2 = \beta_0 + \beta_1 x_{21} + \beta_2 x_{22} + \cdots + \beta_p x_{2p} + e_2$$
$$\vdots \qquad\qquad \vdots \qquad\qquad \vdots$$
$$y_n = \beta_0 + \beta_1 x_{n1} + \beta_2 x_{n2} + \cdots + \beta_p x_{np} + e_n \tag{11.16}$$

It is possible to write the n equations in (11.16) in matrix notation as

$$\mathbf{y} = \mathbf{X}\boldsymbol{\beta} + \mathbf{e} \tag{11.17}$$

where

$$\mathbf{y} = \begin{pmatrix} y_1 \\ y_2 \\ \vdots \\ y_n \end{pmatrix}, \quad \mathbf{X} = \begin{pmatrix} 1 & x_{11} & x_{12} & \cdots & x_{1p} \\ 1 & x_{21} & x_{22} & \cdots & x_{2p} \\ & & \vdots & & \\ 1 & x_{n1} & x_{n2} & \cdots & x_{np} \end{pmatrix}, \quad \boldsymbol{\beta} = \begin{pmatrix} \beta_0 \\ \beta_1 \\ \vdots \\ \beta_p \end{pmatrix}, \quad \mathbf{e} = \begin{pmatrix} e_1 \\ e_2 \\ \vdots \\ e_n \end{pmatrix} .$$

The letters $\mathbf{y}, \mathbf{X}, \boldsymbol{\beta}, \mathbf{e}$ are written in bold because they refer to vectors and matrices rather than scalars. The capital letter $\mathbf{X}$ makes it clear that $\mathbf{X}$ is a matrix of order $n \times p$ representing the n observations on each of the covariates $X_1, X_2, \ldots, X_p$.

Similarly, $\mathbf{y}$ is a $n \times 1$ vector of n observations on Y, $\boldsymbol{\beta}$ is a $p \times 1$ vector of regression coefficients associated with $X_1, X_2, \ldots, X_p$ and $\mathbf{e}$ is a $n \times 1$ vector of n errors. The lower case letter $\mathbf{x}$ relates to a vector representing a variable which means we can denote the multiple linear model from now on also as

$$\mathbf{y} = \beta_0 \mathbf{1} + \beta_1 \mathbf{x_1} + \cdots + \beta_p \mathbf{x_p} + \mathbf{e} \tag{11.18}$$

where $\mathbf{1}$ is the $n \times 1$ vector of 1's. We assume that $\mathrm{E}(\mathbf{e}) = 0$ and $\mathbf{Cov}(\mathbf{e}) = \sigma^2 I$, see Sect. 11.9 for more details.

We would like to highlight that $\mathbf{X}$ is not the data matrix. The matrix $\mathbf{X}$ is called the **design matrix** and contains both a column of 1's denoting the presence of the intercept term and all explanatory variables which are relevant to the multiple linear model (including possible transformations and interactions, see Sects. 11.6.3 and 11.7.3). The errors $\mathbf{e}$ reflect the deviations of the observations from the regression line and therefore the difference between the observed and fitted relationships. Such deviations may occur for various reasons. For example, the measured values can be affected by variables which are not included in the model, the variables may not be accurately measured, there is unmeasured genetic variability in the study population, all of which are covered in the error $\mathbf{e}$. The estimate of $\boldsymbol{\beta}$ is obtained by using the least squares principle by minimizing $\sum_{i=1}^{n} e_i^2 = e'e$. The **least squares estimate** of $\boldsymbol{\beta}$ is given by

$$\hat{\boldsymbol{\beta}} = (\mathbf{X}'\mathbf{X})^{-1}\mathbf{X}'\mathbf{y}. \tag{11.19}$$

The vector $\hat{\boldsymbol{\beta}}$ contains the estimates of $(\beta_0, \beta_1, \ldots, \beta_p)'$. We can interpret it as earlier: $\hat{\beta}_0$ is the estimate of the intercept which we obtain because we have added the column of 1's (and is identical to α in (11.1)). The estimates $\hat{\beta}_1, \hat{\beta}_2, \ldots, \hat{\beta}_p$ refer to the regression coefficients associated with the variables $\mathbf{x_1}, \mathbf{x_2}, \ldots, \mathbf{x_p}$, respectively. The interpretation of $\hat{\beta}_j$ is that it represents the partial change in y_i when the value of x_i changes by one unit keeping all other covariates fixed.

A possible interpretation of the intercept term is that when all covariates equal zero then

$$\mathrm{E}(y) = \beta_0 + \beta_1 \cdot 0 + \cdots + \beta_p \cdot 0 = \beta_0. \tag{11.20}$$

There are many situations in real life for which there is no meaningful interpretation of the intercept term because one or many covariates cannot take the value zero. For instance, in the pizza data set, the bill can never be €0 and thus there is no need to describe the average delivery time for a given bill of €0. The intercept term here serves the purpose of improvement of the model fit, but it will not be interpreted.

In some situations, it may happen that the average value of y is zero when all covariates are zero. Then the intercept term is zero as well and does not improve the model. For example, suppose the salary of a person depends on two factors— education level and type of qualification. If any person is completely illiterate, even then we observe in practice that his salary is never zero. So in this case, there is a benefit of the intercept term. In another example, consider that the velocity of a car depends on two variables—acceleration and quantity of petrol. If these two variables take values of zero, the velocity is zero on a plane surface. The intercept term will therefore be zero as well and yields no model improvement.

Example 11.6.2 Consider the pizza data described in Appendix A.4.1. The data matrix $\mathcal{X}$ is as follows:

$$\mathcal{X} = \begin{matrix} & \text{Day} & \text{Date} & \text{Time} & \cdots & \text{Discount} \\ & \begin{pmatrix} \text{Thursday} & \text{1-May-14} & 35.1 & \cdots & 1 \\ \text{Thursday} & \text{1-May-14} & 25.2 & \cdots & 0 \\ \vdots & & & & \vdots \\ \text{Saturday} & \text{31-May-14} & 35.7 & \cdots & 0 \end{pmatrix} \end{matrix}$$

Suppose the manager has the hypothesis that the operator and the overall bill (as a proxy for the amount ordered from the customer) influence the delivery time. We can postulate a linear model to describe this relationship as

$$\text{Delivery Time} = \beta_0 + \beta_1 \text{Bill} + \beta_2 \text{Operator} + e \,.$$

The model in matrix notation is as follows:

$$\underbrace{\begin{pmatrix} 35.1 \\ 25.2 \\ \vdots \\ 35.7 \end{pmatrix}}_{\mathbf{y}} = \underbrace{\begin{pmatrix} 1 & 58.4 & 1 \\ 1 & 26.4 & 0 \\ \vdots & \vdots & \vdots \\ 1 & 42.7 & 0 \end{pmatrix}}_{\mathbf{X}} \underbrace{\begin{pmatrix} \beta_0 \\ \beta_1 \\ \beta_2 \end{pmatrix}}_{\boldsymbol{\beta}} + \underbrace{\begin{pmatrix} e_1 \\ e_2 \\ \vdots \\ e_{1266} \end{pmatrix}}_{\mathbf{e}}$$

To understand the associations in the data, we need to estimate $\boldsymbol{\beta}$ which we obtain by the least squares estimator $\hat{\boldsymbol{\beta}} = (\mathbf{X}'\mathbf{X})^{-1}\mathbf{X}'\mathbf{y}$. Rather than doing this tiresome task manually, we use R:

```
lm(time~bill+operator)
```

If we have more than one covariate in a linear regression model, we simply add all of them separated by the $+$ sign when specifying the model formula in the $lm()$ function. We obtain $\hat{\beta}_0 = 23.1$, $\hat{\beta}_1 = 0.26$, $\hat{\beta}_2 = 0.16$. The interpretation of these parameters is as follows:

- For each extra € that is spent, the delivery time increases by 0.26 min. Or, for each extra €10 spent, the delivery time increases on average by 2.6 min.
- The delivery time is on average 0.16 min longer for operator 1 compared with operator 0.
- For operator 0 and a bill of €0, the expected delivery time is $\beta_0 = 23.1$ min. However, there is no bill of €0 and therefore the intercept β_0 cannot be interpreted meaningfully here and is included only for improvement of the model fit.

11.6.2 Categorical Covariates

We now consider the case when covariates are categorical rather than metric/continuous.

Examples

- Region: East, West, South, North,
- Marital status: single, married, divorced, widowed,
- Day: Monday, Tuesday, . . ., Sunday.

We have already described how to treat a categorical variable which consists of two categories, i.e. a binary variable: one category is replaced by 1's, the other category is replaced by 0's subjects who belong to the category $x = 1$ have on average y-values which are $\hat{\beta}$ units higher than those subjects with $x = 0$.

Consider a variable x which has more than 2 categories, say $k > 2$ categories. To include such a variable in a regression model, we create $k - 1$ new variables $x_i, i = 1, 2, \ldots, k - 1$. Similar to how we treat binary variables, each of these variables is a **dummy variable** and consists of 1's for units which belong to the category of interest and 0's for the other category.

$$x_i = \begin{cases} 1 \text{ for category } i, \\ 0 \text{ otherwise.} \end{cases} \qquad (11.21)$$

The category for which we don't create a dummy variable, is called the **reference category** and the interpretation of the parameters of the dummy variables is with respect to this category. For example, for category i, the **y**-values are on average β_i higher than for the reference category. The concept of creating dummy variables and interpreting them is explained in the following example.

Example 11.6.3 Consider again the pizza data set described in Appendix A.4.1. The manager may hypothesize that the delivery times vary with respect to the branch. There are $k = 3$ branches: East, West, Centre. Instead of using the variable $\mathbf{x} =$ branch, we create $(k - 1)$, i.e. $(3 - 1) = 2$ new variables denoting $\mathbf{x_1} =$ East and $\mathbf{x_2} =$ West. We set $\mathbf{x_1} = 1$ for those deliveries which come from the branch in the East and set $\mathbf{x_1} = 0$ for other deliveries. Similarly, we se $\mathbf{x_2} = 1$ for those deliveries which come from the West and $\mathbf{x_2} = 0$ for other deliveries. The data then is as follows:

Delivery	y (Delivery Time)	x (Branch)	x_1 (East)	x_2 (West)
1	35.1	East	1	0
2	25.2	East	1	0
3	45.6	West	0	1
4	29.4	East	1	0
5	30.0	West	0	1
6	40.3	Centre	0	0
⋮	⋮	⋮	⋮	⋮
1266	35.7	West	0	1

Deliveries which come from the East have $x_1 = 1$ and $x_2 = 0$, deliveries which come from the West have $x_1 = 0$ and $x_2 = 1$, and deliveries from the Centre have $x_1 = 0$ and $x_2 = 0$. The regression model of interest is thus

$$y = \beta_0 + \beta_1 \text{East} + \beta_2 \text{West} + e .$$

We can calculate the least squares estimate $\hat{\boldsymbol{\beta}} = (\hat{\beta}_0, \hat{\beta}_1, \hat{\beta}_2)' = (\mathbf{X}'\mathbf{X})^{-1}\mathbf{X}'\mathbf{y}$ via R: either (i) we create the dummy variables ourselves or (ii) we ask R to create it for us. This requires that "branch" is a factor variable (which it is in our data, but if it was not, then we would have to define it in the model formula via `as.factor()`).

```
East <- as.numeric(branch=='East')
West <- as.numeric(branch=='West')
lm(time~East+West)              # option 1
lm(time~branch)                 # option 2a
lm(time~as.factor(branch))      # option 2b
```

We obtain the following results:

$$Y = 36.3 - 5.2\text{East} - 1.1\text{West}.$$

The interpretations are as follows:

- The average delivery time for the branch in the Centre is 36.3 min, the delivery time for the branch in the East is $36.3 - 5.2 = 31.1$ min, the predicted delivery time for the West is $36.3 - 1.1 = 35.2$ min.
- Therefore deliveries arrive on average 5.2 min earlier in the East compared with the centre ($\hat{\beta}_1 = -5.2$); and deliveries in the West arrive on average 1.1 min earlier than in the centre ($\hat{\beta}_2 = -1.1$). In other words, it takes 5.2 min less to deliver a pizza in the East than in the Centre. The deliveries in the West take on average 1.1 min less than in the Centre.

Consider now a covariate with $k = 3$ categories for which we create two new dummy variables, x_1 and x_2. The linear model is $y = \beta_0 + \beta_1 x_1 + \beta_2 x_2 + e$.

- For $x_1 = 1$, we obtain $E(y) = \beta_0 + \beta_1 \cdot 1 + \beta_2 \cdot 0 = \beta_0 + \beta_1 \equiv E(y|x_1 = 1, x_2 = 0)$;
- for $x_2 = 1$, we obtain $E(y) = \beta_0 + \beta_1 \cdot 0 + \beta_2 \cdot 1 = \beta_0 + \beta_2 \equiv E(y|x_1 = 0, x_2 = 1)$;
- for the reference category ($x_1 = x_2 = 0$), we obtain $y = \beta_0 + \beta_1 \cdot 0 + \beta_2 \cdot 0 = \beta_0 \equiv E(y|x_1 = x_2 = 0)$.

We can thus conclude that the intercept $\beta_0 = E(y|x_1 = 0, x_2 = 0)$ describes the average y for the reference category, that $\beta_1 = E(y|x_1 = 1, x_2 = 0) - E(y|x_1 = 0, x_2 = 0)$ describes the difference between the first and the reference category, and that $\beta_2 = \mathrm{E}(\mathbf{y}|\mathbf{x_1} = 0, \mathbf{x_2} = 1) - \mathrm{E}(\mathbf{y}|\mathbf{x_1} = 0, \mathbf{x_2} = 0)$ describes the difference between the second and the reference category.

Remark 11.6.1 There are other ways of re-coding categorical variables, such as effect coding. However we do not describe them in this book.

11.6.3 Transformations

As we have already outlined in Sect. 11.5, if $\mathbf{x}$ is transformed by a function $T(\mathbf{x}) = (T(x_1), T(x_2), \ldots, T(x_n))$ then

$$y = f(\mathbf{x}) = \beta_0 + \beta_1 T(\mathbf{x}) + \mathbf{e} \qquad (11.22)$$

is still a linear model because the model is linear in its parameters $\boldsymbol{\beta}$. Popular transformations $T(\mathbf{x})$ are $\log(\mathbf{x})$, $\sqrt{\mathbf{x}}$, $\exp(\mathbf{x})$, $\mathbf{x}^p$, among others. The choice of such a function is typically motivated by the application and data at hand. Alternatively, a relationship between $\mathbf{y}$ and $\mathbf{x}$ can be modelled via a **polynomial** as follows:

$$y = f(\mathbf{x}) = \beta_0 + \beta_1 \mathbf{x} + \beta_2 \mathbf{x}^2 + \ldots + \beta_p \mathbf{x}^p + \mathbf{e}. \qquad (11.23)$$

Example 11.6.4 Consider again Examples 11.2.1, 11.3.2, 11.3.3, 11.4.1, and 11.5.1 where we analysed the association of intensity of exercising and recovery time after knee surgery. The linear regression line was estimated as

$$\text{Recovery Time} = 100.28 - 1.22 \, \text{Exercising Time}.$$

One could question whether the association is indeed linear and fit the second and third order polynomials:

$$\text{Recovery Time} = \beta_0 - \beta_1 \text{Exercise} + \beta_2 \text{Exercise}^2,$$
$$\text{Recovery Time} = \beta_0 - \beta_1 \text{Exercise} + \beta_2 \text{Exercise}^2 + \beta_3 \text{Exercise}^3.$$

To obtain the estimates we can use R:

```
lm(Y~X+I(X^2))
lm(Y~X+I(X^2)+I(X^3))
```

The results are $\beta_0 = 121.8$, $\beta_1 = -2.4$, $\beta_2 = 0.014$ and $\beta_0 = 12.9$, $\beta_1 = 6.90$, $\beta_2 = -0.23$, $\beta_3 = 0.002$ for the two models respectively. While it is difficult to interpret these coefficients directly, we can simply plot the regression lines on a scatter plot, see Fig. 11.6.

We see that the regression based on the quadratic polynomial visually fits the data slightly better than the linear polynomial. It seems as if the relation between recovery time and exercise time is not exactly linear and the association is better

Fig. 11.6 Scatter plot and regression lines for Example 11.6.4

modelled through a second order polynomial. The regression line based on the cubic polynomial seems to be even closer to the measured points; however, the functional form of the association looks questionable. Driven by a single data point, we obtain a regression line which suggests a heavily increased recovery time for exercising times greater than 65 min. While it is possible that too much exercising causes damage to the knee and delays recovery, the data doesn't seem to be clear enough to support the shape suggested by the model. This example illustrates the tradeoff between fit (i.e. how good the regression line fits the data) and parsimony (i.e. how well a model can be understood if it becomes more complex). Section 11.8 explores this non-trivial issue in more detail.

11.6.3.1 Transformations of the Outcome Variable

It is also possible to apply transformations to the outcome variable. While in many situations this makes interpretations quite difficult, a log transformation is quite common and easy to interpret. Consider the **log-linear model**

$$\log \mathbf{y} = \beta_0 \mathbf{1} + \beta_1 \mathbf{x_1} + \cdots + \beta_p \mathbf{x_p} + \mathbf{e}$$

where y is required to be greater than 0. Exponentiating on both sides leads to

$$\mathbf{y} = e^{\beta_0 \mathbf{1}} \cdot e^{\beta_1 \mathbf{x_1}} \cdot \ldots \cdot e^{\beta_p \mathbf{x_p}} \cdot e^{\mathbf{e}}.$$

It can be easily seen that a one unit increase in $\mathbf{x_j}$ multiplies $\mathbf{y}$ by e^{β_j}. Therefore, if $\mathbf{y}$ is log-transformed, one simply interprets $\exp(\boldsymbol{\beta})$ instead of $\boldsymbol{\beta}$. For example, if $\mathbf{y}$ is the yearly income, $\mathbf{x}$ denotes gender (1 = male, 0 = female), and $\beta_1 = 0.2$ then one can say that a man's income is $\exp(0.2) = 1.22$ times higher (i.e. 22%) than a woman's.

11.7   The Inductive View of Linear Regression

So far we have introduced linear regression as a method which *describes* the relationship between dependent and independent variables through the best fit to the data. However, as we have highlighted in earlier chapters, often we are interested not only in describing the data but also in drawing conclusions from a sample about the population of interest. For instance, in an earlier example, we have estimated the association of branch and delivery time for the pizza delivery data. The linear model was estimated as delivery time $= 36.3 - 5.2\,\text{East} - 1.1\,\text{West}$; this indicates that the delivery time is on average 5.2 min shorter for the branch in the East of the town compared with the central branch, and 1.1 min shorter for the branch in the West compared with the central branch. When considering all pizza deliveries (the population), not only those collected by us over the period of a month (the sample), we might ask ourselves if 1.1 min is a real difference valid for the entire population or just caused by random error involved in the sample we chose. In particular, we would also like to know what the 95% confidence interval for our parameter estimate is. Does the interval cover "zero"? If yes, we might conclude that we can't be very sure that there is an association and don't reject the null hypothesis that $\beta_{\text{West}} = 0$. In reference to Chaps. 9 and 10 we now apply the concepts of statistical inference and testing in the context of regression analysis.

Point and interval estimation in the linear model. We now re-write the model for the purpose of introducing statistical inference to linear regression as

$$
\begin{aligned}
\mathbf{y} &= \beta_0 \mathbf{1} + \beta_1 \mathbf{x_1} + \ldots + \beta_p \mathbf{x_p} + \mathbf{e} \\
&= \mathbf{X}\boldsymbol{\beta} + \mathbf{e} \quad \text{with} \quad \mathbf{e} \sim N(\mathbf{0}, \sigma^2 \mathbf{I}).
\end{aligned}
\tag{11.24}
$$

where $\mathbf{y}$ is the $n \times 1$ vector of outcomes and $\mathbf{X}$ is the $n \times (p+1)$ design matrix (including a column of 1's for the intercept). The identity matrix $\mathbf{I}$ consists of 1's on the diagonal, and 0's elsewhere and the parameter vector is $\boldsymbol{\beta} = (\beta_0, \beta_1, \ldots, \beta_p)'$. We would like to estimate $\boldsymbol{\beta}$ to make conclusions about a relationship in the population. Similarly, $\mathbf{e}$ reflects the random errors in the population. Most importantly the linear model now contains assumptions about the errors. They are assumed to be normally distributed, $N(0, \sigma^2 \mathbf{I})$, which means that the expectation of the errors is 0, $\mathrm{E}(e_i) = 0$, the variance is $\mathbf{Var}(e_i) = \sigma^2$ (and therefore the same for *all* e_i), and it follows from $\mathbf{Var}(\mathbf{e}) = \sigma^2 \mathbf{I}$ that $\mathrm{Cov}(e_i, e_{i'}) = 0$ for all $i \neq i'$. The assumption of a normal distribution is required to construct confidence intervals and test of hypotheses statistics. More details about these assumptions and the procedures to test their validity on the basis of a given sample of data are explained in Sect. 11.9.

The least squares estimate of $\boldsymbol{\beta}$ is obtained by

$$
\hat{\boldsymbol{\beta}} = (\mathbf{X}'\mathbf{X})^{-1}\mathbf{X}'\mathbf{y}.
\tag{11.25}
$$

It can be shown that $\hat{\boldsymbol{\beta}}$ follow a normal distribution with mean $\mathrm{E}(\hat{\boldsymbol{\beta}}) = \boldsymbol{\beta}$ and covariance matrix $\mathbf{Var}(\hat{\boldsymbol{\beta}}) = \sigma^2 \mathbf{I}$ as

$$
\hat{\boldsymbol{\beta}} \sim N(\boldsymbol{\beta}, \sigma^2 (\mathbf{X}'\mathbf{X})^{-1}).
\tag{11.26}
$$

Note that $\hat{\boldsymbol{\beta}}$ is unbiased (since $E(\hat{\boldsymbol{\beta}}) = \boldsymbol{\beta}$); more details about (11.26) can be found in Appendix C.6. An unbiased estimator of σ^2 is

$$\hat{\sigma}^2 = \frac{(\mathbf{y} - \mathbf{X}\hat{\boldsymbol{\beta}})'(\mathbf{y} - \mathbf{X}\hat{\boldsymbol{\beta}})}{n - (p + 1)} = \frac{\hat{\mathbf{e}}'\hat{\mathbf{e}}}{n - (p + 1)} = \frac{1}{n - (p + 1)} \sum_{i=1}^{n} \hat{e}_i^2 . \qquad (11.27)$$

The errors are estimated from the data as $\hat{\mathbf{e}} = \mathbf{y} - \mathbf{X}\hat{\boldsymbol{\beta}}$ and are called **residuals**.

Before giving a detailed data example, we would like to outline how to draw conclusions about the population of interest from the linear model. As we have seen, both $\boldsymbol{\beta}$ and σ^2 are unknown in the model and are estimated from the data using (11.25) and (11.27). These are our point estimates. We note that if $\beta_j = 0$, then $\beta_j \mathbf{x_j} = 0$ and then the model will not contain the term $\beta_j \mathbf{x_j}$. This means that the covariate $\mathbf{x_j}$ does not contribute to explaining the variation in $\mathbf{y}$. Testing the hypothesis $\beta_j = 0$ is therefore equivalent to testing whether $\mathbf{x}_j$ is associated with $\mathbf{y}$ or not in the sense that it helps to explain the variations in $\mathbf{y}$ or not. To test whether the point estimate is different from zero, or whether the deviations of estimates from zero can be explained by random variation, we have the following options:

1. The $100(1 - \alpha)\%$ confidence interval for each $\hat{\beta}_j$ is

$$\hat{\beta}_j \pm t_{n-p-1;1-\alpha/2} \cdot \hat{\sigma}_{\hat{\beta}_j} \qquad (11.28)$$

with $\hat{\sigma}_{\hat{\beta}} = \sqrt{s_{jj}\hat{\sigma}^2}$ where s_{jj} is the jth diagonal element of the matrix $(\mathbf{X}'\mathbf{X})^{-1}$. If the confidence interval does not overlap 0 then we can conclude that β is different from 0 and therefore $\mathbf{x_i}$ is associated with $\mathbf{y}$ and it is a relevant variable. If the confidence interval includes 0, we cannot conclude that there is an association between $\mathbf{x_i}$ and $\mathbf{y}$.

2. We can formulate a formal test which tests if $\hat{\beta}_j$ is different from 0. The hypotheses are:

$$H_0 : \beta_j = 0 \quad \text{versus} \quad H_1 : \beta_j \neq 0 .$$

The test statistic

$$T = \frac{\hat{\beta}_j}{\sqrt{\hat{\sigma}_{\hat{\beta}_j^2}}}$$

follows a t_{n-p-1} distribution under H_0. If $|T| > t_{n-p-1;1-\alpha/2}$, we can reject the null hypothesis (at α level of significance) otherwise we accept it.

3. The decisions we get from the confidence interval and the T-statistic from points 1 and 2 are identical to those obtained by checking whether the p-value (see also Sect. 10.2.6) is smaller than α, in which case we also reject the null hypothesis.

4. An alternative to making a binary decision is to draw p- and S-value functions by varying the null hypothesis $H_0 : \beta_j = b$ with respect to b. Note that the test model is the combination of the null hypothesis and the collection of background assumptions, which includes $\mathbf{e} \sim N(\mathbf{0}, \sigma^2 \mathbf{I})$. How to check those assumptions is explained in Sect. 11.9.

Example 11.7.1 Recall Examples 11.2.1, 11.3.2, 11.3.3, 11.4.1, 11.5.1, and 11.6.4 where we analysed the association of exercising time and time of recovery from knee surgery. We can use R to obtain a full inductive summary of the linear model:

```
summary(lm(Y~X))
```

```
Coefficients:
            Estimate Std. Error t value Pr(>|t|)
(Intercept) 100.1244    10.3571   9.667 2.17e-06 ***
X            -1.2153     0.2768  -4.391  0.00135 **
---
Signif. codes:  0   *** 0.001   ** 0.01   * 0.05   . 0.1      1

Residual standard error: 11.58 on 10 degrees of freedom
```

Now we explain the meaning and interpretation of different terms involved in the output.

- Under "Estimate" the parameter estimates are listed and we read that the linear model is fitted as $\mathbf{y}$ (recovery time) $= 100.1244 - 1.2153 \cdot \mathbf{x}$ (exercise time). The subtle differences to Example 11.2.1 are due to rounding of numerical values.
- The variance is estimated as $\hat{\sigma}^2 = 11.58^2$ ("Residual standard error"). This is easily calculated manually using the residual sum of squares: $\sum \hat{e}_i^2 / (n - p - 1) = 1/10 \cdot \{(90 - 70.84)^2 + ... + (45 - 62.30)^2\} = 11.58^2$.
- The standard errors $\hat{\sigma}_{\hat{\beta}}$ are listed under "Std. Error". Given that $n - p - 1 = 12 - 1 - 1 = 10$, and therefore $t_{10;0.975} = 2.28$ we can construct a confidence interval for age:

$$-1.22 \pm 2.28 \cdot 0.28 = [-1.86; -0.58]$$

The interval does not include 0 and we therefore conclude that there is an association between exercising time and recovery time. The random error involved in the data is not sufficient to explain the deviation of $\hat{\beta}_i = -1.22$ from 0 (given $\alpha = 0.05$).
- We therefore reject the null hypothesis that $\beta_j = 0$. This can also be seen by comparing the test statistic (listed under "t value" and obtained by $(\hat{\beta}_j - 0)/\sqrt{\hat{\sigma}_{\hat{\beta}}^2} = -1.22/0.277$) with $t_{10,0.975}$, $|-4.39| > 2.28$. Moreover, $p = 0.001355 < \alpha = 0.05$. We can say that there is a significant association between exercising and

recovery time. An S-value of $-\log_2(0.001355) \approx 9.5$ confirms this, with almost 10 bits of information against the null hypothesis. All hypotheses $\beta_j = b$ relating to the values of b contained in the compatibility interval $[-1.86; -0.58]$ have at most 4.3 bits of information against them; this illustrates the support for the claim of a negative relationship between exercising and recovery time.

Sometimes one is interested in whether a regression model is useful in the sense that all β_i's are different from zero and we therefore can conclude that there is an association between any x_i and y. The null hypothesis is

$$H_0 : \beta_1 = \beta_2 = ... = \beta_p = 0$$

and can be tested by the **overall F -test**

$$F = \frac{(\hat{y} - \bar{y})'(\hat{y} - \bar{y})/(p)}{(y - \hat{y})'(y - \hat{y})/(n - p - 1)} = \frac{n - p - 1}{p} \frac{\sum_i (\hat{y}_i - \bar{y})^2}{\sum_i e_i^2}.$$

The null hypothesis is rejected if $F > F_{1-\alpha; p, n-p-1}$. Note that the null hypothesis in this case tests only the equality of slope parameters and does not include the intercept term.

Example 11.7.2 In this chapter, we have already explored the associations between branch, operator and bill with delivery time for the pizza data (Appendix A.4.1). If we fit a multiple linear model including all of the three variables, we obtain the following results:

```
summary(lm(time~branch+bill+operator))
```

```
Coefficients:
                  Estimate Std. Error t value Pr(>|t|)
(Intercept)       26.19138    0.78752  33.258  < 2e-16 ***
branchEast        -3.03606    0.42330  -7.172 1.25e-12 ***
branchWest        -0.50339    0.38907  -1.294   0.196
bill               0.21319    0.01535  13.885  < 2e-16 ***
operatorMelissa    0.15973    0.31784   0.503   0.615
---
Signif. codes:  0   *** 0.001   ** 0.01   * 0.05   . 0.1      1

Residual standard error: 5.653 on 1261 degrees of freedom
Multiple R-squared:  0.2369,     Adjusted R-squared:  0.2345
F-statistic: 97.87 on 4 and 1261 DF,  p-value: < 2.2e-16
```

By looking at the p-values in the last column, one can easily see (without calculating the confidence intervals or evaluating the t-statistic) that there is a significant association between the bill and delivery time; also, it is evident that the average delivery time in the branch in the East is significantly different ($\approx$ 3 min less) from the Central branch (which is the reference category here). However, the estimated difference in delivery times for both the branches in the West and the operator was not found to be significantly different from zero. We conclude that some variables in the model are associated with delivery time, while for others we could not show the existence of such an association. The last line of the output confirms this: the overall F-test has a test statistic of 97.87 which is larger than $F_{1-\alpha;\,p,n-p-1} = F_{0.95;4,1261} = 2.37$; therefore the p-value is smaller 0.05 (2.2×10^{-16}) and the null hypothesis is rejected. The test suggests that there is at least one variable which is associated with delivery time.

11.7.1 Properties of Least Squares and Maximum Likelihood Estimators

The least squares estimator of $\boldsymbol{\beta}$ has several properties:

1. The least squares estimator of $\boldsymbol{\beta}$ is *unbiased*, $E(\hat{\boldsymbol{\beta}}) = \boldsymbol{\beta}$, see Appendix C.6 for the proof.
2. The estimator $\hat{\sigma}^2$ as introduced in Eq. (11.27) is also *unbiased*, i.e. $E(\hat{\sigma}^2) = \sigma^2$.
3. The least squares estimator of $\boldsymbol{\beta}$ is *consistent*, i.e. $\hat{\boldsymbol{\beta}}$ converges to $\boldsymbol{\beta}$ as n approaches infinity.
4. The least squares estimator of $\boldsymbol{\beta}$ is *asymptotically normally distributed*.
5. The least squares estimator of $\boldsymbol{\beta}$ has the smallest variance among all linear and unbiased estimators (*best linear unbiased estimator*) of $\boldsymbol{\beta}$.

We do not discuss the technical details of these properties in detail. It is more important to know that the least squares estimator has good features and that is why we choose it for fitting the model. Since we use a "good" estimator we expect that the model will also be "good". One may ask whether it is possible to use a different estimator. We have already made distributional assumptions in the model: we require the errors to be normally distributed. Given that, it is indeed possible to apply the maximum likelihood principle to obtain estimates for $\boldsymbol{\beta}$ and σ^2 in our setup.

Theorem 11.7.1 *For the linear model (11.24), the least squares estimator and the maximum likelihood estimator for β are identical. However, the maximum likelihood estimator of σ^2 is $\hat{\sigma}_{ML}^2 = 1/n(\hat{\mathbf{e}}'\hat{\mathbf{e}})$ which is a biased estimator of σ^2, but it is asymptotically unbiased.*

How to obtain the maximum likelihood estimator for the linear model is presented in Appendix C.6. The important message of Theorem 11.7.1 is that no matter whether we apply the least squares principle or the maximum likelihood principle, we always obtain $\hat{\boldsymbol{\beta}} = (\mathbf{X}'\mathbf{X})^{-1}\mathbf{X}'\mathbf{y}$; this does not apply when estimating the variance, but it is an obvious choice to go for the unbiased estimator (11.27) in any given analysis.

11.7.2 The ANOVA Table

A table that is frequently shown by software packages when performing regression analysis is the analysis of variance (ANOVA) table. This table can have several meanings and interpretations, and may look slightly different depending on the context in which it is used. We focus here on its interpretation (i) as a way to summarize the effect of categorical variables and (ii) as a hypothesis test to compare k means. This is illustrated in the following example.

Example 11.7.3 Recall Example 11.6.3 where we established the relationship between branch and delivery time as

$$\text{Delivery Time} = 36.3 - 5.2\text{East} - 1.1\text{West}.$$

The R output for the respective linear model is as follows:

```
             Estimate Std. Error t value Pr(>|t|)
(Intercept)   36.3127    0.2957 122.819  < 2e-16 ***
branchEast    -5.2461    0.4209 -12.463  < 2e-16 ***
branchWest    -1.1182    0.4148  -2.696  0.00711 **
```

We see that the average delivery time of the branches in the East and the Centre (reference category) are different, and that the average delivery time of the branches in the West and the Centre are different (because $p < 0.05$ and $S = -\log_2(0.00711) > 7$). This is useful information but it doesn't answer the question if branch as a whole influences the delivery time. It seems that this is the case, but the hypothesis we may have in mind may be

$$H_0 : \mu_{\text{East}} = \mu_{\text{West}} = \mu_{\text{Centre}}$$

which corresponds to

$$H_0 : \beta_{\text{East}} = \beta_{\text{West}} = \beta_{\text{Centre}}$$

in the context of the regression model. These are two identical hypotheses because in the regression setup, we are essentially comparing three conditional means $E(Y|X = x_1) = E(Y|X = x_2) = E(Y|X = x_3)$. The ANOVA table summarizes the corresponding F-Test which tests this hypothesis:

```
m1 <- lm(time~branch)
anova(m1)
```

```
Response: time
            Df Sum Sq Mean Sq F value   Pr(>F)
branch       2   6334  3166.8   86.05 < 2.2e-16 ***
Residuals 1263  46481    36.8
```

We see that the null hypothesis of 3 equal means is rejected because p is close to zero.

What does this table mean more generally? If we deal with linear regression with one (possibly categorical) covariate the table will be as follows:

	df	Sum of squares	Mean squares	F-statistic
Var	p	$SQ_{Reg.}$	$MSR = SQ_{Reg.}/p$	MSR/MSE
Res.	$n - p - 1$	SQ_{Error}	$MSE = SQ_{Error}/(n - p - 1)$	

The table summarizes the sum of squares regression and residuals (see Sect. 11.3 for the definition), standardizes them by using the appropriate degrees of freedom (df) and uses the corresponding ratio as the F-statistic. Note that in the above example this corresponds to the overall F-test introduced earlier. The overall F-test tests the hypothesis that any β_j is different from zero which is identical to the hypothesis above. Thus, if we fit a linear regression model with one variable, the ANOVA table will yield the same conclusions as the overall F-test which we obtain through the main summary. However, if we consider a multiple linear regression model, the ANOVA table may give us more information.

Example 11.7.4 Suppose we are not only interested in the branch, but also in how the pizza delivery times are affected by operator and driver. We may for example hypothesize that the delivery time is identical for all drivers given they deliver for the same branch and speak to the same operator. In a standard regression output, we would get 4 coefficients for 5 drivers which would help us to compare the average delivery time of each driver to the reference driver; it would however not tell us if on an average, they all have the same delivery time. The overall F-test would not help us either because it would test if any β_j is different from zero which includes coefficients from branch and operator, not only driver. Using the anova command yields the results of the F-test of interest:

```
m2 <- lm(time~branch+operator+driver)
anova(m2)
```

```
Response: time
            Df Sum Sq Mean Sq F value    Pr(>F)
branch       2   6334  3166.8 88.6374 < 2.2e-16 ***
operator     1     16    16.3  0.4566   0.4994
driver       4   1519   379.7 10.6282 1.798e-08 ***
Residuals 1258  44946    35.7
```

We see that the null hypothesis of equal delivery times of the drivers is rejected. We can also test other hypotheses with this output: for instance, the null hypothesis of equal delivery times for each operator is not rejected because $p \approx 0.5$ (and $S = -\log_2(0.4994) \approx 1$).

11.7.3 Interactions

It may be possible that the joint effect of some covariates affects the response. For example, drug concentrations may have a different effect on men, woman, and children; a fertilizer could work differently in different geographical areas; or a new education programme may show benefit only with certain teachers. It is fairly simple to target such questions in linear regression analysis by using **interactions**. Interactions are measured as the product of two (or more) variables. If either one or both variables are categorical, then one simply builds products for each dummy variable, thus creating $(k - 1) \times (l - 1)$ new variables when dealing with two categorical variables (with k and l categories respectively). These product terms are called interactions and estimate how an association of one variable differs with respect to the values of the other variable. Now we give examples for metric–categorical, categorical–categorical, and metric–metric interactions for two variables x_1 and x_2.

Categorical–Metric Interaction.
Suppose one variable x_1 is categorical with k categories, and the other variable x_2 is metric. Then $k - 1$ new variables have to be created, each consisting of the product of the metric variable and a dummy variable, $x_2 \times x_{1_i}, i \in 1, \ldots, k - 1$. These variables are added to the regression model in addition to the *main effects* relating to x_1 and x_2 as follows:

$$y = \beta_0 + \beta_1 x_{1_1} + \cdots + \beta_{k-1} x_{1_{k-1}} + \beta_k x_2$$
$$+ \beta_{k+1} x_{1_1} x_2 + \cdots + \beta_p x_{1_{k-1}} x_2 + e.$$

It follows that for the reference category of x_1 the effect of x_2 is just β_k (because each $x_2 x_{1_i}$ is zero since each x_{1_j} is zero). However, the effect for all other categories is $\beta_2 + \beta_j$ where β_j refers to $x_{1_j} x_2$. Therefore the association between x_2 and the outcome y differs by β_j between category j and the reference category. Testing $H_0 : \beta_j = 0$ thus helps to identify whether there is an interaction effect with respect to category j or not.

Example 11.7.5 Consider again the pizza data described in Appendix A.4.1. We may be interested in whether the association of delivery time and temperature varies with respect to branch. In addition to time and branch, we therefore need additional interaction variables. Since there are 3 branches we need $3 - 1 = 2$ interaction variables which are essentially the product of (i) time and branch "East" and (ii) time and branch "West". This can be achieved in R by either using the "$\star$" operator (which will create both the main and interaction effects) or the ":" operator (which only creates the interaction term).

```
int.model.1 <- lm(temperature~time*branch)
int.model.2 <- lm(temperature~time+branch+time:branch)
summary(int.model.1)
summary(int.model.2)
```

```
Coefficients:
                  Estimate Std. Error t value Pr(>|t|)
(Intercept)       70.718327   1.850918  38.207  < 2e-16   ***
time              -0.288011   0.050342  -5.721 1.32e-08   ***
branchEast        10.941411   2.320682   4.715 2.69e-06   ***
branchWest         1.102597   2.566087   0.430  0.66750
time:branchEast   -0.195885   0.066897  -2.928  0.00347   **
time:branchWest    0.004352   0.070844   0.061  0.95103
```

The main effects of the model tell us that the temperature is almost 11° higher for the eastern branch compared to the central branch (reference) and about 1° higher for the western branch. However, only the former difference is significantly different from 0 (since the p-value is smaller than $\alpha = 0.05$). Moreover, the longer the delivery time, the lower the temperature (0.29° for each minute). The parameter estimates related to the interaction imply that this association differs by branch: the estimate is indeed $\beta_{time} = -0.29$ for the reference branch in the Centre, but the estimate for the branch in the East is $-0.29 - 0.196 = -0.486$ and the estimate for the branch in the West is $-0.29 + 0.004 = -0.294$. However, the latter difference in the association of time and temperature is not significantly different from zero. We therefore conclude that the delivery time and pizza temperature are negatively associated and this is more strongly pronounced in the eastern branch compared to the other two branches. It is also possible to visualize this by means of a separate regression line for each branch, see Fig. 11.7.

- Centre: Temperature $= 70.7 - 0.29 \times$ time,
- East: Temperature $= 70.7 + 10.9 - (0.29 + 0.195) \times$ time,
- West: Temperature $= 70.7 + 1.1 - (0.29 - 0.004) \times$ time.

One can see that the pizzas delivered by the branch in the East are overall hotter but longer delivery times level that benefit off. One might speculate that the delivery boxes from the eastern branch are not properly closed and therefore–despite the overall good performance–the temperature falls more rapidly over time for this branch.

Categorical–Categorical Interaction. For two categorical variables x_1 and x_2, with k and l categories respectively, $(k - 1) \times (l - 1)$ new dummy variables $x_{1i} \times x_{2j}$ need to be created as follows:

Fig. 11.7 Interaction of delivery time and branch in Example 11.7.5

$$y = \beta_0 + \beta_1 \mathbf{x}_{1_1} + \ldots + \beta_{k-1} \mathbf{x}_{1_{k-1}} + \beta_k \mathbf{x}_{2_1} + \ldots + \beta_{k+l-2} \mathbf{x}_{2_{l-1}}$$
$$+ \beta_{k+l-1} \mathbf{x}_{1_1} \mathbf{x}_{2_1} + \ldots + \beta_p \mathbf{x}_{1_{k-1}} \mathbf{x}_{2_{l-1}} + \mathbf{e} \, .$$

The interpretation of the regression parameters of interest is less complicated than it looks at first. If the interest is in $\mathbf{x}_1$, then the estimate referring to the category of interest (i.e. $\mathbf{x}_{1_i}$) is interpreted as usual—with the difference that it relates only to the reference category of $\mathbf{x}_2$. The effect of $\mathbf{x}_{1_i}$ may vary with respect to $\mathbf{x}_2$ and the sum of the respective main and interaction effects for each category of $\mathbf{x}_2$ yields the respective estimates. These considerations are explained in more detail in the following example.

Example 11.7.6 Consider again the pizza data. If we have the hypothesis that the delivery time depends on the operator (who receives the phone calls), but the effect is different for different branches, then a regression model with branch (3 categories, 2 dummy variables), operator (2 categories, one dummy variable), and their interaction (2 new variables) can be used.

```
lm(time~operator*branch)
```

```
Coefficients:
                          Estimate Std. Error t value Pr(>|t|)
(Intercept)                36.4203     0.4159  87.567   <2e-16 ***
operatorMelissa            -0.2178     0.5917  -0.368   0.7129
branchEast                 -5.6685     0.5910  -9.591   <2e-16 ***
branchWest                 -1.3599     0.5861  -2.320   0.0205 *
operatorMelissa:branchEast  0.8599     0.8425   1.021   0.3076
operatorMelissa:branchWest  0.4842     0.8300   0.583   0.5598
```

The interaction terms can be interpreted as follows:

- If we are interested in the operator, we see that the delivery time is on average 0.21 min shorter for operator "Melissa". When this operator deals with a branch other than the reference (Centre), the estimate changes to $-0.21 + 0.86 = 0.64$ min longer delivery in the case of branch "East" and $-0.21 + 0.48 = 0.26$ min for branch "West".
- If we are interested in the branches, we observe that the delivery time is shortest for the eastern branch which has on average a 5.66 min shorter delivery time than the central branch. However, this is the estimate for the reference operator only; if operator "Melissa" is in charge, then the difference in delivery times for the two branches is only $-5.66 + 0.86 = -4.8$ min. The same applies when comparing the western branch with the central branch: depending on the operator, the difference between these two branches is estimated as either -1.36 or $-1.36 + 0.48 = 0.88$ min respectively.
- The interaction terms are not significantly different from zero (and the respective S-values are at most 1.7). We therefore conclude that the hypothesis of different delivery times for the two operators, possibly varying by branch, could not be confirmed.

Metric–Metric Interaction. It is also possible to add an interaction of two metric variables. This is done by adding the product of these two variables to the model. If x_1 and x_2 are two mtric variables, then $x_1 x_2$ is the new variable added in the model as an interaction effect as

$$\mathbf{y} = \beta_1 \mathbf{x}_1 + \beta_2 \mathbf{x}_2 + \beta_3 \mathbf{x}_1 \mathbf{x}_2 + \mathbf{e}.$$

Example 11.7.7 If we again consider the pizza data, with pizza temperature as an outcome, we may wish to test for an interaction of bill and time.

```
Coefficients:
              Estimate Std. Error t value Pr(>|t|)
(Intercept) 92.555943   2.747414  33.688  < 2e-16  ***
bill        -0.454381   0.068322  -6.651 4.34e-11  ***
time        -0.679537   0.086081  -7.894 6.31e-15  ***
bill:time    0.008687   0.002023   4.294 1.89e-05  ***
```

The R output above reveals that there is a significant interaction effect. The interpretation is more difficult here. It is clear that a longer delivery time and a larger bill decrease the pizza's temperature. However, for a large product of bill and time (i.e., when both are large) these associations are less pronounced because the negative coefficients become more and more outweighed by the positive interaction term. On the contrary, for a small bill and short delivery time, the temperature can be expected to be quite high.

Combining Estimates. As we have seen in the examples in this section it can make sense to combine regression estimates when interpreting interaction effects. While it is simple to obtain the point estimates, it is certainly also important to report their 95% confidence intervals. As we know from Theorem 7.7.1 the variance of the combination of two random variables can be obtained by $\text{Var}(X \pm Y) = \sigma^2_{XY} = \text{Var}(X) + \text{Var}(Y) \pm 2\text{Cov}(X, Y)$. We can therefore estimate a confidence interval for the combined estimate $\hat{\beta}_i + \hat{\beta}_j$ as

$$(\hat{\beta}_i + \hat{\beta}_j) \pm t_{n-p-1;1-\alpha/2} \cdot \hat{\sigma}_{(\hat{\beta}_i + \hat{\beta}_j)} \tag{11.29}$$

where $\hat{\sigma}_{(\hat{\beta}_i + \hat{\beta}_j)}$ is obtained from the estimated covariance matrix $\widehat{\text{Cov}}(\hat{\boldsymbol{\beta}})$ via

$$\hat{\sigma}_{(\hat{\beta}_i + \hat{\beta}_j)} = \sqrt{\widehat{\text{Var}}(\beta_i) + \widehat{\text{Var}}(\beta_j) + 2\widehat{\text{Cov}}(\beta_i, \beta_j)}.$$

Example 11.7.8 Recall Example 11.7.5 where we estimated the association between pizza temperature, delivery time, and branch. There was a significant interaction effect for time and branch. Using R, we obtain the covariance matrix as follows:

```
mymodel <- lm(temperature~time*branch)
vcov(mymodel)
```

	(Int.)	time	East	West	time:East	time:West
(Int.)	3.4258	-0.09202	-3.4258	-3.4258	0.09202	0.09202
time	-0.0920	0.00253	0.0920	0.0920	-0.00253	-0.00253
branchEast	-3.4258	0.09202	5.3855	3.4258	-0.15232	-0.09202
branchWest	-3.4258	0.09202	3.4258	6.5848	-0.09202	-0.17946
time:East	0.0920	-0.00253	-0.1523	-0.0920	0.00447	0.00253
time:West	0.0920	-0.00253	-0.0920	-0.1794	0.00253	0.00501

The point estimate for the association between time and temperature in the eastern branch is $-0.29 - 0.19$ (see Example 11.7.5). The standard error is calculated as $\sqrt{0.00253 + 0.00447 - 2 \cdot 0.00253} = 0.044$. The confidence interval is therefore $-0.48 \pm 1.96 \cdot 0.044 = [-0.56; -0.39]$.

Remark 11.7.1 If there is more than one interaction term then it is generally possible to test whether there is an overall interaction effect, such as $\beta_1 = \beta_2 = \beta_3 = \beta_4 = 0$ in the case of four interaction variables. These tests belong to the general class of "linear hypotheses" and they are not explained in this book. It is also possible to create interactions between more than two variables. However the interpretation then becomes difficult.

11.8 Comparing Different Models

There are many situations where different multiple linear models can be fitted to a given data set, but it is unclear which is the best one. This relates to the question of which variables should be included in a model and which ones can be removed.

For example, when we modelled the association of recovery time and exercising time by using different transformations, it remained unclear which of the proposed transformations was the best. In the pizza delivery example, we have about ten covariates: does it make sense to include all of them in the model or do we understand the associations in the data better by restricting ourselves to the few most important variables? Is it necessary to include interactions or not?

There are various model selection criteria to compare the quality of different fitted models. They can be useful in many situations but there are also a couple of limitations which make model selection a difficult and challenging task.

Coefficient of Determination (R^2). A possible criterion to check the goodness of fit of a model is the coefficient of determination (R^2). We discussed its development, interpretation, and philosophy in Sect. 11.3, see page 275 for more details.

Adjusted R^2. Although R^2 is a reasonable measure for the goodness of fit of a model it also has some limitations. One limitation is that if the number of covariates increases, R^2 increases too (we omit the theorem and the proof for this statement). The added variables may not be relevant but R^2 will still increase, wrongly indicating an improvement in the fitted model. This criterion is therefore not a good measure for model selection. An adjusted version of R^2 is defined as

$$R^2_{adj} = 1 - \frac{SQ_{\text{Error}}/(n - p - 1)}{SQ_{\text{Total}}/(n - 1)} . \tag{11.30}$$

It can be used to compare the goodness of fit of models with a different number of covariates. Please note that both R^2 and R^2_{adj} are only defined when there is an intercept term in the model (which we assume throughout the chapter).

Example 11.8.1 In Fig. 11.6, the association of exercising time and recovery time after knee surgery is modelled linearly, quadratically, and cubically; in Fig. 11.5 this association is modelled by means of a square-root transformation. The model summary in R returns both R^2 (under "Multiple R-squared") and the adjusted R^2 (under "adjusted R-squared"). The results are as follows:

	R^2	R^2_{adj}
Linear	0.6584	0.6243
Quadratic	0.6787	0.6074
Cubic	0.7151	0.6083
square root	0.6694	0.6363

It can be seen that R^2 is larger for the models with more variables, i.e. the cubic model (which includes three variables) has the largest R^2. The adjusted R^2, which takes the different model sizes into account, favours the model with the square-root transformation. This model provides therefore the best fit to the data among the four models considered using R^2.

Akaike's Information Criterion (AIC) is another criterion for model selection. The AIC is based on likelihood methodology and has its origins in information theory. We don't give a detailed motivation of this criterion here. It can be written for the linear model as

$$\text{AIC} = n \log \left(\frac{SQ_{\text{Error}}}{n} \right) + 2(p+1).\tag{11.31}$$

The smaller the AIC, the better the model. The AIC takes not only the fit to the data via SQ_{Error} into account but also the parsimony of the model via the term $2(p+1)$. It is in this sense a more mature criterion than R^2_{adj} which considers only the fit to the data via SQ_{Error}. Akaike's Information Criterion can be calculated in R via the extractAIC() command. There are also other commands but the results differ slightly because the different formulae use different constant terms. However, no matter what formula is used, the results regarding the model choice are always the same.

Example 11.8.2 Consider again Example 11.8.1. R^2_{adj} preferred the model which includes exercising time via a square-root transformation over the other options. The AIC value for the model where exercise time is modelled linearly is 60.59, when modelled quadratically 61.84, when modelled cubically 62.4, and 60.19 for a square-root transformation. Thus, in line with R^2_{adj}, the AIC also prefers the square-root transformation.

Backward selection. Two models, which differ only by one variable x_j, can be compared by simply looking at the test result for $\beta_j = 0$: if the null hypothesis is rejected, the variable is kept in the model; otherwise, the other model is chosen. If there are more than two models then it is better to consider a systematic approach to comparing them. For example, suppose we have 10 potentially relevant variables and we are not sure which of them to include in the final model. There are $2^{10} = 1024$ possible different combinations of variables and in turn so many choices of models! The inclusion or deletion of variables can be done systematically with various procedures, for example with **backward selection** (also known as **backward elimination**) as follows:

1. Start with the full model which contains all variables of interest, $\Upsilon = \{x_1, x_2, \ldots, x_p\}$.
2. Remove the variable $x_i \in \Upsilon$ which optimizes a criterion, i.e. leads to the smallest AIC (the highest R^2_{adj}, the highest test statistic, or the smallest significance) among all possible choices.
3. Repeat step 2. until a stop criterion is fulfilled, i.e. until no improvement regarding AIC, R^2 or the p-value can be achieved.

There are several other approaches. Instead of moving in the backward direction, we can also move in the forward direction. **Forward selection** starts with a model with no covariates and adds variables as long as the model continues to improve with respect to a particular criterion. The **stepwise selection** approach combines forward selection and backward selection: either one does backward selection and checks in between whether adding variables yields improvement; or, one does forward selection and continuously checks whether removing the already added variables improves the model.

Example 11.8.3 Consider the pizza data: if delivery time is the outcome, and branch, bill, operator, driver, temperature and number of ordered pizzas are potential covariates, we may decide to include only the relevant variables in a final model. Using the stepAIC function of the library(MASS) allows us implementing backward selection with *R*.

```
library(MASS)
ms <- lm(time~branch+bill+operator+driver+temperature+pizzas)
stepAIC(ms, direction='back')
```

At the first step, the *R* output states that the AIC of the full model is 4277.56. Then, the AIC values are displayed when variables are removed: 4275.9 if operator is removed, 4279.2 if driver is removed, and so on. One can see that the AIC is minimized if operator is excluded.

```
Start:  AIC=4277.56
time ~ branch+bill+operator+driver+temperature+pizzas

              Df Sum of Sq   RSS    AIC
- operator     1      9.16 36508 4275.9
<none>                      36499 4277.6
- driver       4    279.71 36779 4279.2
- branch       2    532.42 37032 4291.9
- pizzas       1    701.57 37201 4299.7
- temperature  1   1931.50 38431 4340.8
- bill         1   2244.28 38743 4351.1
```

Now *R* fits the model without operator. The AIC is 4275.88. Excluding further variables does not improve the model with respect to the AIC.

```
Step:  AIC=4275.88
time ~ branch + bill + driver + temperature + pizzas

               Df Sum of Sq   RSS    AIC
<none>                       36508 4275.9
- driver        4    288.45 36797 4277.8
- branch        2    534.67 37043 4290.3
- pizzas        1    705.53 37214 4298.1
- temperature   1   1923.92 38432 4338.9
- bill          1   2249.60 38758 4349.6
```

We therefore conclude that the final "best" model includes all variables considered, except operator. Using stepwise selection (with option both instead of back) yields the same results. We could now interpret the summary of the chosen model.

Limitations and further practical considerations. The above discussions of variable selection approaches makes it clear that there are a variety of options to compare fitted models but there is no unique best way to do so. Different criteria are based on different philosophies and may lead to different outcomes. There are a few considerations that should however always be taken into account:

- If categorical variables are included in the model then all respective dummy variables need to be considered as a whole: all dummy variables of this variable should either be removed or kept in the model. For example, a variable with 3 categories is represented by two dummy variables in a regression model. In the process of model selection, both of these variables should either be kept in the model or removed, but not just one of them.
- A similar consideration refers to interactions. If an interaction is added to the model, then the main effects should be kept in the model as well. This enables straightforward interpretations as outlined in Sect. 11.7.3.
- The same applies to polynomials. If adding a cubic transformation of x_i, the squared transformation as well as the linear effect should be kept in the model.
- When applying backward, forward or stepwise regression techniques, the results may vary considerably depending on the method chosen! This may give different models for the same data set. Depending on the strategy used to add and remove variables, the choice of the criterion (e.g. AIC versus p-values), and the detailed setup (e.g. if the strategy is to add/remove a variable if the p value is smaller than α, then the choice of α is important), one may obtain different results.
- All the above considerations show that model selection is a non-trivial task which should always be complemented by substantial knowledge of the subject matter. If there are not too many variables to choose from, simply reporting the full model can be meaningful too.

11.9   Checking Model Assumptions

The assumptions made for the linear model mostly relate to the error terms: $\mathbf{e} \sim N(\mathbf{0}, \sigma^2 \mathbf{I})$. This implies a couple of things: (i) the difference between $\mathbf{y}$ and $\mathbf{X}\boldsymbol{\beta}$ (which is $\mathbf{e}$) needs to follow a normal distribution, (ii) the variance for each error e_i is constant as σ^2 and therefore does not depend on i, and (iii) there is no correlation between the error terms, $\mathrm{Cov}(e_i, e_{i'}) = 0$ for all i, i'. We also assume that $\mathbf{X}\boldsymbol{\beta}$ is adequate in the sense that we included all relevant variables and interactions in the model, and that the estimator $\hat{\boldsymbol{\beta}}$ is stable and adequate due to influential observations or highly correlated variables. Now we demonstrate how to check assumptions (i) and (ii).

Normality assumption. The errors $\mathbf{e}$ are assumed to follow a normal distribution. We never know the errors themselves. The only possibility is to estimate the errors by means of the residuals $\hat{\mathbf{e}} = \mathbf{y} - \mathbf{X}\hat{\boldsymbol{\beta}}$. To ensure that the estimated residuals are comparable to the scale of a *standard* normal distribution, they can be standardized via

$$\hat{e}_i^* = \frac{\hat{e}_i}{\hat{\sigma}^2 \sqrt{1 - P_{ii}}}$$

where P_{ii} is the ith diagonal element of the hat matrix $\mathbf{P} = \mathbf{X}(\mathbf{X}'\mathbf{X})^{-1}\mathbf{X}'$. Generally the estimated **standardized residuals** are used to check the normality assumption. To obtain a first impression about the error distribution, one can simply plot a histogram of the standardized residuals. The normality assumption is checked with a QQ-plot where the theoretical quantiles of a standard normal distribution are plotted against the quantiles from the standardized residuals. If the data approximately matches the bisecting line then we have evidence for fulfilment of the normality assumption (Fig. 11.8a), otherwise we do not (Fig. 11.8b).

Small deviations from the normality assumption are not a major problem as the least squares estimator of $\boldsymbol{\beta}$ remains unbiased. However, confidence intervals and tests of hypothesis rely on this assumption and particularly for small sample sizes and/or stronger violations of the normality assumptions, these intervals and conclusions from tests of hypothesis no longer remain valid. In some cases these problems can be fixed by transforming $\mathbf{y}$.

Heteroscedasticity. If errors have a constant variance, we say that the errors are homoscedastic and we call this phenomenon **homoscedasticity**. When the variance of errors is not constant, we say that the errors are heteroscedastic. This phenomenon is also known as **heteroscedasticity**. If the variance σ^2 depends on i then the variability of the e_i will be different for different groups of observations. For example, the daily expenditure on food ($\mathbf{y}$) may vary more among persons with a high income ($\mathbf{x_1}$), so fitting a linear model yields stronger variability of the e_i among higher income groups. Plotting the fitted values $\hat{y}_i$ (or alternatively x_i) against the standardized residuals (or a transformation thereof) can help to detect whether or not problems exist. If there is no pattern in the plot (random plot) then there is likely no violation of the assumption, see Fig. 11.10a. However, if there is a systematic trend,

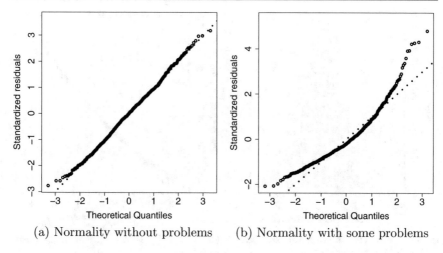

(a) Normality without problems   (b) Normality with some problems

Fig. 11.8 Checking the normality assumption with a QQ-plot

i.e. higher/lower variability for higher/lower $\hat{y}_i$, then this indicates heteroscedasticity (Fig. 11.10b, trumpet plot). The consequences are again that confidence intervals and tests may no longer be correct; again, in some situations, a transformation of **y** can help.

Example 11.9.1 Recall Example 11.8.3 where we identified a good model to describe the delivery time for the pizza data. Branch, bill, temperature, number of pizzas ordered, and driver were found to be associated with delivery time. To explore whether the normality assumption is fulfilled for this model, we can create a histogram for the standardized residuals (using the *R* function rstandard()). A QQ-plot is contained in the various model diagnostic plots of the plot() function.

```
fm <- lm(time~branch+bill+driver+temperature+pizzas)
hist(rstandard(fm))
plot(fm, which=2)
```

Figure 11.9 shows a reasonably symmetric distribution of the residuals, maybe with a slightly longer tail to the right. As expected for a standard normal distribution not many observations are larger than 2 or smaller than -2 (which are close to the 2.5 and 97.5% quantiles). The QQ-plot also points towards a normal error distribution because the observed quantiles lie approximately on the bisecting line. It seems that the lowest residuals deviate a bit from the expected normal distribution, but not extremely. The normality assumption doesn't seem to be violated.

Plotting the fitted values $\hat{y}_i$ against the square root of the absolute values of the standardized residuals ($= \sqrt{|\hat{e}_i^*|}$, used by *R* for stability reasons) yields a plot with no visible structure, see Fig. 11.10a. There is no indication of heteroscedasticity.

```
plot(fm, which=3)
```

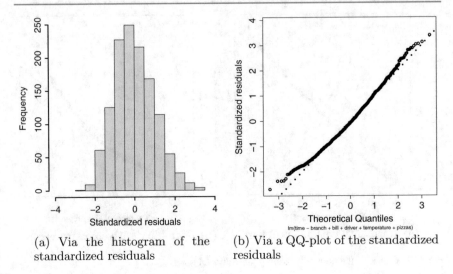

(a) Via the histogram of the standardized residuals

(b) Via a QQ-plot of the standardized residuals

Fig. 11.9 Checking the normality assumption in Example 11.9.1

(a) Example 11.9.1: chaos plot

(b) Violation: σ^2 depends on i

Fig. 11.10 Checking the heteroscedasticity assumption

Figure 11.10b shows an artificial example where heteroscedasticity occurs: the higher the fitted values, the higher is the variation of residuals. This is also called a **trumpet plot**.

11.10   Association Versus Causation

There is a difference between association and causation: association says that higher values of X relate to higher values of Y (or vice versa) whereas causation says that *because* values of X are higher, values of Y are higher too. Regression analysis, as introduced in this chapter, establishes associations between X_i's and Y, not causation, unless additional assumptions are made.

For example, recall the pizza delivery example from Appendix A.4.1. We have seen in several examples and exercises that the delivery time varies by driver. In fact, in a multiple linear regression model where Y is the delivery time, and the covariates are branch, bill, operator, driver and temperature, we get significant differences of the mean delivery times of some drivers (e.g. Domenico and Bruno). Does this mean that *because* Domenico is the driver, the pizzas are delivered faster? Not necessarily. If Domenico drives the fastest scooter then this may be the cause of his speedy deliveries. However, we haven't measured the variable "type of scooter", and thus can't take it into account. The result we obtain from the regression model is still useful because it provides the manager with useful hypotheses and predictions about his delivery times, but the interpretation that the driver's abilities cause shorter delivery times may not necessarily be appropriate.

This example highlights that one of the most important assumptions to interpret regression parameters in a causal way, is to have measured (and used) all variables X_i which affect both the outcome Y and the variable of interest A (e.g. the variable "driver" above). Another assumption is that the relationship between all X_i's and Y is modelled correctly, for example non-linear if appropriate. Moreover, we need some *a priori* assumptions about how the variables relate to each other, and some technical assumptions need to be met as well. More details are given in Chapter 14.

11.11   Key Points and Further Issues

Note:

✓ If X is a metric variable, the parameter β in the linear regression model $Y = \alpha + \beta X + e$ can be interpreted as follows:

"For each unit of increase in X, there is an increase of β units in $\mathrm{E}(Y)$."

✓ If X can take a value of zero then it can be said that

"For $X = 0$, $\mathrm{E}(Y)$ equals α."

If $X = 0$ is out of the data range or implausible it does not make sense to interpret α.

✓ A model is said to be linear when it is linear in its parameters.

✓ It is possible to include many different covariates in the regression model: they can be binary, log-transformed, or take any other form, it is still a *linear* model. The interpretation of these variables is always conditional on the other variables, i.e. the reported association assumes that all other variables in the model are held fixed.

Note:

✓ An important assumption in the multiple linear model $\mathbf{y} = \mathbf{X}\boldsymbol{\beta} + \mathbf{e}$ relates to the errors:

$$\mathbf{e} \sim N(\mathbf{0}, \sigma^2 \mathbf{I}).$$

To check whether the errors are (approximately) normally distributed, one can plot the standardized residuals in a histogram or a QQ-plot. The heteroscedasticity assumption (i.e. σ^2 does not depend on i) is tested by plotting the fitted values ($\hat{y}_i$) against the standardized residuals ($\hat{e}_i$). The plot should show no pattern (random plot).

✓ To evaluate whether $\mathbf{x_j}$ is associated with $\mathbf{y}$, it is possible to test whether $\beta_j = 0$. If the null hypothesis $H_0 : \beta_j = 0$ is rejected (i.e. the p-value is smaller than α), one says that there is a *significant* association of $\mathbf{x_j}$ and $\mathbf{y}$. However, note that a non-significant result does not necessarily mean there is no association, it just means we couldn't show an association.

✓ To avoid dichotomies, a graph of p- or S-values against possible parameter values of β_j shows which parameter values are most compatible with the data under the background assumptions; see Rafi and Greenland (2020), Section 10.9 and Exercise 11.5.

✓ Different models can be compared by systematic hypothesis testing, R^2_{adj}, AIC, and other methods. Different methods may yield different results and there is no unique best option.

11.12 Exercises

Exercise 11.1 The body mass index (BMI) and the systolic blood pressure of 6 people were measured to study a cardiovascular disease. The data are as follows:

Body mass index	26	23	27	28	24	25
Systolic blood pressure	170	150	160	175	155	150

(a) The research hypothesis is that a high BMI relates to a high blood pressure. Estimate the linear model where blood pressure is the outcome and BMI is the covariate. Interpret the coefficients.
(b) Calculate R^2 to judge the goodness of fit of the model.

Exercise 11.2 A psychologist speculates that spending a lot of time on the internet has a negative effect on children's sleep. Consider the following data on hours of deep sleep (Y) and hours spent on the internet (X) where x_i and y_i are the observations on internet time and deep sleep time of the ith($i = 1, 2, \ldots, 9$) child respectively:

Child i	1	2	3	4	5	6	7	8	9
Internet time x_i (in h)	0.3	2.2	0.5	0.7	1.0	1.8	3.0	0.2	2.3
Sleep time y_i (in h)	5.8	4.4	6.5	5.8	5.6	5.0	4.8	6.0	6.1

(a) Estimate the linear regression model for the given data and interpret the coefficients.
(b) Calculate R^2 to judge the goodness of fit of the model.
(c) Reproduce the results of (a) and (b) in R and plot the regression line.
(d) Now assume that we only distinguish between spending more than 1 h on the internet ($X = 1$) and spending less than (or equal to) one hour on the internet ($X = 0$). Estimate the linear model again and compare the results. How can $\hat{\beta}$ now be interpreted? Describe how $\hat{\beta}$ changes if those who spend more than one hour on the internet are coded as 0 and the others as 1.

Exercise 11.3 Consider the following data on weight and height of 17 female students:

Student i	1	2	3	4	5	6	7	8	9
Weight y_i	68	58	53	60	59	60	55	62	58
Height x_i	174	164	164	165	170	168	167	166	160

Student i	10	11	12	13	14	15	16	17
Weight y	53	53	50	64	77	60	63	69
Height x	160	163	157	168	179	170	168	170

(a) Calculate the correlation coefficient of Bravais–Pearson (use $\sum_{i=1}^{n} x_i y_i = 170, 821, \bar{x} = 166.65, \bar{y} = 60.12, \sum_{i=1}^{n} y_i^2 = 62, 184, \sum_{i=1}^{n} x_i^2 = 472, 569$). What does this imply with respect to a linear regression of height on weight?
(b) Now estimate and interpret the linear regression model where "weight" is the outcome.
(c) Predict the weight of a student with a height 175 cm.
(d) Now produce a scatter plot of the data (manually or by using R) and interpret it.
(e) Add the following two points to the scatter plot $(x_{18}, y_{18}) = (175, 55)$ and $(x_{19}, y_{19}) = (150, 75)$. Speculate how the linear regression estimate will change after adding these points.
(f) Re-estimate the model using all 19 observations and $\sum x_i y_i = 191, 696$ and $\sum x_i^2 = 525, 694$.
(g) Given the results of the two regression models: what are the general implications with respect to the least squares estimator of β?

Exercise 11.4 To study the association of the monthly average temperature (in °C, X) and hotel occupation (in %, Y), we consider data from three cities: Polenca (Mallorca, Spain) as a summer holiday destination, Davos (Switzerland) as a winter skiing destination and Basel (Switzerland) as a business destination.

Month	Davos		Polenca		Basel	
	X	Y	X	Y	X	Y
Jan	−6	91	10	13	1	23
Feb	−5	89	10	21	0	82
Mar	2	76	14	42	5	40
Apr	4	52	17	64	9	45
May	7	42	22	79	14	39
Jun	15	36	24	81	20	43
Jul	17	37	26	86	23	50
Aug	19	39	27	92	24	95
Sep	13	26	22	36	21	64
Oct	9	27	19	23	14	78
Nov	4	68	14	13	9	9
Dec	0	92	12	41	4	12

(a) Interpret the following regression model output where the outcome is "hotel occupation" and "temperature" is the covariate:

```
            Estimate Std. Error t value Pr(>|t|)
(Intercept) 50.33459    7.81792   6.438 2.34e-07 ***
LLX          0.07717    0.51966   0.149    0.883
```

(b) Interpret the following output where "city" is treated as a covariate and "hotel occupation" is the outcome:

```
            Estimate Std. Error t value Pr(>|t|)
(Intercept)  48.3333    7.9457    6.083 7.56e-07 ***
cityDavos     7.9167   11.2369    0.705    0.486
cityPolenca   0.9167   11.2369    0.082    0.935
```

(c) Interpret the following output and compare it with the output from (b):

```
Analysis of Variance Table
Response: Y
            Df  Sum Sq Mean Sq F value Pr(>F)
city         2   450.1  225.03   0.297  0.745
Residuals   33 25001.2  757.61
```

(d) In the following multiple linear regression model, both "city" and "temperature" are treated as covariates. How can the coefficients be interpreted?

```
                Estimate Std. Error t value Pr(>|t|)
(Intercept)     44.1731    10.9949   4.018 0.000333 ***
X                0.3467     0.6258   0.554 0.583453
cityDavos        9.7946    11.8520   0.826 0.414692
cityPolenca     -1.1924    11.9780  -0.100 0.921326
```

(e) Now consider the regression model for hotel occupation and temperature, fitted separately for each city: how can the results be interpreted and what are the implications with respect to the models estimated in (a)–(d)? How can the models be improved?

```
Davos:
                Estimate Std. Error t value Pr(>|t|)
(Intercept)     73.9397     4.9462  14.949 3.61e-08 ***
X               -2.6870     0.4806  -5.591 0.000231 ***
```

```
Polenca:
                Estimate Std. Error t value Pr(>|t|)
(Intercept)    -22.6469    16.7849  -1.349  0.20701
X                3.9759     0.8831   4.502  0.00114 **
```

```
Basel:
               Estimate Std. Error t value Pr(>|t|)
(Intercept)     32.574     13.245   2.459   0.0337 *
X                1.313      0.910   1.443   0.1796
```

(f) Describe what the design matrix will look like if city, temperature and the inter-action between them are included in a regression model.

(g) If the model described in (f) is fitted the output is as follows:

```
                 Estimate Std. Error t value Pr(>|t|)
(Intercept)      32.5741    10.0657   3.236 0.002950 **
X                 1.3133     0.6916   1.899 0.067230 .
cityDavos        41.3656    12.4993   3.309 0.002439 **
cityPolenca     -55.2210    21.0616  -2.622 0.013603 *
X:cityDavos      -4.0003     0.9984  -4.007 0.000375 ***
X:cityPolenca     2.6626     1.1941   2.230 0.033388 *
```

Interpret the results.

(h) Summarize the association of temperature and hotel occupation by city–including 95% confidence intervals–using the interaction model. The covariance matrix is as follows:

	(Int.)	X	Davos	Polenca	X:Davos	X:Polenca
(Int.)	101.31	-5.73	-101.31	-101.31	5.73	5.73
X	-5.73	0.47	5.73	5.73	-0.47	-0.47
Davos	-101.31	5.73	156.23	101.31	-9.15	-5.73
Polenca	-101.31	5.73	101.31	443.59	-5.73	-22.87
X:Davos	5.73	-0.47	-9.15	-5.73	0.99	0.47
X:Polenca	5.73	-0.47	-5.73	-22.87	0.47	1.42

Exercise 11.5 The theatre data (see Appendix A.4.3) describes the monthly expenditure on theatre visits of 699 residents of a Swiss city. It captures not only the expenditure on theatre visits (in SFR) but also age, gender, yearly income (in 1000 SFR), expenditure on cultural activities in general as well as expenditure on theatre visits in the preceding year.

(a) The summary of the multiple linear model where expenditure on theatre visits is the outcome is as follows:

	Estimate	Std. Error	t value	Pr(>|t|)
(Intercept)	-127.22271	19.15459	-6.642	6.26e-11 ***
Age	0.39757	0.19689	[1]	[2]
Sex	22.22059	5.22693	4.251	2.42e-05 ***
Income	1.34817	0.20947	6.436	2.29e-10 ***
Culture	0.53664	0.05053	10.620	<2e-16 ***
Theatre_ly	0.17191	0.11711	1.468	0.1426

How can the missing values [1] and [2] be calculated?

(b) Interpret the model diagnostics in Fig. 11.11.
(c) Given the diagnostics in (b), how can the model be improved? Plot a histogram of theatre expenditure in R if you need further insight.
(d) Consider the model where theatre expenditure is log-transformed:

Coefficients:

	Estimate	Std. Error	t value	Pr(>|t|)
(Intercept)	2.9541546	0.1266802	23.320	< 2e-16 ***
Age	0.0038690	0.0013022	2.971	0.00307 **
Sex	0.1794468	0.0345687	5.191	2.75e-07 ***
Income	0.0087906	0.0013853	6.346	4.00e-10 ***
Culture	0.0035360	0.0003342	10.581	< 2e-16 ***
Theatre_ly	0.0013492	0.0007745	1.742	0.08197 .

How can the coefficients be interpreted?

(e) Judge the quality of the model from d) by means of Fig. 11.12a, b. What do they look like compared with those from (b)?
(f) Draw and interpret a S-value function for $\beta_{Theatre\_ly}$ based on the model from d). Hint: Use the function curve_gen of the package concurve, which is explained in Section 10.9.3.

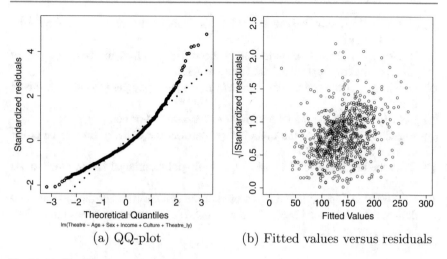

Fig. 11.11 Checking the model assumptions

Fig. 11.12 Checking the model assumptions

Exercise 11.6 Consider the pizza delivery data described in Appendix A.4.1.

(a) Read the data into R. Fit a multiple linear regression model with delivery time as the outcome and temperature, branch, day, operator, driver, bill, number of ordered pizzas, and discount customer as covariates. Give a summary of the coefficients.

(b) Use R to calculate the 95% confidence intervals of all coefficients. Hint: the standard errors of the coefficients can be accessed either via the covariance matrix or the model summary.

(c) Reproduce the least squares estimate of σ^2. Hint: use `residuals()` to obtain the residuals.

(d) Now use R to estimate both R^2 and R^2_{adj}. Compare the results with the model output from a).

(e) Use backward selection by means of the stepAIC function from the library MASS to find the best model according to AIC.

(f) Obtain R^2_{adj} from the model identified in (e) and compare it to the full model from a).

(g) Identify whether the model assumptions are satisfied or not.

(h) Are all variables from the model in (e) causing the delivery time to be either delayed or improved?

(i) Test whether it is useful to add a quadratic polynomial of temperature to the model.

(j) Use the model identified in (e) to predict the delivery time of the last captured delivery (i.e. number 1266). Use the predict() command to ease the calculation of the prediction.

$\rightarrow$ Solutions of all exercises in this chapter can be found in Appendix B

Logistic Regression

12

In Chap. 11, we introduced the linear regression model to describe the association between a metric response variable y and a metric covariate X, assuming a linear relationship between X and y at first. Note that a random variable is typically denoted in uppercase, say Y, and its realization is denoted in lowercase, say y. In the context of regression models, it is however common to not necessarily distinguish between the random variable Y and the observed outcome y. Therefore, in this chapter we use y to denote the random variable, and $y_1, y_2, \ldots, y_n$ to denote a set of n observations on y. Graphically, we can plot the observations of two metric variables in a scatter plot to visualize their relationship; analytically, we can use the method of least squares or maximum likelihood estimation (Sect. 11.2 and Appendix C.6) to fit a linear regression model. We also discussed more general cases in which X is binary as in Sect. 11.4, a transformed covariate as in Sect. 11.5 and where more than one covariate is used to explain the variation in the response variable (the multiple linear regression model, Sect. 11.6). The covariates in the multiple linear regression model can also be categorical with more than two possible categories while y remains a metric response variable. In practice, however, there can be situations where the response variable can also be categorical and its outcome is binary. For example, the outcomes of y may be 'absent' and 'present', indicating if a drug led to a desired response or not. The outcomes 'absent' and 'present' can then be coded as 0 and 1, respectively. In this chapter, we extend the linear regression model to the case when y is a binary response variable with only the two possible categories whose coded values are denoted by 0 and 1. Some other examples of binary outcomes are as follows:

- A patient has a certain disease ($y = 1$) or not ($y = 0$),
- a customer buys a certain product ($y = 1$) or not ($y = 0$) after viewing it online,
- a bank lends a loan to a customer ($y = 1$) or not ($y = 0$).

© Springer Nature Switzerland AG 2022
C. Heumann et al., *Introduction to Statistics and Data Analysis*,
https://doi.org/10.1007/978-3-031-11833-3_12

In Chap. 8, we introduced the Bernoulli distribution, which is suitable to model problems of type $0 - 1$ where the outcomes are categorized into two categories (i.e., 0 and 1). If the response variable in a regression model is binary, then we use the **logistic regression model** which describes the association between one or more categorical and/or metric covariates $(X_1, X_2, \ldots, X_p)$ with the binary outcome y. Recall that such a relationship was formulated in the linear regression model (11.15) for metric outcomes. In Chap. 8 we introduced the probabilities $p = P(y = 1)$ and $1 - p = P(y = 0)$. We also found the mean of y as $E(y) = p$ in (8.4) and the variance of y as $Var(y) = p(1 - p)$ in (8.5). A subtle difference between the logistic regression model and the linear regression model is as follows: In the linear regression model, we link the covariates directly to the response y, whereas in the logistic regression model, we link them to the expected value of y with $E(y) = p$ (of course, we also model $E(y)$ in the linear model, but the model equation usually includes the error term and thus the model is formulated with the observations y, and not the expectation, on the left-hand side of the model equation). Since the expectation is a probability and a probability lies in the interval $(0, 1)$ (we exclude the cases when the probability is exactly 0 or 1), this information can be utilized in the formulation of a logistic regression model. The covariates are linked to $E(y)$, requiring that its values have to lie in the interval $(0, 1)$. Consider n observations, which follow

$$\eta_i = \beta_0 + \beta_1 x_{i1} + \cdots + \beta_p x_{ip} = \mathbf{x}_i'\boldsymbol{\beta} \, , i = 1, 2, \ldots, n, \qquad (12.1)$$

where η_i is called the **linear predictor for observation** i and $\boldsymbol{\beta} = (\beta_0, \beta_1, \ldots, \beta_p)'$ is, as in the linear regression model, a column vector of unknown coefficients. The logistic regression model is then defined as

$$p_i = P(y_i = 1|\mathbf{x}_i) = h(\eta_i) = \frac{\exp(\eta_i)}{1 + \exp(\eta_i)} = \frac{1}{1 + \exp(-\eta_i)} \, . \qquad (12.2)$$

The function $h(\eta_i)$ is called the **response function**. Note that the probabilities p_i are now dependent on the observation index i, i.e. each observation $y_i, i = 1, 2, \ldots, n$, can have a different expected value p_i, depending on the values of the covariates $\mathbf{x}_i = (x_{i1}, \ldots, x_{ip})$ of observation i. Therefore, we say that we model the **conditional probabilities** in a logistic regression model, that is, conditional on the observation specific covariates $\mathbf{x}_i$.

The function on the right-hand side of (12.2) is called the (standard) logistic function (see Fig. 12.1). This is why we call the model a logistic regression model, or sometimes logit model:

The model can be expressed as

$$p_i = P(y_i = 1|\mathbf{x}_i) = \frac{\exp(\mathbf{x}_i'\boldsymbol{\beta})}{1 + \exp(\mathbf{x}_i'\boldsymbol{\beta})} = \frac{1}{1 + \exp(-\mathbf{x}_i'\boldsymbol{\beta})} \, . \qquad (12.3)$$

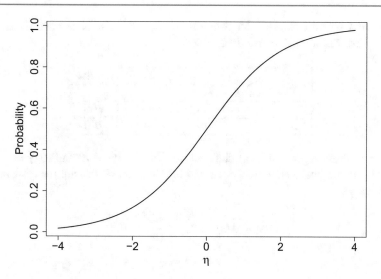

Fig. 12.1 The logistic function, which is the response function of the logistic regression model

Since the function $\frac{\exp(\eta)}{1+\exp(\eta)}$ is strictly monotone increasing in η it can be reformulated based on its inverse function, called the **logit link function**:

$$\log\left(\frac{p_i}{1-p_i}\right) = \log\left(\frac{P(y_i = 1|\mathbf{x}_i)}{P(y_i = 0|\mathbf{x}_i)}\right) = \eta_i = \mathbf{x}_i'\boldsymbol{\beta} = \beta_0 + \beta_1 x_{i1} + \cdots + \beta_p x_{ip}.$$
(12.4)

The terms

$$\log\left(\frac{P(y_i = 1|\mathbf{x}_i)}{P(y_i = 0|\mathbf{x}_i)}\right), \quad i = 1, \ldots, n,$$
(12.5)

are called **logarithmic chances**, **log-odds** or simply the **logits**. For the chances themselves, we have therefore a **multiplicative model**:

$$\frac{P(y_i = 1|\mathbf{x}_i)}{P(y_i = 0|\mathbf{x}_i)} = \exp(\beta_0) \cdot \exp(\beta_1 x_{i1}) \cdot \cdots \cdot \exp(\beta_p x_{ip}).$$
(12.6)

It is important to note that the interpretation of the logistic regression model is not as straightforward as the linear regression model because the response and the link function of the logistic regression model are non-linear functions. Before discussing appropriate interpretations of the regression coefficients $\boldsymbol{\beta}$ further, we first consider the following example.

Example 12.0.1 Consider a study where two treatments A (standard treatment) and B (new treatment) are to be compared. A group of 200 eligible patients are selected, half of them are randomly selected to receive treatment A and the other half receives treatment B. The binary outcome y describes whether the treatment was effective ($y = 1$) in treating the disease or not ($y = 0$). The treatment is the corresponding covariate X with the two possible values A and B, and is also binary in this case. We can thus represent the data in a 2×2-table:

	$y = 1$	$y = 0$	Total
Treatment X			
B	80	20	100
A	40	60	100

The conditional probabilities of interest are $P(y_i = 1 | x_i = A)$ and $P(y_i = 1 | x_i = B)$. The maximum likelihood estimators of these probabilities are the relative frequencies, i.e.

$$\widehat{P}(y_i = 1 | x_i = B) = \frac{80}{100} = 0.8 \,,$$

$$\widehat{P}(y_i = 1 | x_i = A) = \frac{40}{100} = 0.4 \,.$$

It is clear that treatment B has a higher proportion of successful treatment outcomes than treatment A.

Now we formulate this example in terms of a logistic regression model where we represent the treatment covariate x_i as a binary variable:

$$x_i = \begin{cases} 1 & \text{if the assigned treatment is the new treatment } B \\ 0 & \text{if the assigned treatment is the standard treatment } A. \end{cases}$$

The logistic regression model can then be written as

$$\log \left(\frac{P(y_i = 1 | x_i)}{P(y_i = 0 | x_i)} \right) = \beta_0 + \beta_1 x_i \,.$$

Now we relate the relative frequencies to the coefficients. An explicit solution for the estimators $\hat{\beta}_0$ and $\hat{\beta}_1$ can be derived as follows:

$$\log \left(\frac{0.4}{1 - 0.4} \right) = \beta_0 + \beta_1 \cdot 0 \,,$$

$$\log \left(\frac{0.8}{1 - 0.8} \right) = \beta_0 + \beta_1 \cdot 1 \,.$$

We get the estimates

$$\hat{\beta}_0 = \log \left(\frac{0.4}{0.6} \right) = -0.405465 \,,$$

$$\hat{\beta}_1 = \log \left(\frac{0.8}{0.2} \right) - \log \left(\frac{0.4}{0.6} \right) = \log \left(\frac{0.8}{0.2} \right) - \hat{\beta}_0 = 1.791759 \,.$$

As we can see, $\hat{\beta}_1$ is the difference of the two logarithmic chances or logits. Using the calculation rules for the logarithm, we derive

$$\hat{\beta}_1 = \log\left(\frac{0.8 \cdot 0.6}{0.2 \cdot 0.4}\right),$$

$$\exp(\hat{\beta}_1) = \frac{0.8 \cdot 0.6}{0.2 \cdot 0.4} = 6.0, \tag{12.7}$$

where the right-hand side of (12.7) is equal to the odds ratio of the corresponding 2×2-table (i.e., $(80 \cdot 60)/(40 \cdot 20) = 6$; see (4.14) for the definition of the odds ratio in a 2×2-table). We thus have the interpretation of the coefficient β_1 as a log odds ratio, or $\exp(\beta_1)$ as an odds ratio. In this case, the model fits the data perfectly, and such models are called **saturated models**, see also Sect. 14.5.1 for a precise definition.

12.1 Parameter Interpretation

As we have already seen in Example 12.0.1, the coefficients of a logistic regression model have an interpretation as log odds ratios. For the general case of a logistic regression model with p covariates, we can see this with the following ceterus paribus calculations, where we increase the value of covariate x_{ij} by one measurement unit to $x_{ij} + 1$, keeping all other covariates at the same value:

$$\log\left(\frac{P(y_i = 1 | x_{i1}, \dots, x_{ij} + 1, \dots, x_{ip})}{P(y_i = 0 | x_{i1}, \dots, x_{ij} + 1, \dots, x_{ip})}\right)$$
$$- \log\left(\frac{P(y_i = 1 | x_{i1}, \dots, x_{ij}, \dots, x_{ip})}{P(y_i = 0 | x_{i1}, \dots, x_{ij}, \dots, x_{ip})}\right) = \beta_j, \tag{12.8}$$

$$\frac{P(y_i = 1 | x_{i1}, \dots, x_{ij} + 1, \dots, x_{ip})}{P(y_i = 0 | x_{i1}, \dots, x_{ij} + 1, \dots, x_{ip})} \bigg/ \frac{P(y_i = 1 | x_{i1}, \dots, x_{ij}, \dots, x_{ip})}{P(y_i = 0 | x_{i1}, \dots, x_{ij}, \dots, x_{ip})} = \exp(\beta_j). \tag{12.9}$$

For (12.9), we have used the multiplicative representation of Eq. (12.6). We can summarize the possible interpretations as follows:

1. β_j is the additive change of the logits (logarithmic chances), if we increase a unit in x_j by one unit, i.e. from x_{ij} to $x_{ij} + 1$.
2. $\exp(\beta_j)$ is the multiplicative change in the chances (odds), if we increase a unit in x_j by one unit, i.e. from x_{ij} to $x_{ij} + 1$.

3. $\exp(\beta_j)$ is the odds ratio, i.e. the ratio of the chances at $x_{ij} + 1$ and at x_{ij}. Alternatively, we can say that the chance at $x_{ij} + 1$ is "higher" than the chance at x_{ij} by the factor $\exp(\beta_j)$. Note that we have used quotation marks in the last statement since it depends on the sign of the coefficient β_j: if $\beta_j > 0$, the chance increases if x_{ij} is increases and if $\beta_j < 0$, the chance decreases if x_{ij} increases. If $\beta_j = 0$, the chance does not change in either direction.

12.2   Estimation of Parameters and Predictions

In the introductory Example 12.0.1 we used a one-to-one correspondence between the relative frequencies as maximum likelihood estimates and the coefficients of the respective logistic regression. In this case, we had a saturated model which perfectly fits the data. The invariance property of the maximum likelihood estimator ensured that the calculated coefficients $\hat{\beta}_0$ and $\hat{\beta}_1$ are also the maximum likelihood estimators. The invariance property is a nice property of maximum likelihood estimators that says that one gets the maximum likelihood estimate of a function of a parameter by applying the function to the maximum likelihood estimator of the parameter, i.e., for a function g and a parameter θ, $\widehat{g}(\theta)_{ML} = g(\hat{\theta}_{ML})$. In general, an iterative procedure has to be used to find the estimates. The likelihood, as well as more general information on inference in the logistic regression model, is given in Appendix C.7.

Once we have calculated the maximum likelihood estimator $\hat{\beta}$ for the coefficient vector β, the predictions can be calculated through the response function as

$$\hat{p}_i = \frac{\exp(\mathbf{x}_i'\hat{\beta})}{1 + \exp(\mathbf{x}_i'\hat{\beta})} = \frac{1}{1 + \exp(-\mathbf{x}_i'\hat{\beta})} . \tag{12.10}$$

Sometimes, logistic regression is used for classification purposes in two-class (0 or 1 type) classification problems. Once the estimated probabilities $\hat{p}_i$ have been calculated, a classification procedure can be implemented by introducing a constant threshold c such that the predicted class is 1 if $\hat{p}_i > c$ and 0 if $\hat{p}_i \leq c$.

12.3   Logistic Regression in R

The logistic regression model is well integrated into R through the `glm`, a general command for **generalized linear models (GLM)** which not only includes the logistic model but also models for other situations, for example count data. The function has the usual parameters known from the linear regression module in R such as `formula` and `data`, but also an additional `family` parameter where we specify the distribution and the link function. In logistic regression, the `family` parameter has to be specified as `family=binomial(link='logit')` or just `family=binomial`. As an example we use the theatre data (see Appendix A.4.3).

Example 12.3.1 To create a binary response variable, we categorize the `Theatre` variable (expenditure on theatre visits in SFR per month) as 1 if the expenditure exceeds 150 SFR and 0 otherwise and add it to the data frame:

```
theatre$Theatre.bin <- ifelse(theatre$Theatre>150,1,0)
```

Then, a logistic regression model is specified as follows:

```
model.log <- glm(Theatre.bin~Age+Sex+Income+Culture+
Theatre_ly, family=binomial(link='logit'),
data=theatre)
```

The model summary

```
summary(model.log)
```

yields the following output (regression coefficients only):

```
             Estimate Std. Error z value Pr(>|z|)
(Intercept) -8.512305   0.845365 -10.069  < 2e-16 ***
Age          0.013594   0.007274   1.869  0.06164 .
Sex          0.634535   0.196549   3.228  0.00124 **
Income       0.039834   0.007830   5.087 3.63e-07 ***
Culture      0.016468   0.002177   7.564 3.90e-14 ***
Theatre_ly   0.002134   0.004219   0.506  0.61299
```

The output can be interpreted in a similar way to the linear model. It shows the estimates of the regression coefficients, their standard errors, the test statistics (z-values) and the two-sided p-values. The test statistics or z-values are based on the asymptotic normal distribution of the maximum-likelihood estimator, see Appendix C.7 for more details.

We see that the variables `Age`, `Sex`, `Income` and `Culture` have a significant influence on whether a visitor will spend more than 150 SFR per month (or not) if we choose a significance level of $\alpha = 0.1$ (since their p-values are lower than α, see Sect. 10.2.6). The coefficients are all positive (except the intercept). This means that the odds for females (`Sex=1`) to spend more than 150 SFR (versus less than or equal to 150 SFR) is (ceterus paribus) $\exp(0.634535) = 1.886145$ higher than the odds for males. Also, visitors who are older or have a higher income, or have higher general expenditures for culture have higher odds of spending more than 150 SFR for a theatre visit. For example, for each SFR that is spent more on general expenditures for `Culture`, the odds for spending more than 150 SFR for theatre visits is higher by a factor of $\exp(0.016468) = 1.016604$, i.e. 1.66% higher. Of course, all odds ratios can be calculated in one step, applying the exponential function to the whole vector of regression coefficients:

```
print(exp(coef(model.log)), digits=2)
```

```
(Intercept)    Age Sex    Income Culture Theatre_ly
      0.0002 1.0137 1.8861 1.0406 1.0166  1.0021
```

Unfortunately, the summary output of the logistic regression model contains only the coefficients and not the odds ratios themselves or the confidence intervals for the odds ratios. Fortunately, the function `confint` in the standard `MASS` package can be applied directly on a `glm` object, providing confidence intervals in one line. These are intervals for the regression coefficients. Taking the exponential of the lower and upper bounds of these confidence intervals provides confidence intervals for the odds ratios, where the default is a 95%-confidence interval.

```
exp(confint(model.log))
```

In our example the result is

```
              2.5 % 97.5 %
(Intercept) 0.0000 0.0010
Age         0.9993 1.0283
Sex         1.2863 2.7822
Income      1.0250 1.0570
Culture     1.0124 1.0211
Theatre_ly  0.9939 1.0105
```

12.4   Model Selection and Goodness-of-Fit

In Sect. 11.8 we introduced Akaike's Information Criterion (AIC) for the linear model. It can be used for model selection also in the logistic regression case (more technical details are given in Appendix C.7). Note that a model is considered to be "better" than another model, based on the AIC, if its AIC value is lower. The `stepAIC` function from the library `MASS` in R can also be used for GLMs.

Example 12.4.1 Consider again the theatre data analyzed in Example 12.3.1. The `summary` function applied to the `glm` object gives us AIC $= 743.17$. A stepwise (backward) selection (or elimination) can be employed as follows:

```
stepAIC(model.log)
```

```
Start:  AIC=743.17
Theatre.bin ~ Age + Sex + Income + Culture + Theatre_ly
```

```
             Df Deviance    AIC
- Theatre_ly  1    731.42 741.42
<none>             731.17 743.17
- Age         1    734.66 744.66
- Sex         1    741.78 751.78
- Income      1    758.46 768.46
- Culture     1    802.78 812.78


Step:  AIC=741.42
Theatre.bin ~ Age + Sex + Income + Culture

             Df Deviance    AIC
<none>             731.42 741.42
- Age         1    735.51 743.51
- Sex         1    741.89 749.89
- Income      1    761.56 769.56
- Culture     1    803.23 811.23


Call:  glm(formula = Theatre.bin ~ Age + Sex + Income
         + Culture,family = binomial(link = "logit"),
           data = theatre)

Coefficients:
(Intercept)      Age     Sex   Income  Culture
   -8.31499  0.01437 0.62872  0.04071  0.01648
```

The variable `Theatre_ly` is removed in the first step, and then the backward selection process stops. The final model contains the variables `Age`, `Sex`, `Income`, `Culture` with AIC = 741.4, which is lower than the AIC value of the original model, including `Theatre_ly`. Thus, this model could be considered to be better. The point estimates of the final model are also given, but to get the full information, we would have to re-estimate the logistic regression with the reduced number of variables (see also Exercise 12.3).

It is important to reiterate the limitations of model selection, which we have discussed in Sect. 11.8 in detail. When applying backward, forward or stepwise regression techniques, the results may vary considerably depending on the chosen setup, i.e. the specific criterion and inclusion/exclusion rules. The results of the model, e.g. the estimated confidence intervals and tests statistics of the regression parameters, are *conditional* on the model choice and the assumption that the chosen model is appropriate.

Once a model has been selected, one often wishes to check the model's goodness of fit. This means that we want to check whether the model can "explain the data" well. Generally, this is a difficult task, which cannot be solved easily. The discussion about type I error and type II error in Sect. 10.2.4 may be recalled: there, we argued and explained why only a rejection of H_0 gives us a statistically valid decision while not rejecting H_0 does not mean that H_0 is true. The same problem occurs here. We can only reject the hypothesis that the model fits the data well using some test statistic (and corresponding p-value), but we cannot prove that the model fits the data well. What we can say at most is that there is no evidence against the model. In logistic regression, establishing goodness-of-fit measures is even more difficult than for linear regression because the response variable can only take the values 0 and 1. Therefore, Hosmer and Lemeshow (2013) introduced a test statistic based on grouping the data using quantiles of the predicted probabilities. Often, $g = 10$ groups are used. We demonstrate the method in the following example.

Example 12.4.2 Returning to our previous example, the selected model is

```
model.log.2 <- glm(Theatre.bin~Age+Sex+Income+Culture,
family=binomial(link='logit'), data=theatre)
```

The idea is the following: after having estimated the regression coefficients, a model prediction is possible for each observation i. In our data set, we have $n = 699$ observations and the predictions are the model probabilities p_i, $i = 1, \ldots, 699$. These probabilities are then sorted in ascending order together with the associated observed response values (which can only be 0 or 1). If $g = 10$, the percentiles of the predicted probabilities are calculated and the data is accordingly grouped (aggregated) into g groups. In our example the percentiles are

```
quantile(predict(model.log.2, type='response'),
probs=seq(0,1,0.1))
```

10%	20%	30%	40%	50%	60%	70%	80%	90%	100%
0.0921	0.135	0.176	0.229	0.293	0.368	0.448	0.548	0.684	0.995

Note that we have removed the 0%-quantile (the minimum of all predicted probabilities) 0.0131 and rounded all quantiles.

For each of the g groups,

- the number of observations in this group,
- the sum of the predicted probabilities p_i in this group (i.e., the expected number of 1's), denoted as yhat1 in the table below,
- the observed number of 1's, denoted as y1,
- the expected number of 0's (sum of the counter probabilities $(1 - p_i)$) denoted as yhat0 and
- the observed number of 0's denoted as y0

are recorded. Then a χ^2-test, which corresponds to the Hosmer–Lemeshow test with $g - 2$ (here $10 - 2 = 8$) degrees of freedom, is applied to the aggregated data. The command `logitgof` from the package `generalhoslem` can be used to proceed with the test.

The number of observations per group may slightly differ (e.g. because of ties in the predicted probabilities).

```
goodness.of.fit <- logitgof(theatre$Theatre.bin,
predict(model.log.2, type='response') )

print(goodness.of.fit)
print(cbind(goodness.of.fit$observed,
goodness.of.fit$expected))
```

```
> goodness.of.fit <- logitgof(theatre$Theatre.bin,
          predict(model.log.2, type="response") )
> print(goodness.of.fit)
Hosmer and Lemeshow test (binary model)
data:  theatre$Theatre.bin, predict(model.log.2,
                          type = "response")
X-squared = 4.334, df = 8, p-value = 0.8258

> print(cbind(goodness.of.fit$observed,
              goodness.of.fit$expected))
                    y0 y1 yhat0 yhat1
[0.0131,0.0921] 65  5  66.0  4.04
(0.0921,0.135]  64  6  62.3  7.70
(0.135,0.176]   56 14  59.0 11.05
(0.176,0.229]   59 11  55.9 14.13
(0.229,0.293]   53 17  51.7 18.31
(0.293,0.368]   49 20  46.3 22.72
(0.368,0.448]   41 29  41.3 28.68
(0.448,0.548]   33 37  35.2 34.79
(0.548,0.684]   24 46  27.7 42.29
(0.684,0.995]   16 54  14.7 55.30
```

The χ^2-statistic is calculated as

$$\chi^2 = \left(\frac{(65-66)^2}{66} + \frac{(5-4.04)^2}{4.04} \right) + \cdots + \left(\frac{(16-14.7)^2}{14.7} + \frac{(54-55.30)^2}{55.30} \right).$$

We summarize the Hosmer–Lemeshow procedure again:

The command `predict(model.log.2, type='response')` returns the (unsorted) vector of predicted probabilities of the model. Note that the specification of the argument `type='response'` is important since it is also possible to get the predictions on the scale of the linear predictors η_i. The `quantile` function sorts them and calculates the desired quantiles (minimum or 0%-quantile, 10%-quantile, ..., 90%-quantile, maximum or 100%-quantile). It can be seen that exactly 70 observations are in each of the 10 intervals, except the interval (0.293, 0.368] where there are 69 observations ($n = 699$) of predicted probabilities. This shows the idea of using quantiles to create the grouped data. Here y1 and y0 are the observed frequencies of 1s and 0s of the observations falling into the respective interval while y1hat and y0hat are the corresponding expected frequencies based on the logistic model. These are simply the sum of all predicted probabilities (and counter probabilities) in the interval. The test statistic is the χ^2-statistic based on these observed and expected frequencies. In our example, the value of the test statistic is 4.334 with 8 degrees of freedom showing no significant result ($p \approx 0.8$), i.e. the expected frequencies based on the model and the observed frequencies in the data are somehow compatible and the model is not rejected.

12.5   Key Points and Further Issues

> **Note:**
>
> ✓ Logistic regression is typically used if the outcome variable is binary, in which case a linear regression model cannot be used.
>
> ✓ In logistic regression, no explicit error term such as **e** in linear regression is necessary. Instead, it directly models the conditional expectations or probabilities $p_i = P(y_i = 1|\mathbf{x}_i) = \mathrm{E}(y_i|\mathbf{x}_i)$.
>
> ✓ Instead of the method of least squares, maximum likelihood is used to get the estimated coefficient vector $\hat{\boldsymbol{\beta}}$. The function glm can be used for that purpose.
>
> ✓ The estimated coefficients $\hat{\beta}_j$ can be interpreted in three different ways: as an additive change of the logits (logarithmic chances), as a log odds ratio, or as a multiplicative change of the chances, or odds ratio, when applying the exponential function.
>
> ✓ Models can be selected and compared using AIC as in the linear regression model, which assumes normally distributed outcomes. The reason is that AIC (only) needs the likelihood. In the logistic regression model, the likelihood is a product of Bernoulli probability functions, where the probabilities are subject-specific, i.e. may depend on the covariates of observation i.
>
> ✓ The goodness-of-fit can be judged using the Hosmer–Lemeshow statistic, which groups the data according to the quantiles of predicted probabilities.

12.6 Exercises

Exercise 12.1 Consider the data from the 2×2 table of Example 12.0.1:

	$y = 1$	$y = 0$	Total
Treatment X			
B	80	20	100
A	40	60	100

(a) Calculate and interpret the odds ratio.
(b) Estimate a logistic regression model with the `glm` function. Note that you can either expand the data to 200 observations using the `weights` argument in the `glm` function, or use the number of 1s and 0s for each treatment directly. Interpret the estimated coefficient and compare your results with those from (a).

Exercise 12.2 Consider the pizza data from Appendix A.4.1.

(a) We define a long delivery to last longer than 30 minutes. If we fit a logistic regression model with "long delivery" as the outcome and branch, operator and number of pizzas as covariates, we obtain the following output.

	Estimate	Std. Error	t-value	Pr(>|t|)
(Intercept)	0.588917	0.035967	16.374	<2e-16
operatorMelissa	0.011009	0.022535	0.489	0.625
branchEast	-0.199329	0.029211	-6.824	1.37e-11
branchWest	-0.027212	0.027542	-0.988	0.323
pizzas	0.076823	0.008087	9.500	<2e-16

Interpret the exponentiated coefficients.
(b) *R* does not return confidence intervals by default (one has to use the `confint` command). Calculate the 95% confidence interval for the estimated coefficient of the number of pizzas manually using the model output.
(c) Now consider a competing logistic regression model (with lower AIC), which—in addition to the variables from (a)—also contains the variable "driver" and an interaction between number of pizzas and branch:

	Estimate	Std. Err.	t value	Pr(>|t|)
(Intercept)	0.701036	0.055653	12.596	< 2e-16
operatorMelissa	0.004115	0.022483	0.183	0.854815
branchEast	-0.343767	0.065758	-5.228	2.01e-07
branchWest	-0.068525	0.068128	-1.006	0.314692
pizzas	0.051539	0.013574	3.797	0.000154
driverDomenico	-0.242016	0.047572	-5.087	4.18e-07

```
driverLuigi          -0.039195    0.040024   -0.979 0.327637
driverMario          -0.001957    0.030402   -0.064 0.948685
driverSalvatore      -0.023442    0.030671   -0.764 0.444823
branchEast:pizzas     0.058587    0.020475    2.861 0.004289
branchWest:pizzas     0.011414    0.018772    0.608 0.543287
```

Interpret the estimated interaction between branch and number of pizzas.

Exercise 12.3 Consider the theatre data described in Appendix A.4.3.

(a) Estimate a logistic regression model using the `glm` function in R, where the binarized Theatre variable (`Theatre.bin`) is the outcome (`Theatre.bin` equals 1 if `Theatre`> 150, and 0 else) and `Age`, `Sex`, `Income`, `Culture`, `Theatre_ly` are the covariates.
(b) Plot the predicted probabilities $\hat{p}_i$ on the y-axis against the linear predictors η_i on the x-axis.
(c) In the following, we exclude `Theatre_ly` from the analysis. Show how the AIC values change when you estimate the following models:

   (i) all possible models with two variables (without any interactions),
   (ii) all possible models with three variables (without any interactions).

(d) Consider the model with `Age`, `Sex`, `Income`, `Culture` as covariates. Investigate whether the Hosmer–Lemeshow test shows a different result if you vary the number of groups, e.g. if you increase the number of groups from $g = 10$ to $g = 20$, or if you reduce the number of groups to $g = 5$.
(e) Using again the model with `Age`, `Sex`, `Income` and `Culture` as covariates, calculate the confusion matrix. The confusion matrix is the 2×2-matrix containing the pairs of true and predicted outcomes. To classify the predicted outcomes as 1 or 0, you can use the decision rule "$\hat{y}_i = 1$ if $\hat{p}_i > c$" and "$\hat{y}_i = 0$ if $\hat{p}_i \leq c$" for a number $c \in (0, 1)$. Use the default $c = 0.5$. How "accurate" is your model, i.e. what is the proportion of correct predictions? Try different values for c and observe whether the *accuracy* changes.

Predicted	$y = 1$	$y = 0$	total
True outcome y			
1	?	?	239
0	?	?	460

→ Solutions to all exercises in this chapter can be found in Appendix B

Part IV
Additional Topics

Simple Random Sampling and Bootstrapping

13

13.1 Introduction

The statistical inferences drawn about a population of interest are dependent on the actual sample of data. An underlying assumption in many statistical analyses is that this sample is a good representation of the population. By representative we mean that the sample contains all the salient features and characteristics present in the population. If the sample is representative, the statistical inferences drawn from the sample will remain valid for the population too. Also, many statistical methods depend upon several assumptions: for example, we often assume that the sample observations are independently and identically distributed (i.i.d.). The success of applying statistical methods, and obtaining the correct statistical inferences from them, crucially depends on whether the respective assumptions are satisfied and the sample is representative. An important question that arises is how to draw the sample from the population such that all such assumptions are satisfied as far as possible, and the sample is representative.

Another point of concern relates to the size of the population from where the sample is drawn. The population size can be finite or infinite. For example, a frequently used statement is "Let $y_1, y_2, \ldots, y_n$ be a random sample from a given population, say $N(\mu, \sigma^2)$". A silent assumption in all such cases is that a sample of finite size is drawn from a population of infinite or very large size. In many situations however, the finite sample is drawn from a population of finite (possibly small) size. We need to understand how statistical inferences will change when a sample is drawn from a population of finite size instead of a population of infinite size. Population parameters represent the characteristics of the population. Different populations have different features, and therefore the methods for drawing representative samples from such populations also differ.

Various sampling schemes enable the drawing of samples in different ways, considering the different structures and characteristics of the populations. For example,

© Springer Nature Switzerland AG 2022
C. Heumann et al., *Introduction to Statistics and Data Analysis*,
https://doi.org/10.1007/978-3-031-11833-3_13

suppose the population is homogeneous with respect to the variable under study. The variance of all the units in the population is the same across the entire population. Then the sample can be drawn from any part of the population using the simple random sampling methodology. On the other hand, suppose the population is heterogeneous with respect to the variable under study, i.e. different parts of the population have different variances. Then, instead of drawing the sample from any part of the population, it is better to first divide the entire population into different homogeneous subpopulations. One can then draw the sample from each subpopulation to make the whole sample representative. This is called stratified random sampling. The drawn sample will exhibit the same level of heterogeneity as in the population in this case. Similarly, there exist other useful sampling schemes to obtain representative samples from populations with different features, for example cluster sampling, two-stage sampling, multistage sampling, two-phase sampling, multi-phase sampling, varying probability sampling and others. More details can be found in the book of Rao (2000).

In most statistical analyses, an investigator estimates a multitude of statistical parameters based on a sample of data, e.g. the mean, median, variance or the correlation coefficient. Both the estimation of population parameters and their standard errors and confidence intervals are then of interest. However, while the exact analytical expressions of standard errors and confidence intervals are readily known for some estimators (e.g., the estimator of the mean), the exact forms of other estimators is unknown (e.g., estimators of the median or the correlation coefficient). Under such conditions, there is an alternative computational approach to derive them: the **bootstrap methodology**. It is based on repeatedly drawing simple random samples with replacement from the given sample of data to calculate standard errors, confidence intervals and other quantities.

Among the available sampling schemes, the simple random sampling scheme may be the most common one in many applications. The bootstrap methodology is based on simple random sampling (with replacement). This is the motivation for this chapter, where we discuss the details of sampling as well as bootstrap methods for computing estimators, their standard errors, confidence intervals, manually and with R.

13.2 Methodology of Simple Random Sampling

There are numerous different study designs and ways of collecting data: through randomized experiments (Sect. 14.4.1), web searching, medical registries, surveys, a census, and many others. We consider data collection processes where we either conduct a survey to obtain a sample of data or obtain information through a census.

Sample surveys collect information on a fraction of the total population, whereas a census collects information on the whole population. Some surveys, for example economic (household) surveys, are conducted regularly by government agencies. Other surveys are need-based and are conducted whenever their need arises: consumer satisfaction surveys are, for example, conducted at newly opened shopping malls to assess the satisfaction level with the amenities provided. Complete enumeration or a census selects the entire population and collects the data from every unit in the population. For example, a population census is conducted in different countries at a given time: the national census in India is conducted every ten years. Sampling theory provides the basis, fundamentals, methods and techniques for data collection such that the sample is representative while fulfilling the objectives of the study and ensuring that the assumptions for the chosen statistical methods are satisfied.

Simple random sampling is the easiest way of drawing samples. We first consider an example to understand the methodology of simple random sampling. Suppose a box contains ten nearly identically-sized balls marked with the numbers 1, 2, . . . , 10. Each ball is a sampling unit on which the observations on any of its characteristics, e.g., the weight of the ball, is recorded. The population size is ten balls. The balls are identified by the numbers marked on them. The collection of these ten balls with identification marks is called the sampling frame. Now suppose a sample of three balls is required. We draw the three balls according to the following procedure: the box is well shaken, randomly rearranging the balls to such an extent that it is no longer possible to know their exact locations. Then, without looking into the box, we draw the first ball. Now there are two options to draw the second ball. The first option is to keep the drawn ball out of the box and draw the second ball from the remaining balls in the box. The second option is to return the drawn ball into the box and then draw the second ball—which is equivalent to drawing a ball from all the balls available at the beginning. The box is then well shaken again so that the locations of all the balls are random. Then the third ball is drawn without looking into the box. Once again, there are two ways to draw the third ball: without placing the second ball back into the box or after placing the second ball back into the box. Finally, these three drawn balls constitute a random sample of size three from a population of ten balls. This is the method of **simple random sampling (SRS)**. When the earlier drawn ball is not replaced in the box before drawing the next ball, the method is called **simple random sampling without replacement (SRSWOR)**. When the earlier drawn ball is placed back into the box before drawing the next ball, the method is called **simple random sampling with replacement (SRSWR)**. Note that no ball has identical numbers in the sample in the case of SRSWOR whereas the sample drawn by SRSWR may have balls with identical numbers. Since all the balls are nearly identically sized, the population of the balls can be considered homogeneous. The balls are well shaken so that it is not known before the draw which of the balls will be selected in the next draw, so the balls are considered to be drawn independently. All the balls have an equal chance of being selected. Each ball is a sampling unit. Suppose we want to measure the weight of the balls; then the weight of the ball is the characteristic under study and the measurements of weights of the three selected balls will constitute a random sample of three observations. One can study more than one characteristic with the same sampling unit, for example, the radius of the ball, or the colour of the ball.

Definition 13.2.1 An element or a group of elements from which the observations can be taken is called a sampling unit. The objective of the survey helps in determining the definition of the sampling unit.

For example, if the objective is to determine the total household income of all persons in a household, then the sampling unit is a household. If the objective is to determine the income of any particular person in the household, then the sampling unit is the specific person in the household whose income is recorded. So, the definition of a sampling unit depends on and varies with the objective of the survey. Similarly, in another example, if the aim is to study someone's blood sugar level, then the sampling unit is the person whose blood sample is collected and the value of the blood sugar level is obtained. On the other hand, if the objective is to study someone's health conditions, then the sampling unit is the person on whom the blood sugar level, blood pressure, and other factors are obtained. These values together classify the person as healthy or unhealthy.

Definition 13.2.2 The collection of all the sampling units in a given region at a particular time or during a particular period is called the population. The total number of sampling units in the population is the population size, generally denoted by N. The population size can be finite or infinite (N is large).

For example, suppose we want to learn about the medical facility in a hospital through a survey conducted over the patients. In that case, the total number of patients registered in the medical facility during the time period of the survey is the population and the total number of registered patients is the population size. Similarly, if the production of a crop in a state is to be studied, then all the fields cultivating the crop in that state will constitute the population, and the total number of fields is the population size.

Definition 13.2.3 The complete count of the population is called a census. The observations of all the sampling units in the population are collected in the census.

For example, the census is conducted in India every ten years, in which observations on all the persons staying in the country are collected.

Definition 13.2.4 A sample consists only of a portion of the population units in which one or more sampling units are selected from the population according to some specified procedure. Such a collection of units is called the sample.

A census is a 100% sample, and it is a complete count of the population.

Definition 13.2.5 When all the relevant features of the population are present in the sample, it is called a representative sample.

It is an underlying assumption in many statistical analyses that every sample is a representative sample. For example, suppose we are interested in the gender distribution with respect to some medical condition. Then, if in the population of people who have the particular disease 30% are male and 70% are female, we would expect the sample to be close to 30% male and 70% female. In another example, we would expect a handful of wheat taken from a 100 kilogram bag to have the same quality as the entirety of the wheat inside the bag. Similarly, a drop of blood is expected to give the same information as all of the blood in the body from which it has been taken. For example, a person's hemoglobin level is often determined on the basis of a single drop of blood.

Definition 13.2.6 The list of all the units of the population to be surveyed constitutes the sampling frame. All the sampling units in the sampling frame have identification particulars.

For example, all the students in a particular university listed along with their enrolment numbers constitute the sampling frame. Similarly, the list of households with the name of the head of the family or house address constitutes the sampling frame. In another example, the residents of a city area may be listed in more than one sampling frame—as per automobile registration or the listing in the telephone directory.

There are two possible ways to ensure that the selected sample is representative based on the law of probability.

Definition 13.2.7 When the laws of chance or probability govern the selection of units in the sample from a population, then such a sample is called a random sample or probability sample. The probability of selection of a unit can be equal as well as unequal in a random sample. When the probability laws do not govern the selection of units in the sample from the population, it is called a non-random sample or purposive sample.

An example of the latter is when the units are selected based on the personal judgment of the investigator. If persons have volunteered to take some medical product or to drink a new type of coffee, they also constitute a sample based on non-random laws.

Definition 13.2.8 When the survey is continued until a predetermined number of units with the characteristic under study is enrolled, it is called quota sampling.

For example, to conduct an experiment for a rare type of disease, the study is continued until the required number of patients with the disease is collected.

Suppose a population has N sampling units and a sample of n sampling units is to be selected.

Definition 13.2.9 Simple random sampling (SRS) is a method for selecting sampling units out of a population such that every sampling unit has an equal probability of being chosen.

The samples can be drawn in two possible ways:

- The sampling units are selected without replacement , i.e. the units, once drawn, are not replaced in the population.
- The sampling units are selected with replacement , i.e. the drawn units are replaced in the population.

Now we define simple random sampling without replacement (SRSWOR) and simple random sampling with replacement (SRSWR) as follows:

Definition 13.2.10 SRSWOR is a method of selection of n sampling units out of the N sampling units one by one such that at any stage of selection, any one of the remaining sampling units has the same probability of being selected.

Definition 13.2.11 SRSWR is a method of selection of n sampling units out of the N sampling units one by one such that at each stage of selection, each sampling unit has an equal probability of being selected.

13.2.1 Procedure of Selection of a Random Sample

The procedure of selection of a simple random sample is as follows:

1. Identify the sampling units in the population and mark them with the numbers 1 to N.
2. Choose a number randomly from any of those numbers.
3. Choose the sampling unit whose serial number corresponds to the random number.
4. Choose another number from 1 to N randomly.
5. In the case of SRSWR, all the random numbers are accepted even if repeated more than once. In the case of SRSWOR, if any random number is repeated, then it is ignored, and more numbers are drawn.
6. Continue with the procedure until a sample of size n is collected.

One way to implement such a process through programming is by using the discrete uniform distribution, see Sect. 8.1.1 in Chap. 8. The integers $1, 2, \ldots, N$ are indexed to the N sampling units in the population. Any number between 1 and N is generated from the discrete uniform distribution and is treated as an index. The sampling unit corresponding to the generated index in the population is selected in the sample.

13.2.2 Probabilities of Selection

First, we consider an example to understand the concept and role of selecting a sample and the probability of selecting a sampling unit.

Example 13.2.1 Suppose a population has three balls which are numbered as $1, 2$ and 3. Suppose we want to draw a sample of two balls. A total of $\binom{3}{2}$ samples—i.e. the sampling units $\{1, 2\}, \{2, 3\}$, and $\{1, 3\}$—can be drawn from this population using SRSWOR, and each sample has an equal probability of selection, that is $1/\binom{3}{2}$.

Similarly, a total of 3^2 samples with sampling units—$\{1, 1\}, \{2, 2\}, \{3, 3\}, \{1, 2\}$, $\{2, 3\}, \{1, 3\}, \{2, 1\}, \{3, 2\}, \{3, 1\}$—can be drawn from this population using SRSWR and each sample has an equal probability of selection (=$1/3^2$).

The probability of selection of any sampling unit from the population is $1/3$ and this remains the same in both SRSWOR and SRSWR.

Based on this example, we can extend these results to the general situation when a sample of size n is to be drawn from a population of size N.

13.2.2.1 SRSWOR

If n sampling units are selected by SRSWOR from a population of size N, the total number of possible samples that can be chosen is $\binom{N}{n}$. So the probability of selecting any one of these samples is $1/\binom{N}{n}$. The probability of selection of a sampling unit from the population under SRSWOR is $1/N$.

13.2.2.2 SRSWR

If n sampling units are selected with SRSWR from a population of size N, the total number of possible samples is N^n. The probability of drawing a sample is $1/N^n$. The probability of selection of a sampling unit from the population under SRSWR is $1/N$.

The proofs for the calculation of the above probabilities for SRSWR and SRSWOR are presented in Appendix C.8.

Now we review R-commands that can be used to draw a sample by the SRSWOR and SRSWR schemes.

13.2.2.3 Drawing Samples in R

Let y be a vector that contains all the sampling units, i.e. the population from which a sample of n sampling units is to be selected. Here y acts as the sampling frame where the position of a unit in the data vector is its index and is considered to be an identification tag.

SRSWOR

The command `sample` in *R* is used to draw a sample of the specified size by SRSWOR:

```
sample(y, n, replace = FALSE)    # SRSWOR
sample(y, n)                     # SRSWOR
```

The above commands return a sample of size n from the elements of y. The option `replace = FALSE` is used so that drawing is done without replacement.

The command

```
sample(y)
```

generates a random permutation of *all* elements of y, i.e. n is set to the actual sample size.

Another command in *R* to generate a sample of size n from a population of size N by SRSWOR is based on the package `sampling`. After its installation the following command

```
srswor(n, N)
```

generates a data vector of the same size *N* as the sampling frame and consisting of '1's and '0's. Since there are only two values ('0' and '1') in the data vector, we will call it a binary data vector in the text below. These numbers, i.e. '1' and '0', indicate whether or not a sampling unit has been selected into the sample from the population. The interpretation is as follows: each element *k* of this outcome data vector indicates the status of sampling unit *k*. If the outcome at the kth place is 1, it means the kth sampling unit is selected, and if the outcome at the kth place is 0, it means the kth sampling unit is not selected in the sample. So, the position of a particular number in the outcome data vector identifies the selection of the corresponding sampling unit number into the sample. Obviously, the number of times a particular sampling unit can be repeatedly drawn into the sample drawn with SRSWOR is at the most one; hence, the outcome contains only two values—0 and 1. Note that the data vector y is not used in the `srswor()` command as compared to the earlier command `sample()`.

SRSWR

The command `sample` in *R* is used to draw a sample of the specified size by SRSWR as follows:

```
sample(y, n, replace = TRUE)
```

This provides us with a sample of size n from the elements of y and `replace = TRUE` indicates that the sample is drawn with replacement.

Another command in R to generate a sample of size n from a population of size N by SRSWR is based on the package `sampling`. It works slightly differently. After its installation, the following command

```
srswr(n, N)
```

generates an outcome data vector (like the 'binary vector' in the case of `srswor()`) of size N, the same size as the sampling frame, where every element can take any value between 0 and n such that the sum of all the numbers in the generated outcome data vector is equal to n. The numbers in this outcome data vector indicate the frequency with which the sampling unit is selected, i.e. the number of times the unit is entered into the sample. The position of the values in the outcome data vector identifies the corresponding sampling unit in the population which is now selected into the sample. Thus the number 0 (zero) in the binary outcome vector indicates that the corresponding unit is not selected. But if, for example, the number 3 appears at the 5th place in the outcome data vector, it indicates that the 5th unit of the sampling frame is selected three times into the sample. Note that the outcome data vector y is not used in the `srswr()` command as compared to the earlier command `sample()`.

Now we illustrate all the above concepts related to SRSWOR and SRSWR through the following example.

Example 13.2.2 Consider a set of ten balls and suppose a sample of five balls is to be drawn. First we create the sampling frame by assigning and marking all the balls in the population with the unique numbers $21, 22, \ldots, 30$ and store them in a data vector y which creates the sampling frame as

```
y <- 21:30
y
[1] 21 22 23 24 25 26 27 28 29 30
```

We can summarize these data as follows:

Marking on ball	21	22	23	24	25	26	27	28	29	30
Index assigned to ball	1	2	3	4	5	6	7	8	9	10

We have intentionally not marked the balls with the numbers $1, 2, \ldots, 10$ so that we can discriminate between the balls and ball numbers, making the explanation clearer.

We consider the drawing of a sample of five balls by SRSWOR as follows:

```
set.seed(1234)
sample(y, size = 5, replace = FALSE)
[1] 30 26 25 24 21
```

This indicates that the balls bearing the numbers 30, 26, 25, 24, and 21 are selected into the sample. Note that the use of the command set.seed() makes the results reproducible. Unless the set.seed() command is used, repeating the same command gives another outcome each time. Different seeds give different results, for example:

```
set.seed(4321)
sample(y, size = 5, replace = FALSE)
[1] 25 30 29 22 26
```

which indicates that the balls bearing the numbers 25, 30, 29, 22, and 26 are selected into the sample.

Note that the probability of selecting a sample with $N = 10$ and $n = 5$ is

$$1/\binom{N}{n} = 1/\binom{10}{5} = 1/\text{choose}(10,5) = 0.003968.$$

It is thus clear that the chances of obtaining a repeated sample is low. As N and n increase, such a probability will become smaller.

Alternatively, the following R command can be used after installing the package sampling to draw a sample of size $n = 5$ from a population of size $N = 10$.

```
set.seed(1234)
srswor(5, 10)
[1] 1 0 0 1 1 1 0 0 0 1
```

Note that the 1's in this binary vector are at the $1^{st}, 4^{th}, 5^{th}, 6^{th}$ and 10th positions. Based on this, the balls located at these positions, bearing the numbers 21, 24, 25, 26 and 30, are selected into the sample.

Suppose we are interested only in determining the selected units. In that case, one can remove the "0" values from the selected sample, choose the nonzero outcomes and relate them to the corresponding values in the population. For example, suppose the population has ten values 21, 22, ..., 30, which are stored in the variable y. Now one can store all the generated values in a outcome data vector, say z, and select the relevant sample values from the population using the logical operator "!=" ('not equal to') as follows:

```
y <- 21:30
y
[1] 21 22 23 24 25 26 27 28 29 30
set.seed(1234)
z <- srswor(5, 10)
z
[1] 1 0 0 1 1 1 0 0 0 1
y[z != 0]
[1] 21 24 25 26 30
```

Next, we draw a sample of five numbers by SRSWR as

```
set.seed(1234)
sample(y, size = 5, replace = TRUE)
[1] 30 26 25 29 25
```

This indicates that the balls bearing the numbers 30, 26, 25, 29, and 25 are selected into the sample. Note that ball number 25 has been selected twice. It is also possible that no number is selected twice, or three times. Repeating the same command gives another outcome unless the command set.seed() is used

```
set.seed(2212)
sample(y, size = 5, replace = TRUE)
[1] 30 29 29 28 28
```

which indicates that the balls bearing the numbers 30, 29, 29, 28 and 28 are selected into the sample, and that ball numbers 28 and 29 are selected twice. Note that the probability of selecting a particular sample is $1/N^n = 1/10^5 = 1 \times 10^{-5}$. It is therefore clear that the probability of getting a repeated sample is low and becomes even lower with a further increase of N and n.

Alternatively, the following R command can also be used (after installing the package sampling) to draw a sample of size $n = 5$ from a population of size $N = 10$.

```
set.seed(2222)
srswr(5, 10)
[1] 0 0 1 1 2 0 1 0 0 0
```

Observe that the number 1 occurs at the 3rd, 4th and 7th positions, whereas the number 2 occurs at the 5th position in the returned outcome data vector. Now, we have to match the positions and values of the nonzero numbers with the respective values and positions in the sampling frame y. The sampling units corresponding to the the 3rd, 4th and 7th position in y are selected. Hence ball numbers 23, 24, and 27 are selected once each. Similarly, the next nonzero value is 2, which is located at the 5th position in the outcome, and hence the 5th ball in y (bearing the number 25) is selected twice. Note that the sum of all the numbers in the outcome vector is $0 + 0 + 1 + 1 + 2 + 0 + 1 + 0 + 0 + 0 = 5$, which is the required sample size.

Repeating the command with another seed gives different results:

```
set.seed(5555)
srswr(5, 10)
[1] 0 0 0 0 0 0 2 2 1 0
```

Observe that the numbers 2, 2 and 1 occur at the 7th, 8th, and 9th positions in the outcome data vector. Thus, ball numbers 27, 28, and 29, which are located at the 7th, 8th, and 9th positions respectively in y, are selected twice, twice, and once, respectively. Note that the sum of all the numbers is again 5, which is the required sample size.

13.3   Estimation of the Population Mean and Population Variance

One of the main objectives after selecting a sample is to learn about the tendency of the data to cluster around the central value and its scatteredness around it. There are various popular estimators for estimating the population mean and population variability. One may consider estimators for measuring the central tendency, such as the mean, median, mode, geometric mean, harmonic mean, and others, and estimators for the population variability, like the variance, mean deviation, absolute deviation, and others. We consider the arithmetic mean and variance to estimate the population mean and population variance, respectively, as they possess nice statistical properties. The sample arithmetic mean and sample variance possess nice statistical properties too, and we consider them for estimating the population mean and population variance, respectively.

Suppose Y denotes the characteristic under study and Y_i denotes the value of the characteristic for the ith, $i = 1, 2, \ldots, N$, sampling unit in the population. Suppose y_i denotes the value of the characteristic for the ith, $i = 1, 2, \ldots, n$, sampling unit in the sample.

Let $\bar{Y} = \frac{1}{N} \sum_{i=1}^{N} Y_i$ and $\bar{y} = \frac{1}{n} \sum_{i=1}^{n} y_i$ denote the population mean and sample mean respectively. Also, let

$$S^2 = \frac{1}{N-1} \sum_{i=1}^{N} (Y_i - \bar{Y})^2 = \frac{1}{N-1} \left(\sum_{i=1}^{N} Y_i^2 - N\bar{Y}^2 \right),$$

$$\sigma^2 = \frac{1}{N} \sum_{i=1}^{N} (Y_i - \bar{Y})^2 = \frac{1}{N} \left(\sum_{i=1}^{N} Y_i^2 - N\bar{Y}^2 \right),$$

$$s^2 = \frac{1}{n-1} \sum_{i=1}^{n} (y_i - \bar{y})^2 = \frac{1}{n-1} \left(\sum_{i=1}^{n} y_i^2 - n\bar{y}^2 \right).$$

First we clarify the above notation through an example.

Example 13.3.1 Consider again Example 13.2.2 where a set of ten balls were marked with the numbers $21, 22, \ldots, 30$. Suppose we now also want to study their weights. The weights (in grams) of each of these ten balls are denoted by $Y_1, Y_2, \ldots, Y_{10}$ as given in the following table:

Ball	21	22	23	24	25	26	27	28	29	30
Index of ball	1	2	3	4	5	6	7	8	9	10
Population values	Y_1	Y_2	Y_3	Y_4	Y_5	Y_6	Y_7	Y_8	Y_9	Y_{10}
Weight of ball	11	12	13	14	15	16	17	18	19	20

We have chosen the values in a friendly manner to make the understanding of the introduced concepts easy and straightforward.

Suppose a sample of five balls is drawn and the following set of five numbers 25, 21, 28, 23, and 26 is obtained using the command `sample(x, size = 5, replace = FALSE)`. This indicates that the balls bearing the numbers 25, 21, 28, 23 and 26 are selected into the sample. So we consider the weight of selected balls, i.e. Y_5, Y_1, Y_8, Y_3 and Y_6 to be in the sample and the sample values are denoted as $y_1 = Y_5 = 15, y_2 = Y_1 = 11, y_3 = Y_8 = 18, y_4 = Y_3 = 13$ and $y_5 = Y_6 = 16$. Now $\bar{Y} = \frac{1}{10} \sum_{i=1}^{10} Y_i = 15.5$, $\bar{y} = \frac{1}{5} \sum_{i=1}^{5} y_i = 14.6$, $S^2 = \frac{1}{9} \sum_{i=1}^{10} (Y_i - \bar{Y})^2 = \frac{1}{9} \left(\sum_{i=1}^{10} Y_i^2 - 10\bar{Y}^2 \right) = 9.17, \sigma^2 = \frac{1}{10} \sum_{i=1}^{10} (Y_i - \bar{Y})^2 = \frac{1}{10} \left(\sum_{i=1}^{10} y_i^2 - 10\bar{y}^2 \right) = 8.25, s^2 = \frac{1}{4} \sum_{i=1}^{5} (y_i - \bar{y})^2 = \frac{1}{4} \left(\sum_{i=1}^{5} y_i^2 - 5\bar{y}^2 \right) = 7.3$.

Estimation of the Population Mean

Suppose $y_1, y_2, \ldots, y_n$ is a sample of size n drawn by SRS from a population of size N. Let us consider the sample arithmetic mean $\bar{y} = \frac{1}{n} \sum_{i=1}^{n} y_i$ as an estimator of the population mean $\bar{Y} = \frac{1}{N} \sum_{i=1}^{N} Y_i$. Let $\bar{y}_{WOR}$ and $\bar{y}_{WR}$ denote the sample means based on the samples drawn by SRSWOR and SRSWR, respectively. Next, we investigate its bias.

SRSWOR Consider the setup of SRSWOR and let $P_i(j)$ denote the probability of selection for the jth unit at the ith draw, $i = 1, 2, \ldots, n, j = 1, 2, \ldots, N$. Then, $P_i(j) = \frac{1}{N}$ and

$$E(\bar{y}_{WOR}) = \frac{1}{n} \sum_{i=1}^{n} E(y_i) = \frac{1}{n} \sum_{i=1}^{n} \left[\sum_{j=1}^{N} Y_j P_i(j) \right] = \frac{1}{n} \sum_{i=1}^{n} \bar{Y} = \bar{Y}.$$

Thus $\bar{y}_{WOR}$ is an unbiased estimator of $\bar{Y}$ under SRSWOR.

SRSWR Consider the setup of SRSWR and let $P_j = \frac{1}{N}$, $j = 1, 2, \ldots, N$, denote the probability of selection of the jth unit. Then,

$$E(\bar{y}_{WR}) = \frac{1}{n} \sum_{i=1}^{n} E(y_i) = \frac{1}{n} \sum_{i=1}^{n} \left[\sum_{j=1}^{N} Y_j P_j \right] = \frac{1}{n} \sum_{i=1}^{n} \bar{Y} = \bar{Y}.$$

Thus $\bar{y}_{WR}$ is an unbiased estimator of $\bar{Y}$ under SRSWR too.

Thus, the sample mean is an unbiased estimator of the population mean under both SRSWOR and SRSWR.

Computation of the sample mean in R can thus be done using the `mean(y)` command for both SRSWOR and SRSWR.

Variance of the Sample Mean

Since the sample mean is a function of the random sample, we can determine the variance of the sample mean under SRSWOR and SRSWR. Assume that each observation has the same variance σ^2.

The variance of $\bar{y}_{WOR}$ under SRSWOR is

$$\text{Var}(\bar{y}_{WOR}) = \frac{N - n}{Nn} S^2. \tag{13.1}$$

The variance of $\bar{y}_{WR}$ under SRSWR is

$$\text{Var}(\bar{y}_{WR}) = \frac{N - 1}{Nn} S^2 = \frac{\sigma^2}{n}. \tag{13.2}$$

The proofs of the results (13.1)–(13.2) are presented in Appendix C.8.

Note that if N is infinite (large enough), then

$$\text{Var}(\bar{y}) \approx \frac{S^2}{n}$$

for both SRSWOR and SRSWR. The factor $\frac{N-n}{N}$ is therefore responsible for changing the variance of $\bar{y}$ when the sample is drawn from a finite population in comparison to when the sample is drawn from an infinite population. That is why the factor $\frac{N-n}{N}$ is called a finite population correction (fpc). It may be noted that $\frac{N-n}{N} = 1 - \frac{n}{N}$, so $\frac{N-n}{N}$ is close to 1 if the ratio of sample size to population $\frac{n}{N}$ is very small or negligible. The term $\frac{n}{N}$ is called the sampling fraction. In practice, the fpc can be ignored whenever $\frac{n}{N} < 5\%$, and for many purposes, even if it is as high as 10%. Since the fpc lies between 0 and 1, ignoring it will result in the overestimation of the variance of the sample mean $\bar{y}$.

Efficiency of the Sample Mean

A natural question is which sampling scheme, SRSWOR or SRSWR, will provide a more efficient estimate when estimating the population mean by the sample mean. We therefore now compare the variances of $\bar{y}$ under SRSWOR and SRSWR. Consider

$$\mathrm{Var}(\bar{y}_{WR}) = \frac{N-1}{Nn}S^2$$

$$= \frac{N-n}{Nn}S^2 + \frac{n-1}{Nn}S^2$$

$$= \mathrm{Var}(\bar{y}_{WOR}) + \frac{n-1}{Nn}S^2.$$

Hence

$$\mathrm{Var}(\bar{y}_{WOR}) \le \mathrm{Var}(\bar{y}_{WR})$$

because $\frac{n-1}{Nn}S^2 > 0$ when $n > 1$, which is obviously always the case. Thus, a sample drawn by SRSWOR yields a more efficient estimate of the population mean than a sample drawn by SRSWR.

Estimation of Variance of the Sample Mean

Since the expressions of the variances of the sample mean (13.1) and (13.2) involve S^2, which is based on population values, these expressions cannot be used to estimate the variance on the basis of a sample of data in real-life applications. In order to estimate the variance of $\bar{y}$ on the basis of a sample, an estimator of S^2 (or equivalently σ^2) is required. Consider s^2 as an estimator of S^2 (or σ^2). We find that

$$\mathrm{E}(s^2) = \begin{cases} S^2 & \text{in SRSWOR} \\ \sigma^2 & \text{in SRSWR.} \end{cases} \tag{13.3}$$

Thus we replace S^2 by its unbiased estimator in (13.1) and (13.2). Then, under SRSWOR, an unbiased estimator of $\mathrm{Var}(\bar{y}_{WOR})$ is

$$\widehat{\mathrm{Var}}(\bar{y}_{WOR}) = \frac{N-n}{Nn}s^2. \tag{13.4}$$

Under SRSWR, an unbiased estimator of $\mathrm{Var}(\bar{y}_{WR})$ is

$$\widehat{\mathrm{Var}}(\bar{y}_{WR}) = \frac{N-1}{Nn}\frac{N}{N-1}s^2 = \frac{s^2}{n}. \tag{13.5}$$

The proofs of the results (13.3) to (13.5) are provided in Appendix C.8.

Computation of the estimate of the variance of the sample mean in R can be done for the SRSWOR and SRSWR cases using the var() command, as introduced in Sect. 3. Let the sample of size n be drawn from a population of size N and the sample values be stored in a data vector y. If the sample is drawn by SRSWOR, then an estimate of the variance of the sample mean is obtained as `((N- length(y))/(N*length(y)))*var(y)`. If the sample is drawn by SRSWR, then an estimate of the variance of the sample mean is obtained as `var(y)/length(y)`.

Standard Errors of the Sample Mean

Standard errors are commonly used to measure variability. The standard deviation depends upon the population parameters whereas the standard error depends only on the sample values. An advantage is that standard errors have the same unit of measurement as the actual variable and so it is appealing to practitioners. The standard error of $\bar{y}$ is defined as $\sqrt{\widehat{\text{Var}}(\bar{y})}$. Thus under SRSWOR, the standard error of $\bar{y}$ is

$$se(\bar{y}_{WOR}) = \sqrt{\widehat{\text{Var}}(\bar{y}_{WOR})} = \sqrt{\frac{N-n}{Nn}} s, \tag{13.6}$$

and under SRSWR, the standard error of $\bar{y}$ is

$$se(\bar{y}_{WR}) = \sqrt{\widehat{\text{Var}}(\bar{y}_{WR})} = \frac{s}{\sqrt{n}}. \tag{13.7}$$

Note that the standard errors do not possess the same statistical properties as $\widehat{\text{Var}}(\bar{y})$: we know, for example, that under SRSWOR $E(s^2) = S^2$ but $E(s) \neq S$.

The standard errors of the sample means in R can be obtained by computing the square root of the estimate of variance under SRSWOR and SRSWR. For example, if the sample of size n is drawn from a population of size N and the sample values are stored in a data vector y, then the standard errors of the sample mean under SRSWOR and SRSWR are obtained as `sqrt(((N-length(y))/(N*length(y)))*var(y))` and `sqrt(var(y)/length(y))` respectively.

13.3.1 Estimation of the Population Total

Sometimes, it is also of interest to estimate the population total, for example the total household income or total expenditures. The population total of $Y_1, Y_2, \ldots, Y_N$ is defined as

$$Y_T = \sum_{i=1}^{N} Y_i = N\bar{Y}, \tag{13.8}$$

which can be estimated in the case of both SRSWOR and SRSWR by

$$\hat{Y}_T = N\hat{\bar{Y}} = N\bar{y}. \tag{13.9}$$

Obviously,

$$E(\hat{Y}_T) = NE(\bar{y}) = N\bar{Y},\qquad(13.10)$$

for both SRSWOR and SRSWR.

In R, we can estimate the population total under SRWOR and SRSWR for the sample data vector y as $N*mean(y)$.

The variances of $\hat{Y}_T$ are obtained under SRSWOR as

$$\mathrm{Var}(\hat{Y}_T)_{WOR} = N^2\mathrm{Var}(\bar{y}_{WOR}) = N^2\left(\frac{N-n}{Nn}\right)S^2 = \frac{N(N-n)}{n}S^2$$

and under SRSWR as

$$\mathrm{Var}(\hat{Y}_T)_{WR} = N^2\mathrm{Var}(\bar{y}_{WR}) = N^2\left(\frac{N-1}{Nn}\right)S^2 = \frac{N(N-1)}{n}S^2.$$

The estimates of the variances of $\hat{Y}_T$ are obtained under SRSWOR as

$$\widehat{\mathrm{Var}}(\hat{Y}_T)_{WOR} = \frac{N(N-n)}{n}s^2\qquad(13.11)$$

and under SRSWR as

$$\widehat{\mathrm{Var}}(\hat{Y}_T)_{WR} = \frac{N^2}{n}s^2.\qquad(13.12)$$

The standard errors of the estimator of population totals under SRSWOR and SRSWR are obtained by taking the positive square roots of the estimates of variances in (13.11) and (13.12), respectively. In R, they can be found for a data vector y for a given population of size N under SRSWOR and SRSWR as $sqrt(((N*(N- length(y)))/length(y))*var(y))$ and $N*sqrt(var(y)/length(y))$, respectively.

13.3.2 Confidence Interval for the Population Mean

Now we construct the $100(1-\alpha)\%$ confidence interval for the population mean. It may be noted that while estimating the population mean and population variance, no assumption has been made about the probability distribution of the population. In order to construct the confidence interval, we now need to assume a probability distribution. Assume that the population associated with Y is normally distributed $N(\bar{Y}, \sigma^2)$ with mean $\bar{Y}$ and variance σ^2; then $\bar{y} - \bar{Y}/\sqrt{\mathrm{Var}(\bar{y})}$ follows a $N(0, 1)$ distribution when σ^2 is known. If σ^2 is unknown and is estimated from the sample then $\bar{y} - \bar{Y}/\sqrt{\mathrm{Var}(\bar{y})}$ follows a t-distribution with $(n-1)$ degrees of freedom.

When σ^2 is known, then the $100(1-\alpha)\%$ confidence interval is obtained as

$$P\left[-Z_{\frac{\alpha}{2}} \le \frac{\bar{y}-\bar{Y}}{\sqrt{\mathrm{Var}(\bar{y})}} \le Z_{\frac{\alpha}{2}}\right] = 1-\alpha$$

or

$$P\left[\bar{y} - Z_{\frac{\alpha}{2}}\sqrt{\mathrm{Var}(\bar{y})} \le \bar{Y} \le \bar{y} + Z_{\frac{\alpha}{2}}\sqrt{\mathrm{Var}(\bar{y})}\right] = 1 - \alpha.$$

The confidence limits of the $100(1 - \alpha)\%$ confidence interval are obtained as

$$\left[\bar{y} - Z_{\frac{\alpha}{2}}\sqrt{\mathrm{Var}(\bar{y})},\ \bar{y} + Z_{\frac{\alpha}{2}}\sqrt{\mathrm{Var}(\bar{y})}\right], \tag{13.13}$$

where $Z_{\frac{\alpha}{2}}$ denotes the $100\left(1 - \frac{\alpha}{2}\right)\%$ quantile of the $N(0, 1)$ distribution. The respective variances of $\bar{y}$ under SRSWOR and SRSWR, as given by (13.1) and (13.2)
respectively, can be substituted into (13.13) to obtain suitable confidence intervals
under SRSWOR and SRSWR, respectively.

In R, the command qnorm() is used to find the quantiles $Z_{\frac{\alpha}{2}}$ of a normal distribution. For example, the quantile of a standard normal distribution for a 5% level
of significance, i.e. $\alpha = 0.05$, is obtained by qnorm(0.95), see also Sect. 8.4 and
Appendix A. Suppose the sample is drawn by SRSWOR and the sample is stored in
a data vector y. Then, for example, the lower limit of the 95% confidence interval
$(\bar{y} - Z_{\frac{\alpha}{2}}\sqrt{\mathrm{Var}(\bar{y}_{WOR})})$ is obtained through

```
mean(y)
 - qnorm(0.975)*sqrt(((N-length(y))/(length(y)*N))*var(y))
```

The upper limit of the confidence interval can be calculated accordingly.

Similarly, when σ^2 is unknown, the $100(1 - \alpha)\%$ confidence interval is obtained
as

$$P\left[-t_{\frac{\alpha}{2}} \le \frac{\bar{y} - \bar{Y}}{\sqrt{\widehat{\mathrm{Var}}(\bar{y})}} \le t_{\frac{\alpha}{2}}\right] = 1 - \alpha$$

or

$$P\left[\bar{y} - t_{\frac{\alpha}{2}}\sqrt{\widehat{\mathrm{Var}}(\bar{y})} \le \bar{Y} \le \bar{y} + t_{\frac{\alpha}{2}}\sqrt{\widehat{\mathrm{Var}}(\bar{y})}\right] = 1 - \alpha.$$

The respective confidence interval is

$$\left[\bar{y} - t_{\frac{\alpha}{2}}\sqrt{\widehat{\mathrm{Var}}(\bar{y})},\ \bar{y} + t_{\frac{\alpha}{2}}\sqrt{\widehat{\mathrm{Var}}(\bar{y})}\right],$$

where $t_{\frac{\alpha}{2}}$ denotes the $100\left(1 - \frac{\alpha}{2}\right)\%$ quantile of a t-distribution with $(n - 1)$ degrees
of freedom.

In R, the value of the quantile of the t-distribution with df degrees of freedom is
found by the function qt(·, df). For example, the quantile for a 5% level of significance, i.e. $\alpha = 0.05$, and $df = 4$ degrees of freedom is obtained by qt(0.95,
4). Let the sample be drawn by SRSWOR and the sample values be stored in the
data vector y; then, for example, the lower limit of the 95% confidence interval
$(\bar{y} - t_{\frac{\alpha}{2}}\sqrt{\widehat{\mathrm{Var}}(\bar{y})})$ is obtained as

```
mean(y)
- qt(0.975, df)*sqrt(((N-length(y))/(length(y)*N))*var(y))
```

The upper limit of the confidence interval can be calculated accordingly.

Suppose the sample is drawn by SRSWR and the sample values are stored in a data vector y. When σ^2 is known, for example, the lower limit of a 95% confidence interval, i.e. $\bar{y} - Z_{\frac{\alpha}{2}}\sqrt{\widehat{\text{Var}}(\bar{y})}$, is obtained as

```
mean(y) - qnorm(0.975)*sqrt(var(y)/length(y))
```

and the upper limit of a 95% confidence interval ($\bar{y} + Z_{\frac{\alpha}{2}}\sqrt{\widehat{\text{Var}}(\bar{y})}$) can be calculated accordingly.

Similarly, when σ^2 is unknown, then the lower limit of a $100(1 - \alpha)\%$ confidence interval ($\bar{y} - t_{\frac{\alpha}{2}}\sqrt{\widehat{\text{Var}}(\bar{y})}$) is obtained as

```
mean(y) - qt(0.975, df)* sqrt(var(y)/length(y))
```

and the upper limit of a 95% confidence interval ($\bar{y} + t_{\frac{\alpha}{2}}\sqrt{\widehat{\text{Var}}(\bar{y})}$) can be similarly calculated.

A comprehensive example on estimating the population mean, population variance, and confidence interval for the population mean is given below in Example 13.4.1.

13.4 Sampling for Proportions

The SRS methodology discussed in Sect. 13.2 assumes that the random variable under consideration is quantitative. In sample surveys the response variable is often binary. For example, in marketing surveys, customers may be asked to reply in categories such as 'yes' or 'no', 'agree' or 'disagree', 'happy' or 'unhappy' etc. In such situations, one may ask how to do the sampling and, secondly, how to estimate population parameters like population proportions and their variance.

Sampling Procedure

The SRS procedures like SRSWOR and SRSWR used for drawing a sample in the case of quantitative characteristics can also be used for drawing a sample for binary characteristics. The sampling procedures therefore remain the same, irrespective of the nature of the characteristic under study—either binary or quantitative.

Estimation of Population Proportion

The population proportion in the case of a binary characteristic can be estimated similarly to the estimation of the population mean for a quantitative characteristic.

Consider a binary variable based on which the population can be divided into two mutually exclusive classes, say C and C^*. For example, if C is the part of the population comprising those who agreed with the proposal, then C^* is the part of the population who disagreed with the proposal. Let A be the number of units in C in a population of size N. Then $(N - A)$ units will be in C^*. The proportion of units in C is

$$P = \frac{A}{N}$$

and the proportion of units in C^* is

$$Q = \frac{N - A}{N} = 1 - P.$$

An indicator variable Y can be associated with the characteristic under study as follows and then for $i = 1, 2, \ldots, N$,

$$Y_i = \begin{cases} 1 & \text{if } i^{th} \text{ unit belongs to } C, \\ 0 & \text{if } i^{th} \text{ unit belongs to } C^*. \end{cases}$$

Now the population total is given as

$$Y_T = \sum_{i=1}^{N} Y_i = A$$

and the population mean is obtained as

$$\bar{Y} = \frac{\sum_{i=1}^{N} Y_i}{N} = \frac{A}{N} = P.$$

Thus the population mean in the case of a quantitative variable is the same as the population proportion in the case of a binary variable.

Suppose a sample of size n is drawn from a population of size N by simple random sampling.

Let a be the number of units in the sample which fall into class C so that $(n - a)$ units fall into class C^*; then the sample proportion of the units in C is

$$p = \frac{a}{n},$$

which can be written as

$$p = \frac{a}{n} = \frac{\sum_{i=1}^{n} y_i}{n} = \bar{y}.$$

Hence the sample proportion in the case of a binary characteristic is the same as the sample mean in the case of a quantitative characteristic. Similarly, we can define

$$q = 1 - p,$$

which is the sample proportion of the units in C^*.

We can now write S^2 and s^2 in terms of P, Q, p and q as follows. Since $\sum_{i=1}^{N} Y_i^2 = A = NP$, we have

$$S^2 = \frac{1}{N-1} \sum_{i=1}^{N} (Y_i - \bar{Y})^2 = \frac{1}{N-1} \left(\sum_{i=1}^{N} Y_i^2 - N\bar{Y}^2 \right)$$

$$= \frac{1}{N-1} (NP - NP^2) = \frac{N}{N-1} PQ.$$

Similarly, $\sum_{i=1}^{n} y_i^2 = a = np$ and

$$s^2 = \frac{1}{n-1} \sum_{i=1}^{n} (y_i - \bar{y})^2 = \frac{1}{n-1} \left(\sum_{i=1}^{n} y_i^2 - n\bar{y}^2 \right)$$

$$= \frac{1}{n-1} (np - np^2) = \frac{n}{n-1} pq.$$

Note that the quantities $\bar{y}$, $\bar{Y}$, s^2, and S^2 have been expressed as functions of the sample and population proportions. Since the sample has been drawn by simple random sampling and the sample proportion is the same as the sample mean, the properties of the sample proportion in SRSWOR and SRSWR can be derived directly by using the properties of the sample mean. We now illustrate this both under SRSWOR and SRSWR.

SRSWOR Suppose a sample of size n is drawn from a population of size N by SRSWOR. Since the sample mean $\bar{y}_{WOR}$ for a quantitative variable Y is an unbiased estimator of the population mean $\bar{Y}$, i.e. $E(\bar{y}_{WOR}) = \bar{Y}$, it follows that

$$E(p_{WOR}) = E(\bar{y}_{WOR}) = \bar{Y} = P.$$

The sample proportion p is thus an unbiased estimator of the population proportion P.

Using the expression of S^2 in terms of P and Q in $\text{Var}(\bar{y}_{WOR}) = \frac{N-n}{Nn} S^2$, the variance of p is found as

$$\begin{aligned} \text{Var}(p_{WOR}) &= \text{Var}(\bar{y}_{WOR}) \\ &= \frac{N-n}{Nn} \frac{N}{N-1} PQ \\ &= \frac{N-n}{N-1} \frac{PQ}{n}. \end{aligned}$$

Similarly, using the expression of s^2 in terms of p and q in $\widehat{\text{Var}}(\bar{y}_{WOR}) = \frac{N-n}{Nn}s^2$, an estimate of the variance of p can be derived as

$$
\begin{aligned}
\widehat{\text{Var}}(p_{WOR}) &= \widehat{\text{Var}}(\bar{y}_{WOR}) \\
&= \frac{N-n}{Nn}\frac{n}{n-1}pq \\
&= \frac{N-n}{N(n-1)}pq .
\end{aligned}
$$

The standard error of p can be obtained as $\sqrt{\widehat{\text{Var}}(p_{WOR})}$.

In R, one approach is to convert the categories of binary variables by using the logical operators TRUE and FALSE (or equivalently as T or F), see also Appendix A. For example, the categories 'good' or 'bad' can be represented by TRUE or FALSE. Similarly the categories 'satisfied' or 'not satisfied' can be represented by TRUE or FALSE too. These categories can then be converted into numerical values through the command as.numeric() into 1 and 0, respectively. The commands sample() or srswor() can then be used to draw samples by SRSWOR. The outcome can be stored in a data vector, say y, and commands like mean(), var() etc. can be used. The population proportion is estimated by mean(y) and the population variance and standard error of p is estimated by

```
# variance
((N-length(y))/(N*(length(y)-1)))*mean(y)*(1-mean(y))
# standard error
sqrt(((N-length(y))/(N*(length(y)-1)))*mean(y)*(1-mean(y)))
```

SRSWR Since the sample mean $\bar{y}_{WR}$ for a quantitative variable Y is an unbiased estimator of the population mean $\bar{Y}$, i.e. $E(\bar{y}_{WR}) = \bar{Y}$ in the case of SRSWR, it follows that

$$
E(p_{WR}) = E(\bar{y}_{WR}) = \bar{Y} = P .
$$

The sample proportion p is therefore an unbiased estimator of the population proportion P.

Using the expression of S^2 in terms of P and Q in $\text{Var}(\bar{y}_{WR}) = \frac{N-1}{Nn}S^2$, the variance of p is found as

$$
\begin{aligned}
\text{Var}(p_{WR}) &= \text{Var}(\bar{y}_{WR}) \\
&= \frac{N-1}{Nn}\frac{N}{N-1}PQ = \frac{PQ}{n} .
\end{aligned}
$$

Similarly, using the expression of s^2 in terms of p and q in $\widehat{\text{Var}}(\bar{y}_{WR}) = \frac{s^2}{n}$, an estimate of the variance of p can be derived as

$$
\begin{aligned}
\widehat{\text{Var}}(p_{WR}) &= \widehat{\text{Var}}(\bar{y}_{WR}) \\
&= \frac{n}{n-1}\frac{pq}{n} = \frac{pq}{n-1} .
\end{aligned}
$$

In R, the same commands `sample()` or `srswr()` can be used to draw the samples by SRSWR. The categories are denoted by `TRUE` or `FALSE` (or equivalently `T` or `F`). These categories are converted into numerical values by using the command `as.numeric()`, which converts `TRUE` and `FALSE` into 1 and 0, respectively. If the outcome consists of '0's and '1's, which are stored in a data vector `y`, then the population proportion can be estimated by `mean(y)`, the population variance through `(mean(y)*(1-mean(y)))/(length(y)-1)` and the standard error by `sqrt((mean(y)*(1-mean(y)))/(length(y)-1))`.

13.4.1 Estimation of the Total Count

Just like we estimated the population total in the case of simple random sampling for quantitative variables, an estimate of the population total A in the case of a binary variable or the total count is obtained as

$$\hat{A} = Np = \frac{Na}{n}.$$

Its variance is obtained as

$$\mathrm{Var}(\hat{A}) = N^2 \mathrm{Var}(p).$$

The estimate of its variance is

$$\widehat{\mathrm{Var}}(\hat{A}) = N^2 \widehat{\mathrm{Var}}(p)$$

and the standard error of p can be obtained as $\sqrt{\widehat{\mathrm{Var}}(\hat{A})}$.

13.4.2 Confidence Interval Estimation of P

If N and n are large, then $p - P/\sqrt{\mathrm{Var}(p)}$ approximately follows a $N(0, 1)$ distribution. With this approximation, we can write

$$P\left[-Z_{\frac{\alpha}{2}} \le \frac{p - P}{\sqrt{\mathrm{Var}(p)}} \le Z_{\frac{\alpha}{2}}\right] = 1 - \alpha$$

and the $100(1 - \alpha)\%$ confidence interval of P is obtained as

$$\left[p - Z_{\frac{\alpha}{2}}\sqrt{\mathrm{Var}(p)},\ p + Z_{\frac{\alpha}{2}}\sqrt{\mathrm{Var}(p)}\right].$$

It may be noted that in this case, a discrete random variable is being approximated by a continuous random variable and a normal distribution is used. Because of this, a continuity correction of $1/2n$ can be introduced in the confidence interval, in which case the limits become

$$\left[p - Z_{\frac{\alpha}{2}} \sqrt{\text{Var}(p)} + \frac{1}{2n}, p + Z_{\frac{\alpha}{2}} \sqrt{\text{Var}(p)} - \frac{1}{2n} \right].$$

We now extend Example 11.2.1 from Chap. 11 to illustrate the concepts of Sects. 13.3 and 13.4.

Example 13.4.1 A physiotherapist selects 12 randomly selected patients out of 100, all of whom had the same knee surgery done, and advises them to regularly perform a set of exercises. She asks them to record how long they exercise. She then summarizes the time they exercised every day (X, time in minutes) and how long it took them to regain their full range of motion again (Y, time in days). Based on the average time of exercise (X), the patients are classified into two categories as follows: $Z =$ '1' if $X > 30$ min and $Z =$ '0' otherwise. The data are as follows:

i	1	2	3	4	5	6	7	8	9	10	11	12
x_i	24	35	64	20	33	27	42	41	22	50	36	31
y_i	90	65	30	60	60	80	45	45	80	35	50	45
z_i	0	1	1	0	1	0	1	1	0	1	1	1

Suppose the physiotherapist is interested in the average time the patients took to regain their full range of motion, based on the data from the selected sample. She may also be interested in the proportion of patients who exercised for more than 30 min, together with the respective standard error and confidence interval. As the purpose of this example is to illustrate the various developed concepts under both SRSWOR and SRSWR, we first assume that the obtained observations are a sample of size 12 drawn from a population of 100 patients selected by SRSWOR. Thereafter, we consider the same sample but assume SRSWR. We store the data on y and z as follows: $y = c(90, 65, 30, 60, 60, 80, 45, 45, 80, 35, 50, 45)$ and $z = c(0, 1, 1, 0, 1, 0, 1, 1, 0, 1, 1, 1)$. We also have $N = 100$ and $n = \text{length}(y) = 12$

Case 1: SRSWOR
Suppose the sample is obtained by SRSWOR and consider the data on y. The estimate of the average time for the patients in the population to regain their full range of motion is obtained by the sample mean as $\text{mean}(y) = 57.08$ min. An estimate of the variance of the sample mean is obtained as $((N-\text{length}(y))/(N*\text{length}(y)))*\text{var}(y) = 26.15$ min^2 and the standard error is $\text{sqrt}(((N- \text{length}(y))/(N*\text{length}(y)))*\text{var}(y)) = 5.11$ min. The lower and upper limits of the 95% confidence interval of population mean of time

to regain the full range of motion are obtained as `mean(y) - qt(0.975, n-1)*sqrt(((N-length(y))/(length(y)*N))*var(y)) = 45.83` min and `mean(y) + qt(0.975, n-1)*sqrt(((N-length(y))/ (length(y)*N))*var(y)) = 68.34` min, respectively.

Consider now the data on z. The estimate of the population proportion of patients who exercised for more than 30 min is obtained by the sample proportion as `mean(z) = 0.667`, which means 66.7% of the patients exercised for more than 30 min and 33.3 % of the patients exercised for less than 30 min. An estimate of the variance of the sample proportion is obtained as `((N- length(z))/ (N*length(z)))*var(z)` or `((N-length(z))/(N*(length(z)-1)))*mean(z)*(1-mean(z)) = 0.018`.

With `sqrt(((N-length(z))/(N*(length(z)-1))) * mean(z) * (1-mean(z))) = 0.13`, we obtain the standard error. The lower and upper limits of the 95% confidence interval of the population proportion of patients who exercised for more than 30 min are obtained using the continuity correction through `mean(z) ± qnorm(0.975)*sqrt(((N-length(z))/(N* (length(z)-1)))* mean(z)*(1-mean(z))) ∓ 1/(2*n)`, which yields 0.45 and 0.89, respectively.

Case 2: SRSWR

Now suppose that the sample had been obtained by SRSWR and consider the data on y. The estimate of the average time to regain a full range of motion for the patients in the population is obtained by the sample mean as `mean(y) = 57.08` min. An estimate of the variance of the sample mean is obtained as `var(y)/length(y) = 29.72` min² and the standard error is `sqrt(var(y)length(y))= 5.45` min. The lower and upper limits of the 95% confidence interval of population mean of time to regain the full range of motion are obtained as `mean(y) ± qt(0.975,n-1)* sqrt(var(y)/length(y))`, which yields 45.08 min and 69.08 min, respectively.

Consider the data on z. The estimate of the population proportion of patients who exercised for more than 30 min is obtained by the sample proportion as `mean(z) = 0.667`, which means that 66.7% of the patients exercised for more than 30 min and 33.3% of the patients exercised for less than 30 min. An estimate of the variance of the sample proportion is obtained as `var(z)length(z) = 0.02` and the standard error is `sqrt(var(z)/length(z))` or `(mean(z)* (1-mean(z)))(length(z)-1) = 0.14`. The lower and upper limits of the 95% confidence interval of population proportion of patients who exercised for more than 30 min are `mean(z) - qnorm(0.975)*sqrt((mean(z)* (1-mean(z)))(length(z)-1)) + 1/(2*n)` and `mean(z) + qnorm (0.975)*sqrt((mean(z)*(1-mean(z)))(length(z)-1)) - 1/(2*n)`, which leads to confidence limits of 0.43 and 0.90, respectively.

It is clear from the results under Cases 1 and 2 that, although the estimates of the population mean and population proportion remain the same under SRSWOR and SRSWR, their standard errors as well as confidence intervals differ. The standard errors of the sample mean and sample proportion are smaller under SRSWOR than SRSWR. The confidence intervals are wider under SRSWR compared to SRSWOR.

13.5   Bootstrap Methodology

The classical approaches to drawing statistical inference are often based on ideal models and assumptions: it is, for example, often assumed that samples are drawn from a given population where the form of the distribution is known or the sample size is assumed to be large. Analysts frequently prefer to choose a simple form of an estimator, as it avoids mathematical complexities in deriving its statistical properties. Often, the choice of an estimator is governed by the availability of expressions to derive its standard error. For example, the arithmetic mean is a popular estimator of the population mean because, besides possessing appealing statistical properties, the expression to derive its standard error is also known. We know that the sample arithmetic mean of a random sample $y_1, y_2, \ldots, y_n$ is $\bar{y} = \sum_{i=1}^{n} y_i/n$. If the variance of each observation is σ^2, then the variance of $\bar{y}$ is σ^2/n, which is unbiasedly estimated by s^2/n where $s^2 = \frac{1}{n-1} \sum_{i=1}^{n}(y_i - \bar{y})^2$ and its standard error is $s/\sqrt{n}$. Expressions for its confidence interval are also available. Suppose we are interested in estimating the central tendency by the sample median, sample mode, sample geometric mean, etc., then finding or deriving the exact form of their exact standard errors is difficult. The unavailability of exact forms of standard errors of these estimators does not undermine the importance and utility of the respective estimators. Asymptotic theory can be used to derive the standard errors in such cases, but such derived expressions may not provide a valid statistical inference for small or finite samples. In another example, the exact standard errors and/or confidence intervals of many important quantities are not available, for example the sample-based coefficient of variation $v = s/\bar{y}$ in the univariate case or the correlation coefficient in a bivariate case. However, this does not undermine the utility and importance of these estimators. Hence the question is how to obtain standard errors and confidence intervals using the available data only.

A modern alternative to the traditional approach is the bootstrap method, see Efron and Tibshirani (2014). The bootstrap is a powerful, computer-based resampling method for deriving statistical inference without relying on too many assumptions. Among many other applications, it is widely used in finding statistical inferences for coefficients of regression models, machine learning estimators, and causal inference (see Chap. 14), and others. The bootstrap methodology is based on simple random sampling with replacement (SRSWR). We discussed in Sect. 13.2.2 that the total number of ways of choosing n sampling units out of N sampling units by SRSWR is N^n, and the probability of drawing a sample by SRSWR is $1/N^n$. One may observe that if $N = n$, then for

- $N = n = 5$, we obtain $N^n = 3125$ and the probability of drawing a particular sample is 0.00032,
- $N = n = 10$, we obtain $N^n = 1 \times 10^{10}$ and the probability of drawing a particular sample is 1×10^{-10},
- $N = n = 20$, we obtain $N^n = 1.04857 \times 10^{26}$ and the probability of drawing a particular sample is 9.536743×10^{-27},

and so on. Thus, as $N = n$ increases, the number of possible samples increases rapidly, and the probability of their selection declines sharply, becoming very small. Suppose a sample of n observations is available: the bootstrap method consists of drawing a large number of samples of size n by SRSWR from the given sample. Note that the size of each sample is the same as the size of the original sample itself. The probability of drawing each sample is expected to be very low for a reasonably large n. These samples are used to study the statistical properties of the estimators, e.g., standard errors, nonparametric confidence intervals, tests of hypotheses, etc.

13.6 Nonparametric Bootstrap Methodology

When applying a statistical method, one usually relies on the assumed distribution. If the information about the distribution is not available, it is reasonable to assume that the observed sample contains all the available information about the underlying distribution. Because we assume that the sample is representative, resampling the sample can provide similar information which can be retrieved from resampling the distribution.

Let θ be the unknown parameter. Let $\hat{\theta} \equiv \hat{\theta}(Y_1, Y_2, \ldots, Y_n)$ be the statistic based on the sample $Y_1, Y_2, \ldots, Y_n$ to estimate the parameter θ. To draw any statistical inference on θ we need the sampling distribution of $\hat{\theta}$. Finding the exact sampling distribution of $\hat{\theta}$ may be difficult or even too complicated to handle in many situations. Such information is nevertheless necessary in order to find the standard error of an estimator or the confidence intervals for an estimate of θ. Suppose the true distribution of $\hat{\theta}$ was known. Then different samples can be drawn from it, and the empirical sampling distribution of $\hat{\theta}$ can be constructed. Since the true distribution of $\hat{\theta}$ is unknown, the samples cannot be drawn from it. Instead, the samples are resampled from the originally drawn sample, and an empirical distribution is constructed from them. The bootstrapping method mimics the data-generating process by means of the samples which are drawn from the unknown true distribution.

Below, we look into standard errors and confidence intervals, but there are many more applications of the bootstrap method. Now, we introduce a few basic concepts to understand the implementation of the steps in the bootstrap methodology.

13.6.1 The Empirical Distribution Function

We have already learnt about the cumulative distribution function (CDF), which is obtained through a probability distribution (or mass) function. When we do not have any information about the probability distribution, we attempt to find it based on a given set of data using the concept of the empirical distribution function (EDF). The EDF usually approximates the CDF well for large sample sizes.

Suppose we have a sample of size n denoted as $y_1, y_2, \ldots, y_n$. The EDF is a discrete distribution that gives equal weight to each data point. It assigns the probability $\frac{1}{n}$ to each $y_i, i = 1, 2, \ldots, n$, forming a CDF that is a step function. The step jumps up by $\frac{1}{n}$ at each of the n data points.

Example 13.6.1 Suppose a die is rolled $n = 1000$ times and its outcomes 1, 2, 3, 4, 5, and 6 are obtained as

Points on upper face (k)	1	2	3	4	5	6
Number of $\{y_i = k\}$	160	240	180	100	200	120
Relative frequency $\{y_i = k\}$	0.16	0.24	0.18	0.10	0.20	0.12

If A is a set in the sample space of Y which has the six possible realizations 1, 2, 3, 4, 5 and 6, then the proportion of $y_1, y_2, \ldots, y_6$ in A is given as

$$\hat{f}_k = \frac{\text{Number of } \{y_i = k\}}{n}, k = 1, 2, 3, 4, 5, 6; i = 1, 2, \ldots, 1000.$$

The EDF of $y_1, y_2, \ldots, y_{1000}$ puts equal probabilities $(1/1000)$ on each of the possible outcomes and then $\hat{F}$ can be expressed as a vector of proportions $\hat{f}_k, k = 1, 2, 3, 4, 5, 6$, as

$$\hat{F} = \left(\hat{f}_1, \hat{f}_2, \hat{f}_3, \hat{f}_4, \hat{f}_5, \hat{f}_6 \right).$$

The EDF is a vector of realizations taken on by the sample along with the respective proportions or relative frequencies.

In other words, $\hat{F}$ assigns the empirical probabilities

$$\hat{P}(A) = \frac{\text{Number of } \{y_i \in A\}}{n}, i = 1, 2, \ldots, n,$$

which are the proportions of $y_1, y_2, \ldots, y_n$ in A, to the given set A in the sample space of Y.

It can be proven that the vector of observed frequencies $\hat{F} = (\hat{f}_1, \hat{f}_2, \ldots)$ is a sufficient statistic for the true distribution $F = (f_1, f_2, \ldots)$. This means that all the information about F contained in $y_1, y_2, \ldots$ is also contained in $\hat{F}$.

13.6.2 The Plug-in Principle

The plug-in principle is a simple method of estimation of parameters based on the empirical distribution function. The plug-in estimator of a parameter $\theta = T(F)$ is defined to be $\hat{\theta} = T(\hat{F})$. In other words, we estimate the function $\theta = T(F)$ of the probability distribution F by the same function of the empirical distribution $\hat{\theta} = T(\hat{F})$ and $\hat{\theta}$ is called a plug-in estimate for θ. For example, the population mean is $\mu = \int_\mu y f(y) dy$ and the sample mean is $\bar{y} = \frac{1}{n} \sum_{i=1}^n y_i$.

13.6.3 Steps in Applying the Bootstrap

Let $Y_1, Y_2, \ldots, Y_n$ be a random sample from some probability function with population parameter θ. The parameter θ may represent any population characteristic, for example the population mean or the population variance, or others. Some statistic, a function of sample observations $Y_1, Y_2, \ldots, Y_n$, estimates the parameter θ. For example, the sample mean estimates the population mean and the sample variance estimates the population variance. The bootstrap methodology is a resampling method and is used to draw inferences about an estimate of θ based on the sample data. For this, simple random sampling with replacement (SRSWR) is used. Suppose a sample of size n is available; then the samples are drawn independently by SRSWR from the existing sample data. The size of the drawn samples is the same as the original data, i.e. n, and the statistical inferences are drawn from these re-sampled data sets.

The following steps are, in general, involved in the bootstrap methodology:

1. An original sample from the population of size n is available.
2. Draw a sample of size n from the original sample by SRSWR. Note that the size of the sample is the same as the size of the original sample.
3. Repeat the process B times. Each re-sampled sample is called a bootstrap sample. There are now B bootstrap samples available.
4. Evaluate the statistic estimating θ for each bootstrap sample. Thus, there will be a total of B estimates of θ.
5. Construct a sampling distribution with these B bootstrap estimates. The sampling distribution constructed from these B bootstrap estimates can be used to make further statistical inferences. For example, the standard error of the statistic used for estimating θ can be computed, or a confidence interval for θ can be obtained using this constructed sampling distribution.

In fact, any statistic of suitable form, irrespective of its complicated structure, can be considered for estimating the parameter. We will discuss how to find the standard error and confidence interval for such a statistic.

We now illustrate the details of the bootstrap methodology. Suppose we have a population of size N for the variable Y, denoted as $\underline{Y} = (Y_1, Y_2, \ldots, Y_N)$. No assumption is made about the associated probability distribution of Y.

Suppose a sample of size n is available and is denoted as $\underline{y} = (y_1, y_2, \ldots, y_n)$. Let $\hat{\theta} = \hat{\theta}(\underline{y})$ be an estimator of θ. For example, $\theta = \theta(\underline{Y}) = \bar{Y} = \frac{1}{N} \sum_{i=1}^{N} Y_i$ is the population mean and $\hat{\theta} = \hat{\theta}(\underline{y}) = \bar{y} = \frac{1}{n} \sum_{i=1}^{n} y_i$ is the sample mean. If we choose a sample by SRSWR of size n from $\underline{y}$, then this sample is going to be a rearrangement of all the sample units in $\underline{y}$ with the possibility that one or more units may be included more than once in the sample, or may not be included at all. This rearrangement of units obtained through SRSWR is denoted as $\underline{y}_1^*$ and this generated sample is the first bootstrap sample. This process is repeated B times, to obtain B bootstrap samples. For example, suppose a sample of size four is available as $\underline{y} = (y_1, y_2, y_3, y_4)$. Now a sample of size four is drawn by SRSWR and is denoted as $\underline{y}_1^* = (y_2, y_4, y_3, y_1)$. This is the first bootstrap sample. Similarly, the second bootstrap sample is drawn as $\underline{y}_2^* = (y_3, y_1, y_3, y_2)$. Eventually we have drawn B bootstrap samples independently as $\underline{y}_1^*, \underline{y}_2^*, \ldots, \underline{y}_B^*$, where B is a large number. Typically, B is chosen to be as large, for example $B = 200$ or 1000 or even larger. While computationally expensive, in many applications bootstrap sampling is still relatively fast on modern computers. Obviously, the probability of selection of a sample is $1/n^n$, which is very small for a reasonably large n. We calculate the statistic $\hat{\theta}$ for each bootstrap sample as $\hat{\theta}(\underline{y}_1^*), \hat{\theta}(\underline{y}_2^*), \ldots, \hat{\theta}(\underline{y}_B^*)$. These B values of $\hat{\theta}(\underline{y}_b^*), b = 1, 2, \ldots, B$, can then be used to construct, among other things, the empirical distribution and various statistical functions.

13.6.4 Bootstrap Estimator and Bootstrap Variance

We calculate the mean and variance of the realizations $\hat{\theta}(\underline{y}_1^*), \hat{\theta}(\underline{y}_2^*), \ldots, \hat{\theta}(\underline{y}_B^*)$ as

$$\hat{\theta}_B = \frac{1}{B} \sum_{b=1}^{B} \hat{\theta}(\underline{y}_b^*) \tag{13.14}$$

and

$$v_B = \frac{1}{B-1} \sum_{b=1}^{B} \left[\hat{\theta}(\underline{y}_b^*) - \hat{\theta}_B \right]^2 \tag{13.15}$$

respectively. The estimator $\hat{\theta}_B$ is the bootstrap estimator of θ and v_B is the bootstrap variance of $\hat{\theta}_B$.

It can be shown that for large n and B:

- $\hat{\theta}_B$ approximates $\hat{\theta}(\underline{y})$;
- v_B approximates the variance of $\hat{\theta}(\underline{y})$; and
- the empirical distribution of $\left[\hat{\theta}(\underline{y}_b^*) - \hat{\theta}(\underline{y}) \right]$, $b = 1, 2, \ldots, B$, approximates the distribution of $\left[\hat{\theta}(\underline{Y}) - \theta \right]$, where $\underline{Y} = (Y_1, Y_2, \ldots, Y_N)$.

We note that $\hat{\theta}_B$ is not used as an estimator of θ but of $\hat{\theta}$. Moreover, $\hat{\theta}_B$ is not a bias corrected estimator of θ.

13.6.5 Bootstrap Estimate of the Bias and Standard Error

The bootstrap estimate of the bias of $\hat{\theta}$ is

$$\widehat{Bias}_{boot}(\hat{\theta}) = \hat{\theta}_B - T(\hat{F}) \tag{13.16}$$

where $\theta = T(F)$ is the parameter of interest, and $\hat{\theta} = T(\hat{F})$ is the plug-in estimate of θ based on the empirical distribution function, i.e. the vector of observed frequencies $\hat{F} = (\hat{f}_1, \hat{f}_2, \ldots)$ is a sufficient statistic for the true distribution $F = (f_1, f_2, \ldots)$. The bootstrap standard error of $\hat{\theta}$ is $\sqrt{v_B}$.

Now we consider an example with $n = 4$ to explain and illustrate the steps involved in the computation of a bootstrap estimate of the unknown parameter and its standard error.

Example 13.6.2 Suppose four children receive chocolate boxes as a present. These children receive 2 boxes, 4 boxes, 6 boxes, and 8 boxes of chocolates respectively. These observations create the original sample of observations, denoted as $\underline{y} = (2, 4, 6, 8)$. The sample median is thus 5 boxes.

Now we draw the first bootstrap sample, which is the same as drawing a sample of size $n = 4$ from $\underline{y}$ by SRSWR. Suppose it is $\underline{y}_1^* = (2, 8, 6, 4)$. Next, we draw another bootstrap sample of size $n = 4$ from $\underline{y}$. Suppose it turns out to be $\underline{y}_2^* = (2, 8, 8, 4)$, which is the second bootstrap sample. This process can be repeated B times so that B bootstrap samples $\underline{y}_1^*, \underline{y}_2^*, \ldots, \underline{y}_B^*$ are generated.

Now we illustrate how to find the bootstrap estimate of the population median and population coefficient of variation. Note that no assumption is made about the associated probability distribution of Y, and it is not easy to find the exact expression of standard errors of the estimators of population median and population coefficient of variation.

First, we consider θ to be the population median. Suppose it is estimated by the sample median. So $\hat{\theta}(\underline{y})$ is the median of the sample observations in the original sample $\underline{y}$. We find the bootstrap sample median of $\underline{y}_1^* = (2, 8, 6, 4)$, using the R command `median(c(2,8,6,4))`, which yields $\hat{\theta}(\underline{y}_1^*) = 5$. Next, we find the median of the second bootstrap sample $\underline{y}_2^* = (2, 8, 8, 4)$ using the R command `median(c(2,8,8,4))` as $\hat{\theta}(\underline{y}_2^*) = 6$. This process is repeated a large number of times and we obtain $\hat{\theta}(\underline{y}_1^*), \hat{\theta}(\underline{y}_2^*), \ldots, \hat{\theta}(\underline{y}_B^*)$. Suppose $B = 10$ and we obtain

$$\hat{\theta}(\underline{y}_3^*) = 2, \hat{\theta}(\underline{y}_4^*) = 6, \hat{\theta}(\underline{y}_5^*) = 6, \hat{\theta}(\underline{y}_6^*) = 5,$$
$$\hat{\theta}(\underline{y}_7^*) = 5, \hat{\theta}(\underline{y}_8^*) = 8, \hat{\theta}(\underline{y}_9^*) = 8, \hat{\theta}(\underline{y}_{10}^*) = 2.$$

We can then calculate the mean of $\hat{\theta}(\underline{y}_1^*), \hat{\theta}(\underline{y}_2^*), \ldots, \hat{\theta}(\underline{y}_{10}^*)$ using the R command `mean(c(5,6,2,6,6,5,5,8,8,2))` as

$$\hat{\theta}_B = \frac{1}{10} \sum_{b=1}^{10} \hat{\theta}(\underline{y}_b^*) = 5.3,$$

which is the bootstrap estimate of the population median (but again we note that the bootstrap estimate is never directly used as an estimate for the population median, only the median of the original sample). The bootstrap variance estimate of the sample median $\hat{\theta}(y)$ is obtained by computing the variance of $\hat{\theta}(\underline{y}_1^*), \hat{\theta}(\underline{y}_2^*), \ldots, \hat{\theta}(\underline{y}_{10}^*)$ using the R command var(c(5,6,2,6,6,5,5,8,8,2)) as

$$v_B = \frac{1}{9} \sum_{b=1}^{B} \left[\hat{\theta}(\underline{y}_b^*) - \hat{\theta}_B \right]^2 \approx 4.23.$$

The bootstrap standard error is approximately $\sqrt{4.23} = 2.06$ boxes and is an estimate of the standard error of the sample median based on the original sample $\underline{y} = (2, 4, 6, 8)$.

Example 13.6.3 Suppose we want to estimate the coefficient of variation v (see Sect. 3.2.3), which is defined, for the population, as the ratio of the standard deviation of the population and the population mean. We consider an estimate of v as the ratio of the standard deviation of the sample and the sample mean. Suppose we want to find the bootstrap estimate of v and the standard error of its estimate. Now, $\theta = S/\bar{Y}$ is v and $\hat{\theta}(y) = s/\bar{y}$ is its estimate based on the original sample $\underline{y} = (2, 4, 6, 8)$. We proceed as in the previous example and draw the bootstrap samples $\underline{y}_1^*, \underline{y}_2^*, \ldots, \underline{y}_B^*$. For each bootstrap sample, we compute $\hat{\theta}(\underline{y}_b^*) = s_{b*}/\bar{y}_{b*}$, $b = 1, 2, \ldots, B$, where $\bar{y}_{b*}$ and s_{b*} are the values of sample mean and $\sqrt{s^2}$ respectively, computed from the bth sample. Then the mean and variance based on all the B values of sample based vs is obtained using (13.14) and (13.15), which provides the bootstrap estimate of v. The square root of its variance will provide us with the bootstrap standard error, which is an estimate of the standard error of $\hat{\theta}(\underline{y})$.

Bootstrap Estimate of Bias and Standard Error in R

First, we consider the R-package boot, which can be used to find the bootstrap estimate of any parameter, as well as its bootstrap bias and bootstrap standard error. The R command boot generates B bootstrap replicates of a statistic applied to a data set. Note that the package boot uses the symbol R to denote the number of bootstrap samples, whereas we have used the symbol B to denote the same. The general arguments and options are as follows:

```
boot(data, statistic, R,...)
```

where

- data inputs the data as a data frame, vector or matrix. If it is a matrix or data frame then each row is considered as one multivariate observation,
- statistic provides a function, which when applied to the data, returns a vector containing the statistic(s) of interest,
- R denotes the number of bootstrap replicates, i.e. a single positive integer.

The function `boot()` returns the following objects, among others:

- `t0` denotes the observed value of the `statistic` argument applied to the `data` argument,
- `R` denotes the number of bootstrap samples as passed to `boot`,
- `t` denotes a matrix with `sum(R)` rows each of which is a bootstrap replicate of the result of calling `statistic`.

The `boot` package works as follows: the function `boot` facilitates the resampling of the given data set by calculating the given statistic on the bootstrap samples. We can thus do bootstrapping by means of the following procedure:

- Before calling `boot`, define a function to compute the desired statistic.
- The function returns the statistic which is to be bootstrapped.
- The first argument passed to the function should be the data set.
- The second argument can be an index vector of the observations in the data set, a frequency, or weight vector that informs the sampling probabilities.

It is easier to understand the implementation of the `boot` package through the following example.

Example 13.6.4 Suppose the following sample of observations is available on the electricity consumption (in kilowatts), denoted as 'Power', and day temperature (in degrees Celsius), denoted as 'Temp', in a house over a period of 20 days:

Day	1	2	3	4	5	6	7	8	9	10
Power	198	184	245	223	263	246	206	216	191	237
Temp	30	25	37	38	27	36	33	29	26	34

Day	11	12	13	14	15	16	17	18	19	20
Power	208	244	221	209	256	276	226	208	198	207
Temp	24	35	37	28	37	36	33	31	26	34

The variables 'Power' and 'Temp' are correlated. Suppose we want to obtain the bootstrap estimate of the Pearson correlation coefficient between 'Power' and 'Temp', its bias and standard error. We have to enter the data and create a data frame `power_temp_data` as

```
day <- c(1:20)

power <- c(198, 184, 245, 223, 263, 246, 206,
216, 191, 237, 208, 244, 221, 209, 256, 276, 226,
208, 198, 207)

temp <- c(30, 25, 37, 38, 27, 36, 33, 29, 26, 34,
24, 35, 37, 28, 37, 36, 33, 31, 26, 34)

power_temp_data <- data.frame(day, power, temp)
```

Now, we have to define a function to compute the desired statistic, i.e. the correlation coefficient between power and temp. *R* already has a function available for this: cor(), as introduced in Sect. 4.3.2. The following generic function takes both the data (through the argument d) and an index (through the argument i) as input and returns the correlation coefficient of Bravais–Pearson:

```
corr_power_temp <- function(d, i){
power_temp_data = d[i,]
return(cor(power_temp_data$power,
power_temp_data$temp))}
```

Then the bootstrap ($B = 5000$) can be executed as follows:

```
set.seed(1234)
corrboot <- boot(data = power_temp_data,
statistic = corr_power_temp, R = 5000)
```

We used the command set.seed(1234) to ensure that the outcome is reproducible. Without it, different random samples, and thus different results, would be obtained every time the code is executed. The above code returns the following:

```
Bootstrap Statistics :
     original              bias     std. error
t1* 0.5838149 0.002973371    0.1772656
```

This provides us with the summary statistics of the bootstrap parameters, saved in the object corrboot. Here the value 0.5838149 under original is the estimated value $\hat{\theta}$ of the correlation coefficient based on the original sample, the value 0.002973371 under bias is the bootstrap bias estimate of the estimator of the correlation coefficient, and the value 0.1772656 under std. error is the bootstrap standard error (which is an estimate of the standard error) of the correlation coefficient estimator $\hat{\theta}$.

Using the returned vectors t and t0, we can calculate the bootstrap estimator of the correlation coefficient, its bias and standard error separately as follows:

```
mean(corrboot$t)
[1] 0.5867883          # estimate

mean(corrboot$t) - corrboot$t0
[1] 0.002973371        # bias

sd(corrboot$t)
[1] 0.1772656          # standard error
```

Note that the bootstrap bias is calculated as $0.5867883 - 0.5838149 = 0.0029734$ and is identical to the bias returned by the boot function as shown above.

13.6.6 Bootstrap Confidence Intervals

Reporting the confidence intervals along with the point estimate of any parameter and its standard error are important, as we have learnt in Sect. 9. The key ingredients of any confidence interval are the estimate of the parameter, its standard error and its probability distribution, which allows us to find the critical values at a given level of significance. In the following sections we discuss how confidence intervals can be constructed with the bootstrap methodology based on three approaches:

1. the percentile of the distribution of bootstrap statistics,
2. the percentile of the standard normal distribution $N(0, 1)$,
3. the studentized form of bootstrap statistics.

13.6.6.1 The Percentile Method for Confidence Intervals
The percentile method for a bootstrap confidence interval is based on the percentile of the empirical distribution of bootstrap statistics. Since B is a large number, it is useful to construct a histogram based on the values $\hat{\theta}(y_b^*)$, $b = 1, 2, \ldots, B$. The bootstrap histogram, i.e. the histogram constructed by B values of $\hat{\theta}(y_b^*)$, is a close approximation to the true distribution of $\hat{\theta}(y)$. Suppose $B = 1000$, then we have $\hat{\theta}(y_b^*) = \hat{\theta}_b^*$, $b = 1, 2, \ldots, 1000$, such that the ordered values are

$$\hat{\theta}_{(1)}^* \leq \hat{\theta}_{(2)}^* \leq \ldots \leq \hat{\theta}_{(1000)}^*.$$

For example, the 90% bootstrap confidence interval is

$$[\hat{\theta}_{(50)}, \hat{\theta}_{(950)}] = [\hat{\theta}_{(\alpha/2=0.05),l}, \hat{\theta}_{(\alpha/2=0.05),u}].$$

Let $(100 \cdot \frac{\alpha}{2})\%$ of the histogram area be below $\hat{\theta}_{(\alpha/2),l}$ and above $\hat{\theta}_{(\alpha/2),u}$. Then, $[\hat{\theta}_{(\alpha/2),l}, \hat{\theta}_{(\alpha/2),u}]$ is taken as the $100(1-\alpha)\%$ confidence interval for θ where $0 \leq \alpha \leq 1$. We use percentiles to create the bootstrap confidence interval. If $\alpha = 0.05$ then for a 95% confidence interval, we need to identify the middle 95% of the distribution. To do this, we follow the steps below:

- Use the 97.5^{th} percentile and 2.5^{th} percentile so that the middle part between these two percentiles is 95%.
- If all the sample statistics $\hat{\theta}(\underline{y}_b^*)$ are ordered from low to high, then chop off the lowest 2.5% and the highest 2.5% points of the $\hat{\theta}(\underline{y}_b^*)$, $b = 1, 2, \ldots, B$.
- The middle 95% of $\hat{\theta}(\underline{y}_b^*)$ is the remaining range and [2.5^{th} percentile, 97.5^{th} percentile] is the bootstrap percentile confidence interval.

13.6.6.2 Bootstrap Confidence Intervals Based on the Standard Normal Distribution

In the bootstrap method for a confidence interval based on the standard normal distribution we consider a form of statistic

$$\frac{\hat{\theta}(\underline{y}) - \theta}{\sqrt{v_B}}, b = 1, 2, \ldots, B, \tag{13.17}$$

based on the $N(0, 1)$ distribution, where v_B is the bootstrap variance estimate of $\hat{\theta}(\underline{y})$. Thus we find

$$P\left[-Z_{\alpha/2} \leq \frac{\hat{\theta}(\underline{y}) - \theta}{\sqrt{v_B}} \leq Z_{\alpha/2}\right] = 1 - \alpha$$

where $\hat{\theta}(\underline{y})$ is based on the original sample data $\underline{y}$. The $100(1-\alpha)\%$ confidence interval for θ, where $0 \leq \alpha \leq 1$, is of the form

$$\left[\hat{\theta}(\underline{y}) - Z_{\alpha/2}\sqrt{v_B}, \hat{\theta}(\underline{y}) + Z_{\alpha/2}\sqrt{v_B}\right], \tag{13.18}$$

where $Z_{\alpha/2}$ denotes the upper $(\alpha/2)\%$ points on the $N(0, 1)$ distribution.

13.6.6.3 The Bootstrap t-Method for Confidence Intervals

The bootstrap t-method for confidence intervals is based on the studentized form of bootstrap statistics. It uses the standard error of the statistic which is obtained by computing the standard deviation of the bootstrap distribution.

In detail, this approach considers a form of Student's t-statistic:

$$\frac{\hat{\theta}(\underline{y}_b^*) - \hat{\theta}(\underline{y})}{\sqrt{v_B^*}} = t_b^*, b = 1, 2, \ldots, B, \tag{13.19}$$

where v_B^* is a variance estimate of $\hat{\theta}(\underline{y}_b^*)$. Since B is a large number, it is useful to construct a histogram based on the values $t_b^*, b = 1, 2, \ldots, B$. This bootstrap histogram is a close approximation to the true t-distribution of $\hat{\theta}(\underline{y}) - \theta/\sqrt{\text{Var}(\hat{\theta}(\underline{y}))}$ and $\text{Var}(\hat{\theta}(\underline{y}))$ is approximated by v_B.

This approach is based on B bootstrap samples $\underline{y}_{b*}, b = 1, 2, \ldots, B$, and considers a form of statistic

$$t_{b*} = \frac{\hat{\theta}(\underline{y}_{b*}) - \hat{\theta}(\underline{y})}{\text{se}(\underline{y}_{b*})}, b = 1, 2, \ldots, B \qquad (13.20)$$

where $\text{se}(\underline{y}_b^*)$ is the estimated standard error of $\hat{\theta}(\underline{y}_b^*)$ for the bth bootstrap sample. It may again be obtained by bootstrapping the bootstrap sample $\underline{y}_b^*$, which is a so-called nested bootstrap.

Given $t_b^*, b = 1, 2, \ldots, B$, the αth quantile is estimated as $\hat{t}_a$

$$\frac{\text{Number of } [t_b^* \leq \hat{t}_\alpha]}{B} = \alpha,$$

where $0 \leq \alpha \leq 1$. For example, if $B = 1000$, the estimate of the 5% percentile is the 50th largest value of the t_b^*'s and the estimate of the 95% point is the 950th largest value of the t_b^*'s.

Let $100\frac{\alpha}{2}\%$ of the histogram area be below $\hat{t}_{\alpha/2}$ and above $\hat{t}_{1-\alpha/2}$. Then a $100(1 - \alpha)\%$ "bootstrap-t" confidence interval for θ is

$$\left[\hat{\theta}(\underline{y}) - \hat{t}_{1-\alpha/2}\sqrt{v_B}, \hat{\theta}(\underline{y}) + \hat{t}_{\alpha/2}\sqrt{v_B}\right]. \qquad (13.21)$$

In brief, compared to a confidence interval based on the standard normal distribution, the normal quantiles are substituted by the estimated quantiles based on the bootstrap.

Some Considerations about Bootstrap Confidence Intervals

Now we discuss some important points about bootstrap confidence intervals which help us in choosing the appropriate bootstrap confidence interval.

- When the bootstrap distribution is symmetric and bell-shaped, any of the two percentile methods or the t-method can be used.
- When the bootstrap distribution is not symmetric or not bell-shaped, then the percentile method is better to use.
- Bootstrapping usually does not give reliable estimates when the sample size is small, say less than 10.
- The bootstrap method does not give reliable results in those distributions that have infinite second moments.
- Bootstrap results are not reliable when estimating extreme values (such as those introduced in the box plot in Sect. 3.3).

- The explained methods for constructing the confidence intervals work satisfactorily if the bootstrap distribution of the statistic is smooth and symmetric.
- It is a good idea to first look at a plot of the bootstrap distribution and even compare it with a normal probability plot. If the bootstrap distribution is highly skewed or too many spikes and gaps appear, then it may not be advisable to use a bootstrap confidence interval.

13.6.6.4 Bootstrap Confidence Intervals in R Using boot and bootstrap

We use the boot package to construct various bootstrap confidence intervals. The command boot.ci provides four types of confidence intervals from the bootstrap samples, including the normal and percentile intervals which we introduced above.

Example 13.6.5 Consider again Example 13.6.4. We want to construct the bootstrap confidence interval for the Pearson correlation coefficient between power and temp. For this, we use the object corrboot from the previous example, apply the function boot.ci to it and pick the option type = c('norm', 'perc'):

```
boot.ci(boot.out = corrboot, type = c('norm', 'perc'))
```

This yields the following output:

```
BOOTSTRAP CONFIDENCE INTERVAL CALCULATIONS
Based on 5000 bootstrap replicates

Intervals :
Level        Normal              Percentile
95%    ( 0.2334,  0.9283 )    ( 0.1959,  0.8578 )
Calculations and Intervals on Original Scale
```

We can interpret this output as follows.

- The confidence interval of the correlation coefficient based on the percentile of the standard normal distribution $N(0, 1)$ can be seen under Normal and is (0.2334, 0.9283).
- The confidence interval of the correlation coefficient based on the percentile of the distribution of the bootstrap statistic can be seen under Percentile and is (0.1959, 0.8578).

Alternatively, the package bootstrap can also be used for calculating bootstrap confidence intervals. We discuss its role in finding the bootstrap confidence interval by the t-method. For this, the command boott can be used after installing and loading the package `bootstrap`. More details about this command can be found in `help(boott)`. We discuss here only a few relevant options briefly.

The *R* command is

```
boott(x, theta, nboott,...)
```

where x denotes a vector containing the data or sometimes the row indices, `theta` denotes the function to be bootstrapped, and `nboott` denotes the number of bootstrap samples used to estimate the distribution of the bootstrap *t*-statistic.

We first have to define a function which calculates the correlation coefficient:

```
corboott <- function(x,xdata){
cor(xdata[x,1],xdata[x,2])
}
```

Note that the above function takes x as the argument to define the row indices and xdata is a second argument which relates to the actual data. With $n = 20$ and nboott = 5000 we can apply `boott` and access the object `confpoints` (which stores the confidence interval estimates) as follows:

```
set.seed(1234)
boott(1:n, theta = corboott, xdata =
power_temp_data, nboott = 5000)$confpoints
```

Again, we use the command `set.seed(1234)` so that the results are reproducible. The returned results are:

```
          0.001      0.01     0.025      0.05        0.1
[1,] -1.42142 -0.809075 -0.59909 -0.452103 -0.324847
            0.5       0.9      0.95     0.975      0.99    0.999
[6,] 0.0511655 0.414774 0.559325 0.732107 0.989264 1.99366
```

The first row of the output indicates the percentile of bootstrap estimates of the Pearson correlation coefficient, the second row the respective bootstrap estimates. For example, the 95th percentile is indicated by 0.95 and its value is 0.559325. If we need a 95% bootstrap confidence interval of the correlation coefficient, then it can be found by obtaining the 2.5^{th} and 97.5^{th} percentiles as $(-0.59909, 0.732107)$. The same results can also be obtained by directly using the option `perc = c(0.025, 0.975)` in the `boott` command.

13.7   Key Points and Further Issues

Note:

✓ The sample mean is an unbiased estimator of the population mean under both SRSWR and SRSWOR. However, SRSWOR yields a more efficient estimate of the population mean than a sample drawn by SRSWR.

✓ Similarly, the sample proportion is an unbiased estimator of the population proportion under both SRSWR and SRSWOR. SRSWOR yields a more efficient estimate of the population proportion than a sample drawn by SRSWR.

✓ The bootstrap is a resampling method where the samples are generated under SRSWR from the original data.

✓ The bootstrap allows estimation of the standard error, distribution and bias of an estimator $\hat{\theta}$ without making strong assumptions about the distribution of the sample.

✓ The bootstrap is never used to directly estimate population parameters, but only those of the sample.

✓ With bootstrapping, confidence intervals can be constructed based on the percentiles of the distribution of the bootstrap statistic, the percentile of a standard normal distribution or Student's t-statistic.

13.8   Exercises

Exercise 13.1 Consider the general situation where a sample of size n is drawn from a population of size N. Now, suppose a sample of size $n/3$ is drawn by SRSWOR and another sample of size $2n/3$ is drawn by SRSWR. Determine the ratio of the variances of the sample means, i.e. $\mathrm{Var}(\bar{y})_{WOR}/\mathrm{Var}(\bar{y})_{WR}$.

Exercise 13.2 The following data represent the amounts spent (in €) by 60 customers in a restaurant:

60.80, 41.59, 61.01, 59.72, 27.02, 55.43, 26.74, 37.21, 47.29, 40.90, 48.01, 34.04, 65.24, 58.71, 31.85, 31.64, 64.00, 55.10, 58.91, 45.70, 35.90, 58.86, 43.83, 49.62, 31.69, 32.35, 35.20, 50.34, 63.33, 45.67, 36.69, 44.52, 63.13, 50.34, 54.24, 58.55, 27.64, 49.44, 42.79, 21.67, 37.80, 56.37, 42.14, 33.02, 58.30, 50.22, 44.31, 44.09, 56.31, 48.18, 55.76, 46.26, 64.41, 55.95, 70.15, 44.95, 64.95, 49.22, 50.29, 39.16

(a) Determine the total number of different samples that can be drawn by SRSWOR.

(b) Draw a sample of 10 customers by SRSWOR in R (using the seed value 4321). Which sampling units are selected into the sample?

(d) Use R to calculate an estimate of the average amount spent, as well as an estimate of its variance and its 95% confidence interval using the sampled values.

(d) Find an estimate of the total spending of the customers and an estimate of its variance using the sampled values.

Exercise 13.3 Consider again the data from Exercise 13.2.

(a) Determine the total number of different samples that can be drawn by SRSWR.
(b) Draw a sample of 15 customers by SRSWR in R (using the seed value 1234). Which sampling units are selected into the sample?
(c) Use R to calculate an estimate of the average amount spent, as well as an estimate of its variance and its 95% confidence interval using the sampled values.
(d) Find an estimate of the total spending of the customers and an estimate of its variance using the sampled values.

Exercise 13.4 Suppose that in an examination 100 points is the highest possible score. We have a sample of scores from 50 students:

29.1, 69.8, 68.7, 69.9, 88.9, 71.2, 20.8, 38.6, 73.3, 61.1, 75.5, 63.6, 42.6, 93.9, 43.4, 87.0, 42.9, 41.3, 34.9, 38.6, 45.3, 44.2, 32.7, 23.2, 37.5, 84.8, 62.1, 93.2, 86.5, 23.7, 56.5, 41.2, 44.4, 60.6, 34.5, 80.8, 36.1, 40.7, 99.4, 84.6, 64.3, 71.7, 44.9, 69.7, 46.4, 60.2, 74.2, 58.8, 39.5, 81.2

(a) Draw a sample by SRSWOR of those 10 students whose marks are greater than 30 points (using R and the seed value 4321). Which sampling units are selected into the sample?
(b) Find an estimate of the average mark among those students securing more than 30 points, as well as its standard error and 95% confidence interval using the sampled values.
(c) Calculate an estimate of the total count among those students securing more than 30 points, as well as its standard error, using the sampled values.

Exercise 13.5 Consider again the data from Exercise 13.4.

(a) Draw a sample by SRSWR of those 15 students whose points are less than 60 (using R and the seed value 1234). Which sampling units are selected into the sample?
(b) Find an estimate of the average score of those students securing less than 60 points, as well as its standard error and 95% confidence interval using the sampled values.
(c) Find an estimate of the total count among those students securing less than 60 points, as well as its its standard error using the sampled values.

Exercise 13.6 A total of 50 persons were asked about their opinion of a new, organic soft drink. Their responses were collected as good or bad as follows:
good, bad, good, good, bad, bad, bad, good, bad, good, bad, good, good, bad, good,
bad, bad, good, bad, good, bad, bad, bad, bad, good, bad, good, good, bad, bad,

good, good, good, good, good, good, good, bad, bad, bad, bad, good, good, good, good, good, bad, bad, good, bad.

(a) A sample of size 15 was drawn by SRSWOR using the seed value 1234. How many sampling units are selected into the sample who answered 'good'?
(b) Find an estimate of the proportion of people responding as 'good', as well as the corresponding standard error.

Exercise 13.7 Consider again the data from Exercise 13.6.

(a) A sample of size 10 was drawn by SRSWR using the seed value 4321. How many sampling units are selected into the sample who answered 'bad'?
(b) Find an estimate of the proportion of people responding as 'bad', as well as the corresponding standard error.

Exercise 13.8 A psychologist speculates that spending a lot of time on the internet affects children's sleep. Consider the following data on hours of deep sleep (Y) and hours spent on the internet (X) where x_i and y_i are the observations on internet time and deep sleep time of the ith ($i = 1, 2, \ldots, 10$) child respectively (it may be noted that the sample size of 10 may not be a desirable sample size in a real study):

i	1	2	3	4	5	6	7	8	9	10
x_i	0.3	2.2	0.5	0.7	1.0	1.8	3.0	0.2	2.3	2.2
y_i	5.8	4.4	6.5	5.8	5.6	5.0	4.8	6.0	6.1	5.9

(a) Compute the correlation coefficient between X and Y using bootstrapping with 5000 bootstrap samples (seed value: 1234). Determine the bootstrap standard error.
(b) Calculate the bootstrap confidence intervals of the correlation coefficient based on the normal and percentile methods, respectively.

Exercise 13.9 Consider again the data from Exercise 13.8.

(a) Compute the bootstrap bias of the coefficient of variation based on 5000 bootstrap samples (seed value: 1234). Determine the bootstrap standard error.
(b) Calculate the bootstrap confidence intervals of the coefficient of variation based on the normal and percentile methods, respectively.

→ Solutions to all exercises in this chapter can be found in Appendix B

Causality

14

In the previous chapters, we have learnt about methods to describe the observed data (Part I) and how to make probabilistic statements about the characteristics of the underlying populations generating the data (Parts II and III). We introduced concepts of association, correlation, dependence and independence, conditional and marginal probabilities, likelihood, odds ratios and other notions to understand the joint behaviour of variables. What do all these concepts have in common? They can be defined based on a joint distribution of variables. In this chapter, we discuss questions which are more ambitious, i.e. questions that relate to actions which may modify distributions of variables. For example, we may be interested in the probability of developing a particular disease within one year if receiving a new drug against the disease—*counter to the fact* that not everyone would actually receive the drug. That is, we are interested in what would happen if *everyone* in the population received the drug. If we then compare these **potential outcomes** to the hypothetical situation where *everyone* did *not* receive the drug, we evaluate the effect of the treatment. More generally, we would like to know how variables change if we (hypothetically) intervene on other variables. For example, how would rates of unemployment change if a new social grant policy is implemented?; how would sales go up if a new marketing strategy is implemented?; and would banning cigarettes have an impact on lung cancer, i.e. would rates of lung cancer decrease if cigarettes were banned? All these questions are causal in nature and are reflected in day-to-day language through words like "because", "due to", "impact", "effect", "what if". In this chapter, we introduce concepts of causality. We will show that we need to make structural assumptions about the process generating the data, in addition to statistical techniques, to make causal inferences. As discussed in Sect. 11.10, correlation is not necessarily causation; but answering causal questions may be our scientific goal and with a structural framework, we can often answer them.

© Springer Nature Switzerland AG 2022
C. Heumann et al., *Introduction to Statistics and Data Analysis*,
https://doi.org/10.1007/978-3-031-11833-3_14

14.1   Potential Outcomes

Suppose we are interested in the effect of a treatment T (intervention, exposure) on an outcome Y. The potential outcome $Y^{t'}$ is the random variable Y that would have been observed if everyone in the population had received $T = t'$. This variable is a *counterfactual* as not every unit necessarily has $T = t'$ as a realization. If T is binary, the relationship between the observed and the counterfactual outcome is

$$Y = Y^1 \quad \text{if} \quad T = 1$$
$$Y = Y^0 \quad \text{if} \quad T = 0$$

and therefore

$$Y = TY^1 + (1 - T)Y^0. \tag{14.1}$$

Example 14.1.1 Consider a sample of 8 participants in a study for whom an outcome Y has been recorded. Out of these participants, suppose four participants are not given the treatment. Their treatment status is $T = 0$, and their outcomes are recorded as $Y = Y^0$. The remaining four participants have received the treatment ($T = 1$) and their outcomes are recorded as $Y = Y^1$. For those participants who were not given the treatment, it is unknown what the outcome would have been had they been treated. Similarly, for those participants who were treated, it can not be known what the outcome would have been if they were not given the treatment. Such unknown, counterfactual outcomes are denoted by '?' in the following data set:

i	T	Y	Y^0	Y^1
1	0	2.3	2.3	?
2	0	4.8	4.8	?
3	0	0.5	0.5	?
4	0	1.2	1.2	?
5	1	5.1	?	5.1
6	1	3.9	?	3.9
7	1	4.6	?	4.6
8	1	2.0	?	2.0

14.2   Causal Questions

We define causal estimands with respect to counterfactuals, and when we speak of *causal effects*, we refer to differences in counterfactual outcomes with respect to different levels of the treatment. That is, we are comparing the outcomes of the entire population in two hypothetical worlds. If treatment is binary, individual and average causal effects can be defined as follows:

Definition 14.2.1 An individual causal effect exists if

$$y_i^1 \neq y_i^0. \tag{14.2}$$

An *Average Treatment Effect (ATE)* exists if

$$E(Y^1) \neq E(Y^0) \tag{14.3}$$

and is quantified by

$$ATE = E(Y^1) - E(Y^0). \tag{14.4}$$

The measures of causal effect that are used if both T and Y are binary are the causal marginal risk difference, causal marginal relative risk and causal marginal odds ratio given as follows:

$$\theta_{MRD} = P(Y^1 = 1) - P(Y^0 = 1) \tag{14.5}$$
$$\theta_{MRR} = P(Y^1 = 1)/P(Y^0 = 1) \tag{14.6}$$
$$\theta_{MOR} = \frac{P(Y^1 = 1)/P(Y^1 = 0)}{P(Y^0 = 1)/P(Y^0 = 0)} \tag{14.7}$$

These measures are not necessarily identical to measures of association such as the risk ratio (4.13) or odds ratio (4.14) introduced in Sect. 4. Following our definitions on conditional and marginal probability distributions from Sect. 7.5, and related to the conditional expectations and probabilities defined in the regression models in Chaps. 11 and 12, we can write the associational (conditional) risk difference, (conditional) risk ratio and (conditional) odds ratio as follows:

$$\theta_{RD} = P(Y = 1|T = 1) - P(Y = 1|T = 0) \tag{14.8}$$
$$\theta_{RR} = P(Y = 1|T = 1)/P(Y = 1|T = 0) \tag{14.9}$$
$$\theta_{OR} = \frac{P(Y = 1|T = 1)/P(Y = 0|T = 1)}{P(Y = 1|T = 0)/P(Y = 0|T = 0)} \tag{14.10}$$

Therefore, the above associational measures are defined with respect to differences in risks between two disjoint subsets of the population, defined through the actual treatment value of the unit. Causal and associational estimands (effect measures) are not necessarily identical, and this is where the famous phrase "correlation is not causation" comes from.

Example 14.2.1 Recall Example 4.2.5 where we saw the following 2×2 table:

		Smoking (T)		
		Yes (1)	No (0)	Total (row)
	Yes (1)	34	66	100
Disease (Y)	No (0)	22	118	140
	Total (column)	56	184	240

The (associational) odds ratio was calculated as

$$\theta_{OR} = \frac{P(Y=1|T=1)/P(Y=0|T=1)}{P(Y=1|T=0)/P(Y=0|T=0)} = \frac{(34/56)/(22/56)}{(66/184)/(118/184)} = 2.76 \, .$$

The interpretation was given that the chances of having the particular disease are 2.76 times higher among smokers compared to non-smokers. Note that this interpretation compares the risks between two disjoint subsets of the population, i.e. smokers and non-smokers.

Is smoking the *cause* of the disease? What would be the odds of developing the disease if everyone had smoked compared to if everyone had not smoked? The calculated odds ratio cannot answer this question because θ_{OR} is not necessarily identical to θ_{MOR}. To answer it, further knowledge and more assumptions are needed, as explained below.

14.3 The Causal Model: Directed Acyclic Graphs

For causal inference, we need a structural model that summarizes subject matter knowledge about the data-generating process. One way to express this knowledge is by using directed acyclic graphs (DAGs). In a DAG, arrows are used to encode assumptions about causes: an arrow from variable A to variable B ($A \to B$) means that we assume that A potentially causes B, whereas the absence of an arrow means we assume no causal relationship between the two respective variables. An arrow does not say anything about the size of an effect. A DAG summarizes our prior knowledge about causal dependencies and independencies and may contain both measured and unmeasured variables. It is acyclic, meaning that no variable can cause itself, i.e. an arrow $A \to A$ is not allowed. No parametric assumptions are encoded in the DAG, it is entirely nonparametric.

Commonly, the assumed time-ordering of variables is reflected in the graph, so that the variable furthest to the left precedes all other variables. This allows us to encode a basic but important piece of knowledge: events in the past can affect those in the future, but not vice versa.

Example 14.3.1 Consider the following DAG:

We assume that X precedes T, which precedes the outcome Y. The graph visualizes our knowledge that X affects both T and Y, but that T does *not* cause Y. The graph does not make any assumptions about effect sizes or parametric relationships between variables.

Typically kinship terminology is used to describe relationships between variables in graphs. For example, in the DAG in Example 14.3.1, T is a *child* of X and X is a *parent* of T. For a path $X \rightarrow T \rightarrow Y$, T and Y are *descendants* of X, and X and T are *ancestors* of Y.

14.3.1 Confounders and Confounding

In a DAG, a back-door path is defined as a path from T to Y that starts with an arrow into T. Consider Example 14.3.1: here $T \leftarrow X \rightarrow Y$ is a back-door path. A back-door path generates **confounding** for the effect of T on Y, and X is the confounder. Confounding means that association is not causation and associational measures, such as (14.8)–(14.10), are not equivalent to the causal measures (14.5)–(14.7). Conditioning on the confounder, for example using linear or logistic regression, removes the association created by the confounder and thus the bias introduced with respect to the causal effect measure. It is also said that conditioning on a confounder *blocks* the respective back-door path.

Example 14.3.2 Consider the DAGs shown in Fig. 14.1. Suppose we are interested in knowing whether a larger stork population causes more births (indicated by the arrow storks → births). It may happen that there are more storks in certain regions of a country (region → storks) and that the birth rate differs by region as well (region →

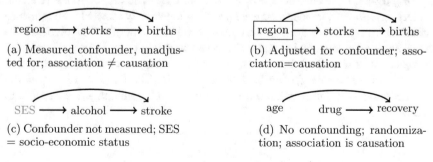

(a) Measured confounder, unadjusted for; association ≠ causation

(b) Adjusted for confounder; association=causation

(c) Confounder not measured; SES = socio-economic status

(d) No confounding; randomization; association is causation

Fig. 14.1 Examples for confounding. A rectangle indicates that the variable has been adjusted for by means of stratification, regression or the g-formula (see Sect. 14.5.1)

(a) No association of T and Y, the path $T \to X \leftarrow Y$ is blocked

(b) Conditioning on collider X; $\to$ opens path and creates an association between T and Y

Fig. 14.2 Examples for colliders

births). We might observe an association between number of storks and births, e.g., E(Births|Many Storks) $\neq$ E(Births|Few Storks), but this does not imply that there is a causal relationship between number of storks and births because of the confounder "region" that has not been considered (note the back-door path storks $\leftarrow$ region $\to$ births). Adjusting for the region, as in Fig. 14.1b, would show if the association is really causation (under the assumption that there are no other relevant confounders). If confounders are unmeasured, as in Fig. 14.1c, causation cannot be established.

Figure 14.1d shows a situation in which no back-door path exists: assignment to a drug or placebo may not depend on anything and is entirely random. Such a situation may arise, for example, in a randomized study. This knowledge is encoded by omitting the arrow from age to drug. With no back-door path, there is no confounding, and thus causation can be established from the association, see also Sect. 14.4.1.

14.3.2 Colliders

In the previous section, we explained that conditioning on a confounder blocks a (back-door) path, but conditioning on a *collider* opens a path. A variable is a collider on a given path if the path enters and leaves the respective variable via arrowheads, such as the variable X in Fig. 14.2a. In this case, adjustment on X would not be needed because the path $T \to X \leftarrow Y$ would be blocked anyway (due to the arrowheads colliding in X). Conditioning on X, for example, via regression or stratification, opens the path and introduces association where there was none before. This situation is illustrated in Fig. 14.2b.

Example 14.3.3 We consider the example from Cole et al. (2010). Suppose that at a meeting with 100 attendees, 10 were pre-symptomatically infected with influenza at arrival. At lunch, everyone gets a random sandwich. Out of 100 lunch boxes, 50 lunch boxes contained sandwiches with egg and 50 without egg. There are 5 expected influenza cases in each of the groups, i.e. 5 cases among those who ate sandwiches with and without eggs respectively. This equates to a risk of $P(Y = 1) = \frac{5}{50} = 0.1$ in each group and thus a risk difference of $\theta_{RD} = 0$. In summary, the data are as follows:

	Influenza		Sum	Risk	θ_{RD}
	Yes	No			
No egg	5	45	50	0.1	0.0
With egg	5	45	50	0.1	

(a) DAG for Example 14.3.3

(b) Conditioning on fever creates an association between sandwich type and influenza

Fig. 14.3 Sandwich example for collider bias

Now, let's say in the evening, 55 people develop a fever. Let's assume a simplified world where one could get a fever only through influenza or eating the egg sandwich. Conditioning on the variable fever means conditioning on the common effect of both influenza and sandwich-type, as visualized in Fig. 14.3b.

If one additionally knows that someone ate a sandwich without egg, then it is clear that the fever is developed due to influenza. The following tables summarize the data when conditioning on fever.

Fever	Influenza				
	Yes	No	Sum	Risk	θ_{RD}
No egg	5	0	5	1.0	0.9
With egg	5	45	50	0.1	

No Fever	Influenza				
	Yes	No	Sum	Risk	θ_{RD}
No egg	0	45	45	0	–
With egg	0	0	0	–	

Among those with a fever, the influenza risk for the attendees who ate a sandwich without egg is $5/5 = 1$, compared with $5/50$ for those who ate a sandwich with egg. This equates to a risk difference of 0.9. This association has been created by conditioning on fever.

Now, suppose one recalls that sandwich-type was randomly assigned, as shown in the DAG in Fig. 14.3a. In that case, it is clear that there is no association between sandwich and influenza (as demonstrated in the overall cross-table above). Conditioning on fever status created an association, e.g., $\theta_{RD} = 0.9 \neq 0$ among those with fever, that was not present before. This association represents no causation.

14.3.3 Mediators

A mediator is a variable M that lies on a path between intervention on the outcome: $T \rightarrow M \rightarrow Y$. For example, the effect of gender (T) on income (Y) may be mediated through years of education (M), among many other variables. The path $T \rightarrow M \rightarrow Y$ then describes the *indirect* effect of gender through education and the path $T \rightarrow Y$ the *direct effect*. This implies that there are two paths through which gender can affect the outcome, and both together describe the *total effect*. Figure 14.4a shows a DAG that describes such a situation.

(a) M is a mediator   (b) Conditioning on the mediator blocks the path $T \to M \to Y$ and introduces bias with respect to the total effect of T on Y

Fig. 14.4 Examples for the role of mediators

Mediation analysis is concerned with disentangling the different pathways through which an intervention can affect an outcome and is an area on its own. However, in this chapter, we are concerned with total effects, and it is relevant to ask: what should we do with the mediators? Should we condition on them, as with confounders or as with colliders? The answer is the latter. If we are interested in the effect of T on Y, conditioning on M blocks the path of the indirect effect and thus creates *overadjustment bias* with respect to the causal quantity of interest (Fig. 14.4b). To understand this better and in more detail, we now need to embark on the more general issue of identification, allowing us to illustrate this section's concepts through examples.

14.4   Identification

Before continuing further, it is important at this point to reflect where we are so far and what we are aiming for. We have a causal question, formally represented in a causal estimand, such as (14.5)–(14.7). We know that if we are interested in a binary intervention, we can never observe both counterfactual outcomes for both the interventions. But we may have data measured and ask: can we use these data for causal inference? Alternatively, more formally, can our causal question be answered by a function of the observed data distribution? The answer is affirmative under structural, partly untestable assumptions, some of which can be visualized in a DAG. The relevant components of these assumptions have already been introduced through concepts such as confounders, colliders, and mediators. However, to answer the general question of identifiability of a causal effect, i.e. if it can be identified and subsequently estimated from the observed data, requires a more formal treatment, summarized in Judea Pearl's back-door theorem.

To start with, consider the concept of d-separation:

Definition 14.4.1 (*d-separation*) A set of variables $\tilde{X}$ is said to block, or *d-separate*, a path p in a DAG if either

(i) p contains a chain $T \to X \to Y$ or a fork $T \leftarrow X \to Y$ such that X is in $\tilde{X}$ or
(ii) p contains an inverted fork $T \to X \leftarrow Y$ such that neither X nor descendant of X is in $\tilde{X}$.

If $\tilde{X}$ blocks all back-door paths from T to Y, then T and Y are independent conditional on $\tilde{X}$ and we write $T \perp\!\!\!\perp Y | \tilde{X}$.

The d-separation definition implies that T and Y are conditionally independent given $\tilde{X}$ if all back-door paths between T and Y are blocked, which can be achieved by conditioning on a confounder or non-conditioning on a collider, mediator or descendant of a mediator/collider. However, it is important to note that lack of d-separation does not imply conditional dependence.

Example 14.4.1 Consider the following DAG:

The path $X_1 \rightarrow X_2 \rightarrow X_4 \rightarrow Y$ is blocked by $\tilde{X} = \{X_2\}$ and $\tilde{X} = \{X_4\}$, i.e. $X_1 \perp\!\!\!\perp Y | X_2$ and $X_1 \perp\!\!\!\perp Y | X_4$, which implies that X_1 and Y are d-separated by either X_2 or X_4. This is immediately evident from the first statement in Definition 14.4.1. The path $X_1 \rightarrow X_2 \rightarrow X_4 \leftarrow X_3$ is blocked by the null set $\tilde{X} = \{\emptyset\}$, but *not* by $\tilde{X} = \{Y\}$, as Y is a descendent of the collider X_4; that is, $X_1 \perp\!\!\!\perp X_3$, but we can't say whether $X_1 \perp\!\!\!\perp X_3 | Y$. This follows from the second statement of Definition 14.4.1.

With the concept of d-separation, we are able to establish when paths in a DAG are blocked and when they are not. This knowledge can be used for the following crucial definition that paves the road to establishing when and how causal effects can be identified:

Definition 14.4.2 (*back-door criterion*) A set of variables $\tilde{X}$ satisfies the back-door criterion with respect to two variables T and Y in a DAG if

(i) $\tilde{X}$ blocks all back-door paths from T to Y and
(ii) no element of $\tilde{X}$ is a descendant of T.

If $\tilde{X}$ satisfies the back-door criterion, then *conditional exchangeability*, defined as follows, is achieved:

$$Y^t \perp\!\!\!\perp T | \tilde{X} \quad \forall T = t, \tilde{X} = \tilde{x} . \tag{14.11}$$

This is immediately clear when looking at the definition of d-separation. If T is binary, then conditional exchangeability means that $P(Y^t = 1 | T = 1, \tilde{X} = \tilde{x}) = P(Y^t = 1 | T = 0, \tilde{X} = \tilde{x})$. For *exchangeability*, it is required that

$$Y^t \perp\!\!\!\perp T \quad \forall T = t \tag{14.12}$$

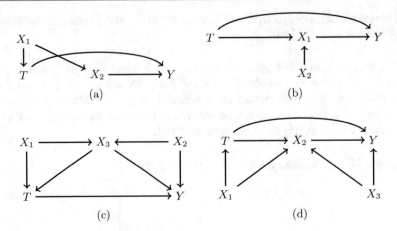

Fig. 14.5 Directed Acyclic Graphs for Example 14.4.2

which for binary interventions says that $P(Y^t = 1|T = 1) = P(Y^t = 1|T = 0) = P(Y^t = 1)$, i.e. the potential outcomes are conditionally independent of the actual, observed interventions.

The above concepts are helpful to understand the *back-door theorem*:

Theorem 14.4.1 (Back-door adjustment) *If a set of variables $\tilde{X}$ satisfies the back-door criterion with respect to T and Y, then the causal effect of T on Y is identifiable and is given in the discrete case by*

$$P(Y^{T=t}) = \sum_{\tilde{x}} P(Y|T = t, \tilde{X} = \tilde{x}) \times P(\tilde{X} = \tilde{x}). \qquad (14.13)$$

This key theorem tells us when and how causal effects can be identified from the observed data distribution. It gives us an idea of which variables are needed to represent a counterfactual quantity (the left side of the equation) as a function of the actual data (the right side of the equation): simply by closing *all* back-door paths, for example, through the inclusion of confounders but not colliders, and by not conditioning on consequences of the intervention of interest (which includes, but is not restricted to, mediators and their descendants). Before delving deeper into understanding formula (14.13) in Sect. 14.5.1, we first illustrate the back-door adjustment theorem using some examples.

Example 14.4.2 Consider four data generating processes illustrated in the DAGs of Fig. 14.5. In each of them the effect of T on Y is of interest. In Fig. 14.5a, there is one back-door path $T \leftarrow X_1 \rightarrow X_2 \rightarrow Y$. The path can be blocked by inclusion of either X_1 or X_2 or both. Without inclusion of either X_1 or X_2, the causal effect of T on Y cannot be identified, and confounding would occur.

In Fig. 14.5b, the (total) effect of T on Y can be identified without inclusion of X_1 or X_2 because there does not exist any back-door path. Inclusion of X_1 would be incorrect and harmful, as it is a mediator, but the inclusion of X_2 would be possible (though not needed) as this variable is not a descendant of T or X_1.

There are several back-door paths in Fig. 14.5c, including: $T \leftarrow X_3 \rightarrow Y$ and $T \leftarrow X_1 \rightarrow X_3 \leftarrow X_2 \rightarrow Y$. The first one can be closed by conditioning on X_3; however, X_3 is a collider on the path $T \leftarrow X_1 \rightarrow X_3 \leftarrow X_2 \rightarrow Y$ and the sub-path $X_1 \rightarrow X_3 \leftarrow X_2$ would therefore be opened. It follows that either X_1 or X_2 need also to be included in the adjustment set, to close the path again. This leads to the following valid adjustment sets: $\tilde{X}_1 = \{X_1, X_3\}$ and $\tilde{X}_2 = \{X_2, X_3\}$.

Figure 14.5d has two back-door paths: $T \leftarrow X_1 \rightarrow X_2 \rightarrow Y$ and $T \leftarrow X_1 \rightarrow X_2 \leftarrow X_3 \rightarrow Y$. The first path could be closed by inclusion of either X_1 or X_2; however, X_2 acts as both a mediator and collider on the second back-door path. It should therefore not be included due to the back-door criterion. Adjustment with respect to X_3 is possible, but not necessarily needed.

We now return to the question posed at the beginning of this section: can our causal question be answered by a function of the observed data distribution, and if yes, how? The back-door theorem tells us that this is possible if the correct variables are chosen (and measured for estimation). Equation (14.11) says that if $\tilde{X}$ satisfies the back-door criterion then conditional exchangeability is achieved. To ensure that causal inference is possible, the literature typically states that the following three assumptions need to be met:

(1) *conditional exchangeability*: this has been defined in (14.11); if and how the assumption is satisfied can be checked in the assumed DAG by application of the back-door theorem.

(2) *Positivity*:

$$P(T = t | \tilde{X} = \tilde{x}) > 0 \quad \forall \tilde{x} \quad \text{with} \quad P(\tilde{X} = \tilde{x}) \neq 0. \qquad (14.14)$$

(3) *Consistency*:

$$\text{If } T = t, \text{ then } Y^t = Y \text{ for } \forall t. \qquad (14.15)$$

The positivity assumption says that the probability of receiving the treatment $T = t$ (e.g. for $t = 1$ and $t = 0$) is greater than zero for all possible combinations of covariate values. This assumption may be violated for two reasons:

(i) it may not be possible to receive $T = t$ in $\tilde{X} = \tilde{x}$ by definition, or
(ii) $P(T = t | \tilde{X} = \tilde{x})$ may be zero, or close to zero, in a finite sample.

For example, in Fig. 14.1a, the probability of finding any (i.e., $t = 1$) storks in a particular region $\tilde{X} = \tilde{x}$ may be zero, simply because no storks live in this region. More commonly, however, we would have settings where there are multiple covariates, and hence the assumption is practically violated (at least to some degree) because of the limited sample size.

To understand what the consistency assumption implies, recall (14.1), which followed from our definition of a counterfactual under a binary intervention: $Y = TY^1 + (1 - T)Y^0$. It is easy to see that this is identical to the consistency assumption. So, how is it even possible that this assumption can be violated at all? The answer is: it can be violated if the treatment is not well-defined, as the link between actual treatment and the counterfactual outcome may then not be clear. For example, suppose one is interested in the effect of being a "single parent" ($T = 1$) on an outcome such as final graduation mark at age 18; then, it is not clear what "single parent" actually means. It could be someone taking care of a child on her/his own for one month, or 18 years, by turns with a partner, or with support from a grandparent. In each case, the potential outcome Y^1, may be different; therefore, the effect of $T = 1$ on the outcome may be different for different versions of the treatment. Any intervention that can be interpreted and implemented in different ways is potentially prone to bias arising from the violation of the consistency assumption.

Example 14.4.3 Consider the example from Luque-Fernandez, Schomaker, Rachet and Schnitzer (2018): we are interested in the effect of dual treatment therapy ($T = 1$), compared to single therapy ($T = 0$), on a binary outcome Y among colorectal cancer patients. We know that there are confounders that affect both treatment assignment and the outcome, namely age (X_1), socioeconomic status (SES, X_2), co-morbidities (X_3) and clinical stage (X_4). The detailed assumptions are given in the DAG below:

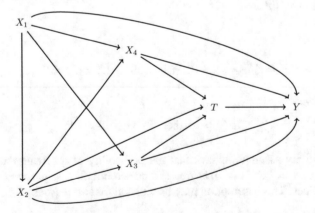

Suppose we have measured all variables represented in the DAG: would conditional exchangeability be met? Yes, because all back-door paths from T to Y are closed. In fact, it would be sufficient to measure $\tilde{X} = \{X_2, X_3, X_4\}$ as they already block all back-door paths. Consistency would be satisfied provided dual and single treatment are well-defined, which is likely the case because the intervention of interest is a medical treatment that can unambiguously be defined and given to the patients. Whether positivity is violated could be checked by looking into cross tables of T and X_j, stratified by the other covariates:

```
, , x2 = 2_middle, x3 = none, x4 = stage 3

    x1
t   young    old
0   13       17
1   25       26
```

For example, in the above R-table we see that $P(T = t | X1, X2 = 2, X3 =$ no comorbidities, $X4 = 3) > 0$ for both $t = 0$ and $t = 1$. This is reassuring in terms of the positivity assumption and could be repeated for other covariate combinations.

14.4.1 Randomization

The most popular approach to establish causality is based on *randomization* or, more specifically, *randomized experiments*. With an experiment, we describe a setting where the system under study is under full control of the investigator studying it. We have already described simple experiments in Chap. 6, such as tossing a coin or rolling a die. In other experiments, where humans are involved, the assignment and manipulation of treatment T, the materials used, and the measurement procedures and instruments are controlled by the investigator. If one assigns the treatment randomly in such an experiment, it means that each unit i receives $T = t$ (e.g. $T = 1$ and $T = 0$) based on a prior specified probability $P(T = t) = p_t$. This can be facilitated by drawing random numbers from a Binomial distribution as described in Sect. 8.1.4 (there are other options too).

Example 14.4.4 Suppose we are interested in knowing the effect of a drug A ($T = 1$) compared to a drug B ($T = 0$) on a medical outcome Y, for example by estimating the difference between $E(Y^1)$ and $E(Y^0)$, i.e. the ATE. We may therefore select $n = 100$ patients who are eligible to receive the drug. We could then randomly draw 100 times from a $B(1, 0.5)$ distribution using the R command `rbinom(100, size=1, p=0.5)`. This gives us a random sequence of 100 numbers R_i ($i = 1, 2, \ldots, 100$) which are either 0 or 1. Patient i then receives drug A if $R_i = 1$, and drug B otherwise. This is a possible randomization procedure in an experiment.

Why does this procedure help us to establish causality? Consider again the DAG of Fig. 14.1d, which looks as follows:

There is no arrow that goes into T (i.e., drug): this encodes our *knowledge* that treatment assignment does not depend on any third variables (e.g., age), only on randomness. Hence there is no back-door path that can be open, and thus we have achieved conditional exchangeability by design! In fact, in the case described, we have even achieved exchangeability as introduced in (14.12), i.e. $Y^t \coprod T \quad \forall T = t$. Thus, the first assumption for causal inference is met. There are also so-called blocked randomization procedures where treatment assignment is random within blocks such as age groups or regions, for example, when one is interested in knowing whether a vaccine works for both younger and older people. If this is done, then blocked randomization ensures exchangeability conditional on the respective block; according to the back-door criterion, we then need to adjust for the respective blocking variable(s).

What about positivity and consistency? The positivity assumption is satisfied by design too: as one is under control $P(T = t) = p_t$, for example, by setting $p_t = 0.5$ for a binary treatment, randomization has helped again. If then the treatment of interest is well-defined and unambiguous, which is under full control of the study investigator, consistency will also be satisfied. This explains *why* randomized experiments are used to establish causal conclusions.

More formally, we can write the average treatment effect as

$$E(Y^1) - E(Y^0) = E(Y^1|T = 1) - E(Y^0|T = 0)$$
$$= E(Y|T = 1) - E(Y|T = 0)$$

where the first equality follows from exchangeability and positivity and the second equality follows from consistency. It follows that after randomization, the effect of a binary treatment on an outcome can be estimated by comparing the estimated means in the two disjoint groups (i.e., those units with $T = 1$ and those with $T = 0$) by using an appropriate test, for example, a t-test (i.e., Welch-test), or the Fisher-test (or Two-sample binomial test) for binary outcomes.

Example 14.4.5 As an example, consider the following R-code that simulates data from a data-generating process in line with the above DAG: that is, treatment assignment is random (based on a Binomial distribution), and the outcome depends both on the treatment and a third variable X through a normal distribution with mean $1 \times X + 1 \times T$ and variance 1. It is obvious that the true average treatment effect is thus 1. Drawing the observations 1000 times from the respective process and evaluating the differences in means for the two treatment groups leads to approximately the true effect, showing that unbiased effect estimation with a simple comparison in means is possible.

```
results <- rep(NA,1000)
for(i in 1:1000){
X <- rnorm(1000,mean=0,sd=1)
T <- rbinom(1000, size=1, prob=0.5)
Y <- rnorm(1000,mean=1*X+1*T,sd=1) # true ATE=1
results[i]<-mean(Y[T==1])-mean(Y[T==0])
}
mean(results)
[1] 0.9985847 # approximately unbiased
```

14.5   The Statistical Model: Estimation

So far, we have focused on the causal question and the causal model (i.e., the DAG). Once the estimand of interest is unambiguously defined, and identification has been established, causal effect estimation with appropriate statistical techniques is the next step. We focus on standardization with the g-formula and regression. References for other methods, such as weighting, are given at the end of this section.

14.5.1 The g-formula

Theorem 14.5.1 (g-formula) *Under consistency, conditional exchangeability and positivity it holds that*

$$E(Y^t) = \sum_x E(Y|T = t, \mathbf{X} = \mathbf{x}) \times P(\mathbf{X} = \mathbf{x}).$$

If $\mathbf{X}$ is continuous, the sum can be replaced with an integral.

A simple proof is as follows:

$$E(Y^t) = \sum_x E(Y^t|\mathbf{X} = \mathbf{x}) \times P(\mathbf{X} = \mathbf{x})$$

$$= \sum_x E(Y^t|T = t, \mathbf{X} = \mathbf{x}) \times P(\mathbf{X} = \mathbf{x})$$

$$= \sum_x E(Y|T = t, \mathbf{X} = \mathbf{x}) \times P(\mathbf{X} = \mathbf{x}).$$

The first equality follows by definition and is intuitive when recalling the law of total probability from Theorem 6.4.2. The second equality follows directly from the assumption of conditional exchangeability, and the third equality is based on the consistency assumption. Positivity is required to guarantee that the conditional expectations are well-defined.

The above theorem gives us the method for estimating the counterfactual quantity $E(Y^t)$:

Example 14.5.1 Consider the following data:

i	X	T	Y	i	X	T	Y
1	0	0	0	7	0	0	0
2	0	0	1	8	1	0	0
3	0	0	0	9	1	1	0
4	0	1	0	10	1	1	1
5	0	1	1	11	1	0	1
6	0	1	1	12	1	0	1

Suppose we are interested in the causal risk difference $\theta_{MRD} = P(Y^1 = 1) - P(Y^0 = 1)$. If the assumptions of conditional exchangeability, positivity and consistency are met, we can calculate $E(Y^1) = P(Y^1 = 1) = p$ based on the g-formula:

$$
\begin{aligned}
E(Y^1) &= \sum_{x=0}^{1} P(Y = 1 | T = 1, X = x) \times P(X = x) \\
&= P(Y = 1 | T = 1, X = 0) \times P(X = 0) + \\
&\quad P(Y = 1 | T = 1, X = 1) \times P(X = 1) \\
&= \frac{2}{3} \cdot \frac{7}{12} + \frac{1}{2} \cdot \frac{5}{12} = \frac{43}{72} \approx 0.5972 \, .
\end{aligned}
$$

The values in this calculation are obtained from the table. For example, there are three observations with $T = 1$ and $X = 0$, of which two have the observed outcome $Y = 1$. This yields $P(Y = 1 | T = 1, X = 0) = \frac{2}{3}$. Similarly, we can estimate $E(Y^0) = P(Y^0 = 1) = 0.42361$. Therefore,

$$
\hat{\theta}_{MRD} = 0.5972 - 0.42361 = 0.17359 \, .
$$

However, in more complex cases, probabilities and expectations are typically calculated with regression models. We illustrate this point in the following example:

Example 14.5.2 Consider again Example 14.5.1. We use logistic regression to model $P(Y = 1 | T = t, X = x)$:

```
summary(glm(Y ~ T+X+T:X, family=binomial))
```

This gives us the following summary:

```
Coefficients:
              Estimate Std. Error z value Pr(>|z|)
(Intercept)    -1.099     1.155   -0.951    0.341
T               1.792     1.683    1.064    0.287
X               1.792     1.683    1.064    0.287
T:X            -2.485     2.517   -0.987    0.323
```

We can now estimate $P(Y = 1|T = t, X = x)$ by means of the fitted regression model. For, example, for $T = 1$ and $X = 1$ we can use (12.3) to obtain

$$P(Y = 1|T = 1, X = 1) = (1 + \exp(-[-1.099 + 1.792 + 1.792 - 2.485]))^{-1}$$
$$= 0.5.$$

Using the fitted model, we can calculate $P(Y = 1|T = 1, X = 0) = \frac{2}{3}$, $P(Y = 1|T = 0, X = 0) = \frac{1}{4}$ and $P(Y = 1|T = 0, X = 1) = \frac{2}{3}$ in the same manner. Given that $P(X = 1) = \frac{5}{12}$ and $P(X = 0) = \frac{7}{12}$ we can calculate:

$$
\begin{aligned}
E(Y^0) = P(Y^0 = 1) &= P(Y = 1|T = 0, X = 0) \times P(X = 0) + \\
&\quad P(Y = 1|T = 0, X = 1) \times P(X = 1) \\
&\quad \frac{1}{4} \cdot \frac{7}{12} + \frac{2}{3} \cdot \frac{5}{12} = 0.42361 \\
E(Y^1) = P(Y^1 = 1) &= P(Y = 1|T = 1, X = 0) \times P(X = 0) + \\
&\quad P(Y = 1|T = 1, X = 1) \times P(X = 1) \\
&= \frac{2}{3} \cdot \frac{7}{12} + \frac{1}{2} \cdot \frac{5}{12} = 0.59722
\end{aligned}
$$

$$\theta_{MRD} = 0.5972 - 0.42361 = 0.17359.$$

The above example shows that using a *parametric model*, like logistic regression, yields the same results as the non-parametric approach used in Example 14.5.1. This is because we used a *saturated* model. In a saturated model, the number of parameters in the model is equal to the number of population quantities (parameters) that can be estimated by using the model. For example, in our example there is only one binary covariate X, as well as a binary treatment and binary outcome. A regression model $P(Y = 1|T, X) = h(\beta_0 + \beta_1 T + \beta_2 X + \beta_3 T X)$ contains four parameters, and it estimates four quantities: $P(Y = 1|T = 0, X = 0)$, $P(Y = 1|T = 1, X = 0)$, $P(Y = 1|T = 0, X = 1)$, $P(Y = 1|T = 1, T = 1)$; thus, the model is saturated and imposes no modeling restrictions. In most practical examples, there are no saturated models due to the number of covariates and their nature, for example there might be many metric covariates. In such a case, model mis-specification may yield biased effect estimates.

We now describe the general approach to implement the *g*-formula in *R* using regression modeling:

(i) Estimate $P(\mathbf{X} = \mathbf{x})$ through the empirical distribution, i.e. the measured data.
(ii) Model the conditional expectation $E(Y|T = t, \mathbf{X} = \mathbf{x})$. For example, with linear or logistic regression as described in Chaps. 11 and 12.
(iii) Replace T with $T = t$. For example, set $T = 1$ for every observation if this is the intervention of interest.
(iv) Obtain the predicted outcome (under the post-intervention distribution) from the fitted model of step (ii) by using $\mathbf{X}$ from step (i) and T from step (iii). Estimate $E(Y|T = t, \mathbf{X} = \mathbf{x})$ by calculating the arithmetic mean of the predicted outcome.
(v) Repeat steps (iii) and (iv) for other t if estimands such as (14.5), (14.6) or (14.7) are of interest.

Example 14.5.3 Consider again Examples 14.5.1 and 14.5.2. We can implement the steps (i)–(v) as follows:

```
# Step i) keep X
# Step ii)
m1 <- glm(Y ~ T+X+T:X, family=binomial)

# Step iii) overwrite T with T=1
newdata1 <- as.data.frame(cbind(T,X))
newdata1$T <- c(rep(1,12))

# Step iv) Predict the outcome under intervention
# and take arithmetic mean
Y1 <- mean(predict(m1, type='response',
newdata=newdata1)) # E(Y^1)

# Step v): repeat iii) and iv) for T=0
newdata0 <- as.data.frame(cbind(T,X))
newdata0$T <- c(rep(0,12))
Y0 <- mean(predict(m1, type='response',
newdata=newdata0)) # E(Y^0)

# ATE
Y1-Y0
> 0.1736 # correct result
```

Again, the results are correct. However, under a mis-specified model this may not be the case. For example, when omitting the interaction term, we get $\hat{\theta}_{\text{MRD}} = 0.1764$.

Confidence Intervals and Bootstrapping

When using the g-formula, confidence intervals are obtained by bootstrapping as explained in Sect. 13.6. Exercise 14.4 illustrates its practical application.

14.5.2 Regression

Instead of using the g-formula, estimated regression parameters can often be interpreted as causal effects too. The key message of the above sections remains valid: if the correct set of variables is included, i.e. those that close back door-paths and are not colliders or descendants of the intervention, we can make causal inference if the additional assumptions of positivity and consistency are satisfied. However, it is crucial to reflect what estimand is then implicitly chosen. As an example, consider the linear model with two covariates, treatment T and a confounder X:

$$\mathrm{E}(Y) = \beta_0 + \beta_1 T + \beta_2 X .$$

Suppose T is binary. Then,

$$\begin{aligned}
\beta_1 &= \mathrm{E}(Y|T = 1, X) - \mathrm{E}(Y|T = 0, X) \\
&= \mathrm{E}(Y^1|T = 1, X) - \mathrm{E}(Y^0|T = 0, X) \\
&= \mathrm{E}(Y^1|X) - \mathrm{E}(Y^0|X),
\end{aligned}$$

where the first equality follows by definition of the linear model, the second from consistency and the third by conditional exchangeability. This means the regression parameter associated with the intervention T relates to an average treatment effect, *conditional* on X. This is a valid causal effect, but it may not necessarily be identical to the ATE, which is $\mathrm{E}(Y^1) - \mathrm{E}(Y^0)$. Recall that in Sect. 11.7.3 we introduced the concept of interactions to model situations in which the association between a variable and the outcome differed by a third variable. In the context of causal inference, we speak of *effect modification* when the effect of T on Y varies with respect to a third variable X. In this case, the conditional effect measure will not necessarily be identical to the marginal effect measure.

   To better understand which causal parameters can be estimated with regression, it is important to understand the concept of collapsibility, and reflect more on both effect modification and confounding. Consider the following data of 3 binary variables, with $n = 52$:

	$X = 1$		$X = 0$		Marginal	
	$T = 1$	$T = 0$	$T = 1$	$T = 0$	$T = 1$	$T = 0$
$Y = 1$	5	8	15	12	20	20
$Y = 0$	3	4	3	2	6	6
OR	$\frac{5}{6}$		$\frac{5}{6}$		1	

   The marginal odds ratio between T and Y is $OR = 1$, and the conditional odds ratios for $X = 1$ and $X = 0$ are equal to $\frac{5}{6}$. Thus, conditional and marginal odds ratios are different. Also, T and Y are marginally independent and conditionally dependent.

If there is no effect modification (as in the given example), one could speculate that the two conditional ORs are identical to the marginal OR; however, this is obviously not the case. If effect modification is actually present, one might expect that the marginal OR is between the two conditional ORs; again, this is not the case, as seen in the table. The reason is that the odds ratio is not *collapsible*. An effect measure is said to be collapsible if the marginal measure is equal to a weighted average of the stratum-specific measures of association. Both risk ratio and the risk difference are collapsible measures.

What are the practical consequences when using a regression parameter for causal effect estimation? First, the marginal causal odds ratio,

$$\theta_{MOR} = \frac{P(Y^1 = 1)/P(Y^1 = 0)}{P(Y^0 = 1)/P(Y^0 = 0)},$$

cannot be obtained by interpreting logistic regression parameters from models that contain one or more covariates $\mathbf{X}$. A logistic regression model, which contains multiple variables, for example to adjust for confounding, will only be able to estimate conditional causal odds ratios, i.e.,

$$\theta_{COR} = \frac{P(Y^1 = 1|\mathbf{X})/P(Y^1 = 0|\mathbf{X})}{P(Y^0 = 1|\mathbf{X})/P(Y^0 = 0|\mathbf{X})}. \tag{14.16}$$

Whether this makes sense depends on the scientific question, see also Sect. 14.6. It is, however, important to note that *linear* regression coefficients, which target the differences in expectations, don't suffer from the problem of non-collapsibility.

Second, when using multiple logistic regression, the chosen estimands are implicitly conditional. Those estimands may have a causal interpretation if the assumptions outlined above are satisfied, but they are not necessarily identical to marginal causal estimands. The differences between marginal and conditional effect measures may be due to confounding, effect modification, the inclusion of colliders/mediators, and non-collapsibility.

Example 14.5.4 Consider again Examples 14.5.1, 14.5.2 and 14.5.3. We can use R to calculate the marginal odds ratio based on the g-formula and the conditional odds ratio using regression:

```
MOR <- (Y1/(1-Y1)) / (Y0/(1-Y0))
MOR
> 2.017524
m1 <- glm(Y ~ T+X, family='binomial')
COR <- exp(coef(m1))[2]
COR
> 2.090139
```

Clearly, MOR and COR are different. Both may have a causal interpretation if conditional exchangeability, positivity, and consistency are met. The MOR will be useful if the average treatment effect of T on Y is of interest. To evaluate effect modification, an interaction can be added to the logistic regression model, and the relevant CORs can be interpreted:

```
> m2 <- glm(Y~T*X, family=binomial)
> coef(m2)
(Intercept) T           X           T:X
-1.098612   1.791759    1.791759    -2.484907
```

The output suggests a treatment effect of approximately $\exp(1.79) = 6$ for $X = 0$ and of approximately $\exp(1.79 - 2.48) = \exp(-0.69) \approx 0.5$ for $X = 1$.

14.6 Roadmap

To apply the methods from this chapter in practice, it is recommended to broadly follow a structured roadmap:

1. Translate the scientific question of interest into a causal target quantity (defined as some parameter of the distribution of counterfactual random variables, for example (14.4), (14.5), (14.6) or (14.7)).
2. Develop a causal model (i.e., a DAG) that summarizes the relevant the background knowledge.
3. Specify which data has been measured.
4. Assess identifiability, i.e. whether the causal target quantity can be represented as a parameter of the observed data distribution (that is, the estimand), and if yes, under which assumptions.
5. Estimate the target quantity using appropriate statistical techniques. In many instances, g-methods, for example the g-formula can be used.
6. Interpret the results.

Example 14.6.1 Recall Example 14.4.3. The scientific question of interest was what the chances of experiencing the outcome would have been had every patient received dual therapy compared to if every patient had received monotherapy. The causal marginal odds ratio (MOR), as given in (14.7), is a target quantity that answers this question (step 1). It is known from the literature that treatment decisions are based on co-morbidities (X_3), clinical-stage (X_4), and sometimes also on features that relate to socioeconomic status (SES, X_2). The DAG shown in Example 14.4.3 encodes this knowledge (step 2). Suppose we have actually measured all data (Y, T, X_1, X_2, X_3, X_4; step 3); then, we know that the target quantity (a counterfactual) can be represented as a function of the observed data distribution using the g-formula, if the assumptions of conditional exchangeability, positivity and consistency are satisfied. The DAG tells us that conditional exchangeability is achieved by the inclusion of the variable set $\tilde{X} = (X_2, X_3, X_4)$ because the back-door theorem says that those variables are sufficient to block all back-door paths, and no descendants of T are then included either. As explained in Example 14.4.3, positivity and consistency would also likely not be violated in the given setting (step 4). This gives us a recipe for estimation (step 5): we can use (14.13) or Theorem 14.5.1 to identify $P(Y^{T=t} = 1)$ and thus the estimator for the MOR, i.e. $P(Y^1 = 1)/P(Y^1 = 0)/P(Y^0 = 1)/P(Y^0 = 0)$ is:

$$\frac{\sum_{\mathbf{x}} P(Y|T = 1, \mathbf{X} = \mathbf{x}) \cdot P(\mathbf{X} = \mathbf{x})/(1 - \sum_{\mathbf{x}} P(Y|T = 1, \mathbf{X} = \mathbf{x}) \cdot P(\mathbf{X} = \mathbf{x}))}{\sum_{\mathbf{x}} P(Y|T = 0, \mathbf{X} = \mathbf{x}) \cdot P(\mathbf{X} = \mathbf{x})/(1 - \sum_{\mathbf{x}} P(Y|T = 0, \mathbf{X} = \mathbf{x}) \cdot P(\mathbf{X} = \mathbf{x}))}$$

The notation of the above estimator assumes that X_1, X_2, X_3 and X_4 are categorical variables. For estimation, this means we simply use the g-formula to estimate $P(Y^{T=0} = 1)$ and $P(Y^{T=1} = 1)$, which is sufficient to calculate our estimand. To implement the steps in the g-formula algorithm, we use logistic regression. This means following the R-code of Example 14.5.4 and the top lines of Example 14.5.3. The result is $\widehat{\text{MOR}} = 0.38$. We interpret that the chances of experiencing the outcome are 0.38 times lower if everyone had received dual therapy than if everyone had received monotherapy. As we assume that all assumptions have been met, this interpretation is causal and relates to the population of interest.

14.7   Key Points and Further Issues

Note:

✓ To answer causal questions, both a statistical model and a causal model (i.e., a DAG) are needed.

✓ Causal inference requires measurement and inclusion of the right variables. The back-door criterion can be used to determine those variables.

✓ For identification, i.e. to represent a counterfactual quantity as a function of the actual data the assumptions of consistency, positivity and conditional exchangeability need to be met. Sometimes those assumptions can be satisfied by design, for example when using randomization.

✓ Many causal target quantities are marginal with respect to (a subset) of the covariates. Thus, the g-formula is often preferred over using regression parameters for effect estimation.

✓ A comprehensive introduction to causal inference, and more details, can be found in Hernan and Robins (2021).

14.8   Exercises

Exercise 14.1 Define appropriate target quantities for the following scientific questions:

(a) What is the average weight gain (in kg) of primary school children if they follow a given physical exercise plan for 1 year (strategy A) compared to not following any exercise plan (strategy B)?
(b) How much more likely is it to fully recover from a knee surgery after 6 months (i.e. obtaining full range of motion) with surgery technique A compared to technique B?
(c) What would be the expected body-mass-index (Y, BMI) of obese patients after one month if they had exercised (T_1) at least 30 min a day and not smoked (T_2) at all?
(d) What is the effect of drug dose (in mg/L) on expected recovery time (in days)?
(e) Would the chances of reemployment 1 year after the start of unemployment be different for younger ($\leq$40 years) and older ($>$40 years) jobseekers if a particular training program had been attended, compared to not?

Exercise 14.2 Use the back-door criterion to determine the smallest valid adjusment sets for the data-generating processes visualized in the following DAGs:

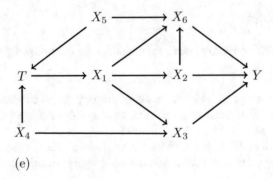

(e)

Exercise 14.3 Consider the following data:

i	X	T	Y	i	X	T	Y
1	0	1	0	5	0	1	0
2	0	1	1	6	1	1	1
3	0	0	0	7	1	0	0
4	0	0	1	8	1	1	0

Assume that conditional exchangeability, positivity and consistency are met.

(a) Calculate the marginal causal risk ratio manually using the g-formula.
(b) Reproduce the result from (a) using R

Exercise 14.4 Consider the Cattaneo data described in Appendix A.4.4. We are interested in the effect of smoking on birth weight.

(a) The following R-output comes from a linear regression model where birth weight (in grams) is the outcome and smoking status during pregnancy (mbsmoke) as well as an indicator variable whether it is the mothers first child (fbaby) are the covariates:

```
                Estimate Std. Error t value Pr(>|t|)
(Intercept)      3445.18      11.98  287.56  < 2e-16 ***
mbsmokesmoker    -281.06      21.46  -13.10  < 2e-16 ***
fbabyYes          -71.20      16.83   -4.23 2.38e-05 ***
```

For simplicity, assume fbaby is the only confounder (which it is not). Calculate the ATE for the effect of smoking on birth weight using the g-formula. Note that 43.8% of mothers gave birth to their first child.
(b) Compare the estimate of the ATE obtained from the g-formula with the estimated regression coefficient related to mbsmoke. If they are identical, explain why; if they are different, again explain why this is the case.
(c) Use R to compute 95% confidence intervals of the ATE based on the bootstrap percentile method (1000 samples) and the g-formula.

(d) Is the assumption of consistency met in the given example?
(e) Suppose that all captured variables are added to the linear model from (a). Would conditional exchangeability be achieved and a causal interpretation of the effect be possible?

$\rightarrow$ Solutions to all exercises in this chapter can be found in Appendix B

Introduction to R

A

A.1 Background

The open source software R was developed as a free implementation of the language S which was designed as a language for statistical computation, statistical programming, and graphics. The main intention was to allow users to explore data in an easy and interactive way, supported by meaningful graphical representations. The statistical software R was originally created by Ross Ihaka and Robert Gentleman (University of Auckland, New Zealand).

A.2 Installation and Basic Functionalities

- The "base" R version, i.e. the software with its most relevant commands, can be downloaded from https://www.r-project.org/. After installing R, it is recommended to install an editor too. An editor allows the user to conveniently save and display R-code, submit this code to the R Console (i.e. the R software itself), and to control the settings and the output. A popular choice of editor is RStudio (free of charge) which can be downloaded from https://www.rstudio.com/, see Fig. A.1 for a screenshot of the software. There are alternative good editors, for example "Tinn-R" (http://sourceforge.net/projects/tinn-r/).
- A lot of additional user-written packages are available online and can be installed within the R console or using the R menu. Within the console, the `install.packages("package to install")` function can be used. Please note that an internet connection is required.

Statistics has a close relationship to algebra: data sets can be seen as matrices, and variables as vectors. R makes use of these structures and this is why we first intro-

© Springer Nature Switzerland AG 2022

C. Heumann et al., *Introduction to Statistics and Data Analysis*,

https://doi.org/10.1007/978-3-031-11833-3

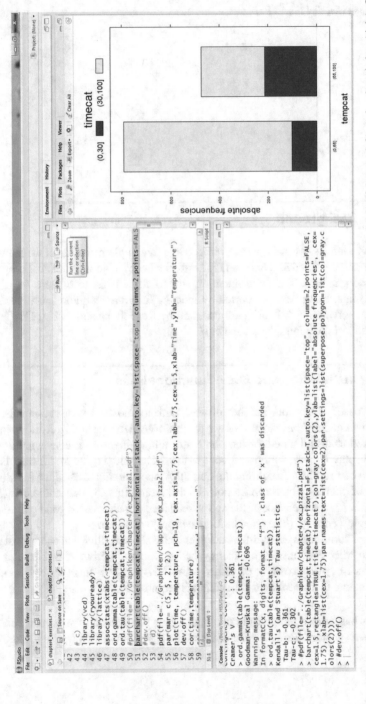

Fig. A.1 Screenshot of RStudio with the command window (top left), the output window ("console", i.e. *R* itself, bottom left), and a plot window (right). Other windows (e.g. about the environment or package updates) are closed. "RStudio and Shiny are trademarks of RStudio, PBC"

duce data structure functionalities before explaining some of the most relevant basic statistical commands.

R as a Calculator, Basic Data Structures and Arithmetic Operations

- The character # marks the beginning of a comment. All characters until the end of the *line* are ignored by *R*. We use # to comment on our *R*-code.
- If we know the name of a command we would like to use, and we want to learn about its functionality, typing ?command in the *R* command line prompt displays a help page, e.g.

```
?mean
```

displays a help page for the arithmetic mean function.
- Using

```
example(mean)
```

shows application examples of the respective function.
- The command c(1,2,3,4,5) combines the numbers 1, 2, 3, 4 and 5 into a vector (e.g. a variable).
- Vectors can be assigned to an "object". For example,

```
X <- c(2,12,22,32)
```

assigns a numeric vector of length 4 to the object X. In general, the arrow sign ($\leftarrow$) is a very important concept to store data, summaries and outputs in objects (that is, the name in front of the $\leftarrow$ sign). Note that *R* is case sensitive, i.e. X and x are two different variable names. Instead of "$\leftarrow$", one can also use "$=$".
- Sequences of numbers can be created using the seq and rep commands. For example,

```
seq(1,10)
```

and

```
rep(1,10)
```

yield

```
[1]  1  2  3  4  5  6  7  8  9 10
```

and

```
[1] 1 1 1 1 1 1 1 1 1 1
```

respectively.
- Basic data structures are **vectors, matrices, arrays, lists**, and **data frames**. They can contain numeric values, logical values or even characters (strings). In the latter case, arithmetic operations are not allowed.
  - A numeric vector of length 5 can be constructed by the command

    ```
    x <- vector(mode="numeric", length=5)
    ```

    The elements can be accessed by squared brackets: [ ]. For example,

    ```
    x[3] <- 4
    ```

    assigns the value 4 to the third element of the object x. Logical vectors containing the values TRUE and FALSE are also possible. In arithmetic operations, TRUE corresponds to 1 and FALSE corresponds to 0 (which is the default). Consider the following example:

    ```
    x.log <- vector(mode="logical", length=4)
    x.log[1] = x.log[3] = x.log[4] = TRUE
    mean(x.log)
    ```

    returns as output 0.75 because the mean of $(1, 0, 1, 1)$ =(TRUE, FALSE, TRUE, TRUE) is 0.75.
  - A matrix can be constructed by the matrix() command:

    ```
    x <- matrix( nrow=4, ncol=2, data=1:8, byrow=T )
    ```

    creates a 4×2 matrix, where the data values are the natural numbers 1, 2, 3, ..., 8 which are stored row-wise in the matrix,

    ```
         [,1] [,2]
    [1,]   1    2
    [2,]   3    4
    [3,]   5    6
    [4,]   7    8
    ```

    because of the parameter byrow=T (which is equivalent to byrow=TRUE). The default is byrow=F which would store the data column-wise.
  - Arrays are more general data structures than vectors and matrices in the sense that they can have more than two dimensions. For instance,

    ```
    x <- array(data=1:12, dim=c(3,2,2) )
    ```

    creates a three-dimensional array with $3 \cdot 2 \cdot 2 = 12$ elements.
  - A list can contain objects of different types. For example, a list element can be a vector or matrix. Lists can be initialized by the command list and can grow dynamically. It is important to understand that list elements should be accessed by the name of the entry via the dollar sign or using double brackets:

    ```
    x <- list(one=c(1,2,3,4,5),two=c("Hello", "world", "!"))
    ```

```
x
$one
[1] 1 2 3 4 5

$two
[1] "Hello" "world" "!"

x[[2]]
[1] "Hello" "world" "!"
x$one
     [1] 1 2 3 4 5
```

- A data frame is the standard data structure for storing a data set with rows as observations and columns as variables. Many statistical procedures in *R* (such as the lm function for linear models), expect a data frame. A data frame is conceptually not much different from a matrix and can either be initialized by reading a data set from an external file or by binding several column vectors. As an example, we consider three variables (age, favourite hobby, favourite animal) each with five observations:

```
age     <- c(25,33,30,40,28)
hobby   <- c("Reading","Sports","Games","Reading","Games")
animal <- c("Elephant", "Giraffe", NA, "Monkey", "Cat")
dat     <- data.frame(age,hobby,animal)
names(dat) <- c("Age","Favourite.hobby",
                "Favourite.animal")
dat
```

The resulting output is

```
> dat
  Age Favourite.hobby Favourite.animal
1  25          Reading         Elephant
2  33           Sports          Giraffe
3  30            Games             <NA>
4  40          Reading           Monkey
5  28            Games              Cat
```

where NA denotes a missing value. With write.table or a specialized version thereof such as write.csv (for writing the data in a file using comma-separated fields), a data frame can be saved in a file. The command sequence

```
write.csv(x=dat,file="toy.csv",row.names=FALSE)
read.dat <- read.csv(file="toy.csv")
read.dat
```

saves the data frame as an external (comma-separated) file, and then loads the data again using read.csv.

Individual elements of the data frame can be accessed using squared brackets, as for matrices. For example dat[1,2] returns the first observation of the second variable column for the data set `dat`. Individual columns (variables) can also be selected using the $ sign:

```
dat$Age
```
returns the age column:

```
[1] 25 33 30 40 28
```

- The `factor` command is very useful to store nominal variables, the command `ordered` is ideal for ordinal variables. Both commands are extremely important since factor variables with more than two categories are automatically expanded into several columns of dummy variables if necessary, e.g. if they are included as covariates in a linear model. In the previous paragraph, two factor variables have already been created. This can be confirmed by typing

```
is.factor(dat$Favourite.hobby)
is.factor(dat$Favourite.animal)
```

which return the value TRUE. Have a look at the following two factor variables:

```
sex <- factor("female","male","male","female","female")
grade <- ordered(c("low", "medium", "low", "high", "high"),
        levels=c("low", "medium","high"))
```

Please note that by default alphabetical order is used to order the categories (e.g. female is coded as 1 and male as 2). However, the mapping of integers to strings can be controlled by the user as seen for the variable "grade":

```
grade
```

returns

```
[1] low     medium low    high    high
Levels: low < medium < high
```

- Basic arithmetic operations can be applied directly to a numeric vector. Basic operations are addition $+$, subtraction $-$, multiplication $*$ and division $/$, integer division $\%/\%$, modulo operation $\%\%$, and exponentiation with two possible notations: $**$ or $\hat{\ }$. Examples are:

```
2^3                    # command
[1] 8                  # output
2**3                   # command
[1] 8                  # output
2^0.5                  # command
```

```
[1] 1.414214            # output
c(2,3,5,7)^2            # command: application to a vector
[1]   4  9 25 49        # output
c(2,3,5,7)^c(2,3)       # command: !!! ATTENTION!
[1]    4   27   25  343  # output
c(1,2,3,4,5,6)^c(2,3,4) # command
[1]    1    8   81    16  125 1296 #output
c(2,3,5,7)^c(2,3,4)     # command: !!! WARNING MESSAGE!
[1]   4  27 625  49
Warning message:
longer object length
        is not a multiple of shorter object length
        in: c(2, 3, 5, 7)^c(2, 3, 4)
```

The last four commands show the "recycling property" of *R*. It tries to match the vectors with respect to the length if possible. In fact,

```
c(2,3,5,7)^c(2,3)
```

is expanded to

```
c(2,3,5,7)^c(2,3,2,3)
```

The last example shows that *R* gives a warning if the length of the shorter vector cannot be expanded to the length of the longer vector by a simple multiplication with a natural number $(2, 3, 4, \ldots)$. Here

```
c(2,3,5,7)^c(2,3,4)
```

is expanded to

```
c(2,3,5,7)^c(2,3,4,2)
```

such that not all elements of

```
c(2,3,4)
```

are "recycled".

More on Indexing

The standard ways of accessing/indexing elements in a vector, matrix, list or data frame have already been introduced above; but *R* allows a lot more flexible accessing of elements.

1. Selecting elements using vectors of positive numbers (`letters` and LETTERS show the 26 letters of the alphabet)

```
letters[1:3]
letters[ c(2,4,6) ]
[1] "a" "b" "c"
[1] "b" "d" "f"
```

2. Selecting elements using logical vectors

```
x <- 1:10              # numbers 1 to 10
x[ (x>5) ]             # selecting any number >5
x[ (x%%2==0) ]         # numbers that are divisible by 2
x[(x%%2==1)]           # numbers that are not divisible by 2
x[5] <- NA             # 5th element of x is defined
                    # to be missing (NA)
x
y <- x[!is.na(x)]  # all x which are not missing
y
```

returns the output

```
[1]   6   7   8   9 10
[1]   2   4   6   8 10
[1]   1   3   5   7   9
[1]   1   2   3   4 NA   6   7   8   9 10
[1]   1   2   3   4   6   7   8   9 10
```

3. Selecting (deleting) elements using negative numbers

```
x <- 1:10
x[-(1:5)]     # x, but delete first five entries of x
```

returns the output

```
[1]   6   7   8   9 10
```

because the first five elements have been removed.

4. Selecting elements using characters

```
x <- c(Water=1, Juice=2, Lemonade=3 )
names(x)
x["Juice"]
```

returns the output

```
[1] "Water"    "Juice"       "Lemonade"
Juice
    2
```

Standard Functions

Some standard functions and their roles in *R* are

abs()	absolute value
sqrt()	square root
round(), floor(), ceiling()	rounding, up and down
sum(), prod()	sum and product
log(), log10(), log2()	logarithms
exp()	exponential function
sin(), cos(), tan(),	trigonometric functions
asin(), acos(), atan()	
sinh(), cosh(), tanh(),	hyperbolic functions
asinh(x), acosh(), atanh(x)	

All functions can again be applied directly to numeric vectors.

Statistical Functions

Some statistical functions and their roles in *R* are

mean(), var()	mean and variance
cov(), cor()	covariance and correlation
min(), max()	minimum and maximum

Note: the arguments of the functions vary depending on the chosen method. For example, the mean() function can be applied to general *R* objects where averaging makes sense (numeric or logical vectors, but also, e.g., matrices). The functions var(), cov(), cor() expect one or two numeric vectors, matrices or data frames. Minimum and maximum functions work also with a comma-separated list of values, i.e.

```
min(2, 6.7, 1.2, 8.0)
```

provides the same result (1.2) as

```
min(c(2, 6.7, 1.2, 8.0))
```

Examples:

```
mean( c(1,2,5,6) )
[1] 3.5
var( c(1,2,5,6) )
[1] 5.666667
```

Note that `var()`, `cov()` use the factor $1/(n-1)$ for the unbiased estimate of the variance instead of $1/n$ for the empirical variance, i.e. $1/3$ in the example above. Both functions can also be applied to several vectors or a matrix. Then the covariance matrix (and correlation matrix in case of `cor()`) are computed. For example, consider the two variables

```
age.v <- c(25,30,35,40)
income.v <- c(2000, 2500, 2800, 3200)
```

Then both commands return the symmetric covariance matrix (with variances as the diagonal entries, and covariances as the non-diagonal entries).

```
var(cbind(age.v, income.v))
                  age.v income.v
age.v        41.66667   3250.0
income.v 3250.00000 255833.3
cov(cbind(age.v, income.v))
                  age.v income.v
age.v        41.66667   3250.0
income.v 3250.00000 255833.3
```

The (Pearson) correlation between the two variables is calculated as 0.9954293.

```
cor(cbind(age.v, income.v))
                age.v   income.v
age.v     1.0000000 0.9954293
income.v 0.9954293 1.0000000
```

The Spearman rank correlation is perfectly 1, since both vectors are in increasing order:

```
cor(cbind(age.v, income.v), method="spearman")
           age.v income.v
age.v          1        1
income.v       1        1
```

More Useful Functions

Some more commonly used standard functions and their roles in *R* are as follows.

- Cumulative sum and product:

```
x <- c( 1,3, 2, 5)
cumsum(x)               # 1, 1+3, 1+3+2, 1+3+2+5
cumprod(x)              # 1, 1*3, 1*3*2, 1*3*2*5
```

  give the output

```
[1]  1  4  6 11
[1]  1  3  6 30
```

- Factorial:

```
factorial(5)
```

  returns 5! as

```
[1] 120
```

- Binomial coefficient $\binom{n}{k}$:

```
choose(4,2)
```

  returns $\binom{4}{2}$ as

```
[1] 6
```

Mathematical Constants

The number π is a mathematical constant, the ratio of a circle's circumference to its diameter, and is commonly approximated as 3.14159. It is directly available in *R* as pi.

```
pi
[1] 3.141593
```

Other "constants" are

```
Inf, -Inf   ∞, −∞
NaN         Not a Number: e.g. 0/0[1] NaN
NA          Not Available: missing values
NULL        empty set
```

Assignment Operator for Creating Functions

The assignment operator $\leftarrow$ ("less than" sign followed by hyphen) has already been introduced above in the context of variables. Alternatively, $=$ (equality sign) can be used. One can create one's own functions: The function is an object with a name which takes values specified in the round brackets and returns what has been specified in the curly braces. For example, the following function `myfunction` returns a polynomial of degree 3 for a given argument x. Note that by default all four coefficients are equal to 1.

```
my.function <- function(x,a=1,b=1,c=1,d=1){
  h <- a+b*x+c*x^2+d*x^3
  return(h)
}

my.function(2)
[1] 15

my.function(x=2, a=4, b=3)
[1] 22
```

Loops and Conditions

The concept of loops is convenient when some operation has to be repeated. Loops can be utilized in various ways, for example via `for` or `while`. Conditions are specified with the `if` statement. For example,

```
x <- 1:10
for(i in 1:10){
if(x[i]>5){x[i] <- x[i]+i}
}
x
```

returns

```
[1]   1   2   3   4   5  12  14  16  18  20
```

In this example, x is a vector with 10 elements: 1, 2, 3, 4, 5, 6, 7, 8, 9, 10. For each element `x[i]`, we replace `x[i]` with `x[i]+i` if the condition `x[i]>5` is true, otherwise we do not.

A.3   Statistical Functions

Now we consider some basic statistical functions in R. For illustration, we use the `painters` data in the following example. This data is available after loading the library MASS (only a subset is shown below). The data lists the subjective assessment, on a 0 to 20 integer scale, of 54 classical painters. The painters were assessed on four characteristics: composition, drawing, colour and expression. The data is due to the eighteenth-century art critic, de Piles. Use `?painters` for more information on the data set.

```
library(MASS)
painters
```

shows

```
            Composition Drawing Colour Expression School
Da Udine             10       8     16          3      A
Da Vinci             15      16      4         14      A
Del Piombo            8      13     16          7      A
Del Sarto            12      16      9          8      A
Fr. Penni             0      15      8          0      A
```

The Summary Function

The summary function allows a quick overview of a data frame. For numeric variables, the five-point summary (which is also used in a simple box plot, see Sect. 3.3) is calculated together with the arithmetic mean. For factor variables, the absolute frequencies of the most frequent categories are printed. If the factor has more than six categories the other categories are summarized in a separate category Other.

```
summary(painters)
```

yields

```
   Composition            Drawing          ...            School
Min.    : 0.00    Min.    : 6.00     ...    A          :10
1st Qu.: 8.25    1st Qu.:10.00       ...    D          :10
Median :12.50    Median :13.50       ...    E          : 7
Mean    :11.56    Mean    :12.46     ...    G          : 7
3rd Qu.:15.00    3rd Qu.:15.00       ...    B          : 6
Max.    :18.00    Max.    :18.00     ...    C          : 6
                                     ...    (Other) : 8
```

The summary function can also be applied to a single variable:

```
summary(painters$School)
```

returns

```
 A  B  C  D  E  F  G  H
10  6  6 10  7  4  7  4
```

Accessing Subgroups in Data Frames

Subgroups, i.e. groups of observations which share the same feature(s) of one or several variables, can be accessed using the subset command.

```
subset(painters, School=="F")
```

accesses all painters for which School=="F" holds.

```
            Composition Drawing Colour Expression School
Durer                 8      10     10          8      F
Holbein               9      10     16         13      F
Pourbus               4      15      6          6      F
Van Leyden            8       6      6          4      F
```

This is a more elegant method than selecting these observations by specifying a condition in squared brackets via the [rows,columns] argument.

```
painters[ painters[["School"]] == "F", ]
```

Note that the explained structure is an important one: we access the rows and columns of a matrix or data set by using the [rows,columns] argument. Here we access all rows for which the variable "school" is "F". If, in addition, we also want to restrict the data set to the first two variables, we can write:

```
painters[ painters[["School"]] == "F", c(1,2)]
```

Similarly,

```
subset(painters, Composition <= 6)
```

gives the output

	Composition	Drawing	Colour	Expression	School
Fr. Penni	0	15	8	0	A
Perugino	4	12	10	4	A
Bassano	6	8	17	0	D
Bellini	4	6	14	0	D
Murillo	6	8	15	4	D
Palma Vecchio	5	6	16	0	D
Caravaggio	6	6	16	0	E
Pourbus	4	15	6	6	F

Uninteresting columns can be eliminated using negative indexing. For instance, in the following example,

```
subset(painters, School=="F", select=c(-3,-5) )
```

	Composition	Drawing	Expression
Durer	8	10	8
Holbein	9	10	13
Pourbus	4	15	6
Van Leyden	8	6	4

the third and the fifth columns (Colour and School) are not shown.

The operator `%in%` allows for more complex searches. For instance

```
subset(painters, Drawing %in% c(6,7,8,9) & Composition==10)
```

returns the following output:

	Composition	Drawing	Colour	Expression	School
Da Udine	10	8	16	3	A
J. Jordaens	10	8	16	6	G
Bourdon	10	8	8	4	H

i.e. those painters with a drawing score between 6 and 9 (= any number which matches 6, or 7, or 8, or 9).

Stratifying a Data Frame and Applying Commands to a List

Sometimes it is of interest to apply statistical commands (such as `summary`) to
several subgroups. If this is the case, the data is partitioned into different groups using
`split` and then `lapply` applies a function to each of these groups. The command
`split` partitions the data set by values of a specific variable. For example, we first
stratify the `painters` data with respect to the painter's school:

```
splitted <- split(painters, painters$School)
splitted
```

```
$A
                Composition Drawing Colour Expression School
Da Udine                 10       8     16          3      A
Da Vinci                 15      16      4         14      A
Del Piombo                8      13     16          7      A
Del Sarto                12      16      9          8      A
Fr. Penni                 0      15      8          0      A
Guilio Romano            15      16      4         14      A
Michelangelo              8      17      4          8      A
Perino del Vaga          15      16      7          6      A
Perugino                  4      12     10          4      A
Raphael                  17      18     12         18      A
```

```
$B
                Composition Drawing Colour Expression School
F. Zucarro               10      13      8          8      B
Fr. Salviata             13      15      8          8      B
Parmigiano               10      15      6          6      B
Primaticcio              15      14      7         10      B
T. Zucarro               13      14     10          9      B
Volterra                 12      15      5          8      B
```

```
$C
...
```

Note, that `splitted` is now a list,

```
is.list(splitted)
```

returns

```
[1] TRUE
```

while the objects `splitted$A` to `splitted$H` are data frames.

```
is.data.frame(splitted$A)
```

returns

```
[1] TRUE
```

Secondly, as indicated above, the command `lapply` allows us to apply a function to a list. For instance

```
lapply(splitted, summary)
```

applies the summary function to all data frames in the list `splitted` (output not shown). See also `?apply`, `?sapply`, `?tapply` and `?mapply` for similar operations.

Sorting, Ranking, Finding Duplicates, and Unique Values

- Sorting a vector:

  ```
      x <- c( 1,3, 2, 5)
  sort(x)
  sort(x, decreasing=TRUE)
  ```

  returns the ordered values in decreasing order as

  ```
  [1] 5 3 2 1
  ```

  See also the command `order` for showing the order of vector elements.

- Calculating ranks;

  ```
      x <- c( 10,30, 20, 50, 20)
  rank(x)
  ```

  returns the following output:

  ```
  [1] 1.0 4.0 2.5 5.0 2.5
  ```

- Finding duplicate values:

```
x <- c( 1,3, 2, 5, 2)
duplicated(x)
```

  indicates which values occur more than once:

```
[1] FALSE FALSE FALSE FALSE  TRUE
```

- Removing duplicates:

```
x <- c( 1,3, 2, 5, 2)
unique(x)
```

  shows the output as

```
[1] 1 3 2 5
```

  This means `unique` finds out how many *different* values a vector has.

Categorizing Numeric Variables

Continuous variables (vectors) can be categorized using the `cut` command.

```
x <- c(1.3, 1.5, 2.5, 3.8, 4.1, 5.9, 7.1, 8.4, 9.0)
xdiscrete <- cut(x, breaks=c(-Inf, 2, 5, 8, Inf) )
is.factor(xdiscrete)
xdiscrete
table(xdiscrete)
```

returns

```
[1] TRUE
[1] (-Inf,2] (-Inf,2] (2,5] (2,5] (2,5] (5,8] (5,8] (8,Inf]
[9] (8,Inf]
Levels: (-Inf,2] (2,5] (5,8] (8,Inf]
(-Inf,2]     (2,5]     (5,8]    (8,Inf]
  2           3          2         2
```

Random Variables

- R has built-in functions for several probability density/mass functions (PMF/PDF), (probability) distribution function (i.e. the CDF), quantile functions and for generating random numbers.
- The function names use the following scheme:

First letter	Function	Further letters
d	density	distribution name
p	probability	distribution name
q	quantiles	distribution name
r	random number	distribution name

- Examples:

```
dnorm(x=0)
[1] 0.3989423
```

returns the value of the density function (i.e. $P(X = x)$) of a $N(0, 1)$-distribution at $x = 0$, which is $1/\sqrt{2\pi}$.

```
pnorm(q=0)
pnorm(q=1.96)
```

```
[1] 0.5
[1] 0.9750021
```

returns the value of the CDF of a $N(0, 1)$-distribution at q, i.e. $\Phi(q) = P(X \leq q)$.

```
qnorm(p=0.95)
```

returns the value

```
[1] 1.644854
```

which is the 95%-quantile of a $N(0, 1)$-distribution.

```
X <- rnorm(n=4)
X
```

returns a vector of four normal random numbers of a $N(0, 1)$ distribution:

```
[1] -0.90826678 -0.09089654 -0.47679821  1.22137230
```

Note that a repeated application of this function leads to different random numbers. To get a reproducible sequence of random numbers, a `seed` value needs to be set:

```
set.seed(89234512)
X <- rnorm(n=4)
X
```

If all three commands are executed, the sequence is (using the standard random generator)

```
[1] -1.07628865  0.37797715  0.04925738 -0.22137107
```

- The following functions for distributions can be used:

Model distributions

Function	Distribution
beta	Beta
binom	Binomial
cauchy	Cauchy
exp	Exponential
gamma	Gamma
geom	Geometric
hyper	Hypergeometric
lnorm	Log–Normal
norm	Normal
pois	Poisson
unif	Uniform
mvnorm	Multivariate Normal (in package mvtnorm)

Test distributions

Function	Distribution
chisq	χ^2
f	F
signrank	Wilcoxon signed rank (1 sample)
t	t
wilcox	Wilcoxon rank sum (2 samples)

- For convenience we list a few important PDF and CDF values in Appendix C.9.

Key Points and Further Issues

> **Note:**
>
> ✓ *R* uses the following data structures: vectors, matrices, arrays, lists, and data frames.
>
> ✓ Entries of matrices and data frames can be accessed using squared brackets. For example `data[1:5,2]` refers to the first 5 observations of the second column (variable) of the data. Variables of data frames can also be accessed via the $ sign, e.g. via `data$variable`.
>
> ✓ If operations such as statistical analyses have to be repeated on several subgroups, using `split` together with `lapply` is a viable option. Alternatively loops (e.g. `for`) together with conditions (such as `if`) can be used.
>
> ✓ *R* contains the most relevant statistical functions needed for descriptive and inductive analyses (as shown throughout the book). User-written packages can be installed using `install.packages("package_name")`.
>
> ✓ Readers who are interested in learning more about *R* are referred to Albert and Rizzo (2012), Crawley (2013), Dalgaard (2008), and Everitt and Hothorn (2011).

A.4   Data Sets

From the data used in this book, we publish some relevant data sets, along with solutions of the *R*-exercises, on https://statsbook.github.io/. The important data sets are explained in the following paragraphs.

A.4.1   Pizza Delivery Data

The pizza delivery data (`pizza_delivery.csv`, see also Table A.1), is a simulated data set. The data refers to an Italian restaurant which offers home delivery of pizza. It contains the orders received during a period of one month: May 2014. There are three branches of the restaurant. The pizza delivery is centrally managed: an operator receives a phone call and forwards the order to the branch which is nearest to the customer's address. One of the 5 drivers (two of whom only work part time at the weekend) delivers the order. The data set captures the number of pizzas ordered as well as the final bill (in €) which may also include drinks, salads, and pasta dishes. The owner of the business observed an increased number of com-

Table A.1 First few rows of the pizza delivery data

```
        day         date  time operator branch    driver temperature
1 Thursday 1 May 2014  35.1    Laura   East     Bruno        68.3
2 Thursday 1 May 2014  25.2  Melissa   East Salvatore        71.0
3 Thursday 1 May 2014  45.6  Melissa   West Salvatore        53.4
4 Thursday 1 May 2014  29.4  Melissa   East Salvatore        70.3
5 Thursday 1 May 2014  30.0  Melissa   West Salvatore        71.5
6 Thursday 1 May 2014  40.3  Melissa Centre     Bruno        60.8
...
```

```
    bill  pizzas free_wine got_wine discount_customer
1  58.4       4         0        0                 1
2  26.4       2         0        0                 0
3  58.1       3         1        0                 0
4  35.2       3         0        0                 0
5  38.4       2         0        0                 0
6  61.8       4         1        1                 0
....
```

plaints, mostly because pizzas arrive too late and too cold. To improve the service quality of his business, the owner wants to measure (i) the time from call to delivery and (ii) the pizza temperature at arrival (which can be done with a special device). Ideally, a pizza arrives within 30 min. of the call; if it takes longer than 40 min., then the customers are promised a free bottle of wine (which is not always handed out though). The temperature of the pizza should be above 65 °C at the time of delivery. The analysis of the data aims to determine the factors which influence delivery time and temperature of the pizzas (Table A.4).

A.4.2 Decathlon Data

This data (decathlon.csv, see also Table A.2), describes the results of the decathlon competition during the 2004 Olympic Games in Athens. The performance of all 30 athletes in the 100 m race (in seconds), long jump (in metres), shot-put (in metres), high jump (in metres), 400 m race (in seconds), 110 m hurdles race (in seconds), discus competition (in metres), pole vault (in metres), javelin competition (in metres) and 1500 m race (in seconds) are recorded in the data set.

A.4.3 Theatre Data

This data (theatre.csv, see also Table A.3) summarizes a survey conducted on 699 participants in a big Swiss city. The survey participants are all frequent visitors to a local theatre and were asked about their age, sex (gender, female=1), annual income (in 1000 SFR), general expenditure on cultural activities ("Culture", in SFR

Table A.2 First few rows of the decathlon data from the 2004 Olympic Games in Athens data

	100m	Long.jump	Shot.put	High.jump	400m
Roman Sebrle	10.85	7.84	16.36	2.12	48.36
Bryan Clay	10.44	7.96	15.23	2.06	49.19
Dmitriy Karpov	10.50	7.81	15.93	2.09	46.81
Dean Macey	10.89	7.47	15.73	2.15	48.97
Chiel Warners	10.62	7.74	14.48	1.97	47.97
Attila Zsivoczky	10.91	7.14	15.31	2.12	49.40
...					

	110m.hurdle	Discus	Pole.vault	Javelin	1500m
Roman Sebrle	14.05	48.72	5.0	70.52	280.01
Bryan Clay	14.13	50.11	4.9	69.71	282.00
Dmitriy Karpov	13.97	51.65	4.6	55.54	278.11
Dean Macey	14.56	48.34	4.4	58.46	265.42
Chiel Warners	14.01	43.73	4.9	55.39	278.05
Attila Zsivoczky	14.95	45.62	4.7	63.45	269.54
...					

Table A.3 First few rows of the theatre data

	Age	Sex	Income	Culture	Theatre	Theatre_ly
1	31	1	90.5	181	104	150
2	54	0	73.0	234	116	140
3	56	1	74.3	289	276	125
4	36	1	73.6	185	75	130
5	24	1	109.0	191	172	140
6	25	0	93.1	273	168	130
...						

per month), expenditure on theatre visits (in SFR per month), and their estimated expenditure on theatre visits in the year before the survey was done (in SFR per month).

A.4.4 Cattaneo Data

This hypothetical data set is based on the study of Cattaneo et al. (Efficient semi-parametric estimation of multi-valued treatment effects under ignorability, *Journal of Econometrics*, 2010, 155:138–154). It has been created originally to illustrate causal inference methods in the Stata statistical software package (StataCorp. 2021. Stata Statistical Software: Release 17. College Station, TX: StataCorp LLC). The

Table A.4 First rows of selected variables from the Cattaneo data

	bweight	mmarried	alcohol	mage	medu	fage	mbsmoke	
1	3459	married	0	24	14	28	nonsmoker	...
2	3260	notmarried	0	20	10	0	nonsmoker	...
3	3572	married	0	22	9	30	nonsmoker	...
4	2948	married	0	26	12	30	nonsmoker	...
5	2410	married	0	20	12	21	nonsmoker	...
6	3147	notmarried	0	27	12	40	nonsmoker	...

. . .

data includes information on women who gave birth, including the infant's birth-weight (in grams, bweight), whether cigarettes have been smoked during pregnancy (msmoke, mbsmoke), if it is the mother's first baby (fbaby), information on whether the mother consumed alcohol during pregnancy (alcohol), the mother's marital status (mmarried), the mother's and father's education attainment (medu, fedu) and age (mage, fage), among other variables.

Solutions to Exercises

B

Chapter 1

Solution to Exercise 1.1

(a) The population consists of all employees of the airline. This may include administration staff, pilots, stewards, cleaning personnel and others. Each single employee relates to an observation in the survey.

(b) The population comprises all students who take part in the examination. Each student represents an observation.

(c) All people suffering high blood pressure in the study area (city, province, country, …), are the population of interest. Each of these persons is an observation.

Solution to Exercise 1.2   The *population* in this study refers to all leopards in the park. Only a few of the leopards are equipped with the GPS devices. This is the *sample* on which the study is conducted in. Each leopard refers to an *observation*. The measurements are taken for each leopard in the sample. The GPS coordinates allow to determine the position during the entire day. Important *variables* to capture would therefore be $X_1 =$"latitude", $X_2 =$"longitude", and $X_3 =$"time". Each variable would take on certain *values* for each observation, for example the first leopard may have been observed at latitude $32°$ at a certain time point, and thus $x_{11} = 32°$.

Solution to Exercise 1.3

Qualitative:	preferred political party, eye colour, gender, blood type, subject line of an email.
Quantitative and discrete:	shoe size, customer satisfaction on a scale from 1 to 10, number of goals in a hockey match.
Quantitative and continuous:	time to travel to work, price of a canteen meal, wavelength of light, delivery time of a parcel, height of a child.

© Springer Nature Switzerland AG 2022
C. Heumann et al., *Introduction to Statistics and Data Analysis*,
https://doi.org/10.1007/978-3-031-11833-3

Solution to Exercise 1.4

(a) The choice of a political party is measured on a nominal scale. The names of the parties don't have a natural order.
(b) Typically the level of a computer game is measured on an ordinal scale: for example level 10 may be more difficult than level 5, but this does not imply that levels 10 is twice as difficult as level 5, or that the difference in difficulty between levels 2 and 3 is the same as the difference between level 10 and 11.
(c) The production time of a car is measured on a metric scale (ratio scale). In practice it may be measured in days from the start of the production.
(d) This variable is measured on a metric scale (ratio scale). Typically the age is captured in years starting from the day of birth.
(e) Calender year is a metric variable which is measured on an interval scale. Note that the year which we define as "zero" is arbitrary, it varies from culture to culture. Because the year zero is arbitrary, and we also have dates before this year, the calender year is measured on an interval scale.
(f) The scale is metric (ratio scale).
(g) The scale of ID numbers is nominal. The ID number may indeed consist of numbers; however, "112233" does not refer to something half as much/good as "224466". The number is descriptive.
(h) The final rank is measured on an ordinal scale. The ranks can be clearly ordered, and the participants can be ranked by using their final results. However, the first winner may not have "double" the beauty of the second winner, it is merely a ranking.
(i) The intelligence quotient is a variable on a metric scale. It is constructed in such a way that differences are interpretative, i.e. being 10 points above or 10 points below the average score of 100 points means the same deviation from the average. However, ratios cannot be interpreted, so the intelligence quotient is measured on an interval scale.

Solution to Exercise 1.5

(a) The data is provided in *.csv* format. We thus read it in with the `read.csv()` command (after we have set a working directory with `setwd()`):

```
setwd('C:/directory')
pizza <- read.csv('pizza_delivery.csv')
```

(b) The data can be viewed by means of the `View()` command or simply being printed:

```
View(pizza)
pizza
```

(c) We can access the data, as for any matrix, by using squared brackets [,], see also Appendix A. The first entry in the brackets refers to the row, the second entry to the columns. Each entry either is empty (referring to every row/column) or consists of a vector or sequence describing the columns/rows we want to select. This means that we obtain the first 5 rows and variables via pizza[1:5,1:5]. If we give the new data the name "pizza2" we simply need to type:

```
pizza2 <- pizza[1:5,1:5]
pizza2
```

We can save this new data either as a *.dat* file (with write.table()), or as a *.csv* file (with write.csv()), or directly as an *R* data file (with save()) which gives us access to our entire *R* session.

```
write.csv(pizza2,file='pizza2.csv')
write.table(pizza2,file='pizza2.dat')
save(pizza2,file='pizza2.Rdata')
```

(d) We can access any variable by means of the $ sign. If we type pizza$ new we create a new variable in the pizza data set called "new". Therefore a simple way to add a variable to the data set is as follows:

```
pizza$NewTemperature <- 32+1.8*pizza$temperature
```

(e)

```
attach(pizza)
NewTemperature
```

(f) We can apply all these commands onto the object "pizza". The command str(pizza) gives us an overview of the data set, see also Fig. B.1. The output shows that the data set contains 1266 observations (deliveries) and 13 variables. We can also see which of these variables are factors (categorical with defined categories) and which are numerical. We also see the first actual numbers for the each variable and the coding scheme used for the categories of the factor variables. The command dim summarizes the dimension of the data, i.e. the

```
> str(pizza)
'data.frame':  1266 obs. of  13 variables:
 $ day               : Factor w/  7 levels "Friday","Monday",..: 5 5 5 5
 $ date              : Factor w/ 31 levels "01-May-14","02-May-14",..: 1
 $ time              : num  35.1 25.2 45.6 29.4 30 ...
 $ operator          : Factor w/  2 levels "Laura","Melissa": 1 2 2 2 2
 $ branch            : Factor w/  3 levels "Centre","East",..: 2 2 3 2 3
 $ driver            : Factor w/  5 levels "Bruno","Domenico",..: 1 5 5 5
 $ temperature       : num  68.3 71 53.4 70.3 71.5 ...
 $ bill              : num  58.4 26.4 58.1 35.2 38.4 61.8 57.9 35.8 36.6
 $ pizzas            : int  4 2 3 3 2 4 3 2 2 5 ...
 $ free_wine         : int  0 0 1 0 0 1 1 0 0 0 ...
 $ got_wine          : int  0 0 0 0 0 1 1 0 0 0 ...
 $ discount_customer : int  1 0 0 0 0 0 0 0 0 0 ...
 $ NewTemperature    : num  155 160 128 159 161 ...
```

Fig. B.1 Applying str() to the pizza data

number of rows and columns of the data matrix. `Colnames` gives us the names of the variables from the data set, and so does `names`. The commands `nrow` and `ncol` give us the number of rows and columns respectively. Applying `head` and `tail` to the data prints the first and last rows of the data, respectively.

Solution to Exercise 1.6

(a) The appropriate study design is a survey. The information would be obtained via a questionnaire given to a sample of parents. It is not a controlled experiment because we do not manipulate one particular variable, while controlling others; we rather collect data on all variables of interest.

(b) There are different options to ask for parents' attitudes: of course one could simply ask "what do you think of immunization?"; however, capturing long answers in a variable "attitude" may make it difficult to summarize and distill the information obtained. A common way to deal with such variables is to translate a concept into a score: for example, one could ask 5 "yes/no" type questions (instead of one general question) which relate to attitudes towards immunization, such as "do you think immunization may be harmful for your child?" or "do you agree that it is a priority to immunize infants in their first year of life?". The number of answers which show a positive attitude towards immunization can be summed up. If there are 5 questions, there are up to 5 points "to earn". Thus, each parent may be asked 5 questions and his/her attitude can be summarized on a scale ranging from 0 to 5, depending on the answers given.

(c) The following variables are needed:

- Attitude: possibly several variables are needed to capture parents' information in a score, see (b) for details. The scale is ordinal because a higher score relates to a more positive attitude towards immunization, but the differences between different score levels cannot be interpreted meaningfully.
- Immunized: a binary ("yes–no" type) variable capturing whether the parent agrees to immunization against chicken-pox for their youngest child or not. This is a nominal variable.
- Gender: to compare "Immunized" for male and female parents. This is a nominal variable.
- Age: to compare the age distribution in the group of parents who would immunize their child with the age distribution in the group who would not immunize their child. Age is measured on a metric scale.

(d) A data set might look as follows:

Parent	A_1	...	A_5	Attitude	Immunization	Gender	Age
1	yes	...	yes	3	yes	male	35
2	no	...	yes	2	no	female	26
$\vdots$	$\vdots$	$\vdots$		$\vdots$			

where $A_1, ..., A_5$ refer to variables capturing attitudes towards immunization and "Attitude" is the score variable summarizing these questions. The questions may be written down as follows:

- What is the average attitude score towards immunization among parents and how much does it vary?
- What percentage of parents answer "yes" when asked whether they would immunize their youngest child against chicken-pox?
- What is the difference in the proportion calculated in (b) when stratified by gender?
- What is the average age of those parents who would immunize their child compared with the average age of those parents who would not immunize their child?

Chapter 2

Solution to Exercise 2.1

(a) The table shows the relative frequencies of each party and not the absolute frequencies. We can thus draw a bar chart where the relative frequencies of votes are plotted on the *y*-axis and different parties are plotted on the *x*-axis. In *R*, we can first type in the data by defining two vectors and then use the "barplot" command to draw the bar chart (Fig. B.2a). Typing "?barplot" and "?par" shows the graphical options to improve the presentation and look of the graph:

```
results2014 <- c(0.6215,0.2223,0.0635,0.0240,0.0067,0.0620)
barplot(results2014)
barplot(results2014,names.arg=c('ANC','DA','EFF','IFP','COPE'
'O.'), col=gray.colors(6),ylim=c(0,0.7),xlab='Parties',ylab
= 'Votes(%)')
```

(b) There are several options to compare the results. Of course, one can simply plot the two bar charts with each bar chart representing one election. It would be important for this solution to ensure that the *y*-axes are identical in both the plots. However, if we want to compare the election results in one graph then we can plot the difference between the two elections, i.e. the win/loss per party. The bar chart for the new variable "difference in proportion of votes between the two elections" is shown in Figure B.2b and is obtained as follows:

```
results2009 <- c(0.6590,0.1666,0,0.0455,0.0742,0.0547)
difference <- results2014-results2009
barplot(difference)
```

Remark Another solution would be to create subcategories on the *x*-axis: for example, there would be categories such as "ANC 2009 results" and "ANC 2014 results", followed by "DA 2009 results" and "DA 2014 results", and so on.

(a) Results of the 2014 elections   (b) Difference in results between the
                                               2014 and 2009 elections

Fig. B.2 Bar charts for national elections in South Africa

Solution to Exercise 2.2

(a) The scale of X is metric. However, please note that the number of values X can
    practically take is limited (90 min. plus extra time, recorded in 1 min. intervals).
(b) It is straightforward to obtain the frequency table, as well as the other information
    needed to obtain the histogram and the ECDF, see Table B.1.
(c) We need to obtain the heights for each category to obtain the histogram using
    $h_j = f_j/d_j$, see Table B.1.
(d) We obtain the histogram and kernel density plot in R (Fig. B.3a) using the com-
    mands

```
goals <- c(6,24,91,...,7)
hist(goals, breaks=c(0,15,30,45,60,75,90,96))
plot(density(goals, adjust=1,kernel='gaussian'))
```

Table B.1 Frequency table and other information for the variable "time until first goal"

j	$[e_{j-1}, e_j)$	n_j	f_j	d_j	h_j	$F(x)$
1	[0, 15)	19	$\frac{19}{55}$	15	$\frac{19}{825}$	$\frac{19}{55}$
2	[15, 30)	17	$\frac{17}{55}$	15	$\frac{17}{825}$	$\frac{36}{55}$
3	[30, 45)	6	$\frac{6}{55}$	15	$\frac{6}{825}$	$\frac{42}{55}$
4	[45, 60)	5	$\frac{5}{55}$	15	$\frac{5}{825}$	$\frac{47}{55}$
5	[60, 75)	4	$\frac{4}{55}$	15	$\frac{4}{825}$	$\frac{51}{55}$
6	[75, 90)	2	$\frac{2}{55}$	15	$\frac{2}{825}$	$\frac{53}{55}$
7	[90, 96)	2	$\frac{2}{55}$	6	$\frac{2}{825}$	1
Total		56	1			

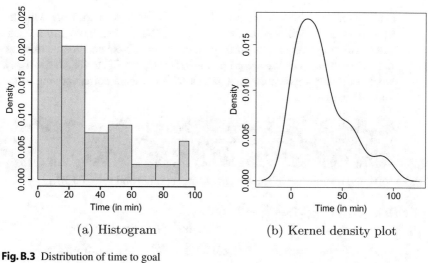

(a) Histogram   (b) Kernel density plot

Fig. B.3 Distribution of time to goal

(a) Original data   (b) Grouped data

Fig. B.4 Empirical cumulative distribution function for the variable "time until first goal"

(e) The ECDF values for $F(x)$ are calculated using the relative frequencies $f(x)$, see Table B.1.

(f) (i) We can easily plot the ECDF (Fig. B.4a) for the original data using the R command

```
plot.ecdf(goals)
```

(ii) Generating the ECDF for the grouped data requires more effort and is not necessarily intuitive: first we categorize the metric variable using the function cut. We use the label option to indicate that the name of each category corresponds to the upper limit of the respective interval. This new variable is a "factor" variable and the `plot.ecdf` function does not work with this type of variable. We need to first change the "factor" variable into a "character" variable with strings corresponding to the labels and coerce this into numeric values. Then we use `plot.ecdf`, see also Fig. B.4b. Alternatively, we can directly replace the raw values with numbers corresponding to the upper interval limits.

```
goals_cat <- cut(goals, breaks=c(0,15,30,45,60,75,90,96),
labels=c(15,30,45,60,75,90,96))
plot.ecdf(as.numeric(as.character(goals_cat)))
```

(g) To solve the exercises, we simply use formula (2.11) and rules (2.3ff.)

(i) $H(X \le 45) = F(45) = \frac{42}{55} \approx 0.76$.

(ii) $H(X > 80) = 1 - F(80) = 1 - (\frac{51}{55} + \frac{2/55}{15}(80 - 75)) \approx 0.085$.

(iii) $H(20 \le X \le 65) = F(65) - F(20) = \frac{47}{55} + \frac{4/55}{15} \cdot (65 - 60) - (\frac{19}{55} + \frac{17/55}{15} \cdot (20 - 15)) \approx 0.43$.

(h) We know from (2.11) that

$$F(x_p) = p = F(e_{j-1}) + h_j(x_p - e_{j-1})$$

with $h_j = f_j/d_j$ which relates to

$$x_p = e_{j-1} + \frac{p - F(e_{j-1})}{h_j}.$$

We are interested in $x_{0.8}$ because 80% of the matches have seen a goal at this time point:

$$x_{0.8} = 45 + \frac{0.8 - \frac{43}{56}}{\frac{1}{168}} = 50.4.$$

We conclude that 80% of the "first goals" happened up to 50.4 min.

Solution to Exercise 2.3

(a) We obtain the relative frequencies for the first and fourth intervals as 0.125 ($h_j \cdot d_j = 0.125 \cdot 1$). Accordingly, for the other two intervals, we obtain frequencies of $f_j = 3 \cdot 0.125 = 0.375$.

(b) We obtain the absolute frequencies for the first and fourth intervals as 250 (2000 · 0.125). For the other intervals, we obtain 750 (2000 · 0.375).

Solution to Exercise 2.4

(a) The absolute frequencies n_j are evident from the following table:

j	e_{j-1}	e_j	$F(e_j)$	f_j	$n_j(= f_j n)$	d_j	a_j
1	8	14	0.25	0.25	$0.25 \cdot 500 = 125$	6	11
2	14	22	0.40	0.15	75	8	18
3	22	34	0.75	0.35	175	12	28
4	34	50	0.97	0.22	110	16	42
5	50	82	1.00	0.03	15	32	66

(b) We obtain $F(X > 34) = 1 - F(34) = 1 - 0.75 = 0.25$.

Solution to Exercise 2.5

(a) The data needed to calculate and draw the ECDF is summarized in Table B.2; the ECDF is plotted in Fig. B.5.
(b) It follows from Table B.2 that $F(3) = 12\%$ and $F(9) = 98\%$.
(c) The grey solid line in Fig. B.5 summarizes the ECDF for the grouped data. It goes from $(0, 0)$ to $(1, 1)$ with a breakpoint at $(5, 0.47)$ since $F(5) = 0.47$ summarizes the information for the group "disagree". Using (2.11) we calculate:

$$F(3) = F(e_{j-1}) + \frac{f_j}{d_j}(x - e_{j-1})$$

Table B.2 Information needed to calculate the ECDF

Score	1	2	3	4	5	6	7	8	9	10
Results	1	3	8	8	27	30	11	6	4	2
f_j	$\frac{1}{100}$	$\frac{3}{100}$	$\frac{8}{100}$	$\frac{8}{100}$	$\frac{27}{100}$	$\frac{30}{100}$	$\frac{11}{100}$	$\frac{6}{100}$	$\frac{4}{100}$	$\frac{2}{100}$
F_j	$\frac{1}{100}$	$\frac{4}{100}$	$\frac{12}{100}$	$\frac{20}{100}$	$\frac{47}{100}$	$\frac{77}{100}$	$\frac{88}{100}$	$\frac{94}{100}$	$\frac{98}{100}$	1

Fig. B.5 Empirical cumulative distribution function for the variable "score"

$$= F(0) + \frac{0.47}{5} \cdot (3 - 0) = 28.2\%$$

$$F(9) = F(5) + \frac{0.53}{5} \cdot (9 - 5) = 89.4\%.$$

(d) The results of (b) and (c) are very different. The formula applied in (c) assumes that the values in each category are uniformly distributed, i.e. that within each category, each value occurs as often as each other value. However, we know from (a) that this is not true: there are more values towards the central score numbers. The assumption used in (c) is therefore inappropriate as also highlighted in Fig. B.5.

Solution to Exercise 2.6   We read in and attach the data as follows:

```
setwd('C:/directory')
pizza <- read.csv('pizza_delivery.csv')
attach(pizza)
```

(a) We need the options ylim, xlim, ylab, xlab, col to adjust the histogram produced with hist(). We then add a dashed (lty=2) line (type='l'), which is thick (lwd=3), from (65, 0) to (65, 400) using the lines() command. See also Fig. B.6a.

```
hist(temperature,xlab='Temperature',xlim=c(40,90),
ylim=c(0,400),col='lightgrey',ylab='Number of deliveries')
lines(c(65,65),c(0,400),type='l',lty=2,lwd=3)
```

(b) We can create the histogram as follows, see also Fig. B.6b:

```
library(ggplot2)
p1 <- ggplot(data=pizza,aes(x=temperature))
p2 <- p1 + geom_histogram(fill='darkgrey',alpha=0.5,binwidth=2.5)
+ scale_y_continuous('Number of deliveries')
plot(p2)
```

(c) A possible solution is as follows, see also Fig. B.6c:

```
barplot(table(driver),ylim=c(0,200),col=gray.colors(7),
ylab='Number of deliveries', xlab='Driver',main='Title'))
```

(d) We can produce the graph (Fig. B.6d) as follows:

```
p3 <- qplot(driver,data=pizza,aes=('bar'),fill=day)
p4 <- p3 + scale_fill_grey() +theme_bw()
plot(p4)
```

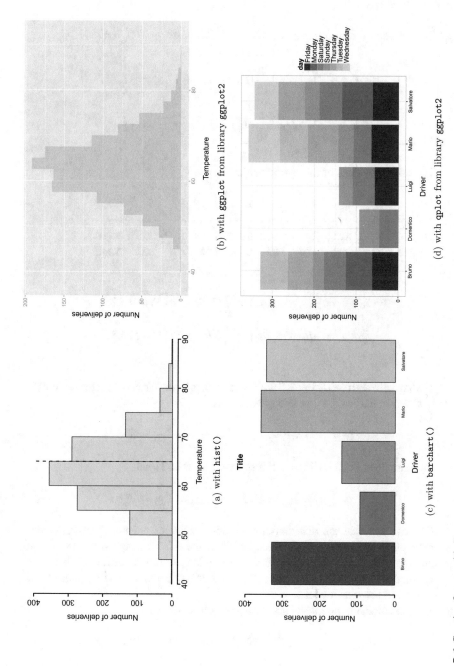

Fig. B.6 Creating figures with *R*

Chapter 3

Solution to Exercise 3.1

(a) The arithmetic means can be calculated as follows:

$$\bar{x}_D = \frac{1}{10} \sum_{i=1}^{10} x_i = \frac{1}{10}(12.5 + \ldots + 17.5) = 17.32,$$

$$\bar{x}_A = \frac{1}{10} \sum_{i=1}^{10} x_i = \frac{1}{10}(342 + \ldots + 466) = 612.4.$$

The ordered values of the two variables are:

```
i: 1     2     3     4     5     6     7     8     9     10
----------------------------------------------------------------
D: 7.6  12.1  12.5  14.8  16.2  16.5  17.5  18.5  27.4  29.9
A: 238   342   398   466   502   555   670   796   912  1245
```

$$\tilde{x}_{0.5,D} = \frac{1}{2}(\tilde{x}_{(5)} + \tilde{x}_{(6)}) = \frac{1}{2}(16.2 + 16.5) = 16.35,$$

$$\tilde{x}_{0.5,A} = \frac{1}{2}(\tilde{x}_{(5)} + \tilde{x}_{(6)}) = \frac{1}{2}(502 + 555) = 528.5.$$

(b) We have $n\alpha = 10 \cdot 0.25 = 2.5$ which is not an integer. We can therefore calculate the 25% quantiles, i.e. the first quartiles, as

$$\tilde{x}_{0.25,D} = \tilde{x}_{(3)} = 12.5; \quad \tilde{x}_{0.25,A} = \tilde{x}_{(3)} = 398.$$

Similarly, $n\alpha = 10 \cdot 0.75 = 7.5$ is not an integer and thus

$$\tilde{x}_{0.75,D} = \tilde{x}_{(8)} = 18.5; \quad \tilde{x}_{0.75,A} = \tilde{x}_{(8)} = 796.$$

One can see that the distributions for both variables are not symmetric. For example, when looking at the distance hiked, the difference between the median and the first quartile $(16.35 - 12.5)$ is much larger than the difference between the median and the third quartile $(18.5 - 16.35)$; this indicates a distribution that is skewed towards the left.

(c) We can calculate the interquartile ranges as follows:

$$d_{Q,A} = 796 - 398 = 398; \quad d_{Q,D} = 18.5 - 12.5 = 6.$$

The mean absolute deviations are:

$$D_D(\tilde{x}_{0.5}) = = \frac{1}{10}(|7.6 - 16.35| + \ldots + |29.9 - 16.35|) = 4.68\,,$$

$$D_A(\tilde{x}_{0.5}) = = \frac{1}{10}(|238 - 528.5| + \ldots + |1245 - 528.5|) = 223.2\,.$$

The variances of both the variables are

$$\tilde{s}_D^2 = \frac{1}{10}([7.6 - 16.35]^2 + \ldots + [29.9 - 16.35]^2) \approx 41.5\,,$$

$$\tilde{s}_A^2 = \frac{1}{10}([238 - 528.5]^2 + \ldots + [1245 - 528.5]^2) \approx 82,314\,.$$

The standard deviations are therefore $\tilde{s}_D = \sqrt{41.5}$ and $\tilde{s}_A = \sqrt{82,314}$.
(d) To answer this question, we can use the rules of linear transformations.

$$\bar{y} \overset{(3.4)}{=} 0 + 3.28\bar{x} = 3.28 \cdot 528.5 = 1722.48\,,$$

$$\tilde{s}_y^2 \overset{(3.29)}{=} b^2\tilde{s}_x^2 = 3.28^2 \cdot 272.2 \approx 2928\,.$$

(e) To draw the box plots, we can use the results from (a)–(c) in this solution. The median, first quartile, third quartile, and the interquartile range have already been calculated. It is also easy to determine the minimum and maximum values from the table of ordered values in (a). What is missing is the knowledge of whether to treat some of the values as extreme values or not. For the distance hiked an extreme value would be any value $> 18.5 + 1.5 \times 6 = 27.5$ or $< 12.5 - 1.5 \times 6 = 3.5$. It follows that there is only one extreme value: 29.9 km. For the maximum altitude there is no extreme value because there is no value $> 796 + 1.5 \times 398 = 1292$ or $< 398 - 1.5 \times 398 = -199$. The box plots are shown in Fig. B.7a, b.
(f) The data can be summarized as follows:

Class intervals	(5;15]	(15;20]	(20;30]
n_j	4	4	2
f_j	4/10	4/10	2/10
$\sum f_j$	4/10	8/10	1

We can calculate the weighted arithmetic mean by using the relative frequencies f_j and the middle of the intervals m_j:

$$\bar{x} = \sum_j f_j m_j = \frac{4}{10} \cdot 10 + \frac{4}{10} \cdot 17.5 + \frac{2}{10} \cdot 25 = 16\,.$$

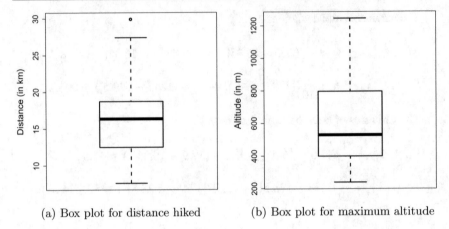

(a) Box plot for distance hiked   (b) Box plot for maximum altitude

Fig. B.7 Box plots for Exercise 3.1

To estimate the weighted median, we need to determine the class for which

$$\sum_{j=1}^{m-1} f_j < 0.5 \quad \text{and} \quad \sum_{j=1}^{m} f_j \geq 0.5$$

holds. This is clearly the case for the second class $K_2 = (15; 20]$. Thus

$$\tilde{x}_{0.5} = e_{m-1} + \frac{0.5 - \sum\limits_{j=1}^{m-1} f_j}{f_m} d_m = 15 + \frac{0.5 - 0.4}{0.4} \cdot 5 = 16.25 \, .$$

(g) If the raw data is known then the variance for the grouped data will be identical to the variance calculated in c). For educational purposes we show the identity here. The variance for grouped data can be calculated as:

$$\tilde{s}^2 = \underbrace{\frac{1}{n} \sum_{j=1}^{k} n_j (\bar{x}_j - \bar{x})^2}_{\text{between}} + \underbrace{\frac{1}{n} \sum_{j=1}^{k} n_j \tilde{s}_j^2}_{\text{within}}$$

Using the arithmetic mean $\bar{x} = 17.32$ as well as the means in each class, $\bar{x}_1 = 11.75$, $\bar{x}_2 = 17.225$, and $\bar{x}_3 = 28.65$, we can calculate the variance between the classes:

$$\tilde{s}_b^2 = \frac{1}{10} ([4 \cdot (11.75 - 17.32)^2] + [4 \cdot (17.225 - 17.32)^2]$$
$$+ [2 \cdot (28.65 - 17.32)^2]) = 38.08735 \, .$$

The variances within each class are:

$$\tilde{s}_1^2 = \frac{1}{4}[(7.6 - 11.75)^2 + \ldots + (14.8 - 11.75)^2] = 6.8025,$$

$$\tilde{s}_2^2 = \frac{1}{4}[(16.2 - 17.225)^2 + \ldots + (18.5 - 17.225)^2] = 0.956875,$$

$$\tilde{s}_3^2 = \frac{1}{2}[(27.4 - 28.65)^2 + (29.9 - 28.65)^2] = 1.5625.$$

We thus get

$$\tilde{s}_w^2 = \frac{1}{10}(4 \cdot 6.8025 + 4 \cdot 0.956875 + 2 \cdot 1.5625) = 3.41625.$$

The total variance is therefore $\tilde{s}^2 = \tilde{s}_w^2 + \tilde{s}_b^2 = 3.41625 + 38.08735 \approx 41.5$. The results will typically differ if we do not know the raw data: we have to replace the arithmetic means within each class, $\bar{x}_j$, with the middle of each class a_j, i.e. $a_1 = 10$, $a_2 = 17.5$, $a_3 = 25$:

$$\tilde{s}_b^2 = \frac{1}{10}([4 \cdot (10 - 17.32)^2] + [4 \cdot (17.5 - 17.32)^2]$$
$$+ [2 \cdot (25 - 17.32)^2]) = 33.2424.$$

We further get

$$\tilde{s}_1^2 = \frac{1}{4}[(7.6 - 10)^2 + \ldots + (14.8 - 10)^2] = 9.865,$$

$$\tilde{s}_2^2 = \frac{1}{4}[(16.2 - 17.5)^2 + \ldots + (18.5 - 17.5)^2] = 0.9225,$$

$$\tilde{s}_3^2 = \frac{1}{2}[(27.4 - 25)^2 + (29.9 - 25)^2] = 1.5625,$$

and

$$\tilde{s}_w^2 = \frac{1}{10}(4 \cdot 9.865 + 4 \cdot 0.9225 + 2 \cdot 14.885) = 7.292.$$

The variance is $\tilde{s}^2 = \tilde{s}_w^2 + \tilde{s}_b^2 = 7.292 + 33.2424 \approx 40.5$. The approximation is therefore good. However, please note that the between-class variance was estimated too low, but the within-class variance was estimated too high; only the combination of the two variance components led to reasonable results in this example. It is evident that the approximation in the third class was not ideal. The middle of the interval, 25, was not a good proxy for the true mean in this class, 28.65.

(h) It is easy to calculate the mean and the median:

```
distance <- c(12.5,29.9,...,17.5)
altitude <- c(342,1245,...,466)
mean(distance)
mean(altitude)
median(distance)
median(altitude)
```

We can use the quantile function, together with the probs option, to get the quantiles:

```
quantile(distance,probs=0.75)
quantile(distance,probs=0.25)
quantile(altitude,probs=0.75)
quantile(altitude,probs=0.25)
```

However, the reader will see that the results differ slightly from our results obtained in (b). As noted in Example 3.1.5, R offers nine different ways to obtain quantiles, each of which can be chosen by the type argument. The difference between these options cannot be understood easily without a background in probability theory. It may however be worth highlighting that we get the same results as in (b) if we choose the type=2 option in this example. The interquartile ranges can be calculated by means of the difference of the quantiles obtained above. To determine the mean absolute deviation we have to program our own function:

```
amd <- function(mv){1/length(mv)*sum(abs(mv-median(mv)))}
amd(distance)
amd(altitude)
```

We can calculate the variance using the var command. However as noted in Example 3.2.4, R uses $1/(n-1)$ rather than $1/n$ when calculating the variance. This important alternative formula for variance estimation is explained in Chap. 9, Theorem 9.2.1. To obtain the results from (c) we hence need to multiply the output from R by $(n-1)/n$:

```
var(altitude)*9/10
var(distance)*9/10
```

The box plots can be drawn by using the boxplot command:

```
boxplot(altitude)
boxplot(distance)
```

The weighted mean is obtained as follows:

```
weighted.mean(c(10,17.5,25),c(4/10,4/10,2/10))
```

Solution to Exercise 3.2

(a) We need to solve the equation that defines the arithmetic mean:

$$\bar{x} = \frac{1}{n} \sum_i x_i$$

$$-90 = \frac{1}{10}(200 + 600 - 200 - 200 - 200 - 100 - 100 - 400 + 0 + R)$$

$$-90 = \frac{1}{10}(-400 + R)$$

$$\Rightarrow R = -500.$$

(b) The mode is $\bar{x}_M = -200$. Using $n\alpha = 2.5$ and $n\alpha = 7.5$ respectively, we can determine the quartiles as follows:

$$\tilde{x}_{0.25} = x_{(3)} = -200,$$
$$\tilde{x}_{0.75} = x_{(8)} = 0.$$

The interquartile range is $d_Q = 0 - (-200) = 200$.

(c) It is not possible to use the coefficient of variation because some of the values are negative.

Solution to Exercise 3.3 We calculate

$$n_m = n - n_w - n_c = 100 - 45 - 20 = 35.$$

Using the formula for the arithmetic mean for grouped data,

$$\bar{x} = \frac{1}{n}(n_w \bar{x}_w + n_m \bar{x}_m + n_c \bar{x}_c),$$

we further get

$$\bar{x}_m = \frac{1}{n_m}(n\bar{x} - n_w \bar{x}_w - n_c \bar{x}_c)$$
$$= \frac{1}{35}(100 \cdot 15 - 45 \cdot 16 - 20 \cdot 7.5) = 18.$$

Similarly, we can use the Theorem of Variance Decomposition, Eq. (3.27), to calculate the variance for the grouped data:

$$s^2 = s_w^2 + s_b^2 = \frac{1}{n}(n_w s_w^2 + n_m s_m^2 + n_c s_c^2)$$

$$+ \frac{1}{n}\left(n_w(\bar{x}_w - \bar{x})^2 + n_m(\bar{x}_m - \bar{x})^2 + n_c(\bar{x}_c - \bar{x})^2\right).$$

This yields

$$s_m^2 = \frac{1}{n_m}\left[ns^2 - n_w s_w^2 - n_c s_c^2 - n_w(\bar{x}_w - \bar{x})^2 - n_m(\bar{x}_m - \bar{x})^2 - n_c(\bar{x}_c - \bar{x})^2\right]$$

$$= \frac{1}{35}(100 \cdot 19.55 - 45 \cdot 6 - 20 \cdot 3 - 45 \cdot 1^2 - 35 \cdot 3^2 - 20 \cdot 7.5^2) = 4.$$

Solution to Exercise 3.4

(a) We evaluate a period of 6 years which means that $T = 0, 1, 2, 3, 4, 5$. To determine the average growth rate we need to first calculate the geometric mean. There are two options to facilitate this:

(i) Solution:

$$B_t/B_{t-1} = (-, 1.04, 1.125, 0.925, 1.2, 0.933)$$
$$\bar{x}_G = (1.04 \cdot 1.125 \cdot 0.925 \cdot 1.2 \cdot 0.933)^{1/5} = 1.04.$$

(ii) Easier solution:

$$\bar{x}_G = (28/23)^{1/5} = 1.04.$$

Since $\bar{x}_G = 1.04$, the average growth rate is $r = 1.04 - 1 = 4\%$.

(b) To predict the number of members in 2018, we could apply the average growth rate to the number of members in 2016 for two consecutive years:

$$B_{2018} = \bar{x}_G B_{2017}, \quad B_{2017} = \bar{x}_G B_{2016}, \quad \Rightarrow B_{2018} = \bar{x}_G^2 B_{2016}$$
$$B_{2018} = 1.04^2 \cdot 28 = 30.28 \approx 31.$$

(c) We could use the approach outlined in (b) to predict the number of members in 2025. However, this assumes that the average growth rate between 2011 and 2016 remains valid until 2025. This is rather unrealistic. The number of members of the club increases in some years, but decreases in other years. There is no guarantee that the pattern observed until 2016 can be generalized to the future. This highlights that statistical methodology should in general be used with care when making long-term future predictions.

Fig. B.8 Lorenz curve

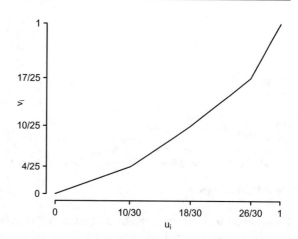

(d) The invested money is $\sum_i x_i = €\ 250$ million. We deal with partially grouped data because the club's members invest in groups, but the invested sum is clearly defined and not part of a group or interval. We can thus summarize the data as follows:

(i)	1	2	3	4
number of members	10	8	8	4
rel. number of members f_j	10/30	8/30	8/30	4/30
$\tilde{u}_i = \sum_j f_j$	10/30	18/30	26/30	1
money invested x_i	40	60	70	80
rel. amount per group	40/250	60/250	70/250	80/250
v_i	40/250	100/250	170/250	1

The Lorenz curve is plotted in Fig. B.8.

(e) The Gini coefficient can be calculated using formula (3.37) as follows:

$$G = 1 - \frac{1}{30}(10(0 + 4/25) + 8(4/25 + 2/5) + 8(2/5 + 17/25) +$$
$$4(17/25 + 1)) = 1 - 268/375 = 107/375 = 0.2853\ .$$

$$G^+ \overset{(3.39)}{=} 30/29 \cdot 107/375 = 214/725 = 0.2952\ .$$

The concentration of investment is thus rather weak.

Solution to Exercise 3.5   Let us look at Fig. B.9b first. The quantiles coincide approximately with the bisection line. This means that the distribution of "length of service" is similar for men and women. They have worked for about the same amount of time for the company. For example, the median service time should be approximately

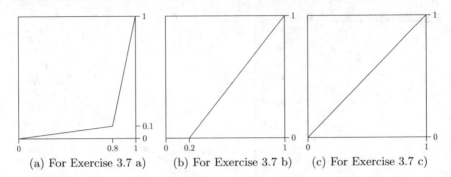

(a) For Exercise 3.7 a)   (b) For Exercise 3.7 b)   (c) For Exercise 3.7 c)

Fig. B.9 Lorenz curves

the same for men and women, the first quartile should be very similar too, the third quartile should be similar too, and so on. However, Fig. B.9a shows a somewhat different pattern: the quantiles for men are consistently higher than those for women. For example, the median salary will be higher for men. In conclusion, we see that men and women have a similar length of service in the company, but earn less.

Solution to Exercise 3.6   There are many ways in which a "mode" function can be programmed. We present a simple and understandable solution, not the most efficient one. Recall that `table` constructs a frequency table. This table shows immediately which value(s) occur(s) most often in the data. How can we extract it? Applying the `names` function on the table returns the names of the values represented in the table. The only thing we need to do is to choose the name which corresponds to the value which has been counted most. This yields the following function:

```
mymode <- function(vec){
mt <- table(vec)
names(mt)[mt == max(mt)]
}
```

This function will work in general, though it returns a character vector. Using `as.numeric` is one option to make the character string numeric, if necessary.

Solution to Exercise 3.7

(a) In this exercise we do not have the individual data, i.e. we do not know how much each inhabitant earns. The summarized data simply tells us about the wealth of two groups. For simplicity and for illustrative purposes, we assume that the wealth is equally distributed in each group. We determine $(\tilde{u}_i, \tilde{v}_i)$ as $(0.8, 0.1)$ and $(1, 1)$ because 80% of the population earn 10% of the wealth and 100% of the population earns everything. The respective Lorenz curve is illustrated in Fig. B.9a.

(b) The upper class lost its wealth. This means that 20% of the population do not own anything at all. However the remaining 80% own the rest. This yields $(\tilde{u}_i, \tilde{v}_i)$ of $(0.2, 0)$ and $(1, 1)$, see also Fig. B.9b.

(c) In this scenario, 20% of the population leave the country. However, the remaining 80%—which are now 100% of the population—earn the rest. The money is equally distributed between the population. Figure B.9c shows this situation.

Solution to Exercise 3.8

(a) It is necessary to use the harmonic mean to calculate the average speed. Using $w_1 = n_1/n = 180/418 \approx 0.43$, $w_2 = 117/418 \approx 0.28$, and $w_3 = 121/418 \approx 0.29$ we get

$$
\bar{x}_H = \frac{1}{\sum_{i=1}^{k} \frac{w_i}{x_i}}
$$

$$
= \frac{1}{0.43/48 + 0.28/37 + 0.29/52} \approx 45.2\,\text{km/h}.
$$

(b) Travelling at 45.2 km/h means travelling about 361 km in 8 h. The bus will not be in time.

Solution to Exercise 3.9

(a) The sum of investment is $\sum_i x_i = €18,020$. To calculate and draw the Lorenz curve we need the following table:

(i)	1	2	3	4
investment x_i	800	2220	4700	10300
f_j	1/4	1/4	1/4	1/4
u_i	1/4	2/4	3/4	1
relative investment	0.044	0.123	0.261	0.572
v_i	0.044	0.168	0.428	1

The curve is presented in Fig. B.10.

(b) We can calculate the Gini coefficient as follows:

$$
G \stackrel{(3.37)}{=} 1 - \frac{1}{4}[(0 + 0.044) + (0.044 + 0.168) + (0.168 + 0.428)
$$

$$
+ (0.428 + 1)] = 1 - \frac{1}{4} \cdot 2.28 = 0.43.
$$

$$
G^+ \stackrel{(3.39)}{=} \frac{n}{n-1} G = \frac{4}{3} \cdot 0.43 = 0.57.
$$

Fig. B.10 Lorenz curve for
investment sum

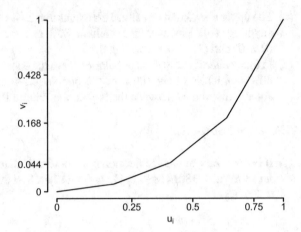

(c) The Gini coefficient remains the same as the relative investment stays the same.
(d) Using the library `ineq` we can easily reproduce the results in *R*:

```
library(ineq)
investment <- c(800,10300,4700,2200)
plot(Lc(investment))
ineq(investment)
```

However, please note that the `ineq` command calculates the unstandardized
Gini coefficient.

Solution to Exercise 3.10

(a) The easiest way to get all these measures is to use the `summary` function and
apply it to the data columns which relate to quantitative variables:

```
setwd('C:/yourpath')
pizza <- read.csv('pizza_delivery.csv')
attach(pizza)
summary(pizza[,c('time','temperature','bill','pizzas')])
```

We then get the following output:

```
          time            temperature           bill              pizzas
Min.   :12.27     Min.    :41.76     Min.    : 9.10     Min.    : 1.000
1st Qu.:30.06     1st Qu.:58.24     1st Qu.:35.50     1st Qu.: 2.000
Median :34.38     Median :62.93     Median :42.90     Median : 3.000
Mean   :34.23     Mean    :62.86     Mean    :42.76     Mean    : 3.013
3rd Qu.:38.58     3rd Qu.:67.23     3rd Qu.:50.50     3rd Qu.: 4.000
Max.   :53.10     Max.    :87.58     Max.    :75.00     Max.    :11.000
```

(b) We can use the quantile function:

```
quantile(time,probs=0.99)
quantile(temperature,probs=0.99)
```

The results are 48.62 min. for delivery time and 79.87 °C for temperature. This means 99% of the delivery times are less than or equal to 48.62 min. and 1% of deliveries are greater than or equal to 48.62 min. Similarly, only 1% of pizzas were delivered with a temperature greater than 79.87 °C.

(c) The following simple function calculates the absolute mean deviation:

```
amdev <- function(mv){1/length(mv)*sum(abs(mv-mean(mv)))}
amdev(temperature)
```

(d) We can use the scale, mean, and var commands respectively.

```
sc.time <- scale(time)
mean(sc.time)
var(sc.time)
```

As one would expect, the mean is zero and the variance is 1 for the scaled variable.

(e) The boxplot command draws a box plot; the range option specifies the range for which extreme values are defined. As specified in the help files, range=0 produces a box plot with no extreme values.

```
boxplot(temperature,range=0)
boxplot(time,range=0)
```

The box plots are displayed in Fig. B.11.

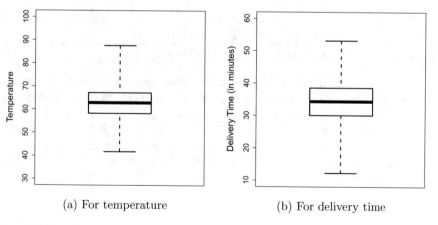

(a) For temperature   (b) For delivery time

Fig. B.11 Box plots

(f) We use the `cut` command to create a variable which has the categories $(10, 20], (20, 30], (30, 40], (40, 50], (50, 60]$ respectively. Using the interval midpoints, as well as the relative frequencies in each class (obtained via the `table` command), we get:

```
tc <- cut(time,breaks=seq(10,60,10))
weighted.mean(c(15,25,35,45,55),table(tc)/sum(table(tc)))
[1] 34.18641
mean(time)
[1] 34.22955
```

The weighted mean is very similar to the mean from the raw data, see output above.

(g) The plots can be reproduced by using the `qqplot` command:

```
qqplot(time[driver=='Luigi'],time[driver=='Domenico'])
qqplot(time[driver=='Mario'],time[driver=='Salvatore'])
```

Chapter 4

Solution to Exercise 4.1

(a) We need the following table to calculate the correlation coefficient R:

Café (i)	x_i	$R(x_i)$	y_i	$R(y_i)$	d_i	d_i^2
1	3	1	6	2	-1	1
2	8	4	7	3	1	1
3	7	3	10	5	-2	4
4	9	5	8	4	1	1
5	5	2	4	1	1	1

$$R = 1 - \frac{6 \sum_{i=1}^{n} d_i^2}{n(n^2 - 1)} = 1 - \frac{6(1 + 1 + 4 + 1 + 1)}{5(25 - 1)} = 1 - 0.4 = 0.6.$$

There is a moderate to strong positive correlation between the ratings of the two coffee enthusiasts. In general, a high rating from one staff member implies a rather high rating from the other staff member.

(b) Above we have assigned ranks in an increasing order, i.e. the lowest x_i/y_i gets the lowest rank (1) and the highest x_i/y_i gets the highest rank (5). If we use decreasing order and assign the lowest rank to the highest values we get the following results:

Café (i)	x_i	$R(x_i)$	y_i	$R(y_i)$	d_i	d_i^2
1	3	5	6	4	1	1
2	8	2	7	3	−1	1
3	7	3	10	1	2	4
4	9	1	8	2	−1	1
5	5	4	4	5	−1	1

As in (a), we have $\sum_i d_i^2 = 8$ and therefore the results are identical: $R = 0.6$. Depending on whether ranks are assigned in an increasing order or a decreasing order, the sign of d_i differs; but the calculation of R is not affected since the squared values of d_i are used for its calculation and the sign of d_i is thus not important.

(c) We can summarize the ratings in a 2×2 table:

		X	Y
Coffee	Bad	2	1
Quality	Good	3	4

The odds ratio is $OR = (2 \times 4)/(3 \times 1) = 2$. The chance of rating a coffee as good is twice as likely for person X compared to person Y.

Solution to Exercise 4.2

(a) The expected frequencies (under independence) are:

	Satisfied	Unsatisfied
Car	$\frac{74 \cdot 58}{150} = 28.61$	$\frac{76 \cdot 58}{150} = 29.39$
Car (diesel engine)	$\frac{74 \cdot 60}{150} = 29.6$	$\frac{76 \cdot 60}{150} = 30.4$
Motorbike	$\frac{74 \cdot 32}{150} = 15.79$	$\frac{76 \cdot 32}{150} = 16.21$

We therefore have

$$\chi^2 = \sum_{i=1}^{k} \sum_{j=1}^{l} \frac{(n_{ij} - \frac{n_{i+}n_{+j}}{n})^2}{\frac{n_{i+}n_{+j}}{n}}$$

$$= \frac{(33 - 28.61)^2}{28.61} + \frac{(25 - 29.39)^2}{29.39} + \frac{(29 - 29.6)^2}{29.6}$$

$$+ \frac{(31 - 30.4)^2}{30.4} + \frac{(12 - 15.79)^2}{15.79} + \frac{(20 - 16.21)^2}{16.21}$$

$$= 0.6736 + 0.6557 + 0.0122 + 0.0112 + 0.9097 + 0.8861 = 3.1485\,.$$

The maximum value χ^2 can take is $150(2 - 1) = 150$ which indicates that there is almost no association between type of vehicle and satisfaction with the insurance. The other measures can be calculated as follows:

$$V = \sqrt{\frac{\chi^2}{n(\min(k, l) - 1)}} = \sqrt{\frac{3.1485}{150(2 - 1)}} = 0.14\,.$$

C_{corr}:

$$C_{corr} = \sqrt{\frac{\min(k,l)}{\min(k,l)-1}} \sqrt{\frac{\chi^2}{\chi^2+n}} =$$

$$= \sqrt{\frac{2}{1}} \sqrt{\frac{3.1485}{3.1485+150}} = \sqrt{2}\sqrt{0.02056} \approx 0.20.$$

The small values of V and C_{corr} confirm that the association is rather weak.

(b) The summarized table looks as follows

	Satisfied	Unsatisfied
Car	62	56
Motorbike	12	20

Using (4.7), we obtain

$$\chi^2 = \frac{n(ad-bc)^2}{(a+d)(c+d)(a+c)(b+d)}$$

$$= \frac{150(1240-672)^2}{118 \cdot 32 \cdot 74 \cdot 76} = \frac{48,393,600}{21,236,224} \approx 2.2788.$$

The maximum value χ^2 can take is $150(2-1)$. The association is therefore weak. The odds ratio is

$$OR = \frac{ad}{bc} = \frac{62 \cdot 20}{12 \cdot 56} = \frac{1240}{672} \approx 1.845.$$

The chances of being satisfied with the insurance are 1.845 times higher among those who drive a car.

(c) All χ^2-based statistics suggest that there is only a small association between the two variables, both for the 2×3 and the 2×2 table. However, the odds ratio gives us a more nuanced interpretation, showing that customers driving a car are somewhat more satisfied with their insurance. The question is whether the additional information from the odds ratio is stable and trustworthy. Confidence intervals for the odds ratio can provide guidance under such circumstances, see Sect. 9.4.4 for more details.

Solution to Exercise 4.3

(a) The scatter plot is given in Fig. B.12. The black circles show the 5 observations. A positive relationship can be discovered: the higher the speed limit, the higher the number of deaths. However, "Italy" (the observation on top right) is the

Fig. B.12 Scatter diagram for speed limit and number of deaths

observation which gives the graph a visible pattern and drives our impression about the potential relationship.

(b) Using $\bar{x} = 61$ and $\bar{y} = 4.86$ we obtain

$$S_{xx} = (55 - 61)^2 + (55 - 61)^2 + \cdots + (75 - 61)^2 = 270$$
$$S_{yy} = (4.1 - 4.86)^2 + (4.7 - 4.86)^2 + \cdots + (6.1 - 4.86)^2 = 2.512$$
$$S_{xy} = (55 - 61)(4.1 - 4.86) + \cdots + (75 - 61)(6.1 - 4.86) = 23.2$$

and therefore

$$r = \frac{S_{xy}}{\sqrt{S_{xx}S_{yy}}} \frac{23.2}{\sqrt{270 \cdot 2.512}} = 0.891 .$$

The correlation coefficient of Spearman is calculated using the following table:

Country (i)	x_i	$R(x_i)$	y_i	$R(y_i)$	d_i	d_i^2
Denmark	55	4.5	4.1	5	−0.5	0.25
Japan	55	4.5	4.7	3	1.5	2.25
Canada	60	2.5	4.3	4	−1.5	2.25
Netherlands	60	2.5	5.1	2	0.5	0.25
Italy	75	1	6.1	1	0	0

This yields

$$R = 1 - \frac{6\sum_{i=1}^{n} d_i^2}{n(n^2 - 1)} = 1 - \frac{6 \cdot 5}{5(25 - 1)} = 0.75 .$$

Please note that above we averaged the ranks for ties. The R function cor uses a more complicated approach; this is why the results differ slightly when using R.

(c) The results stay the same. Pearson's correlation coefficient is invariant with respect to linear transformations which means that it doesn't matter whether we use miles/h or km/h.

(d) (i) The grey square in Fig. B.12 represents the additional observation. The overall pattern of the scatter plot changes with this new observation pair: the positive relationship between speed limit and number of traffic deaths is not so clear anymore. This emphasizes how individual observations may affect our impression of a scatter plot.

(ii) Based on the same approach as in (b) we can calculate $\bar{x} = 62.5$, $\bar{y} = 4.6333$, $S_{xx} = 337.5$, $S_{yy} = 4.0533$, $S_{xy} = 13$, and $r = 0.3515$. A single observation changes our conclusions from a strong positive relationship to a moderate to weak relationship. It is evident that Pearson's correlation coefficient is volatile and may be affected heavily by outliers or extreme observations.

Solution to Exercise 4.4

(a) The contingency table for the absolute frequencies is as follows:

	1. Class	2. Class	3. Class	Staff	Total
Rescued	202	125	180	211	718
Not rescued	135	160	541	674	1510
Total	337	285	721	885	2228

(b) To obtain the conditional relative frequency distributions we calculate the proportion of passengers rescued (X) for each travel class (Y). In the notation of Definition 4.1.1 we determine $f_{i|j}^{X|Y} = f_{ij}/f_{+j} = n_{ij}/n_{+j}$ for all i and j. For example, $f_{\text{rescued}|1.\ \text{class}} = 202/337 = 0.5994$. This yields

	1. Class	2. Class	3. Class	Staff
Rescued	0.5994	0.4386	0.2497	0.2384
Not rescued	0.4006	0.5614	0.7503	0.7616

It is evident that the proportion of passengers being rescued differs by travel class. It can be speculated that there is an association between the two variables pointing towards better chances of being rescued among passengers from higher travel classes.

(c) Using (4.3), we get

	1. Class	2. Class	3. Class	Staff	Total
Rescued	108.6	91.8	232.4	285.2	718
Not rescued	228.4	193.2	488.6	599.8	1510
Total	337	285	721	885	2228

which can be used to calculate χ^2 and V as follows:

$$\chi^2 = \sum_{i=1}^{k} \sum_{j=1}^{l} \frac{(n_{ij} - \frac{n_{i+}n_{+j}}{n})^2}{\frac{n_{i+}n_{+j}}{n}} = \frac{(202 - 108.6)^2}{108.6} + \frac{(125 - 91.8)^2}{91.8}$$

$$+ \frac{(180 - 232.4)^2}{232.4} + \frac{(211 - 285.2)^2}{285.2} + \frac{(135 - 228.4)^2}{228.4}$$

$$+ \frac{(160 - 193.2)^2}{193.2} + \frac{(541 - 488.6)^2}{488.6} + \frac{(674 - 599.8)^2}{599.8}$$

$$= 80.33 + 12.01 + 11.82 + 19.30 + 38.19 + 5.71 + 5.62 + 9.18$$

$$= 182.16.$$

$$V = \sqrt{\frac{\chi^2}{n(\min(k, l) - 1)}} = \sqrt{\frac{182.16}{2228(2 - 1)}} = 0.286.$$

The value of V indicates a moderate or weak relationship between the two variables. This is in contradiction to the hypothesis derived from the conditional distributions in b).

(d) The table is as follows:

	1. Class/2. Class	3. Class/Staff	Total
Rescued	327	391	718
Not rescued	295	1215	1510
Total	622	1606	2228

Using (4.7) we get

$$\chi^2 = \frac{2228(327 \cdot 1215 - 295 \cdot 391)^2}{718 \cdot 1510 \cdot 622 \cdot 1606} = 163.55.$$

and therefore $V = \sqrt{\frac{163.55}{2228}} = 0.271.$

There are several relative risks that can be calculated, for example:

$$\frac{f_{1|1}}{f_{1|2}} = \frac{n_{11}/n_{+1}}{n_{12}/n_{+2}} = \frac{327/622}{391/1606} \approx 2.16,$$

$$\frac{f_{2|1}}{f_{2|2}} = \frac{n_{21}/n_{+1}}{n_{22}/n_{+2}} = \frac{295/622}{1215/1606} \approx 0.63.$$

The proportion of passengers who were rescued was 2.16 times higher in the 1./2. class compared to the 3. class and staff. Similarly, the proportion of passengers who were not rescued was 0.62 times lower in the 1./2. class compared to the 3. class and staff. The odds ratio is $OR = \frac{a \cdot d}{b \cdot c} = \frac{397,305}{115,345} = 3.444$. This is nothing but the ratio of the relative risks, i.e. 2.16/0.63. The chance of being rescued (i.e. the ratio rescued/not rescued) was almost 3.5 times higher for the 1./2. class compared to the 3. class and staff.

(e) While Cramer's V led to rather conservative conclusions regarding a possible relationship of travel class and rescue status, the conditional relative frequency distribution, the relative risks, and the odds ratio support the hypothesis that, at least to some degree, the chances of being rescued were higher in better travel classes. This makes sense because better travel classes were located towards the top of the ship with best access to the lifeboats while both third class passengers and large numbers of the staff were located and working in the bottom of the ship, where the water came in first.

Solution to Exercise 4.5

(a) Using (4.17) and (4.18), we get

$$r = \frac{S_{xy}}{\sqrt{S_{xx}S_{yy}}} = \frac{\sum_{i=1}^{36} x_i y_i - 36\bar{x}\bar{y}}{\sqrt{n\tilde{s}_x^2 n\tilde{s}_y^2}} = \frac{22776 - 36 \cdot 12.22 \cdot 51.28}{n\sqrt{\tilde{s}_x^2 \tilde{s}_y^2}}$$

$$= \frac{216.9}{36\sqrt{76.95 \cdot 706.98}} \approx 0.026.$$

This indicates that there is no linear relationship between temperature and hotel occupancy.

(b) The scatter plot shows no clear pattern. This explains why the correlation coefficient is close to 0. However, if we look only at the points for each city separately, we see different structures for different cities: a possible negative relationship for Davos (D), a rather positive relationship for Polenca (P) and no visible relationship for Basel (B). This makes sense because for winter holiday destinations hotel occupancy should be higher when the temperature is low and for summer holiday destinations occupancy should be high in the summer months.

(c) We type in the data by specifying three variables: temperature (X), occupancy (Y) and city (Z). We then simply use the `cor` command to calculate the correlation—and condition on the values of Z which we are interested in:

```
X <- c(-6,-5,2,...,9,4)
Y <- c(91,89,76,...,9,12)
Z <- c(rep('Davos',12),rep('Polenca',12),rep('Basel',12))
cor(X[Z=='Davos'],Y[Z=='Davos'])
cor(X[Z=='Basel'],Y[Z=='Basel'])
cor(X[Z=='Polenca'],Y[Z=='Polenca'])
```

This yields correlation coefficients of -0.87 for Davos, 0.42 for Basel and 0.82 for Polenca. It is obvious that looking at X and Y only indicates no correlation, but the information from Z shows strong linear relationships in subgroups. This example shows the limitations of using correlation coefficients and the necessity to think in a multivariate way. Regression models offer solutions. We refer the reader to Chap. 11, in particular Sect. 11.7.3 for more details.

Solution to Exercise 4.6

(a) We use the visual rule of thumb and work from the top left to the bottom right for the concordant pairs and from the top right to the bottom left for the discordant pairs:

$$
\begin{aligned}
K &= 82 \cdot 43 + 82 \cdot 9 + 82 \cdot 2 + 82 \cdot 10 + 8 \cdot 2 + 8 \cdot 10 + 4 \cdot 10 + 4 \cdot 9 \\
&\quad + 43 \cdot 10 = 5850 \\
D &= 4 \cdot 8 + 9 \cdot 2 = 50 \\
\gamma &= \frac{K - D}{K + D} = \frac{5800}{5900} = 0.98 \,.
\end{aligned}
$$

(b)

$$
\begin{aligned}
\chi^2 &= \sum_{i=1}^{k} \sum_{j=1}^{l} \frac{\left(n_{ij} - \frac{n_{i+}n_{+j}}{n}\right)^2}{\frac{n_{i+}n_{+j}}{n}} = \frac{(82 - \frac{86 \cdot 90}{158})^2}{\frac{86 \cdot 90}{158}} + \frac{(4 - \frac{49 \cdot 86}{158})^2}{\frac{49 \cdot 86}{158}} \\
&\quad + \frac{(0 - \frac{19 \cdot 86}{158})^2}{\frac{19 \cdot 86}{158}} + \frac{(8 - \frac{90 \cdot 60}{158})^2}{\frac{90 \cdot 60}{158}} + \frac{(43 - \frac{49 \cdot 60}{158})^2}{\frac{49 \cdot 60}{158}} + \frac{(9 - \frac{60 \cdot 19}{158})^2}{\frac{60 \cdot 19}{158}} \\
&\quad + \frac{(0 - \frac{12 \cdot 90}{158})^2}{\frac{12 \cdot 90}{158}} + \frac{(2 - \frac{12 \cdot 49}{158})^2}{\frac{12 \cdot 49}{158}} + \frac{(10 - \frac{12 \cdot 19}{158})^2}{\frac{12 \cdot 19}{158}} \\
&= 22.25.19.27 + 10.34 + 20.05 + 31.98 + 0.47 + 6.84 + 0.80 + 50.74 \\
&= 162.74 \,.
\end{aligned}
$$

$$
V = \sqrt{\frac{\chi^2}{n(\min(k, l) - 1)}} = \sqrt{\frac{162.74}{158 \cdot 2}} \approx 0.72 \,.
$$

(c) The table is as follows

		Use a leash		Total
		Agree or n.o.	Disagree	
Use for concerts	Agree or n.o.	137	9	146
	Disagree	2	10	12
	Total	139	19	158

(d) The relative risk can either be summarized as

$$\frac{2/139}{10/19} \approx 0.03 \quad \text{or} \quad \frac{10/19}{2/139} \approx 36.6 \,.$$

The proportion of those who disagree with using the park for summer concerts is 0.03 times lower in the group who agree or have no opinion about using leashes for dogs compared to those who disagree. Similarly, the proportion of those who disagree with using the park for summer concerts is 36.6 times higher in the group who also disagree with using leashes for dogs compared to those who do not disagree.

(e) The odds ratio is $OR = (137 \cdot 10)/(2 \cdot 9) \approx 36.1$.

- The chance of not disagreeing with the concert proposal is 36.1 times higher for those who also do not disagree with the leash proposal.
- The chance of not disagreeing with the leash proposal is 36.1 times higher for those who also do not disagree with the concert proposal.
- In simpler words: The chance of agreeing or having no opinion for one of the questions is 36.1 higher if the person also has no opinion or agrees with the other question.

(f)

$$\gamma = \frac{137 \cdot 10 - 9 \cdot 2}{137 \cdot 10 + 9 \cdot 2} = \frac{1352}{1388} = 0.974 \,.$$

(g) In general, it makes sense to use all the information available, i.e. to use the ordinal structure of the data and all 3 categories. While it is clear that γ is superior to V in our example, one may argue that the relative risks or the odds ratio could be more useful because they provide an intuitive quantification on how the two variables relate to each other rather than just giving a summary of strength and direction of association. However, as we have seen earlier, the interpretations of the relative risks and the odds ratio are quite clumsy in this example. It can be difficult to follow the somewhat complicated interpretation.

A simple summary would be to say that agreement with both questions was strongly associated ($\gamma = 0.98$).

Solution to Exercise 4.7

We use the definition of the correlation coefficient, replace y_i with $a + bx_i$, and replace $\bar{y}$ with $a + b\bar{x}$ to obtain

$$r = \frac{\sum_{i=1}^{n}(x_i - \bar{x})(a + bx_i - (a + b\bar{x}))}{\sqrt{\sum_{i=1}^{n}(x_i - \bar{x})^2 \sum_{i=1}^{n}(a + bx_i - (a + b\bar{x}))^2}}.$$

This equates to:

$$r = \frac{\sum_{i=1}^{n}(x_i - \bar{x})(b(x_i - \bar{x}))}{\sqrt{\sum_{i=1}^{n}(x_i - \bar{x})^2 \sum_{i=1}^{n}(b(x_i - \bar{x}))^2}} = \frac{b\sum_{i=1}^{n}(x_i - \bar{x})^2}{\sqrt{b^2\sum_{i=1}^{n}(x_i - \bar{x})^2 \sum_{i=1}^{n}(x_i - \bar{x})^2}} = 1.$$

Solution to Exercise 4.8

(a) We read in the data, make sure the first column is recognized as containing the row names, attach the data, and obtain the correlation using the `cor` command:

```
decathlon <- read.csv('decathlon.csv', row.names=1)
attach(decathlon)
cor(X.Discus,X.High.jump)
```

The correlation is 0.52984. This means there is a moderate to strong positive correlation, i.e. the longer the distance the discus is thrown, the higher the height in the high jump competition.

(b) There are 10 variables. For the first variable, we can calculate the correlation with 9 other variables. For the second variable, we can also calculate the correlation with 9 other variables. However, we have already calculated one out of the 9 correlations, i.e. when analysing variable number one. So it is sufficient to calculate 8 correlations for the second variable. Similarly, we need another 7 correlations for the third variable, 6 correlations for the fourth variable, and so on. In total, we therefore need to have $9 + 8 + 7 + \cdots + 1 = 45$ correlations. Since the correlation coefficient describes the relationship between two variables, it makes sense to summarize the results in a contingency table, similar to a matrix, see Fig. B.13.

	X.100m	X.Long.jump	X.Shot.put	X.High.jump	X.400m	X.110m.hurdle	X.Discus	X.Pole.vault	X.Javeline	X.1500m
X.100m	1	−0.7	−0.37	−0.31	0.63	0.54	−0.23	−0.26	−0.01	0.06
X.Long.jump	−0.7	1	0.2	0.35	−0.67	−0.54	0.25	0.29	0.09	−0.15
X.Shot.put	−0.37	0.2	1	0.61	−0.2	−0.25	0.67	0.02	0.38	0.13
X.High.jump	−0.31	0.35	0.61	1	−0.17	−0.33	0.52	−0.04	0.2	0
X.400m	0.63	−0.67	−0.2	−0.17	1	0.52	−0.14	−0.12	−0.05	0.55
X.110m.hurdle	0.54	−0.54	−0.25	−0.33	0.52	1	−0.22	−0.15	−0.08	0.18
X.Discus	−0.23	0.25	0.67	0.52	−0.14	−0.22	1	−0.18	0.25	0.22
X.Pole.vault	−0.26	0.29	0.02	−0.04	−0.12	−0.15	−0.18	1	−0.07	0.18
X.Javeline	−0.01	0.09	0.38	0.2	−0.05	−0.08	0.25	−0.07	1	−0.25
X.1500m	0.06	−0.15	0.13	0	0.55	0.18	0.22	0.18	−0.25	1

Fig. B.13 Correlation matrix for the decathlon data

(c) Using `cor(decathlon)` yields the correlation coefficients between all variable pairs. This is much simpler than calculating the correlation coefficient for each of the 45 comparisons. Note that the correlation matrix provides the 45 comparisons both in the upper triangle and the lower triangle of the table. We know that $r(X, Y) = r(Y, X)$, but R still provides us with both, although they are identical. Note that the diagonal elements are 1 because $r(X, X) = 1$.

(d) One way to omit rows with missing data automatically is to use the `na.omit` command:

```
cor(na.omit(decathlon))
```

The results are displayed in Fig. B.13. We see moderate to strong correlations between the 100 m race, 400 m race, 110 m hurdle race and long jump. This may reflect the speed-and-athletic component of the decathlon contest. We also see moderate to strong correlations between the shot put, high jump and discus events. This may reflect the strength-and-technique component of the contest. The other disciplines also show correlations which make sense, but they are rather weak and may reflect the uniqueness of these disciplines.

Solution to Exercise 4.9

(a) A possible code is listed below:

```
pizza <- read.csv('pizza_delivery.csv')
pizza$tempcat <- cut(pizza$temperature, breaks=c(0,65,100))
pizza$timecat <- cut(pizza$time, breaks=c(0,30,100))
attach(pizza)
addmargins(table(tempcat,timecat))
```

```
             timecat
tempcat    (0,30] (30,100]  Sum
  (0,65]      101      691  792
  (65,100]    213      261  474
  Sum         314      952 1266
```

We can see that there is a higher proportion of high temperature ((65, 100]) in the category of short delivery times ((0, 30]) compared to long delivery times ((30, 100]).

(b) Using the data from (a) we can calculate the odds ratio:

```
(101*261)/(213*691)
```

Thus the chances of receiving a cold pizza are 0.18 lower if the delivery time is short.

(c) We use the vcd and ryouready packages to determine the desired measures of association:

```
library(vcd)
library(ryouready)
library(lattice)
assocstats(xtabs(~tempcat+timecat))
ord.gamma(table(tempcat,timecat))
ord.tau(table(tempcat,timecat))
barchart(table(tempcat,timecat),horizontal=F,stack=T)
```

Cramer's V is 0.361, Stuart's τ_c is -0.302, and γ is -0.696. The first two measures indicate a moderate to weak relationship between temperature and delivery time. It is clear from the last two measures that this relationship is negative, i.e. that shorter delivery times imply higher temperatures. However, interestingly, τ_c and γ provide us with a different strengths of association. In any case, it is clear that for shorter delivery times the customers receive warmer pizzas, as evident from the stacked bar chart (Fig. B.14a).

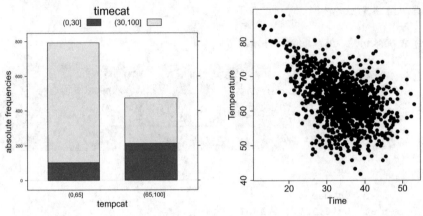

(a) Stacked bar chart for the categorical time and temperature variables

(b) Scatter plot for temperature and time

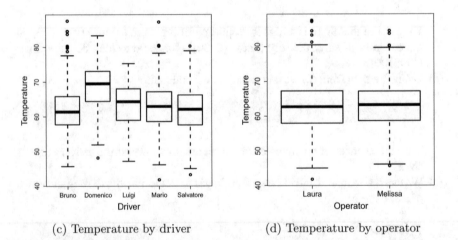

(c) Temperature by driver

(d) Temperature by operator

Fig. B.14 Plots for Exercise 4.9

(d) The scatter plot (Fig. B.14b) shows a decreasing temperature for an increasing delivery time. This is also highlighted in the correlation coefficients which are -0.43 and -0.39 for Bravais–Pearson and Spearman respectively.

```
plot(time,temperature)
cor(time,temperature)
cor(time,temperature,method='spearman')
```

(e) It makes sense to compare metric variables (temperature, number of pizzas, bill) using the correlation coefficient from Bravais–Pearson. The relationships between temperature and driver and operator could be visualized using stratified box plots.

```
boxplot(temperature~driver)
boxplot(temperature~operator)
cor(temperature,pizzas)
cor(temperature,bill)
```

The correlation coefficients are -0.37 and -0.42 for number of pizzas and bill respectively. More pizzas and a higher bill are associated with a lower temperature. The box plots (Fig. B.14c, d) show variation in the delivery times of the different drivers, but almost identical delivery times for the two operators. These results give us a first idea about the relationship in the data. However, they do not tell us the full story. For example: is the pizza temperature for higher bills low because a higher bill means more pizzas, and therefore a less efficient preparation of the food? Is one driver faster because he mainly delivers for a particular branch? Could it be that the operators have a different performance but because they deal with different branches and drivers, these differences are not visible? To address these questions a multivariate perspective is needed. Regression models, as introduced in Chap. 11, provide some answers.

Chapter 5

Solution to Exercise 5.1   There are $n = 10$ guests and $m = 2$ guests shake hands with each other. The order is not important: two guests shake hands no matter who is "drawn first". It is also not possible to shake hands with oneself which means that in terms of the urn model, we have the "no replacement" case. Therefore we obtain the solution as

$$\binom{n}{m} = \binom{10}{2} = \frac{10 \cdot 9}{2 \cdot 1} = 45$$

handshakes in total.

Solution to Exercise 5.2
We assume that it is not relevant to know which student gets tested at which time point. We are thus dealing with combinations without considering the order. We have $n = 25$ and $m = 5$ and therefore obtain:

(a) a total number of $\binom{25}{5} = 53{,}130$ possibilities.
(b) a total number of $\binom{25+5-1}{5} = \binom{29}{5} = 118{,}755$ possibilities.

In R, the commands choose(25,5) and choose(29,5) respectively provide the required results.

Solution to Exercise 5.3   The board consists of $n = 381$ knots. Each knot is either occupied or not. We may thus assume a "drawing" without replacement in the sense that each knot can be drawn (occupied) only once. If we place $m = 64$ counters on the board then we can simultaneously think of "drawing" 64 occupied knots out of a total of 381 knots. The order in which we draw is not relevant—either a knot is occupied or not. The total number of combinations is $\binom{n}{m} = \binom{381}{64} \approx 4.35 \cdot 10^{73}$. We obtain the final number in R using the command choose(381,64).

Solution to Exercise 5.4   We obtain the results (again using the command choose(n,m) in R) as follows:

(a) The customer takes the beers "with replacement" because the customer can choose among any type of beer for each position in the tray. One can also think of an urn model with 6 balls relating to the 6 different beers, where beers are drawn with replacement and placed on the tray. The order in which the beers are placed on the tray is not relevant. We thus have

$$\binom{n+m-1}{m} = \binom{6+20-1}{20} = \binom{25}{20} = 53,130$$

combinations.

(b) If the customer insists on having at least one beer per brewery on his tray, then 6 out of the 20 positions of the tray are already occupied. Based on the same thoughts as in (a), we calculate the total number of combinations as

$$\binom{n+m-1}{m} = \binom{6+14-1}{14} = \binom{19}{14} = 11,628.$$

Solution to Exercise 5.5   Since each team has exactly one final place, we have a "without replacement" situation. Using $n = 32$ and $m = 3$ (and choose(n,m) in R) yields

(a) $\frac{32!}{(32-3)!} = \binom{32}{3}3! = 29,760$ and

(b) $\binom{32}{3} = 4960$.

Solution to Exercise 5.6   There are $n = 12$ different letters for $m = 4$ positions of the membership code. Each letter can be used more than once if desired and we thus obtain $n^m = 12^4 = 20,736$ possible combinations. We therefore conclude that sufficient membership codes are left. However, this may not last long and the book store may still wish to create another solution for its membership codes.

Solution to Exercise 5.7   For each member of the jury there are 61 scoring options:

0	0.1	0.2	0.3	0.4	0.5	0.6	0.7	0.8	0.9
1	1.1	1.2	1.3	1.4	1.5	1.6	1.7	1.8	1.9
.				.				.	
.				.				.	
5	5.1	5.2	5.3	5.4	5.5	5.6	5.7	5.8	5.9
6									

Different jury members are allowed to assign the same scores. We thus deal with combinations "with replacement". To verify this, just think of an urn with 61 balls where each ball refers to one possible score. Now one ball is drawn, assigned to a specific jury member and then put back into the urn. Since each score is "attached" to a particular jury member, we have combinations with consideration of the order and therefore obtain a total of $n^m = 61^9 \approx 1.17 \cdot 10^{16}$ possibilities. If you have difficulties in understanding the role of "replacement" and "order" in this example, recall that each member has 61 scoring options: thus, $61 \times 61 \times \ldots \times 61$ (9 times) combinations are possible.

Solution to Exercise 5.8

(a) We obtain:

$$\binom{2}{0} = 1 \leftrightarrow \binom{n}{2} = \binom{2}{2} = 1; \qquad \binom{3}{1} = 3 \leftrightarrow \binom{n}{2} = \binom{3}{2} = 3;$$

$$\binom{4}{2} = 6 \leftrightarrow \binom{n}{2} = \binom{4}{2} = 6; \qquad \binom{5}{3} = 10 \leftrightarrow \binom{n}{2} = \binom{5}{2} = 10.$$

(b) Based on the observations from (a) we conclude that each entry on the diagonal line can be represented by $\binom{n}{2}$. The sum of two consecutive entries is thus $\binom{n}{2} + \binom{n+1}{2}$. Using the fourth relation in (5.5), it follows that:

$$\binom{n}{2} + \binom{n+1}{2} \overset{(5.5)}{=} \frac{n(n-1)}{2} + \frac{(n+1)n}{2}$$

$$= \frac{n(n-1+n+1)}{2} = \frac{n \cdot 2n}{2} = n^2.$$

Chapter 6

Solution to Exercise 6.1

(a) We obtain

- $A \cap B = \{8\}$
- $B \cap C = \{\emptyset\}$
- $A \cup C = \{0, 4, 8, 9, 15\}$
- $C \setminus A = \{4, 9, 15\}$
- $\Omega \setminus (B \cup A \cup C) = \{6, 7, 11, 13, 14\}$

(b) The solutions are:

- $P(\bar{F}) = 1 - P(F) = 0.5$
- $P(G) = 1 - P(E) - P(F) = 0.3$
- $P(E \cap G) = 0$ because the events are pairwise disjoint

- $P(E \setminus E) = 0$
- $P(E \cup F) = P(E) + P(F) = 0.7$

Solution to Exercise 6.2 We know that the probability of failing the practical examination is $P(PE) = 0.25$, tof failing the theoretical examination is $P(TE) = 0.15$, and of failing both is $P(PE \cap TE) = 0.1$.

(a) If we ask for the probability of failing in *at least* one examination, we imply that either the theoretical examination, or the practical examination, or both are not passed. We can therefore calculate the union of events $P(PE \cup TE) = P(PE) + P(TE) - P(TE \cap PE) = 0.25 + 0.15 - 0.1 = 0.3$.
(b) $P(PE \setminus TE) = P(PE) - P(PE \cap TE) = 0.25 - 0.1 = 0.15$.
(c) $P(\overline{PE \cup TE}) = 1 - P(PE \cup TE) = 1 - 0.3 = 0.7$.
(d) We are interested in the probability of the person failing exactly in one exam. This corresponds to $P(M \setminus C \cup C \setminus M) = P(M \cup C) - P(C \cap M) = 0.3 - 0.1 = 0.2$.

Solution to Exercise 6.3 The total number of possible simple events is $|\Omega| = 12$. The number of favourable simple events is

(a) $|A| = 6$ (i.e. the numbers 2, 4, 6, 8, 10, 12). Hence, $P(A) = \frac{6}{12} = \frac{1}{2}$.
(b) $|B| = 3$ (i.e. the numbers 10, 11, 12). Hence, $P(B) = \frac{3}{12} = \frac{1}{4}$.
(c) $|C| = 2$ (i.e. the numbers 10, 12). Hence, $P(A \cap B) = \frac{2}{12} = \frac{1}{6}$.
(d) $|D| = 7$ (i.e. the numbers 2, 4, 6, 8, 10, 11, 12). Hence, $P(A \cup B) = \frac{7}{12}$.

Solution to Exercise 6.4 The total number of simple events is $\binom{12}{2}$.

(a) The number of favourable simple events is one and therefore $P(\text{right two presents}) = \frac{|A|}{|\Omega|} = \frac{1}{\binom{12}{2}} \approx 0.015$.
(b) The number of favourable simple events is $\binom{10}{2}$ because the person draws two presents out of the ten "wrong" presents:
$P(\text{wrong two presents}) = \frac{|A|}{|\Omega|} = \binom{10}{2}/\binom{12}{2} \approx 0.682$. In Sect. 8.1.8 we explain the hypergeometric distribution which will shed further light on this exercise.

Solution to Exercise 6.5

(a) Let V denote the event that there is too much salt in the soup and let L denote the event that the chef is in love. We know that

$$P(V) = 0.2 \Rightarrow P(\bar{V}) = 0.8.$$

Similarly, we have

$$P(L) = 0.3 \Rightarrow P(\bar{L}) = 0.7.$$

We therefore get:

$$P(V \cap L) = P(V|L) \cdot P(L) = 0.6 \cdot 0.3 = 0.18$$
$$P(\bar{V} \cap L) = P(L) - P(V \cap L) = 0.3 - 0.18 = 0.12$$
$$P(V \cap \bar{L}) = P(V) - P(V \cap L) = 0.2 - 0.18 = 0.02$$
$$P(\bar{V} \cap \bar{L}) = P(\bar{V}) - P(\bar{V} \cap L) = 0.8 - 0.12 = 0.68$$

This yields the following contingency table:

	V	$\bar{V}$	Total
L	0.18	0.12	0.3
$\bar{L}$	0.02	0.68	0.7
Total	0.2	0.8	1

(b) The variables are not stochastically independent since, for example, $P(V) \cdot P(L) = 0.3 \cdot 0.2 = 0.06 \neq 0.18 = P(V \cap L)$.

Solution to Exercise 6.6

(a) We define the following events: $G =$ Basil is treated well, $\bar{G} =$ Basil is not treated well; $E =$ Basil survives, $\bar{E} =$ Basil dies. We know that

$$P(\bar{G}) = \frac{1}{3} \Rightarrow P(G) = \frac{2}{3}; \quad P(E|G) = \frac{1}{2}; \quad P(E|\bar{G}) = \frac{3}{4}.$$

Using the Law of Total Probability, we get

$$P(E) = P(E|G) \cdot P(G) + P(E|\bar{G}) \cdot P(\bar{G})$$

$$= \frac{1}{2} \cdot \frac{2}{3} + \frac{3}{4} \cdot \frac{1}{3} = \frac{1}{3} + \frac{1}{4} = \frac{7}{12} \approx 0.58.$$

(b) We can use Bayes' Theorem to answer the question:

$$P(\bar{G}|E) = \frac{P(E|\bar{G}) \cdot P(\bar{G})}{P(E|\bar{G}) \cdot P(\bar{G}) + P(E|G) \cdot P(G)} = \frac{\frac{3}{4} \cdot \frac{1}{3}}{\frac{7}{12}} = \frac{3}{7} \approx 0.43.$$

Solution to Exercise 6.7   We define the following events and probabilities

- A: Bill never paid, $P(A) = 0.05 \Rightarrow P(\bar{A}) = 0.95$
- M: Bill paid too late, $P(M) = ?$
- $P(M|\bar{A}) = 0.2$

- $P(M|A) = 1$ because someone who never pays will always pay too late.

(a) We are interested in $P(M)$, the probability that someone does not pay his bill in a particular month, either because he is not able to or he pays too late. We can use the Law of Total Probability to obtain the results:

$$P(M) = P(M|A)P(A) + P(M|\bar{A})P(\bar{A})$$
$$= 0.05 \cdot 1 + 0.2 \cdot 0.95 = 0.24.$$

(b) We can use Bayes' Theorem to obtain the results:

$$P(A \mid M) = \frac{P(A)P(M|A)}{P(M)} = \frac{0.05}{0.24} = 0.208.$$

(c) If the bill was not paid in a particular month the probability is 20.8% that it will never be paid, and 78.2% that it will be paid. One could argue that a preventive measure that affects almost 79% of trustworthy customers is not ideal and the bank should therefore not block a credit card if a bill is not paid on time.

Solution to Exercise 6.8

(a) The "and" operator refers to the joint distribution of two variables. In our example, we want to know the probability of being infected *and* having been transported by the truck. This probability can be directly obtained from the respective entry in the contingency table: 40 out of 200 cows fulfill both criteria and thus

$$P(B \cap A) = \frac{40}{200}.$$

(b) We can use $P(A) = \frac{100}{200} = P(\bar{A})$ to obtain:

$$P(B|A) = \frac{P(B \cap A)}{P(A)} = \frac{40/200}{100/200} = \frac{40}{100}.$$

(c) Using these results and $P(B) = \frac{60}{200}$, and $P(\bar{B}) = \frac{140}{200} = 1 - P(B)$, we obtain

$$P(B|\bar{A}) = \frac{P(B \cap \bar{A})}{P(\bar{A})} = \frac{20/200}{100/200} = \frac{20}{100}$$

by using the Law of Total Probability. We can thus calculate

$$P(B) = P(B|A)P(A) + P(B|\bar{A})P(\bar{A})$$
$$= 0.40 \cdot 0.50 + 0.20 \cdot 0.50 = 0.30.$$

This means that the probability of a cow being infected is 30%. Alternatively we could have simply looked at the marginal distribution of the contingency table to get $P(B) = 60/200 = 0.3$.

Solution to Exercise 6.9

(a) The two shots are independent of each other and thus

$$P(A \cap B) = 0.4 \cdot 0.5 = 0.2.$$
$$P(A \cup B) = 0.4 + 0.5 - 0.2 = 0.7.$$

(b) We need to calculate

$$P(A\backslash B \cup B\backslash A) = 0.4 - 0.2 + 0.5 - 0.2 = 0.5.$$

(c)

$$P(B\backslash A) = 0.5 - 0.2 = 0.3.$$

Chapter 7

Solution to Exercise 7.1

(a) The first derivative of the CDF yields the PDF, $F'(x) = f(x)$:

$$f(x) = \begin{cases} 0 & \text{if } x < 2 \\ -\frac{1}{2}x + 2 & \text{if } 2 \leq x \leq 4 \\ 0 & \text{if } x > 4 \end{cases}$$

(b) We know from Theorem 7.2.3 that for any continuous variable $P(X = x_0) = 0$ and therefore $P(X = 4) = 0$. We calculate $P(X < 3) = P(X \leq 3) - P(X = 3) = F(3) - 0 = -\frac{9}{4} + 6 - 3 = 0.75$.

(c) Using (7.15), we obtain the expectation as

$$E(X) = \int_{-\infty}^{\infty} x \, f(x) \, dx = \int_{-\infty}^{2} x 0 \, dx + \int_{2}^{4} x \left(-\frac{1}{2}x + 2\right) dx + \int_{4}^{\infty} x 0 \, dx$$

$$= 0 + \int_{2}^{4} \left(-\frac{1}{2}x^2 + 2x\right) dx + 0$$

$$= \left[-\frac{1}{6}x^3 + x^2\right]_{2}^{4} = (-\frac{64}{6} + 16) - (-\frac{8}{6} + 4) = \frac{8}{3}.$$

Given that we have already calculated $E(X)$, we can use Theorem 7.3.1 to calculate the variance as $\text{Var}(X) = E(X^2) - [E(X)]^2$. The expectation of X^2 is

$$E(X^2) = \int_{2}^{4} x^2 \left(-\frac{1}{2}x + 2\right) dx = \int_{2}^{4} \left(-\frac{1}{2}x^3 + 2x^2\right) dx$$

$$= \left[-\frac{1}{8}x^4 + \frac{2}{3}x^3\right]_{2}^{4} = (-32 + \frac{128}{3}) - (-2 + \frac{16}{3}) = \frac{22}{3}.$$

We thus obtain $\text{Var}(X) = \frac{22}{3} - \left(\frac{8}{3}\right)^2 = \frac{66-64}{9} = \frac{2}{9}$.

Solution to Exercise 7.2

(a) The probability mass function of X is

x_i	1	2	3	4	5	6
$P(X = x)$	$\frac{1}{9}$	$\frac{1}{9}$	$\frac{1}{9}$	$\frac{2}{9}$	$\frac{1}{9}$	$\frac{3}{9}$

Using (7.16), we calculate the expectation as

$$E(X) = 1 \cdot \frac{1}{9} + 2 \cdot \frac{1}{9} + 3 \cdot \frac{1}{9} + 4 \cdot \frac{2}{9} + 5 \cdot \frac{1}{9} + 6 \cdot \frac{3}{9}$$
$$= \frac{1 + 2 + 3 + 8 + 5 + 18}{9} = \frac{37}{9} \approx 4.1.$$

To obtain the variance, we need

$$E(X^2) = 1 \cdot \frac{1}{9} + 4 \cdot \frac{1}{9} + 9 \cdot \frac{1}{9} + 16 \cdot \frac{2}{9} + 25 \cdot \frac{1}{9} + 36 \cdot \frac{3}{9}$$
$$= \frac{1 + 4 + 9 + 32 + 25 + 108}{9} = \frac{179}{9} .$$

Therefore using $\text{Var}(X) = E(X^2) - [E(X)]^2$, we get

$$\text{Var}(X) = \frac{179}{9} - \left(\frac{37}{9}\right)^2 = \frac{1611 - 1369}{81} = \frac{242}{81} \approx 2.98.$$

The manipulated die yields on average higher values than a fair die because its expectation is $4.1 > 3.5$. The variability of is however similar because $2.98 \approx 2.92$.

(b) The probability mass function of $Y = \frac{1}{X}$ is:

$y_i = \frac{1}{x_i}$	1	$\frac{1}{2}$	$\frac{1}{3}$	$\frac{1}{4}$	$\frac{1}{5}$	$\frac{1}{6}$
$P(\frac{1}{X} = y)$	$\frac{1}{9}$	$\frac{1}{9}$	$\frac{1}{9}$	$\frac{2}{9}$	$\frac{1}{9}$	$\frac{3}{9}$

The expectation can hence be calculated as

$$E(Y) = E\left(\frac{1}{X}\right) = 1 \cdot \frac{1}{9} + \frac{1}{2} \cdot \frac{1}{9} + \frac{1}{3} \cdot \frac{1}{9} + \frac{1}{4} \cdot \frac{2}{9} + \frac{1}{5} \cdot \frac{1}{9} + \frac{1}{6} \cdot \frac{3}{9}$$
$$= \frac{1}{9} + \frac{1}{18} + \frac{1}{27} + \frac{1}{18} + \frac{1}{45} + \frac{1}{18} = \frac{91}{270} .$$

Comparing the results from (a) and (b) shows clearly that $E(\frac{1}{X}) \neq \frac{1}{E(X)}$. Recall that $E(bX) = bE(X)$. It is interesting to see that for some transformations $T(X)$ it holds that $E(T(X)) = T(E(X))$, but for some it doesn't. This reminds us to be careful when thinking of the implications of transformations.

Solution to Exercise 7.3

(a) There are several ways to plot the CDF. One possibility is to define the function and plot it with the `curve` command. Since the function has different definitions for the intervals $[\infty, 0)$, $[0, 1]$, $(1, \infty]$, we need to take this into account. Remember that a logical statement in R corresponds to a number, i.e. TRUE=1 and FALSE=0; we can thus simply add the different pieces of the function and multiply them with a condition which specifies if X is contained in the interval or not:

```
cdf <-function(x){
(3*x^2-2*x^3)*(x <= 1 & x >= 0)+1*(x > 1)+0*(x < 0)
}
curve(cdf,from=-0.5,to=1.5)
```

(b) The PDF is

$$\frac{d}{dx}F(x) = F'(x) = f(x) = \begin{cases} 6(x - x^2) & \text{if } 0 \leq x \leq 1 \\ 0 & \text{elsewhere.} \end{cases}$$

(c)

$$P\left(\frac{1}{3} \leq X \leq \frac{2}{3}\right) = \int_{\frac{1}{3}}^{\frac{2}{3}} f(x)dx = F\left(\frac{2}{3}\right) - F\left(\frac{1}{3}\right)$$

$$= \left[3\left(\frac{2}{3}\right)^2 - 2\left(\frac{2}{3}\right)^3\right] - \left[3\left(\frac{1}{3}\right)^2 - 2\left(\frac{1}{3}\right)^3\right]$$

$$= 0.48149.$$

(d) We have already defined the CDF in (a). We can now simply plug in the x-values of interest:

```
cdf(2/3)-cdf(1/3)
```

Fig. B.15 Cumulative distribution function for the proportion of wine sold

(e) The variance can be calculated as follows:

$$E(X) = \int_0^1 x6(x - x^2)dx = 6 \cdot \int_0^1 (x^2 - x^3)dx$$

$$= 6\left[\frac{1}{3}x^3 - \frac{1}{4}x^4\right]_0^1 = 6 \cdot \left(\frac{1}{3} - \frac{1}{4}\right) = 0.5$$

$$E(X^2) = \int_0^1 x^2 6(x - x^2)dx = 6 \cdot \int_0^1 (x^3 - x^4)dx$$

$$= 6\left[\frac{1}{4}x^4 - \frac{1}{5}x^5\right]_0^1 = 6 \cdot \left(\frac{1}{4} - \frac{1}{5}\right) = 0.3$$

$$Var(X) = E(X^2) - [E(X)]^2 = 0.3 - 0.5^2 = 0.05 .$$

Solution to Exercise 7.4

(a) Two conditions need to be satisfied for $f(x)$ to be a proper PDF:

(i) $\int_0^2 f(x)dx = 1$:

$$\int_0^2 f(x)dx = \int_0^2 c \cdot x(2 - x)dx = c \int_0^2 x(2 - x)dx$$

$$= c \int_0^2 (2x - x^2)dx = c \left[x^2 - \frac{1}{3}x^3\right]_0^2$$

$$= c\left[4 - \frac{8}{3} - (0 - 0)\right] = c \cdot \frac{4}{3} \overset{!}{=} 1$$

$$\Longrightarrow c = \frac{3}{4}.$$

(ii) $f(x) \geq 0$:

$$f(x) = \frac{3}{4}x(2 - x) \geq 0 \quad \forall \; x \in [0, 2].$$

(b) We calculate

$$F(x) = P(X \leq x) = \int_0^x f(t)dt = \int_0^x \frac{3}{4}t(2 - t)dt$$

$$= \frac{3}{4}\int_0^x (2t - t^2)dt = \frac{3}{4}\left[t^2 - \frac{1}{3}t^3\right]_0^x$$

$$= \frac{3}{4}\left[x^2 - \frac{1}{3}x^3 - 0\right] = \frac{3}{4}x^2\left(1 - \frac{1}{3}x\right)$$

and therefore

$$F(x) = \begin{cases} 0 & \text{if } x < 0 \\ \frac{3}{4}x^2(1 - \frac{1}{3}x) & \text{if } 0 \leq x \leq 2 \\ 1 & \text{if } 2 < x. \end{cases}$$

(c) The expectation is

$$E(X) = \int_0^2 xf(x)dx = \frac{3}{4}\int_0^2 (2x^2 - x^3)dx$$

$$= \frac{3}{4}\left[\frac{2}{3}x^3 - \frac{1}{4}x^4\right]_0^2 = \frac{3}{4}\left[\frac{2}{3} \cdot 8 - \frac{1}{4} \cdot 16 - 0\right]$$

$$= \frac{3}{4}\left[\frac{16}{3} - \frac{12}{3}\right] = \frac{3}{4} \cdot \frac{4}{3} = 1.$$

Using $\text{Var}(X) = E(X^2) - (E(X))^2$, we calculate the variance as

$$E(X^2) = \int_0^2 x^2 f(x)dx = \frac{3}{4}\int_0^2 (2x^3 - x^4)dx$$

$$= \frac{3}{4}\left[\frac{2}{4}x^4 - \frac{1}{5}x^5\right]_0^2 = \frac{3}{4}\left[\frac{2}{4}\cdot 16 - \frac{1}{5}\cdot 32 - 0\right]$$

$$= 6 - \frac{3\cdot 32}{4\cdot 5} = 6 - \frac{3\cdot 8}{5} = \frac{6}{5}.$$

$$\text{Var}(X) = \frac{6}{5} - 1^2 = \frac{1}{5}.$$

(d)

$$P(|X - \mu| \le 0.5) \ge 1 - \frac{\sigma^2}{c^2} = 1 - \frac{(\frac{1}{5})}{(0.5)^2} = 1 - 0.8 = 0.2.$$

Solution to Exercise 7.5

(a) The marginal distributions are obtained by the row and column sums of the joint PDF respectively. For example, $P(X = 1) = \sum_{j=1}^J p_{1j} = p_{1+} = 1/4$.

X	$P(X = x_i)$
0	3/4
1	1/4

Y	$P(Y = y_i)$
1	1/6
2	7/12
3	1/4

The marginal distribution of X tells us how many customers sought help via the telephone hotline (75%) and via email (25%). The marginal distribution of Y represents the distribution of the satisfaction level, highlighting that more than half of the customers (7/12) were "satisfied".

(b) To determine the 75%-quantile with respect to Y, we need to find the value $y_{0.75}$ for which $F(y_{0.75}) \ge 0.75$ and $F(y) < 0.75$ for $y < y_{0.75}$. Y takes the values $1, 2, 3$. The quantile can't be $y_{0.75} = 1$ because $F(1) = 1/6 < 0.75$. The 75%-quantile is $y_{0.75} = 2$ because $F(2) = 1/6 + 7/12 = 3/4 \ge 0.75$ and for all values which are smaller than 2 we get $F(x) < 0.75$.

(c) We can calculate the conditional distribution using $P(Y = y_j | X = 1) = p_{1j}/p_{1+} = p_{1j}/(1/6 + 1/12 + 0) = p_{1j}/(0.25)$. Therefore,

$$P(Y = 1 | X = 1) = \frac{1/6}{1/4} = \frac{2}{3},$$

$$P(Y = 2 | X = 1) = \frac{1/12}{1/4} = \frac{1}{3},$$

$$P(Y = 3 | X = 1) = \frac{0}{1/4} = 0.$$

Among those who used the email customer service two thirds were unsatisfied, one third were satisfied, and no one was very satisfied.

(d) As we know from (7.27), two discrete random variables are said to be independent if $P(X = x_i, Y = y_j) = P(X = x_i)P(Y = y_j)$. However, in our example, $P(X = 0, Y = 1) = P(X = 0)P(X = 1) = \frac{3}{4} \cdot \frac{1}{6} \neq 0$. This means that X and Y are not independent.

(e) The covariance of X and Y is defined as $\mathrm{Cov}(X, Y) = \mathrm{E}(XY) - \mathrm{E}(X)\mathrm{E}(Y)$. We calculate

$$\mathrm{E}(X) = 0 \cdot \frac{3}{4} + 1 \cdot \frac{1}{4} = \frac{1}{4}$$

$$\mathrm{E}(Y) = 1 \cdot \frac{1}{6} + 2 \cdot \frac{7}{12} + 3 \cdot \frac{1}{4} = \frac{25}{12}$$

$$\mathrm{E}(XY) = 0 \cdot 1 \cdot 0 + 1 \cdot 1 \cdot \frac{1}{6} + 0 \cdot 2 \cdot \frac{1}{2} + 1 \cdot 2 \cdot \frac{1}{12} + 0 \cdot 3 \cdot \frac{1}{4} + 1 \cdot 3 \cdot 0$$

$$= \frac{2}{6}$$

$$\mathrm{Cov}(X, Y) = \frac{2}{6} - \frac{1}{4} \cdot \frac{25}{12} = -\frac{3}{16}.$$

Since $\mathrm{Cov}(X, Y) < 0$ we conclude that there is a negative relationship between X and Y: the higher the values of X, the lower the values of Y—and vice versa. For example, those who use the email-based customer service ($X = 1$) are less satisfied than those who use the telephone customer service ($X = 0$). It is however evident that in this example the values of X have no order and therefore care must be exercised in interpreting the covariance.

Solution to Exercise 7.6 Using Tschebyschev's inequality (7.24)

$$P(|X - \mu| < c) \geq 0.9 = 1 - \frac{\mathrm{Var}(X)}{c^2},$$

we can determine c as follows:

$$1 - \frac{\mathrm{Var}(X)}{c^2} = 0.9$$

$$c^2 = \frac{\mathrm{Var}(X)}{0.1} = \frac{4}{0.1} = 40$$

$$c = \pm\sqrt{40} = \pm 6.325 .$$

Thus, the interval is $[15 - 6.325; 15 + 6.325] = [8.675; 21.325]$.

Solution to Exercise 7.7

(a) The joint PDF is:

		Y		
		0	1	2
X	−1	0.3	0.2	0.2
	2	0.1	0.1	0.1

(b) The marginal distributions are obtained from the row and column sums of the joint PDF respectively:

X	−1	2
$P(X = x)$	0.7	0.3

Y	0	1	2
$P(Y = y)$	0.4	0.3	0.3

(c) The random variables X and Y are independent if

$$P(X = x, Y = y) = P(X = x)P(Y = y) \quad \forall x, y .$$

However, in our example we have, for example,

$$P(X = -1, Y = 0) = 0.3 \neq P(X = -1) \cdot P(Y = 0) = 0.7 \cdot 0.4 = 0.28 .$$

Hence, the two variables are not independent.
(d) The joint distribution of X and Y can be used to obtain the desired distribution of U. For example, If $X = -1$ and $Y = 0$, then $U = X + Y = -1$. The respective probability is $P(U = -1) = 0.3$ because $P(U = -1) = P(X = -1, Y = 0) = 0.3$ and there is no other combination of X and Y values which yields $X + Y = -1$. The distribution of U is therefore as follows:

k	-1	0	1	2	3	4
$P(U = k)$	0.3	0.2	0.2	0.1	0.1	0.1

(e) We calculate

$$E(U) = \sum_{k=-1}^{4} k \cdot P(U = k) = 0.8$$
$$E(X) = (-1)0.7 + 2 \cdot 0.3 = -0.1$$
$$E(Y) = 0 \cdot 0.4 + 1 \cdot 0.3 + 2 \cdot 0.3 = 0.9$$

$$E(U^2) = 0.3 \cdot (-1)^2 + \ldots + 0.1 \cdot 4^2 = 3.4$$
$$E(X^2) = 0.7 \cdot (-1)^2 + 0.3 \cdot 2^2 = 1.9$$
$$E(Y^2) = 0.4 \cdot 0^2 + 0.3 \cdot 1^2 + 0.3 \cdot 2^2 = 1.5$$

$$\text{Var}(U) = E(U^2) - [E(U)]^2 = 3.4 - (0.8)^2 = 2.76$$
$$\text{Var}(X) = E(X^2) - [E(X)]^2 = 1.9 - (-0.1)^2 = 1.89$$
$$\text{Var}(Y) = E(Y^2) - [E(Y)]^2 = 1.5 - (0.9)^2 = 0.69.$$

It can be seen that $E(X) + E(Y) = -0.1 + 0.9 = 0.8 = E(U)$. This makes sense because we know from (7.31) that $E(X + Y) = E(X) + E(Y)$. However, $\text{Var}(U) = 2.76 \neq \text{Var}(X) + \text{Var}(Y) = 1.89 + 0.69$. This follows from (7.7.1) which says that $\text{Var}(X \pm Y) = \text{Var}(X) + \text{Var}(Y) \pm 2\,\text{Cov}(X, Y)$ and therefore $\text{Var}(X \pm Y) = \text{Var}(X) + \text{Var}(Y)$ only if the covariance is 0. We know from (c) that X and Y are not independent and thus $\text{Cov}(X, Y) \neq 0$.

Solution to Exercise 7.8

(a) The random variables X and Y are independent if the balls are drawn with replacement. This becomes clear by understanding that drawing with replacement implies that for both the draws, the same balls are in the urn and the conditions in each draw remain the same. The first draw has no implications for the second draw.
   If we were drawing the balls without replacement then the first draw could possibly have implications for the second draw: for instance, if the first ball drawn was red, then the second one could not be red because there is only one red ball in the urn. This means that drawing without replacement implies dependency of X and Y. This can also be seen by evaluating the independence assumption (7.27):

$$P(X = 2, Y = 2) = 0 \neq P(X = 2) \cdot P(Y = 2) = \frac{1}{8} \cdot \frac{1}{8}.$$

(b) The marginal probabilities $P(X = x_i)$ can be obtained from the given information. For example, 3 out of 8 balls are black and thus $P(X = 1) = 3/8$. The conditional distributions $P(Y|X = x_i)$ can be calculated easily by realizing that under the assumed dependency of X and Y, the second draw is always based on 7 balls (8 balls minus the one drawn under the condition $X = x_i$)—e.g. if the first ball drawn is black, then 7 balls, 2 of which are black, remain in the urn and $P(Y = 1|X = 1) = 2/7$. We thus calculate

$$P(Y = 1, X = 1) = P(Y = 1|X = 1)P(X = 1) = \frac{2}{7} \cdot \frac{3}{8} = \frac{6}{56}$$

$$P(Y = 1, X = 2) = P(Y = 1|X = 2)P(X = 2) = \frac{3}{7} \cdot \frac{1}{8} = \frac{3}{56}$$

$$\ldots$$

$$P(Y = 3, X = 3) = P(Y = 3|X = 3)P(X = 3) = \frac{3}{7} \cdot \frac{4}{8} = \frac{12}{56}$$

and obtain

		Y		
		1	2	3
X	1	$\frac{6}{56}$	$\frac{3}{56}$	$\frac{12}{56}$
	2	$\frac{3}{56}$	0	$\frac{4}{56}$
	3	$\frac{12}{56}$	$\frac{4}{56}$	$\frac{12}{56}$

(c) The expectations are

$$E(X) = 1 \cdot \frac{3}{8} + 2 \cdot \frac{1}{8} + 3 \cdot \frac{4}{8} = \frac{17}{8}$$

$$E(Y) = E(X) = \frac{17}{8}.$$

To estimate $\rho(X, Y)$, we need $\text{Cov}(X, Y)$ as well as $\text{Var}(X)$ and $\text{Var } Y$:

$$E(XY) = 1\frac{6}{56} + 2\frac{3}{56} + 3\frac{12}{56} + 2\frac{3}{56} + 4 \cdot 0 + 6\frac{4}{56} + 3\frac{12}{56} + 6\frac{4}{56} + 9\frac{12}{56}$$

$$= \frac{246}{56}$$

$$E(X^2) = E(Y^2) = 1^2\frac{3}{8} + 2^2\frac{1}{8} + 3^2\frac{4}{8} = \frac{43}{8}$$

$$\text{Var}(X) = E(X^2) - [E(X)]^2 = \frac{43}{8} - \left(\frac{17}{8}\right)^2 = \frac{55}{64}$$

$$\text{Var}(Y) = \text{Var}(X) = \frac{55}{64}.$$

Using (7.38) and $\text{Cov}(X, Y) = E(XY) - E(X)E(Y)$, we obtain

$$\rho = \frac{\text{Cov}(X, Y)}{\sqrt{\text{Var}(X) \cdot \text{Var}(Y)}} = \frac{\frac{246}{56} - \frac{289}{64}}{\sqrt{\frac{55}{64} \cdot \frac{55}{64}}} = -0.143 \,.$$

Solution to Exercise 7.9

(a) The constant c must satisfy

$$\int_{40}^{100} \int_{10}^{100} c \left(\frac{100 - x}{x} \right) dx\, dy = \int_{40}^{100} \int_{10}^{100} \frac{100c}{x} - c\, dx\, dy \overset{!}{=} 1$$

and therefore

$$\int_{40}^{100} [100c \ln(x) - cx]_{10}^{100}\, dy = \int_{40}^{100} 100c \ln 100 - 100c - 100c \ln(10) + 10c\, dy$$

which is

$$\int_{40}^{100} 100c \left(\ln \frac{100}{10} - 1 + \frac{1}{10} \right) dy = [100cy\,(\ln 10 - 9/10)]_{40}^{100}$$

$$= 600c(10 \ln 10 - 9) \quad \rightarrow c \approx 0.00012 \,.$$

(b) The marginal distribution is

$$f_X(x) = \int_{40}^{100} c \left(\frac{100 - x}{x} \right) dy = \left[c \left(\frac{100 - x}{x} \right) y \right]_{40}^{100}$$

$$= 100c \left(\frac{100 - x}{x} \right) - 40c \left(\frac{100 - x}{x} \right) \approx 0.00713 \left(\frac{100 - x}{x} \right)$$

for $10 \leq x \leq 100$.

(c) To determine $P(X > 75)$, we need the cumulative marginal distribution of X:

$$F_X(x) = \int_{-\infty}^{x} f_X(t)dt = \int_{10}^{x} 0.00713 \left(\frac{100 - t}{t} \right) dt$$

$$= \int_{10}^{x} \frac{0.00713}{t} - 0.00713\, dt = [0.713 \ln(t) - 0.00713]_{10}^{x}$$

$$= 0.713 \ln(x) - 0.00713x - 0.00713 \ln(10) + 0.00713 \cdot 10 \,.$$

Now we can calculate

$$P(X > 75) = 1 - P(X \leq 75) = 1 - F_X(75) = 1 - 0.973 \approx 2.7\% \,.$$

(d) The conditional distribution is

$$f_{Y|X}(x, y) = \frac{f(x, y)}{f(x)} = \frac{c\left(\frac{100-x}{x}\right)}{60c\left(\frac{100-x}{x}\right)} = \frac{1}{60}.$$

Solution to Exercise 7.10   If we evaluate the expectation with respect to Y then both μ and σ can be considered to be constants. We can therefore write

$$E(Y) = E\left(\frac{X - \mu}{\sigma}\right) = \frac{1}{\sigma}(E(X) - \mu).$$

Since $E(X) = \mu$, it follows that $E(Y) = 0$. The variance is

$$Var(Y) = Var\left(\frac{X - \mu}{\sigma}\right).$$

Applying $Var(a + bX) = b^2 Var(X)$ to this equation yields $a = \mu$, $b = \frac{1}{\sigma}$ and therefore

$$Var(Y) = \frac{1}{\sigma^2} Var(X) = \frac{\sigma^2}{\sigma^2} = 1.$$

Chapter 8

Solution to Exercise 8.1   The random variable X: "number of packages with a toy" is binomially distributed. In each of $n = 20$ "trials", a toy can be found with probability $p = \frac{1}{6}$.

(a) We thus get

$$P(X = 4) = \binom{n}{k} p^k (1 - p)^{n-k} = \binom{20}{4}\left(\frac{1}{6}\right)^4\left(\frac{5}{6}\right)^{16} \approx 0.20.$$

(b) Similarly, we calculate

$$P(X = 0) = \binom{n}{k} p^k (1 - p)^{n-k} = \binom{20}{0}\left(\frac{1}{6}\right)^0\left(\frac{5}{6}\right)^{20} = 0.026.$$

(c) This question relates to a hypergeometric distribution: there are $N = 20$ packages with $M = 3$ packages with toys and $N - M = 17$ packages without a toy. The daughter gets $n = 5$ packages and we are interested in $P(X = 2)$. Hence we get

$$P(X = 2) = \frac{\binom{M}{x}\binom{N-M}{n-x}}{\binom{N}{n}} = \frac{\binom{3}{2}\binom{17}{3}}{\binom{20}{5}} \approx 0.13.$$

Solution to Exercise 8.2   Given $X \sim N(42.1, 20.8^2)$, we get:

(a)

$$P(X \geq 50) = 1 - P(X \leq 50) = 1 - \phi\left(\frac{x - \mu}{\sigma}\right) = 1 - \phi\left(\frac{50 - 42.1}{20.8}\right)$$
$$= 1 - \phi(0.37) \approx 0.35 .$$

We obtain the same results in R as follows:

```
1-pnorm(50,42.1,20.8)
```

(b)

$$P(30 \leq X \leq 40) = P(X \leq 40) - P(X \leq 30)$$
$$= \phi\left(\frac{40 - 42.1}{20.8}\right) - \phi\left(\frac{30 - 42.1}{20.8}\right)$$
$$= \phi(-0.096) - \phi(-0.577) = 1 - 0.538 - 1 + 0.718$$
$$\approx 18\% .$$

We would have obtained the same results in R using:

```
pnorm(40,42.1,20.8)-pnorm(30,42.1,20.8)
```

Solution to Exercise 8.3   The random variable X follows a discrete uniform distribution because $p_i = \frac{1}{12}$ for each x_i. The expectation and variance are therefore

$$E(X) = \frac{k + 1}{2} = \frac{12 + 1}{2} = 6.5,$$
$$\text{Var}(X) = \frac{1}{12}(12^2 - 1) \approx 11.92 .$$

Solution to Exercise 8.4   Each guess is a Bernoulli experiment where the right answer is given with a probability of 50%. The number of correct guesses therefore follows a Binomial distribution, i.e. $X \sim B(10; 0.5)$. The probability of giving the right answer at least 8 times is identical to the probability of not being wrong more than 2 times. We can thus calculate $P(X \geq 8)$ as $P(X \leq 2)$:

$$P(X = 0) = \binom{10}{0}0.5^0(1 - 0.5)^{10} \approx 0.000977$$

$$P(X = 1) = \binom{10}{1}0.5^1(1 - 0.5)^9 \approx 0.009766$$

$$P(X = 2) = \binom{10}{2}0.5^2(1 - 0.5)^8 \approx 0.043945$$

This relates to

$$P(X \leq 2) = P(X = 0) + P(X = 1) + P(X = 2)$$
$$= 0.000977 + 0.009766 + 0.043945 \approx 0.0547 .$$

We would have obtained the same results in R using:

```
pbinom(2,10,0.5)
1-pbinom(7,10,0.5)
```

Solution to Exercise 8.5

(a) It seems appropriate to model the number of fused bulbs with a Poisson distribution. We assume however that the probabilities of fused bulbs on two consecutive days are independent of each other, i.e. they only depend on λ but not on the time t.

(b) The arithmetic mean is

$$\bar{x} = \frac{1}{30}(0 + 1 \cdot 8 + 2 \cdot 8 + \cdots + 5 \cdot 1) = \frac{52}{30} = 1.7333$$

which means that, on an average, 1.73 bulbs are fused per day. The variance is

$$s^2 = \frac{1}{30}(0 + 1^2 \cdot 8 + 2^2 \cdot 8 + \cdots + 5^2 \cdot 1) - 1.7333^2$$
$$= \frac{142}{30} - 3.0044 = 1.72889 .$$

We see that mean and variance are similar, which is an indication that the choice of a Poisson distribution is appropriate since we assume $E(X) = \lambda$ and $Var(X) = \lambda$.

(c) The following table lists the proportions (i.e. relative frequencies f_j) together with the probabilities $P(X = x)$ from a $Po(1.73)$ distribution. As a reference, we also list the probabilities from a $Po(2)$ distribution since it is not practically possible that 1.73 bulbs stop working and it may hence be an option to round the mean.

	f_i	$Po(1.73)$	$Po(2)$
$P(X = 0)$	0.2	0.177	0.135
$P(X = 1)$	0.267	0.307	0.27
$P(X = 2)$	0.267	0.265	0.27
$P(X = 3)$	0.167	0.153	0.18
$P(X = 4)$	0.067	0.067	0.09
$P(X = 5)$	0.033	0.023	0.036

One can see that observed proportions and expected probabilities are close together which indicates again that the choice of a Poisson distribution was appropriate. Chapter 9 gives more details on how to estimate parameters, such as λ, from data if it is unknown.

(d) Using $\lambda = 1.73$, we calculate

$$P(X > 5) = 1 - P(X \leq 5) = 1 - \sum_{i=0}^{5} \frac{\lambda^i}{i!} \exp(-\lambda)$$

$$= 1 - \exp(-1.73) \left(\frac{1.73^0}{0!} + \frac{1.73^1}{1!} + \cdots + \frac{1.73^5}{5!} \right)$$

$$= 1 - 0.99 = 0.01 .$$

Thus the bulbs are replaced on only 1% of the days.

(e) If X follows a Poisson distribution then, given Theorem 8.2.1, Y follows an exponential distribution with $\lambda = 1.73$.

(f) The expectation of an exponentially distributed variable is

$$E(Y) = \frac{1}{\lambda} = \frac{1}{1.73} = 0.578 .$$

This means that, on average, it takes more than half a day until one of the bulbs gets fused.

Solution to Exercise 8.6

(a) Let X be a random variable describing "the number x of winning tickets among n bought tickets"; then X follows the hypergeometric distribution $X \sim H(n, 500, 4000)$. We need to determine n for the conditions specified. We are interested in

$$P(X \geq 3) = 1 - P(X = 2) - P(X = 1) - P(X = 0) .$$

Using the PMF of the hypergeometric distribution

$$P(X = x) = \frac{\binom{M}{x}\binom{N-M}{n-x}}{\binom{N}{n}} ,$$

this equates to

$$P(X \geq 3) = 1 - \frac{\binom{500}{2}\binom{4000-500}{n-2}}{\binom{4000}{n}} - \frac{\binom{500}{1}\binom{4000-500}{n-1}}{\binom{4000}{1}} - \frac{\binom{500}{0}\binom{4000-500}{n}}{\binom{4000}{n}} .$$

We have the following requirement:

$$1 - \frac{\binom{500}{2}\binom{4000-500}{n-2}}{\binom{4000}{n}} - \frac{\binom{500}{1}\binom{4000-500}{n-1}}{\binom{4000}{1}} - \frac{\binom{500}{0}\binom{4000-500}{n}}{\binom{4000}{n}} \overset{!}{\geq} 0.99.$$

To solve this equation, we can program this function for $P(X > 3; n)$ in R and evaluate it for different numbers of tickets sold, e.g. between 50 and 100 tickets:

```
raffle <- function(n){
p <- 1-((choose(500,2)*choose(3500,n-2))/(choose(4000,n)))
-((choose(500,1)*choose(3500,n-1))/(choose(4000,n)))
-((choose(500,0)*choose(3500,n))/(choose(4000,n)))
return(p)
}
raffle(50:100)
raffle(63:64)
```

The output shows that at least 64 tickets need to be bought to have a 99% guarantee that at least three tickets win. This equates to spending € 96.

(b) We can plot the function as follows:

```
nb <- seq(1:75)
plot(nb,tombola(nb),type='l')
```

Figure B.16 shows the relationship between the number of tickets bought and the probability of having at least three winning tickets.

(c) The solution of (a) shows that it is well worth taking part in the raffle: Marco pays €96 and with a probability of 99% and he wins at least three prizes which are worth €142 · 3 = 426. More generally, the money generated by the raffle is €1.50 × 4000 = 6000, but the prizes are worth €142 · 500 = 71, 000. One may suspect that the company produces the appliances much more cheaply than they are sold for and is thus so generous.

Solution to Exercise 8.7   The probability of getting a girl is $p = 1 - 0.5122 = 0.4878$.

(a) We are dealing with a geometric distribution here. Since we are interested in $P(X \leq 3)$, we can calculate:

$$P(X = 1) = 0.4878$$
$$P(X = 2) = 0.4878(1 - 0.4878) = 0.2498512$$
$$P(X = 3) = 0.4878(1 - 0.4878)^2 = 0.1279738$$
$$P(X \leq 3) = P(X = 1) + P(X = 2) + P(X = 3) = 0.865625.$$

We would have obtained the same result in R using:

Fig. B.16 Probability to have at least three winning tickets given the number of tickets bought

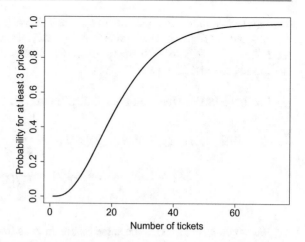

```
pgeom(2,0.4878)
```

Note that we have to specify "2" rather than "3" for x because R takes the number of unsuccessful trials rather the number of trials until success.

(b) Here we deal with a Binomial distribution with $k = 2$ and $n = 4$. We can calculate $P(X = 2)$ as follows:

$$P(X = k) = \binom{n}{k} p^k (1 - p)^{n-k}$$

$$= \binom{4}{2} 0.4878^2 \cdot (1 - 0.4878)^2 = 0.3745536.$$

R would have given us the same result using the `dbinom(2,4,0.4878)` command.

Solution to Exercise 8.8

(a) The random variable Y follows a Poisson distribution, see Theorem 8.2.1 for more details.

(b) The fisherman catches, on average, 3 fish an hour. We can thus assume that the rate λ is 3 and thus $E(Y) = \lambda = 3$. Similarly, $E(X) = \frac{1}{\lambda} = \frac{1}{3}$ which means that it takes, on average, 20 min. to catch another fish.

(c) Using the PDF of the Poisson distribution, we get:

$$P(Y = 5) = \frac{3^5}{5!} \exp(-3) = 0.1 = 10\%$$

$$P(Y < 1) = P(Y = 0) = \frac{3^0}{0!} \exp(-3) \approx 0.0498 \approx 5\%.$$

We would have obtained the same results in R using the `dpois(5,3)` and `dpois(0,3)` commands.

Solution to Exercise 8.9   The random variable $\mathbf{X}$ = "choice of dessert" follows a multinomial distribution. More precisely, X_1 describes whether chocolate brownies were chosen, X_2 describes whether yoghurt was chosen, X_3 describes whether lemon tart was chosen and $\mathbf{X} = \{X_1, X_2, X_3\}$.

(a) Using the PMF of the multinomial distribution, we get

$$P(X_1 = n_1, X_2 = n_2, \ldots, X_k = n_k) = \frac{n!}{n_1! n_2! \cdots n_k!} \cdot p_1^{n_1} \cdots p_k^{n_k}$$

$$P(X_1 = 2, X_2 = 1, X_3 = 2) = \frac{5!}{2!1!2!} \cdot 0.2^2 \cdot 0.3^1 \cdot 0.5^2$$

$$= 9\% \,.$$

We would have obtained the same results in R as follows:

```
dmultinom(c(2,1,2),prob=c(0.2,0.3,0.5))
```

(b) The probability of choosing lemon tart for the first two guests is 1. We thus need to determine the probability that 3 out of the remaining 3 guests order lemon tart:

$$P(X_1 = 0, X_2 = 0, X_3 = 3) = \frac{3!}{0!0!3!} \cdot 0.2^0 \cdot 0.3^0 \cdot 0.5^3$$

$$= 12.5\% \,.$$

Using `dmultinom(c(0,0,3),prob=c(0.2,0.3,0.5))` in R, we get the same result.

(c) The expectation vector is

$$\mathrm{E}(\mathbf{X}) = (np_1, \ldots, np_k) = (20 \cdot 0.2, 20 \cdot 3, 20 \cdot 0.5) = (4, 6, 10) \,.$$

This means we expect 4 guests to order brownies, 6 to order yoghurt, and 10 to order lemon tart. The covariance matrix can be determined as follows:

$$\mathrm{Cov}(X_i, X_j) = \begin{cases} np_i(1 - p_i) & \text{if } i = j \\ -np_i p_j & \text{if } i \neq j \,. \end{cases}$$

Using $n = 20$, $p_1 = 0.2$, $p_2 = 0.3$ and $p_3 = 0.5$, we obtain the covariance matrix as:

$$\begin{pmatrix} 3.2 & -1.2 & -2 \\ -1.2 & 4.2 & -3 \\ -2 & -3 & 5 \end{pmatrix}$$

Solution to Exercise 8.10

$$
\begin{aligned}
P(S \geq 1, W \geq 1) \stackrel{\text{indep.}}{=}\ & P(S \geq 1) \cdot P(W \geq 1) \\
= \ & (1 - P(S = 0)) \cdot (1 - P(W = 0)) \\
= \ & (1 - e^{-3}\frac{3^0}{0!}) \cdot (1 - e^{-4}\frac{4^0}{0!}) \\
\approx \ & 0.93 \, .
\end{aligned}
$$

Solution to Exercise 8.11

(a) Random numbers of a normal distribution can be generated using the `rnorm` command. By default $\mu = 0$ and $\sigma = 1$ (see `?rnorm`), so we don't need to specify these parameters. We simply need to set $n = 1000$. The mean of the 1000 realizations can thus be obtained using `mean(rnorm(1000))`. We can, for example, write a `for` loop to repeat this process 1000 times. An empty (=NA) vector of size 1000 can be used to store and evaluate the results:

```
set.seed(24121980)
R <- 1000
means <- c(rep(NA,R))
for(i in 1:R){means[i] <- mean(rnorm(1000))}
mean(means)
[1] -0.0007616465
var(means)
[1] 0.0009671311
plot(density(means))
```

We see that the mean of the arithmetic means is close to zero, but not exactly zero. The variance is approximately $\sigma^2/n = 1/1000 = 0.001$, as one would expect from the Central Limit Theorem. The distribution is symmetric, similar to a normal distribution, see Fig. B.17. It follows that $\bar{X}_n$ is approximately $N(\mu, \frac{\sigma^2}{n})$ distributed, as one could expect from the Theorem of Large Numbers and the Central Limit Theorem. It is important to understand that $\bar{X}$ is not fixed but a random variable which follows a distribution, i.e. the normal distribution.

(b) We can use the same code as above, except we use the exponential instead of the normal distribution:

```
means2 <- c(rep(NA,R))
for(i in 1:R){means2[i] <- mean(rexp(1000))}
mean(means2)
[1] 1.001321
var(means2)
[1] 0.001056113
plot(density(means))
```

Fig. B.17 Kernel density
plot of the distribution
simulated in Exercise 8.11a

(a) 1000 repetitions   (b) 10,000 repetitions

Fig. B.18 Kernel density plots for Exercises 8.11b and 8.11c

The realizations are i.i.d. observations. Once can see that, as in (a), $\bar{X}_n$ is approximately $N(\mu, \frac{\sigma^2}{n}) = N(1, 1/1000)$ distributed. It is evident that the X_i do not necessarily need to follow a normal distribution for $\bar{X}$ to follow a normal distribution, see also Fig. B.18a.

(c) Increasing the number of repetitions makes the distribution look closer to a normal distribution, see Fig. B.18b. This visualizes that as n tends to infinity $\bar{X}_n$ gets closer to a $N(\mu, \frac{\sigma^2}{n})$ distribution.

Chapter 9

Solution to Exercise 9.1

(a) The exercise tells us that $X_i \overset{iid}{\sim} Po(\lambda)$, $i = 1, 2, \ldots, n$. Let us look at the realizations $x_1, x_2, \ldots, x_n$: under the assumption of independence, which we know is fulfilled because the X_i's are i.i.d., we can write the likelihood function as the product of the n PMF's of the Poisson distribution:

$$L(\theta; x) = \prod_{i=1}^{n} f(x_i; \theta) = \prod_{i=1}^{n} \frac{\lambda^{x_i}}{x_i!} e^{-\lambda} = \frac{\lambda^{\sum x_i}}{\prod x_i!} e^{-n\lambda}.$$

It is better to work on a log-scale because it is easy to differentiate. The results are identical no matter whether we use the likelihood function or the log-likelihood function because the log transformation is monotone in nature. The log-likelihood function is:

$$\ln L = \sum x_i \ln \lambda - \ln(x_1! \cdots x_n!) - n\lambda.$$

Differentiating with respect to λ yields

$$\frac{\partial \ln L}{\partial \lambda} = \frac{1}{\lambda} \sum x_i - n \overset{!}{=} 0$$

which gives us the ML estimate:

$$\hat{\lambda} = \frac{1}{n} \sum x_i = \bar{x}.$$

We need to confirm that the second derivative is < 0 at $\hat{\lambda} = \bar{x}$, otherwise the solution would be a minimum rather than a maximum. We get

$$\frac{\partial^2 \ln L}{\partial \lambda^2} = -\frac{1}{\hat{\lambda}^2} \sum x_i = -\frac{n}{\hat{\lambda}} < 0.$$

It follows that the arithmetic mean $\bar{x} = \hat{\lambda}$ is the maximum likelihood estimator for the parameter λ of a Poisson distribution.

(b) Using the results from (a) we can write the log-likelihood function for $x_1 = 4$, $x_2 = 3$, $x_3 = 8$, $x_4 = 6$, $x_5 = 6$ as:

$$\ln L = 27 \ln \lambda - \ln(4!\, 3!\, 8!\, 6!\, 6!) - 5\lambda.$$

because $\sum x_i = 27$. We can write down this function in R as follows:

Fig. B.19 Illustration of the log-likelihood function

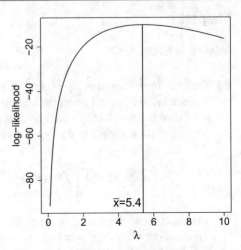

```
MLP <- function(lambda){
27*log(lambda) - log(factorial(4)*...*factorial(6))
- 5*lambda
}
```

The function can be plotted using the `curve` command:

```
curve(MLP, from=0, to=10)
```

Figure B.19 shows the log-likelihood function. It can be seen that the function reaches its maximum at $\bar{x} = 5.4$.

(c) Using (a) we can write the likelihood function as

$$L(\theta; x) = \prod_{i=1}^{n} f(x_i; \theta) = \prod_{i=1}^{n} \frac{\lambda^{x_i}}{x_i!} e^{-\lambda} = \frac{\lambda^{\sum x_i}}{\prod x_i!} e^{-n\lambda} = \underbrace{\lambda^{\sum x_i} e^{-n\lambda}}_{g(t,\lambda)} \underbrace{\frac{1}{\prod x_i!}}_{h(x_1,...,x_n)} .$$

This means $T = \sum_{i=1}^{n} x_i$ is sufficient for λ. The arithmetic mean, which is the maximum likelihood estimate, is a one-to-one function of T and therefore sufficient too.

Solution to Exercise 9.2

(a) The probability density function of a normal distribution equates to

$$f(x) = \frac{1}{\sigma\sqrt{2\pi}} \exp\left(-\frac{(x-\mu)^2}{2\sigma^2}\right)$$

with $\infty < x < \infty$, $-\infty < \mu < \infty$, $\sigma^2 > 0$. The likelihood function is therefore

$$L(x_1, x_2, \ldots, x_n | \mu, \sigma^2) = \left(\frac{1}{\sqrt{2\pi\sigma^2}}\right)^n \exp\left(-\sum_{i=1}^{n} \frac{(x_i - \mu)^2}{2\sigma^2}\right).$$

To find the maximum of $L(x_1, x_2, \ldots, x_n | \mu, \sigma^2)$ it is again easier to work with the *log*-likelihood function, which is

$$l = \ln L(x_1, x_2, \ldots, x_n | \mu, \sigma^2) = -\frac{n}{2} \ln 2\pi - \frac{n}{2} \ln \sigma^2 - \sum_{i=1}^{n} \left(\frac{(x_i - \mu)^2}{2\sigma^2}\right).$$

Assuming σ^2 to be 1, differentiating the log-likelihood function with respect to μ, and equating it to zero gives us

$$\frac{\partial l}{\partial \mu} = 2 \sum_{i=1}^{n} \left(\frac{x_i - \mu}{1^2}\right) = 0 \quad \Leftrightarrow \quad n\mu = \sum_{i=1}^{n} x_i.$$

The ML estimate is therefore $\hat{\mu} = \bar{x}$.

(b) Looking at the differentiated log-likelihood function in (a) shows us that the ML estimate of μ is always the arithmetic mean, no matter whether σ^2 is 1 or any other number.

(c) Differentiating the log-likelihood function from (a) with respect to σ^2 yields

$$\frac{\partial l}{\partial \sigma^2} = -\frac{n}{2} \frac{1}{\sigma^2} + \frac{1}{2\sigma^4} \sum_{i=1}^{n} (x_i - \mu)^2 = 0.$$

Using $\hat{\mu} = \bar{x}$ we calculate $\frac{\partial l}{\partial \sigma^2} = 0$ as

$$\hat{\sigma}^2 = \frac{1}{n} \sum_{i=1}^{n} (x_i - \hat{\mu})^2 = \frac{1}{n} \sum_{i=1}^{n} (x_i - \bar{x})^2.$$

Since the parameter $\theta = (\mu, \sigma^2)$ is two-dimensional, it follows that one needs to solve two ML equations, where $\hat{\mu}_{ML}$ is estimated first, and $\hat{\sigma}^2_{ML}$ second (as we did above). It follows further that one needs to look at the positive definiteness of a matrix (the so-called "information matrix") when checking that the second-order derivatives of the estimates are positive and therefore the estimates yield a maximum rather then a minimum. However, we omit this lengthy and time-consuming task here.

Solution to Exercise 9.3 The probability density function of $U(0, \theta)$ is

$$f(x) = \frac{1}{\theta} \quad \text{if} \ \ 0 < x < \theta \ \ \text{and} \ \ 0 \ \ \text{otherwise}.$$

Note that this equates to the PDF from Definition 8.2.1 for $a = 0$ and $b = \theta$. The likelihood function is therefore

$$L(x_1, x_2, \ldots, x_n | \theta) = \left(\frac{1}{\theta}\right)^n \quad \text{if} \quad 0 < x_i < \theta \quad \text{and} \quad 0 \quad \text{otherwise.}$$

One can see that $L(x_1, x_2, \ldots, x_n | \theta)$ increases as θ decreases. The maximum of the likelihood function is therefore achieved for the smallest valid θ. In particular, θ is minimized when $\theta \geq \max(x_1, x_2, \ldots, x_n) = x_{(n)}$. This follows from the definition of the PDF which requires that $0 < x_i < \theta$ and therefore $\theta > x_i$. Thus, the maximum likelihood estimate of θ is $x_{(n)}$, the greatest observed value in the sample.

Solution to Exercise 9.4

(a) $T_n(X)$ is unbiased, and therefore also asymptotically unbiased, because

$$E(T_n(X)) = E(n X_{\min}) \overset{(7.29)}{=} n \frac{1}{n\lambda} = \frac{1}{\lambda} = \mu.$$

Similarly, $V_n(X)$ is unbiased, and therefore also asymptotically unbiased, because

$$E(V_n(X)) = E\left(\frac{1}{n}\sum_{i=1}^{n} X_i\right) \overset{(7.29)}{=} \frac{1}{n}\sum_{i=1}^{n} E(X_i) = \frac{1}{n} n \frac{1}{\lambda} = \mu.$$

(b) To calculate the MSE we need to determine the bias and the variance of the estimators as we know from Eq. (9.5). It follows from (a) that both estimators are unbiased and hence the bias is 0. For the variances we get:

$$\text{Var}(T_n(X)) = \text{Var}(n X_{\min}) \overset{(7.33)}{=} n^2 \text{Var}(X_{\min}) = n^2 \frac{1}{n^2\lambda^2} = \mu^2.$$

$$\text{Var}(V_n(X)) = \text{Var}\left(\frac{1}{n}\sum_{i=1}^{n} X_i\right) \overset{(7.33)}{=} \frac{1}{n^2}\sum_{i=1}^{n} \text{Var}(X_i) = \frac{1}{n^2} n \frac{1}{\lambda^2} = \frac{1}{n}\mu^2.$$

Since the mean squared error consists of the sum of the variance and squared bias, the MSE for $T_n(X)$ and $V_n(X)$ are μ^2 and $n^{-1}\mu^2$ respectively. One can see that the larger n, the more superior $V_n(X)$ over $T_n(X)$ in terms of the mean squared error. In other words, $V_n(X)$ is more efficient than $T_n(X)$ because its variance is lower for any $n > 1$.

(c) Using the results from (b) we get

$$\lim_{n \to \infty} \text{MSE}(V_n(X)) = \lim_{n \to \infty} \frac{1}{n}\mu^2 = 0.$$

This means the MSE approaches 0 as n tends to infinity. Therefore, $V_n(X)$ is MSE consistent for μ. Since $V_n(X)$ is MSE consistent, it is also weakly consistent.

Solution to Exercise 9.5

(a) The point estimate of μ is $\bar{x}$ which is

$$\hat{\mu} = \bar{x} = \frac{1}{n} \sum_{i=1}^{n} x_i = \frac{1}{24}(450 + \ldots + 790) = 667.92 \,.$$

The variance of σ^2 can be estimated unbiasedly using s^2:

$$\hat{\sigma}^2 = s^2 = \frac{1}{n-1} \sum_{i=1}^{n} (x_i - \bar{x})^2$$

$$= \frac{1}{23}((450 - 667.92)^2 + \ldots + (790 - 667.92)^2) \approx 18,035 \,.$$

(b) The variance is unknown and needs to be estimated. We thus need the t-distribution to construct the confidence interval. We can determine $t_{23;0.975} \approx$ 2.07 using qt(0.975,23) or Table C.2 (though the latter is not detailed enough), $\alpha = 0.05$, $\bar{x} = 667.97$ and $\hat{\sigma}^2 = 18,035$. This yields

$$I_l(X) = \bar{x} - t_{n-1;1-\alpha/2} \cdot \frac{s}{\sqrt{n}} = 667.92 - t_{23;0.975} \cdot \frac{\sqrt{18,035}}{\sqrt{24}} \approx 611.17 \,,$$

$$I_u(X) = \bar{x} + t_{n-1;1-\alpha/2} \cdot \frac{s}{\sqrt{n}} 667.92 - t_{23;0.975} \cdot \frac{\sqrt{18,035}}{\sqrt{24}} \approx 724.66 \,.$$

The confidence interval for μ is thus $[611.17; 724.66]$.

(c) We can reproduce these results in R as follows:

```
eland <- c(450,730,700,600,620,...,790)
t.test(eland)$conf.int
```

Solution to Exercise 9.6

• Let us start with the confidence interval for the "Brose Baskets Bamberg". Using $t_{15;0.975} = 2.1314$ (qt(0.975,15) or Table C.2) and $\alpha = 0.05$, we can determine the confidence interval as follows:

$$I_l(Ba) = \bar{x} - t_{n-1;1-\alpha/2} \cdot \frac{s}{\sqrt{n}} = 199.06 - t_{15;0.975} \cdot \frac{7.047}{\sqrt{16}} = 195.305 \,,$$

$$I_u(Ba) = \bar{x} + t_{n-1;1-\alpha/2} \cdot \frac{s}{\sqrt{n}} = 199.06 + t_{15;0.975} \cdot \frac{7.047}{\sqrt{16}} = 202.815 \,.$$

Thus, we get $[195.305; 202.815]$.

- For the "Bayer Giants Leverkusen", we use $t_{13;0.975} = 2.1604$ to get

$$I_l(L) = \bar{x} - t_{n-1;1-\alpha/2} \cdot \frac{s}{\sqrt{n}} = 196 - t_{13;0.975} \cdot \frac{9.782}{\sqrt{14}} = 190.352 \,,$$

$$I_u(L) = \bar{x} + t_{n-1;1-\alpha/2} \cdot \frac{s}{\sqrt{n}} = 196 + t_{13;0.975} \cdot \frac{9.782}{\sqrt{14}} = 201.648 \,.$$

This leads to a confidence interval of $[190.352; 201.648]$.
- For "Werder Bremen" we need to use the quantile $t_{22,0.975} = 2.0739$ which yields a confidence interval of

$$I_l(Br) = \bar{x} - t_{n-1;1-\alpha/2} \cdot \frac{s}{\sqrt{n}} = 187.52 - t_{22;0.975} \cdot \frac{5.239}{\sqrt{23}} = 185.255 \,,$$

$$I_u(Br) = \bar{x} + t_{n-1;1-\alpha/2} \cdot \frac{s}{\sqrt{n}} = 187.25 + t_{22;0.975} \cdot \frac{5.239}{\sqrt{23}} = 189.786 \,.$$

The interval is therefore $[185.255; 189.786]$.
- The mean heights of the basketball teams are obviously larger than the mean height of the football team. The two confidence intervals of the basketball teams overlap, whereas the intervals of the football team with the two basketball teams do not overlap. It is evident that this indicates that the height of football players is substantially less than the height of basketball players. In Chap. 10 we will learn that confidence intervals can be used to test hypotheses about mean differences.

Solution to Exercise 9.7

(a) Using $n = 98$, we calculate an unbiased point estimate for p using $\bar{x} = \hat{p}$. Note that $x_i = 1$ if the wife has to wash the dishes.

$$\hat{p} = \frac{1}{n} \sum_{i=1}^{n} x_i = \frac{1}{98} \cdot 59 = \frac{59}{98} \approx 0.602 \,.$$

(b) Since $n\hat{p}(1 - \hat{p}) = 98 \cdot 0.602 \cdot 0.398 = 23.48 > 9$ and p is sufficiently large, we can use the normal approximation to calculate a confidence interval for p. Using $z_{1-\alpha/2} = z_{0.975} = 1.96$, we obtain

$$I_l(X) = 0.602 - 1.96\sqrt{\frac{0.602 \cdot 0.398}{98}} = 0.505 \,,$$

$$I_u(X) = 0.602 + 1.96\sqrt{\frac{0.602 \cdot 0.398}{98}} = 0.699 \,.$$

This yields a confidence interval of $[0.505, 0.699]$. Note that the confidence interval does not contain $p = 0.5$ which is the probability that would be expected if the coin was fair. This is a clear sign that the coin might be unfair.

(c) If the coin is fair, we can use $\hat{p} = 0.5$. We would then need

$$n \geq \left[\frac{z_{1-\alpha/2}}{\Delta}\right]^2 \hat{p}(1-\hat{p})$$

$$\geq \left[\frac{1.96}{0.005}\right]^2 0.5^2 = 38,416$$

dinners to get the desired precision—which is not possible as this would constitute a time period of more than 100 years. This shows that the expectation of such a high precision is unrealistic and needs to be modified. However, as the results of (b) show, the coin may not be fair. If the coin is indeed unfair, we may have, for example, $p = 0.6$ and $1 - p = 0.4$ which gives a smaller sample size. We can thus interpret the sample size as a conservative estimate of the highest number of dinners needed.

Solution to Exercise 9.8   If a student fails then $x_i = 1$. We know that $\sum x_i = 11$ and $n = 104$.

(a) Using $\bar{x} = p$ as point estimate, we get $\hat{p} = \frac{11}{104} \approx 0.106 = 10.6\%$. Using $z_{1-\alpha/2}$ $= z_{0.975} = 1.96$, we can calculate the confidence interval as

$$0.106 \pm 1.96 \cdot \sqrt{\frac{0.106 \cdot 0.894}{104}} = [0.047; 0.165] \ .$$

Using R we get:

```
binom.test(11,104)$conf.int
[1]  0.05399514  0.18137316
```

This result is different because the above command does not use the normal approximation. In addition, p is rather small which means that care must be exercised when using the results of the confidence interval with normal approximation in this context.

(b) The point estimate of 10.6% is substantially higher than 3.2%. The lower bound confidence interval is still larger than the proportion of failures at county level. This indicates that the school is doing worse than most other schools in the county.

Solution to Exercise 9.9

(a) Whether the ith household has switched on the TV and watches "Germany's next top model" (GNTM) relates to a random variable X_i with

$$X_i = 1 : \text{if TV switched on and household watching GNTM}$$
$$X_i = 0 : \text{if TV switched off or household watches another show.}$$

It follows that $X = \sum_{i=1}^{2500} X_i$ is the random variable describing the number of TVs, out of 2500 TVs, which are switched on and show GNTM. Since the X_i's can be assumed to be i.i.d., we can say that X follows a Binomial distribution, i.e. $X \sim B(2500; p)$ with p unknown. The length of the confidence interval for p,

$$\left[\hat{p} - z_{1-\alpha/2}\sqrt{\frac{\hat{p}(1-\hat{p})}{n}}, \; \hat{p} + z_{1-\alpha/2}\sqrt{\frac{\hat{p}(1-\hat{p})}{n}} \right],$$

is

$$L = 2z_{1-\alpha/2}\sqrt{\frac{\hat{p}(1-\hat{p})}{n}}.$$

Unfortunately $\hat{p}(1-\hat{p})$ is unknown but $\hat{p}(1-\hat{p}) \le \frac{1}{4}$ because the maximum value for $\hat{p}(1-\hat{p})$ is 0.25 if $\hat{p} = 0.5$. Hence

$$L \le 2z_{1-\alpha/2}\sqrt{\frac{\frac{1}{4}}{n}} = \frac{1.96}{\sqrt{2500}} = 0.0392.$$

This means the precision is half the length, i.e. $\pm 0.0196 = \pm 1.96\%$ in the worst case.

(b) There is the danger of selection bias. A total of 500 households refused to take part in the study. It may be that their preferences regarding TV shows is different from the other 2500 households. For example, it may well be possible that those watching almost no TV refuse to be included; or that those watching TV shows which are considered embarrassing by society are refusing as well. In general, missing data may cause point estimates to be biased, depending on the underlying mechanism causing the absence.

Solution to Exercise 9.10

(a) Using $z_{1-\alpha/2} = 1.96$ and $\hat{\sigma} = 0.233$ we obtain an optimal sample size of at least

$$n_{opt} \ge \left[2\frac{z_{1-\alpha/2}\sigma_0}{\Delta} \right]^2 = \left[2 \cdot \frac{1.96 \cdot 0.233}{0.2} \right]^2 = 20.85.$$

To calculate a confidence interval with a width of not more than $0.2\,s$, the results of at least 21 athletes are needed.

(b) The sample size is 30. Thus, the confidence interval width should be smaller than $0.2\,s$. This is indeed true as the calculations show:

$$[10.93 \pm \underbrace{t_{0.975;29}}_{2.045} \cdot \frac{0.233}{\sqrt{30}}] = [10.84; 11.02].$$

The width is only $0.18\,s$.

(c) If we calculate a 80% confidence interval, then the lower confidence limit corresponds to the running time which is achieved by only 10% of the athletes (of the population). Using $t_{0.9;29} \approx 1.31$ we get

$$[10.93 \pm 1.31 \cdot \frac{0.233}{\sqrt{30}}] = [10.87; 10.99].$$

The athlete's best time is thus below the lower confidence limit. He is among the top 10% of all athletes, using the results of the confidence interval.

Solution to Exercise 9.11

(a) The odds ratio is

$$OR = \frac{163 \cdot 477}{475 \cdot 151} \approx 1.08.$$

This means the chances that a pizza arrives in time are about 1.08 times higher for Laura compared with Melissa. To calculate the 95% confidence interval we need $\hat{\theta}_0 = \ln(1.08) \approx 0.077$, $z_{1-\alpha/2} \approx 1.96$, and

$$\hat{\sigma}_{\hat{\theta}_0} = \left(\frac{1}{163} + \frac{1}{475} + \frac{1}{151} + \frac{1}{477} \right)^{\frac{1}{2}} = 0.13.$$

The interval for the log odds ratio is

$$[\ln(1.08) \pm 1.96 \cdot 0.13] \approx [-0.18; 0.33].$$

Exponentiating the interval gives us the 95% confidence interval for the odds ratio which is [0.84; 1.39]. This indicates that the odds of Laura's pizzas arriving earlier than Melissa's are not much different from one. While the point estimate tells us that Laura's pizzas are delivered 1.08 times faster, the confidence interval tells us that there is uncertainty around this estimate in the sense that it could also be smaller than 1 and Melissa may not necessarily work more slowly than Laura.

(b) We can reproduce the results in *R* by attaching the pizza data, creating a categorical delivery time variable (using `cut`) and then applying the `oddsratio` command from the library `epitools` onto the contingency table:

```
attach(pizza)
timecat <- cut(time, breaks=c(-1,30,150))
library(epitools)
oddsratio(table(timecat,operator), method='wald')
```

Chapter 10

Solution to Exercise 10.1   A type I error is defined as the probability of rejecting H_0 if H_0 is true. This error occurs if A thinks that B does confess, but B does not. In this scenario A confesses, goes free, and B serves a three-year sentence. A type II error is defined as the probability of accepting H_0, despite the fact that H_0 is wrong. In this case, B does confess, but A does not. In this scenario B goes free and A serves a three-year sentence. A type II error is therefore worse for A. With a statistical test we always control the type I error, but not the type II error.

Solution to Exercise 10.2

(a) The hypotheses are

$$H_0 : \mu = 100 \quad \text{versus} \quad H_1 : \mu \neq 100 \,.$$

(b) It is a one sample problem for μ: thus, for known variance, the Gauss test can be used; the t-test otherwise, see also Appendix D.2. Since, in this exercise, σ is assumed to be known we can use the Gauss test, i.e. we can compare the test statistic with the normal distribution (as opposed to the t-distribution when using the t-test). The sample size is small: we must therefore assume a normal distribution for the data.

(c) To calculate the realized test statistic we need the arithmetic mean $\bar{x} = 98.08$. Then we obtain

$$t(x) = \frac{\bar{x} - \mu}{\sigma} \cdot \sqrt{n} = \frac{98.08 - 100}{2} \cdot \sqrt{15} = \frac{-1.92}{2} \cdot \sqrt{15} = -3.72.$$

We reject H_0 if $|t(x)| > z_{1-\frac{\alpha}{2}} = 1.96$. Since $|t(x)| = 3.72 > 1.96$, the null hypothesis can be rejected. The test is significant. There is enough evidence to suggest that the observed weights do not come from a normal distribution with mean 100g.

(d) To show that $\mu < 100$ we need to conduct a one-sided test. Most importantly, the research hypothesis of interest needs to be stated as the *alternative* hypothesis:

$$H_0 : \mu \geq 100 \quad \text{versus} \quad H_1 : \mu < 100 \,.$$

(e) The test statistic is the same as in b): $t(x) = -3.72$. However, the critical region changes. H_0 gets rejected if $t(x) < -z_{1-\alpha} = -1.64$. Again, the null hypothesis is rejected. The producer's hypothesis that the average weight of his chocolate bars was lower than 100 g is confirmed based on the test decision.

Solution to Exercise 10.3

(a) We calculate

	No auction	Auction
$\bar{x}$	16.949	10.995
s^2	2.948	2.461
s	1.717	1.569
v	0.101	0.143

Note that we use the unbiased estimates for the variance and the standard deviation as explained in Chap. 9, i.e. we use $1/(n-1)$ rather than $1/n$. It is evident that the mean price of the auctions (μ_a) is lower than the mean no-auction price (μ_{na}), and also lower than the price from the online book store. There is however a higher variability in relation to the mean for the auction prices. One may speculate that the best offers are found in the auctions, but that there are no major differences between the online store and the internet book store, i.e.

- $\mu_{na} \neq €16.95$,
- $\mu_a < €16.95$,
- $\mu_{na} > \mu_a$.

(b) We can use the t-test to test the hypotheses

$$H_0: \mu_{na} = 16.95 \qquad \text{versus} \qquad H_1: \mu_{na} \neq 16.95 .$$

The test statistic is

$$t(x) = \frac{\bar{x} - \mu}{\sigma} \cdot \sqrt{n} = \frac{16.949 - 16.95}{1.717} \cdot \sqrt{14} = -0.002 .$$

Using the decision rules from Table 10.2 we conclude that H_0 gets rejected if $|t(x)| > t_{13,0.975} = 2.16$. We can calculate $t_{13,0.975}$ either by using Table C.2 or by using R (qt(0.975,13)). Since $|t(x)| = 0.002 \not> 2.16$, we keep the null hypothesis.

(c) Using (9.16) we can calculate the upper and lower limit of the confidence interval as

$$\bar{x} + t_{n-1;1-\alpha/2} \cdot \frac{s_X}{\sqrt{n}} = 16.949 + \underbrace{t_{13;0.975}}_{=2.16} \cdot \frac{1.717}{\sqrt{14}} = 17.94$$

$$\bar{x} - t_{n-1;1-\alpha/2} \cdot \frac{s_X}{\sqrt{n}} = 16.949 - \underbrace{t_{13;0.975}}_{=2.16} \cdot \frac{1.717}{\sqrt{14}} = 15.96$$

respectively. The confidence interval *does* contain $\mu_0 = 16.95$; hence, the null hypothesis $\mu = \mu_0 = 16.95$ would not be rejected. This is the same conclusion as obtained from the one-sample t-test from above. Interpreting the confidence interval as compatibility interval shows that any hypothesized value between 15.96€ and 17.96€ is more compatible with the data as hypothesized values outside this interval, conditional on the background assumptions made.

(d) We test the following hypotheses:

$$H_0: \mu_a \geq 16.95 \qquad \text{versus} \qquad H_1: \mu_a < 16.95 .$$

The realized test statistic is

$$t(x) = \frac{\bar{x} - \mu}{\sigma} \cdot \sqrt{n} = \frac{10.995 - 16.95}{1.569} \cdot \sqrt{14} = -14.201 .$$

Table 10.2 tells us that H_0 gets rejected if $t(x) < -t_{13,0.95}$. Since $-14.201 < -2.16$ we reject H_0. The test confirms our hypothesis that the mean auction prices are lower than €16.95, i.e. the price from the book store.

(e) We need to conduct two tests: the two-sample t-test for the assumption of equal variances and the Welch test for the assumption of different variances.

(i) *Two-sample t-test*:

$$H_0: \mu_{na} \leq \mu_a \qquad \text{versus} \qquad H_1: \mu_{na} > \mu_a .$$

To calculate the test statistic (10.4) we need to determine the pooled sample variance:

$$
\begin{aligned}
s^2 &= \frac{(n_{na} - 1)s_{na}^2 + (n_a - 1)s_a^2}{n_{na} + n_a - 2} \\
&= \frac{(14 - 1)2.948 + (14 - 1)2.461}{14 + 14 - 2} \approx 2.705
\end{aligned}
$$

The test statistic is

$$
\begin{aligned}
t(x) &= \frac{\bar{x}_{na} - \bar{x}_a}{s} \cdot \sqrt{\frac{n_{na} \cdot n_a}{n_{na} + n_a}} = \frac{16.949 - 10.995}{\sqrt{2.705}} \cdot \sqrt{\frac{14 \cdot 14}{14 + 14}} \\
&= \frac{5.954}{1.645} \cdot \sqrt{\frac{196}{28}} = 3.621 \cdot \sqrt{7} = 9.578 .
\end{aligned}
$$

We reject H_0 if $t(x) > t_{n_{na}+n_a-2,0.95} = t_{26,0.95} = 1.71$. Table C.2 does not list the quantile, thus one uses R (qt(0.95,26)) to determine the quantile. Since $9.578 > 1.71$, we reject the null hypothesis. To arrive at a decision, we use the test to conclude that the mean auction prices are lower than the mean non-auction prices.

(ii) *Welch test*: For the Welch test we calculate the test statistic, using (10.6) as:

$$t(x) = \frac{|\bar{x}_{na} - \bar{x}_a|}{\sqrt{\frac{s_{na}^2}{n_{na}} + \frac{s_a^2}{n_a}}} = \frac{|16.949 - 10.995|}{\sqrt{\frac{2.948}{14} + \frac{2.461}{14}}} = 9.578\,.$$

To calculate the critical value we need the degrees of freedom:

$$\begin{aligned}
v &= \left(\frac{s_{na}^2}{n_{na}} + \frac{s_a^2}{n_a}\right)^2 \bigg/ \left(\frac{\left(s_{na}^2/n_{na}\right)^2}{n_{na} - 1} + \frac{\left(s_a^2/n_a\right)^2}{n_a - 1}\right) \\
&= \left(\frac{2.948}{14} + \frac{2.461}{14}\right)^2 \bigg/ \left(\frac{(2.948/14)^2}{13} + \frac{(2.461/14)^2}{13}\right) \\
&\approx 25.79
\end{aligned}$$

We reject H_0 if $t(x) > t_{v,0.95}$. Using $t_{v,0.95} = 1.706$ (in R obtained as qt(0.95,25.79)) and $t(x) = 9.578$ we reject the null hypothesis. The conclusions are identical to using the two-sample t-test (in this example). Interestingly, the two test statistics are similar which indicates that the assumption of equal variances in case (i) was not unreasonable.

(f) The F-test relies on the assumption of the normal distribution and tests the hypotheses:

$$H_0\colon \sigma_{na}^2 = \sigma_a^2 \qquad \text{versus} \qquad H_1\colon \sigma_{na}^2 \neq \sigma_a^2\,.$$

We can calculate the test statistic as described in Sect. C.5:

$$t(x) = \frac{s_{na}^2}{s_a^2} = \frac{2.949}{2.461} = 1.198.$$

The F-distribution is not symmetric around 0; thus we need to calculate two critical values: $f_{n_1-1;n_2-1;1-\alpha/2}$ and $f_{n_1-1;n_2-1;\alpha/2}$. Using R we get $f_{13,13,0.975} = 3.115$ (qf(0.975,13,13)) and $f_{13,13,0.025} = \frac{1}{3.115} = 0.321$ (qf(0.025,13,13)). H_0 is rejected if $t(x) > f_{n_1-1;n_2-1;1-\alpha/2}$ or $t(x) < f_{n_1-1;n_2-1;\alpha/2}$. Since $0.321 < 1.198 < 3.115$ we do not reject H_0. This means there is strong evidence that the two variances are equal. This is also reassuring with respect to using the two-sample t-test above (which assumes equal variances). However, note that testing the equality of variances with the F-test and then using the two-sample t-test or Welch test based on the outcome of the F-test is not ideal and not necessarily correct. In practice, it is best to simply use the Welch test (rather than the t-test), which is also the default option in the R function t.test.

(g) We need the following table to calculate the rank sums:

Value	9.3	9.52	9.54	10.01	10.5	10.5	10.55	10.59	11.02	11.03
Sample	a	a	a	a	a	a	a	a	a	a
Rank	1	2	3	4	5	6	7	8	9	10
Value	11.89	11.99	12	13.79	15.49	15.9	15.9	15.9	15.9	15.9
Sample	a	a	a	na	a	na	na	na	na	na
Rank	11	12	13	14	15	16	17	18	19	20
Value	15.99	16.98	16.98	17.72	18.19	18.19	19.97	19.97		
Sample	na	na	na	na	na	na	na	na		
Rank	21	22	23	24	25	26	27	28		

We can calculate the rank sums as $R_{na+} = 13 + 15 + \ldots + 28 = 300$ and $R_{a+} = 1 + 2 + \ldots + 13 + 15 = 106$ respectively. Thus

$$U_1 = 14^2 + \frac{14 \cdot 15}{2} - 106 = 195; \quad U_2 = 14^2 + \frac{14 \cdot 15}{2} - 300 = 1.$$

With $U = \min(195, 1) = 1$ we can calculate the test statistic, which is approximately normally distributed, as:

$$t(x, y) = \frac{U - \frac{n_1 \cdot n_2}{2}}{\sqrt{\frac{n_1 \cdot n_2 \cdot (n_1 + n_2 + 1)}{12}}} = \frac{1 - \frac{14^2}{2}}{\sqrt{\frac{14 \cdot 14 \cdot 29}{12}}} \approx -4.46$$

Since $|t(x, y)| = 4.46 > z_{1-\alpha/2} = 1.96$ the null hypothesis can be rejected. We conclude that the locations of the two distributions are shifted. Obviously the prices of the auction are smaller in general, and so is the median.

(h) We can type in the data and evaluate the summary statistics using the mean, var, and sd commands:

```
na <- c(18.19,16.98,19.97,...,17.72)
a <- c(10.5,12.0, 9.54,..., 11.02)
mean(na)
mean(a)
var(na)
...
```

The t.test command can be used to answer questions (b)–(e). For (b) and (c) we use

```
t.test(na,mu=16.95)
```

```
One Sample t-test
```

```
data:  na
t = -0.0031129, df = 13, p-value = 0.9976
alternative hypothesis: true mean is not equal to 16.95
95 percent confidence interval:
 15.95714 17.94001
sample estimates:
mean of x
 16.94857
```

The test decision can be made by means of either the p-value ($= 0.9976 > 0.05$) or the confidence interval ([15.95; 17.94], which covers 16.95). To answer (d) and (e) we need to make use of the option `alternative` which specifies the alternative hypothesis:

```
t.test(a,mu=16.95,alternative='less')
t.test(na,a, alternative='greater')
```

Note that the two-sample test provides a confidence interval for the difference of the means. Questions (f) and (g) can be easily solved by using the `var.test` and `wilcox.test` commands:

```
var.test(na,a)
wilcox.test(na,a)
```

```
F test to compare two variances
```

```
data:  na and a
F = 1.198, num df = 13, denom df = 13, p-value = 0.7496
alternative hypothesis: true ratio of variances not equal to 1
95 percent confidence interval:
 0.3845785 3.7317371
sample estimates:
ratio of variances
        1.197976
```

```
Wilcoxon rank sum test with continuity correction
```

```
data:  na and a
W = 195, p-value = 8.644e-06
alternative hypothesis: true location shift is not equal to 0
```

A binary test decision can be made by using the p-value.

Solution to Exercise 10.4   Since the data before and after the diet are dependent (weight measured on the same subjects) we need to use the paired t-test. To calculate the test statistic we need to first determine the weight differences:

Person i	1	2	3	4	5	6	7	8	9	10
Before diet	80	95	70	82	71	70	120	105	111	90
After diet	78	94	69	83	65	69	118	103	112	88
Differences d	2	1	1	−1	6	1	2	2	−1	2

Using $\bar{d} = 1.5$ and

$$
s_d^2 = \frac{1}{10-1} \cdot (0.5^2 + 0.5^2 + 0.5^2 + 2.5^2 + 4.5^2 + \ldots + 0.5^2) = 3.83 ,
$$

we calculate the test statistic as

$$
t(d) = \frac{\bar{d}}{s_d} \sqrt{n} = \frac{1.5}{\sqrt{3.83}} \sqrt{10} = 2.42 .
$$

The null hypothesis is rejected because $|t(d)| = 2.42 > t_{9;0.975} = 2.26$. These results suggest that the mean weight before and after the diet are different. We would have obtained the same results by calculating the confidence interval for the differences:

$$
\left[\bar{d} \pm t_{n-1;1-\alpha/2} \frac{s_d}{\sqrt{n}} \right] \Leftrightarrow \left[1.5 \pm 2.26 \cdot \frac{\sqrt{3.83}}{\sqrt{10}} \right] = [0.1; 2.9] .
$$

Since the confidence interval does not overlap with zero, we reject the null hypothesis; using the confidence interval to arrive at a decision suggests that the mean difference is different (i.e. greater) from zero. While the test is significant, and suggests a weight difference, it is worth noting that the mean benefit from the diet is only 1.5 kg. Whether this is a relevant reduction in weight has to be decided by the ten people experimenting with the diet. Of course, interpreting the interval as a compatibility interval shows that a hypothesized weight loss of up to 2.9 kg is more compatible with the data as hypotheses outside this interval, including no weight difference.

Solution to Exercise 10.5

(a) The production is no longer profitable if the probability of finding a deficient shirt is greater than 10%. This equates to the hypotheses:

$$H_0 : p \leq 0.1 \qquad \text{versus} \qquad H_1 : p > 0.1 .$$

The sample proportion of deficient shirts is $\hat{p} = \frac{30}{230} = \frac{6}{46} \approx 0.13$. Since $np(1 - p) = 230 \cdot \frac{1}{10} \cdot \frac{9}{10} > 9$ we can use the test statistic

$$t(x) = \frac{\hat{p} - p_0}{\sqrt{p_0(1 - p_0)}} \sqrt{n} = \frac{\frac{6}{46} - \frac{1}{10}}{\sqrt{\frac{1}{10} \cdot \frac{9}{10}}} \cdot \sqrt{230}$$

$$= \frac{5}{115} \cdot \frac{10}{3} \cdot \sqrt{230} \approx 2.198 .$$

The null hypothesis gets rejected because $t(x) = 2.198 > z_{0.95} = 1.64$. It seems the production is no longer profitable.

(b) The test statistic is $t(x) = \sum x_i = 30$. The critical region can be determined by calculating

$$P_{H_0}(Y \leq c_l) \leq 0.025 \qquad \text{and} \qquad P_{H_0}(Y \geq c_r) \leq 0.975 .$$

Using R we get

```
qbinom(p=0.975,prob=0.1,size=230)
[1] 32
qbinom(p=0.025,prob=0.1,size=230)
[1] 14
```

The test statistic ($t(x) = 30$) does not fall outside the critical region ([14; 32]); therefore the null hypothesis is *not* rejected. The same result is obtained by using the Binomial test in R: `binom.test(c(30,200), p=0.1, alternative='greater')`. This yields a p-value of 0.095 and a confidence interval covering 10% ([0.09; 0.18]). Interestingly, the approximate Binomial test rejects the null hypothesis whereas the exact test keeps it. This highlights that a binary decision based on a particular test or interval may not be ideal for the controlling department; drawing a p-value function would exhibit that we get the highest support (i.e. $p = 1$) for $H_0 : p = 0.1304348$.

(c) The research hypothesis is that the new machine produces fewer deficient shirts:

$$H_0 : p_{new} \geq p_{old} \qquad \text{versus} \qquad H_1 : p_{new} < p_{old} .$$

To calculate the test statistic, we need the following:

$$\hat{d} = \frac{x_{new}}{n_{new}} - \frac{x_{old}}{n_{old}} = \frac{7}{115} - \frac{7}{46} = -\frac{21}{230}$$

$$\hat{p} = \frac{x_{new} + x_{old}}{n_{new} + n_{old}} = \frac{7 + 35}{230 + 115} = \frac{42}{345} = \frac{14}{115} .$$

This yields:

$$t(x_{new}, x_{old}) = \frac{\hat{d}}{\sqrt{\hat{p}(1 - \hat{p})(\frac{1}{n_{new}} + \frac{1}{n_{old}})}} = \frac{-\frac{21}{230}}{\sqrt{\frac{14}{115} \cdot \frac{101}{115}(\frac{1}{115} + \frac{1}{230})}}$$

$$= \frac{-\frac{21}{230}}{\sqrt{0.1069 \cdot 0.013}} = -\frac{0.0913}{0.0373} = -2.448 .$$

The null hypothesis is rejected because $t(x_{new}, x_{old}) = -2.448 < z_{0.05} = -z_{0.95} = -1.64$. This means we accept the alternative hypothesis that the new machine is better than the old one.

(d) The data can be summarized in the following table:

	Machine 1	Machine 2
Deficient	30	7
Fine	200	112

To test the hypotheses established in (c) we apply the `fisher.test` command onto the contingency table:

```
fisher.test(matrix(c(30,200,7,112),ncol=2))
```

This yields a p-value of 0.0438 suggesting, as the test in (c), that the null hypothesis should be rejected. Note that the confidence interval for the odds ratio, also reported by R, does not necessarily yield the same conclusion as the test of Fisher.

Solution to Exercise 10.6

(a) To compare the two means, we should use the Welch test (since the variances cannot be assumed to be equal). The test statistic is

$$t(x, y) = \frac{|\bar{x} - \bar{y}|}{\sqrt{\frac{s_2^2}{n_2} + \frac{s_1^2}{n_1}}} = \frac{|103 - 101.8|}{\sqrt{\frac{12599.56}{10} + \frac{62.84}{10}}} \approx 0.0337 .$$

The alternative hypothesis is $H_1 : \mu_2 > \mu_1$, i.e. we deal with a one-sided hypothesis. The null hypothesis is rejected if $t(x, y) > t_{v;1-\alpha}$. Calculating

$$v = \left(\frac{s_1^2}{n_1} + \frac{s_2^2}{n_2}\right)^2 \bigg/ \left(\frac{(s_1^2/n_1)^2}{n_1 - 1} + \frac{(s_2^2/n_2)^2}{n_2 - 1}\right)$$

$$= \left(\frac{62.844}{10} + \frac{12599}{10} \right)^2 / \left(\frac{(62.844/10)^2}{9} + \frac{(12599/10)^2}{9} \right) \approx 9.09$$

yields $t_{9.09;0.95} = 1.831$ (using qt(0.95,9.09) in R; or looking at Table C.2 for 9 degrees of freedom). Therefore $t(x, y) < t_{9.09;0.95}$ and the null hypothesis is not rejected. This means player 2 could not prove that she scores higher on average.

(b) We have to first rank the merged data:

Value	6	29	40	47	47	64	87	88	91	98
Sample	2	2	2	2	2	2	2	1	1	2
Rank	1	2	3	4	5	6	7	8	9	10
Value	99	99	101	104	105	108	111	112	261	351
Sample	1	1	1	1	1	1	1	1	2	2
Rank	11	12	13	14	15	16	17	18	19	20

This gives us $R_{1+} = 8 + 9 + \ldots + 18 = 133$, $R_{2+} = 1 + 2 + \ldots + 20 = 77$, $U_1 = 10^2 + (10 \cdot 11)/2 - 133 = 22$, $U_2 = 10^2 + (10 \cdot 11)/2 - 77 = 78$, and therefore $U = 22$. The test statistic can thus be calculated as

$$t(x, y) = \frac{U - \frac{n_1 \cdot n_2}{2}}{\sqrt{\frac{n_1 \cdot n_2 \cdot (n_1 + n_2 + 1)}{12}}} = \frac{22 - \frac{10^2}{2}}{\sqrt{\frac{10 \cdot 10 \cdot 21}{12}}} \approx -2.12 .$$

Since $|t(x, y)| = 2.12 > z_{1-\alpha} = 1.64$, the null hypothesis is rejected. The test supports the alternative hypothesis of higher points for player 1. The U-test has the advantage of not being focused on the expectation μ. The two samples are clearly different: the second player scores with much more variability and her distribution of scores is clearly not symmetric and normally distributed. Since the sample is small, and the assumption of a normal distribution is likely not met, it makes sense to *not* use the t-test. Moreover, because the distribution is skewed the mean may not be a sensible measure of comparison. The two tests yield different conclusions in this example which shows that one needs to be careful in framing the right hypotheses and understanding the assumptions of the test (as highlighted in Sect. 10.9.4). A drawback of the U-test is that it uses only ranks and not the raw data: it thus uses less information than the t-test which could be preferred when comparing means of a reasonably sized sample.

Solution to Exercise 10.7   Otto speculates that the probability of finding a bear of colour i is 0.2, i.e. $p_{white} = 0.2$, $p_{red} = 0.2$, $p_{orange} = 0.2$, $p_{yellow} = 0.2$, and $p_{green} = 0.2$. This hypothesis can be tested by using the χ^2 goodness of fit test. The test statistic is

$$t(x) = \chi^2 = \sum_{i=1}^{k} \frac{(N_i - np_i)^2}{np_i} = \frac{1}{250}(222 - 250)^2 + (279 - 250)^2$$
$$+ (251 - 250)^2 + (232 - 250)^2 + (266 - 250)^2 = 8.824 \,.$$

The null hypothesis cannot be rejected because $t(x) = 8.824 \not> c_{4-1-0;0.95} = 9.49$. While the small number of white bears might be disappointing, the test suggests that there is not enough evidence to reject the hypothesis of equal probabilities.

Solution to Exercise 10.8

(a) To answer this question we need to conduct the χ^2-independence test. The test statistic $t(x, y)$ is identical to Pearson's χ^2 statistic, introduced in Chap. 4. In Exercise 4.4 we have calculated this statistic already, $\chi^2 \approx 182$, for the details. The null hypothesis of independence is rejected if $t(x, y) = \chi^2 > c_{(I-1)(J-1);1-\alpha}$. With $I = 2$ (number of rows), and $J = 4$ (number of columns) we get $c_{3;0.95} = 7.81$ using Table C.3 (or qchisq(0.95,3) in R). Since $182 > 7.81$, we reject the null hypothesis of independence.

(b) The output refers to a χ^2 test of homogeneity: the null hypothesis is that the proportion of passengers rescued is identical for the different travel classes. This hypothesis is rejected because p is smaller than $\alpha = 0.05$. It is evident that the proportions of rescued passengers in the first two classes (60%, 43.9%) is much higher than in the other classes (25%, 23.8%). One can see that the test statistic (182.06) is identical to (a). This is not surprising: the χ^2-independence test and the χ^2 test of homogeneity are technically identical, but the null hypotheses differ. In (a) we showed that "rescue status" and "travel class" are independent; in (b) we have seen that the conditional distributions of rescue status given travel class differ by travel class, i.e. that the proportions of those rescued differ by the categories 1. class/2. class/3. class/staff.

(c) The summarized table is as follows:

	1. Class/2. Class	3. Class/Staff	Total
Rescued	327	391	718
Not rescued	295	1215	1510
Total	622	1606	2228

Using (4.7) we get

$$t(x, y) = \chi^2 = \frac{2228(327 \cdot 1215 - 295 \cdot 391)^2}{718 \cdot 1510 \cdot 622 \cdot 1606} = 163.55 \,.$$

The χ^2-independence test and the χ^2 test of homogeneity are technically identical: H_0 is rejected if $t(x, y) > c_{(I-1)(J-1);1-\alpha}$. Since $163.55 > c_{1;0.95} = 3.84$ the null hypothesis is rejected. As highlighted in (b) this has two interpretations: (i) "rescue status" and "travel class" are independent (independence test) and (ii) the conditional distributions of rescue status given travel class differ by travel class (homogeneity test). The second interpretation implies that the probability of being rescued differs by travel class. The null hypothesis of the same probabilities of being rescued is also rejected using the test of Fisher. Summarizing the data in a matrix and applying the `fisher.test` command yields a p-value smaller than $\alpha = 0.05$.

```
fisher.test(matrix(c(327,295,391,1215),ncol=2,nrow=2))
```

Solution to Exercise 10.9

(a) The hypotheses are:

$$H_0 : \mu_X = \mu_{Y1} \qquad \text{versus} \qquad H_1 : \mu_X \neq \mu_{Y1}\,.$$

The pooled variance,

$$s^2 = \frac{19 \cdot 2.94 + 19 \cdot 2.46}{39} = \frac{102.6}{39} = 2.631\,,$$

is needed to calculate the test statistic:

$$t(x, y) = \frac{4.97 - 4.55}{1.622} \cdot \sqrt{\frac{400}{40}} = \frac{0.42}{1.622} \cdot \sqrt{10} = 0.8188.$$

H_0 is rejected if $|t(x, y)| > t_{38,0.975} \approx 2.02$. Since $|t(x, y)| = 0.8188$ we do not reject the null hypothesis.

(b) The hypotheses are:

$$H_0 : \mu_X = \mu_{Y2} \qquad \text{versus} \qquad H_1 : \mu_X \neq \mu_{Y2}\,.$$

The pooled variance,

$$s^2 = \frac{19 \cdot 2.94 + 19 \cdot 3.44}{39} = 3.108\,,$$

is needed to calculate the test statistic:

$$t(x, y) = \frac{4.97 - 3.27}{1.763} \cdot \sqrt{10} = 3.049.$$

H_0 is rejected because $|t(x, y)| = 3.049 > t_{38,0.975} \approx 2.02$.

(c) In both (a) and (b) there exists a true difference in the mean. However, only in (b) is the t-test able to detect the difference. This highlights that smaller differences can only be detected if the sample size is sufficiently large. However, if the sample size is very large, it may well be that the test detects a difference where there is no difference. Drawing p- and S-value functions are viable options to avoid errors in binary decision making and to visualize which hypotheses are most compatible with the data (given the background assumptions).

Solution to Exercise 10.10

(a) After reading in and attaching the data, we can simply use the t.test command to compare the expenditure of the two samples:

```
theatre <- read.csv('theatre.csv')
attach(theatre)
t.test(Culture[Sex==1],Culture[Sex==0])
```

```
        Welch Two Sample t-test

data:  Culture[Sex == 1] and Culture[Sex == 0]
t = -1.3018, df = 667.43, p-value = 0.1934
alternative hypothesis: true difference not equal to 0
95 percent confidence interval:
 -12.841554    2.602222
sample estimates:
mean of x mean of y
 217.5923   222.7120
```

We see that the null hypothesis is *not* rejected because $p = 0.1934 > \alpha$ (also, the confidence interval overlaps with "0").

(b) A two-sample t-test and a U-test yield the same conclusion. The p-values, obtained with

```
wilcox.test(Culture[Sex==1],Culture[Sex==0])
t.test(Culture[Sex==1],Culture[Sex==0],var.equal=TRUE)
```

are 0.1946 and 0.145 respectively. Interestingly, the test statistic of the two-sample t-test is almost identical to the one from the Welch test in (a) (-1.2983)—this indicates that the assumption of equal variances may not be unreasonable.

(c) We can use the usual t.test command together with the option alternative = 'greater' to get the solution.

```
t.test(Theatre[Sex==1],Theatre[Sex==0],
alternative='greater')
```

Here, p is much smaller than 0.001; hence, the null hypothesis can be rejected. Women spend more on theater visits than men.

(d) We deal with dependent (paired) data because different variables (expenditure this year versus expenditure last year) are measured on the same observations. Thus, we need to use the paired t-test—which we can use in R by specifying the `paired` option:

```
t.test(Theatre,Theatre_ly,paired=TRUE)
```

```
        Paired t-test

data:  Theatre and Theatre_ly
t = 1.0925, df = 698, p-value = 0.275
alternative hypothesis: true difference in means is != 0
95 percent confidence interval:
 -2.481496  8.707533
sample estimates:
mean of the differences
            3.113019
```

Both the p-value (which is greater than 0.05) and the confidence interval (overlapping with "0") state that the null hypothesis should be kept.

(e) We can extract both the confidence interval and the p-value from `t.test` and use `log2()` to calculate the S-value:

```
-log2(t.test(Theatre,Theatre_ly,paired=TRUE)$p.value)
t.test(Theatre,Theatre_ly,paired=TRUE)$conf.int
```

This yields a S-value of ≈ 1.86 and a compatibility interval of $[-2.48; 8.71]$. Thus, 1.86 bits of information can be implied as evidence against the null hypothesis. Rounding to the nearest integer we conclude that the data are no more surprising than getting 2 heads in 2 coin tosses. All hypotheses of a mean difference of expenditure on theatre visits between -2.48SFR and 8.71SFR have more support in the data than hypotheses outside this interval.

We can draw the S-value function, displayed in Fig. B.20, as follows:

```
library(concurve)
pc <- curve_mean(Theatre,Theatre_ly,data=theatre,
paired=T)
ggcurve(pc[[1]],type='s')
```

The function shows the level of evidence against different null hypotheses. It is clear that there is considerable evidence against hypotheses that target practically

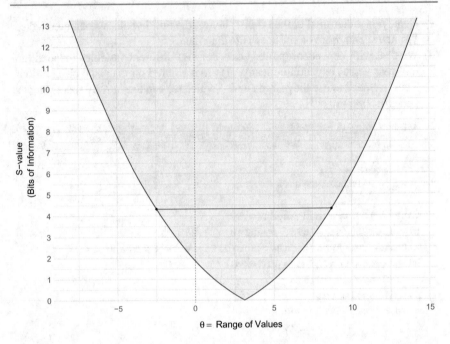

Fig. B.20 *S*-values (surprisals) for a range of hypothesized differences of expenditure between the two years

relevant differences in expenditure (say $5 - 10$ SFR). This complements our findings from (d).

Solution to Exercise 10.11

(a) We can use the one-sample *t*-test to answer both questions:

```
t.test(temperature,mu=65,alternative='greater')
t.test(time,mu=30,alternative='less')
```

```
        One Sample t-test

data:  temperature
t = -11.006, df = 1265, p-value = 1
alternative hypothesis: true mean is greater than 65
...


data:  time
t = 23.291, df = 1265, p-value = 1
alternative hypothesis: true mean is less than 30
```

We cannot confirm that the mean delivery time is less than 30 min. and that the mean temperature is greater than 65 °C. This is not surprising: we have already seen in Exercise 3.10 that the manager should be unsatisfied with the performance of his company.

(b) We can use the exact Binomial test to investigate $H_0 : p \geq 0.15$ and $H_1 : p < 0.15$. For the `binom.test` command we need to know the numbers of successes and failures, i.e. the number of deliveries where a free wine should have been given to the customer. Applying the `table` commands yields 229 and 1037 deliveries respectively.

```
table(free_wine)
binom.test(c(229,1037),p=0.15,alternative='less')
```

```
        Exact binomial test

data:  c(229, 1037)
number of successes = 229, number of trials = 1266,
p-value = 0.9988
alternative hypothesis: true probability
is less than 0.15
95 percent confidence interval:
 0.0000000 0.1996186
sample estimates, probability of success: 0.1808847
```

The null hypothesis cannot be rejected because $p = 0.9988 > \alpha = 0.05$ and because the confidence interval covers $p = 0.15$. We get the same results if we use the variable "got_wine" instead of "free_wine". While the test says that we can't exclude the possibility that the probability of receiving a free wine is less than 15%, the point estimate of 18% suggests that the manager still has to improve the timeliness of the deliveries or stop the offer of free wine.

(c) We first need to create a new categorical variable (using `cut`) which divides the temperatures into two parts: below and above 65 °C. Then we can simply apply the test commands (`fisher.test`, `chisq.test`, `prop.test`) to the table of branch and temperature:

```
hot <- cut(temperature,breaks=c(-Inf,65,Inf))
fisher.test(table(hot,operator))
chisq.test(table(hot,operator))
prop.test(table(hot,operator))
```

We know that the two χ^2 tests lead to identical results. For both of them the p-value is 0.2283 which suggests that we should keep the null hypothesis. There is not enough evidence which would support that the proportion of hot pizzas

differs by operator, i.e. that the two variables are independent! The test of Fisher
yields the same result ($p = 0.2227$).

(d) The null hypothesis is that the proportion of deliveries is the same for each branch:
$H_0 : p_{East} = p_{West} = p_{Centre}$. To test this hypothesis we need a χ^2 goodness of
fit test:

```
chisq.test(table(branch))
```

```
        Chi-squared test for given probabilities

data:   table(branch)
X-squared = 0.74408, df = 2, p-value = 0.6893
```

The null hypothesis of equal proportions is therefore not rejected ($p > \alpha =$
0.05).

(e) To compare three proportions we need to use the χ^2 homogeneity test:

```
prop.test(table(branch, operator))
```

```
X-squared = 0.15719, df = 2, p-value = 0.9244
alternative hypothesis: two.sided
sample estimates:
   prop 1    prop 2    prop 3
0.5059382 0.5097561 0.4965517
```

We can see that the proportions are almost identical and that the null hypothesis
is not rejected ($p = 0.9244$).

(f) To test for this relationship we use the χ^2-independence test:

```
chisq.test(table(driver, branch))
```

```
X-squared = 56.856, df = 8, p-value = 1.921e-09
```

The null hypothesis of independence is rejected.

Solution to Exercise 10.12   To test the hypothesis $p_{Shalabh} = p_{Heumann} = p_{Schomaker} = 1/3$ we use the χ^2 goodness of fit test. The test statistic is

$$\chi^2 = \sum_{i=1}^{k} \frac{(N_i - np_i)^2}{np_i} = \frac{1}{111}[(110 - 111)^2 + (118 - 111)^2 + (105 - 111)^2]$$
$$= (1 + 49 + 36)/111 \approx 0.77$$

H_0 gets rejected if $\chi^2 = 0.77 > c_{3-1-0,1-0.01} = 9.21$. Thus, the null hypothesis is not rejected. We accept that all three authors took about one third of the pictures.

Chapter 11

Solution to Exercise 11.1

(a) Calculating $\bar{x} = \frac{1}{6}(26 + 23 + 27 + 28 + 24 + 25) = 25.5$ and $\bar{y} = \frac{1}{6}(170 + 150 + 160 + 175 + 155 + 150) = 160$, we obtain the following table needed for the estimation of $\hat{\alpha}$ and $\hat{\beta}$:

	Body mass index			Systolic blood pressure			
	x_i	$x_i - \bar{x}$	$(x_i - \bar{x})^2$	y_i	$y_i - \bar{y}$	$(y_i - \bar{y})^2$	v_i
	26	0.5	0.25	170	10	100	5
	23	−2.5	6.25	150	−10	100	25
	27	1.5	2.25	160	0	0	0
	28	2.5	6.25	175	15	225	37.5
	24	−1.5	2.25	155	−5	25	7.5
	25	−0.5	0.25	150	−10	100	5
Total	153		17.5	960		550	80

With $\sum_i v_i = \sum_i (x_i - \bar{x}) \cdot (y_i - \bar{y}) = 80$, it follows that $S_{xy} = 80$. Moreover, we get $S_{xx} = \sum_i (x_i - \bar{x})^2 = 17.5$ and $S_{yy} = \sum_i (y_i - \bar{y})^2 = 550$. The parameter estimates are therefore

$$\hat{\beta} = \frac{S_{xy}}{S_{xx}} = \frac{80}{17.5} \approx 4.57,$$

$$\hat{\alpha} = \bar{y} - \hat{\beta}\bar{x} = 160 - 4.57 \cdot 25.5 = 43.465.$$

A one-unit increase in the BMI therefore relates to a 4.57 unit increase in the blood pressure. The model suggests a positive association between BMI and systolic blood pressure. It is impossible to have a BMI of 0; therefore $\hat{\alpha}$ cannot be interpreted meaningfully here.

(b) Using (11.14), we obtain R^2 as

$$R^2 = r^2 = \left(\frac{S_{xy}}{\sqrt{S_{xx}S_{yy}}} \right)^2 = \left(\frac{80}{\sqrt{17.5 \cdot 550}} \right)^2 \approx 0.66.$$

Thus 66% of the data's variability can be explained by the model. The goodness of fit is good, but not perfect.

Solution to Exercise 11.2

(a) To estimate $\hat{\beta}$, we use the second equality from (11.6):

$$\hat{\beta} = \frac{\sum_{i=1}^n x_i \, y_i - n\bar{x}\bar{y}}{\sum_{i=1}^n x_i^2 - n\bar{x}^2}.$$

Calculating $\bar{x} = 1.333$, $\bar{y} = 5.556$, $\sum_{i=1}^n x_i \, y_i = 62.96$, $\sum_{i=1}^n x_i^2 = 24.24$, and $\sum_{i=1}^n y_i^2 = 281.5$ leads to

$$\hat{\beta} = \frac{62.91 - 9 \cdot 1.333 \cdot 5.556}{24.24 - 9 \cdot 1.333^2} \approx \frac{-3.695}{8.248} \approx -0.45 \,,$$

$$\hat{\alpha} = \bar{y} - \hat{\beta}\bar{x} = 5.556 + 0.45 \cdot 1.333 \approx 6.16 \,,$$

$$\hat{y}_i = 6.16 - 0.45 x_i \,.$$

For children who spend no time on the internet at all, this model predicts 6.16 h of deep sleep. Each hour spent on the internet decreases the time in deep sleep by 0.45 h which is 27 min.

(b) Using the results from (a) and $S_{yy} = \sum_{i=1}^n y_i^2 - n\bar{y}^2 \approx 3.678$ yields:

$$R^2 = r^2 = \frac{S_{xy}^2}{S_{xx}S_{yy}} = \frac{(-3.695)^2}{8.248 \cdot 3.678} \approx 0.45 \,.$$

About 45% of the variation can be explained by the model. The fit of the model to the data is neither very good nor very bad.

(c) After collecting the data in two vectors (c ()), printing a summary of the linear model (summary (lm())) reproduces the results. A scatter plot can be produced by plotting the two vectors against each other (plot ()). The regression line can be added with abline():

```
it <- c(0.3,2.2,...,2.3)
sleep <- c(5.8,4.4,...,6.1)
summary(lm(sleep~it))
plot(it,sleep)
abline(a=6.16,b=-0.45)
```

The plot is displayed in Fig. B.21.

Fig. B.21 Scatter plot and regression line for the association of internet use and deep sleep

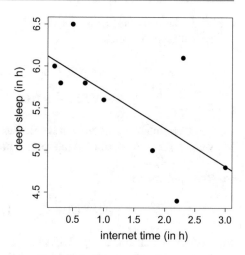

(d) Treating X as a binary variable yields the following values: $0, 1, 0, 0, 0, 1,$ $1, 0, 1$. We therefore have $\bar{x} = 0.444$, $\sum x_i^2 = 4$, and $\sum x_i y_i = 4.4 + 5.0 + 4.8 + 6.1 = 20.3$. Since the Y-values stay the same we calculate

$$\hat{\beta} = \frac{\sum_{i=1}^{n} x_i \, y_i - n\bar{x}\bar{y}}{\sum_{i=1}^{n} x_i^2 - n\bar{x}^2} = \frac{20.3 - 9 \cdot 5.556 \cdot 0.444}{4 - 9 \cdot 0.444^2} \approx -0.85 .$$

Thus, those children who are on the internet for a long time (i.e. > 1 hour) sleep on average $0.85\,\mathrm{h}$ ($= 51$ min.) less. If we change the coding of 1's and 0's, $\hat{\beta}$ will just have a different sign: $\hat{\beta} = 0.85$. In this case, we can conclude that children who spend less time on the internet sleep on average $0.85\,\mathrm{h}$ longer than children who spend more time on the internet. This is the same conclusion and highlights that the choice of coding doesn't affect the interpretation of the results.

Solution to Exercise 11.3

(a) The correlation coefficient is

$$r = \frac{S_{xy}}{\sqrt{S_{yy} S_{xx}}} = \frac{170, 821 - 17 \cdot 166.65 \cdot 60.12}{\sqrt{(62, 184 - 17 \cdot 60.12^2)(472, 569 - 17 \cdot 166.65^2)}}$$

$$= \frac{498.03}{\sqrt{738.955 \cdot 441.22}} = 0.87 .$$

This indicates strong positive correlation: the higher the height, the higher the weight. Since $R^2 = r^2 = 0.87^2 \approx 0.76$, we already know that the fit of a linear regression model will be good (no matter whether height or weight is treated as outcome). From (11.11), we also know that $\hat{\beta}$ will be positive.

(b) We know from (a) that $S_{xy} = 498.03$ and that $S_{xx} = 441.22$. The least squares estimates are therefore

$$\hat{\beta} = \frac{498.03}{441.22} = 1.129,$$
$$\hat{\alpha} = 60.12 - 166.65 \cdot 1.129 = -128.03.$$

Each centimetre difference in height therefore means a 1.129 kg difference in weight. It is not possible to interpret $\hat{\alpha}$ meaningfully in this example.

(c) The prediction is

$$-128.03 + 1.129 \cdot 175 = 69.545 \,\text{kg}.$$

(d) – (g) The black dots in Fig. B.22 show the scatter plot of the data. There is clearly a positive association in that greater height implies greater weight. This is also emphasized by the regression line estimated in (b).

The two additional points appear in dark grey in the plot. It is obvious that they don't match the pattern observed in the original 17 data points. One may therefore speculate that with the inclusion of the two new points $\hat{\beta}$ will be smaller. To estimate the new regression line we need

$$\bar{x} = \frac{1}{19}(17 \cdot 166.65 + 150 + 175) = 166.21,$$

$$\bar{y} = \frac{1}{19}(17 \cdot 60.12 + 75 + 55) = 60.63.$$

This yields

$$\hat{\beta} = \frac{\sum_{i=1}^{n} x_i \, y_i - n\bar{x}\bar{y}}{\sum_{i=1}^{n} x_i^2 - n\bar{x}^2} = \frac{191696 - 19 \cdot 166.21 \cdot 60.63}{525694 - 19 \cdot 166.21^2} = \frac{227.0663}{804.4821} \approx 0.28.$$

This shows that the two added points shrink the estimate from 1.129 to 0.28. The association becomes less clear. This is an insightful example showing that least squares estimates are generally *sensitive to outliers* which can potentially affect the results.

Solution to Exercise 11.4

(a) The point estimate of β suggests a 0.077% increase of hotel occupation for each one degree increase in temperature. However, the null hypothesis of $\beta = 0$ cannot be rejected because $p = 0.883 > 0.05$. We therefore can't show an association between temperature and hotel occupation.

Fig. B.22 Scatter plot and regression line for both 17 and 19 observations

(b) The average hotel occupation is higher in Davos (7.9%) and Polenca (0.9%) compared with Basel (reference category). However, these differences are not significant. Both $H_0 : \beta_{\text{Davos}} = 0$ and $H_0 : \beta_{\text{Polenca}} = 0$ cannot be rejected. The model can't show a significant difference in hotel occupation between Davos/Polenca and Basel.

(c) The analysis of variance table tells us that the null hypothesis of equal average temperatures in the three cities ($\beta_1 = \beta_2 = 0$) cannot be rejected. Note that in this example the overall F-test would have given us the same results.

(d) In the multiple regression model the main conclusions of (a) and (b) don't change: testing $H_0 : \beta_j = 0$ never leads to the rejection of the null hypothesis. We can't show an association between temperature and hotel occupation (given the city); and we can't show an association between city and hotel occupation (given the temperature).

(e) Stratifying the data yields considerably different results compared to (a)–(c): In Davos, where tourists go for skiing, each increase of $1\,°C$ relates to a drop in hotel occupation of 2.7%. The estimate $\hat{\beta} \approx -2.7$ is also significantly different from zero ($p = 0.000231$). In Polenca, a summer holiday destination, an increase of $1\,°C$ implies an increase of hotel occupation of almost 4%. This estimate is also significantly different from zero ($p = 0.00114 < 0.05$). In Basel, a business destination, there is a somewhat higher hotel occupation for higher temperatures ($\hat{\beta} = 1.3$); however, the estimate is not significantly different from zero ($S = -log_2(0.1796) \approx 2.5$). While there is no overall association between temperature and hotel occupation (see (a) and (c)), there is an association between them if one looks at the different cities separately. This suggests that an interaction between temperature and city should be included in the model.

(f) The design matrix contains a column of 1's (to model the intercept), the temperature and two dummies for the categorical variable "city" because it has three categories. The matrix also contains the interaction terms which are both the product of temperature and Davos and temperature and Polenca. The matrix has 36 rows because there are 36 observations: 12 for each city.

	Int.	Temp.	Davos	Polenca	Temp. $\times$ Davos	Temp. $\times$ Polenca
1	1	−6	1	0	−6	0
2	1	−5	1	0	−5	0
⋮	⋮	⋮	⋮	⋮	⋮	⋮
12	1	0	1	0	0	0
13	1	10	0	1	0	10
⋮	⋮	⋮	⋮	⋮	⋮	⋮
24	1	12	0	1	0	12
25	1	1	0	0	0	0
⋮	⋮	⋮	⋮	⋮	⋮	⋮
36	1	4	0	0	0	0

(g) Both interaction terms are significantly different from zero ($p = 0.000375$ and $p = 0.033388$). The estimate of temperature therefore differs by city; and the estimate of city differs by temperature. For the reference city of Basel the association between temperature and hotel occupation is estimated as 1.31; for Davos it is $1.31 - 4.00 = -2.69$ and for Polenca $1.31 + 2.66 = 3.97$. Note that these results are identical to (d) where we fitted three different regressions, they are just summarized in a different way.

(h) From (f) it follows that the point estimates for $b_{\text{temperature}}$ are 1.31 for Basel, −2.69 for Davos, and 3.97 for Polenca. Confidence intervals for these estimates can be obtained via (11.29):

$$(\hat{\beta}_i + \hat{\beta}_j) \pm t_{n-p-1;1-\alpha/2} \cdot \hat{\sigma}_{(\hat{\beta}_i + \hat{\beta}_j)} \, .$$

We calculate $t_{n-p-1;1-\alpha/2} = t_{36-5-1,0.975} = t_{30,0.975} = 2.04$. With $\text{Var}(\beta_{\text{temp.}}) = 0.478$ (obtained via 0.6916^2 from the model output or from the second row and second column of the covariance matrix), $\text{Var}(\beta_{\text{temp:Davos}}) = 0.997$, $\text{Var}(\beta_{\text{Polenca}}) = 1.43$, $\text{Cov}(\beta_{\text{temp.}}, \beta_{\text{temp:Davos}}) = -0.48$, and also $\text{Cov}(\beta_{\text{temp.}}, \beta_{\text{temp:Polenca}}) = -0.48$ we obtain:

$$\hat{\sigma}_{(\hat{\beta}_{\text{temp.}} + \hat{\beta}_{\text{Davos}})} = \sqrt{0.478 + 0.997 - 2 \cdot 0.48} \approx 0.72 \, ,$$

$$\hat{\sigma}_{(\hat{\beta}_{\text{temp.}} + \hat{\beta}_{\text{Polenca}})} = \sqrt{0.478 + 1.43 - 2 \cdot 0.48} \approx 0.97 \, ,$$

$$\hat{\sigma}_{(\hat{\beta}_{\text{temp.}} + \hat{\beta}_{\text{Basel}})} = \sqrt{0.478 + 0 + 0} \approx 0.69 \, .$$

The 95% confidence intervals are therefore:

$$\text{Davos: } [-2.69 \pm 2.04 \cdot 0.72] \approx [-4.2; -1.2],$$
$$\text{Polenca: } [3.97 \pm 2.04 \cdot 0.97] \approx [2.0; 5.9],$$
$$\text{Basel: } [1.31 \pm 2.04 \cdot 0.69] \approx [-0.1; 2.7].$$

Solution to Exercise 11.5

(a) The missing value [1] can be calculated as

$$T = \frac{\hat{\beta}_i - \beta_i}{\hat{\sigma}_{\hat{\beta}_i}} = \frac{0.39757 - 0}{0.19689} = 2.019.$$

Since $t_{699-5-1,0.975} = 1.96$ and $2.019 > 1.96$ it is clear that the p-value from [2] is smaller than 0.05. The exact p-value can be calculated in R via (1-pt(2.019, 693))*2 which yields 0.0439. The pt command gives the probability value for the quantile of 2.019 (with 693 degrees of freedom): 0.978. Therefore, with probability $(1 - 0.978)\%$ a value is right of 2.019 in the respective t-distribution which gives, multiplied by two to account for a two-sided test, the p-value.

(b) – (c) The plot on the left shows that the residuals are certainly not normally distributed as required by the model assumptions. The dots don't approximately match the bisecting line. There are too many high positive residuals which means that we are likely dealing with a right-skewed distribution of residuals. The plot on the right looks alright: no systematic pattern can be seen, it is a random plot. The histogram of both theatre expenditure and log(theatre expenditure) suggests that a log-transformation may improve the model, see Fig. B.23. Log-transformations are often helpful when the outcome's distribution is skewed to the right.

(d) Since the outcome is log-transformed, we can apply the interpretation of a log-linear model:

- Each year increase in age yields an $\exp(0.0038) = 1.0038$ times higher $(= 0.38\%)$ expenditure on theatre visits. Therefore a 10-year age difference relates to an $\exp(10 \cdot 0.0038) = 1.038$ times higher expenditure $(= 3.8\%)$.
- Women (gender= 1) spend on average (given the other variables) $\exp(0.179) \approx 1.20$ times more money on theatre visits.
- Each 1000 SFR more yearly income relates to an $\exp(0.0088) = 1.0088$ times higher expenditure on theatre visits. A difference in 10,000 SFR per year therefore amounts to an 8.8% difference in expenditure.
- Each extra Swiss Franc spent on cultural activities is associated with an $\exp(0.00353) = 1.0035$ times higher expenditure on theatre visits.
- Except for theatre expenditure from the preceding year, all β_j are significantly different from zero.

(a) Expenditure   (b) log(expenditure)

Fig. B.23 Histogram of the outcome

(e) While in (b) the residuals were clearly not normally distributed, this assumption seems to be fulfilled now: the QQ-plot shows dots which lie approximately on the bisecting line. The fitted values versus residuals plot remains a chaos plot. In conclusion, the log-transformation of the outcome helped to improve the quality of the model.

(f) We can draw the S-value function, displayed in Fig. B.24, as follows:

```
library(concurve)
m <- lm(I(log(Theatre))~Age+Sex+Income+Culture+Theatre_ly)
s_fun <- curve_gen(m,'Theatre_ly')
ggcurve(s_fun[[1]],type='s')
```

The function shows the level of evidence against different null hypotheses. There is little or moderate evidence against null hypotheses where $\beta_{\text{Theatre\_ly}} > 0$ and < 0.003 (with less than 4 bits of information against them). This suggests that there could be a relationship between the expenditure from the last and the current year. Interpreting $\hat{\beta}_{\text{Theatre\_ly}} = 0.0013492$ tells us that for every 10 Swiss Francs spent in the preceding year the expenditure on theatre visits in the current year increase by a factor of $\exp(0.0013492)^{10} \approx 1.014$. While is is possible, that there is some relationship, it is rather small.

Solution to Exercise 11.6

(a) The multiple regression model is obtained by using the `lm()` command and separating the covariates with the $+$ sign. Applying `summary()` to the model returns the comprehensive summary.

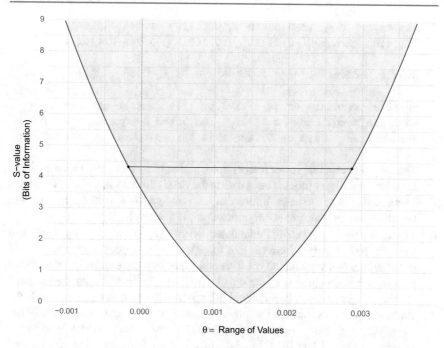

Fig. B.24 *S*-values (surprisals) for a range of hypothesized parameter values for the expenditure from the previous year

```
mp <- lm(time ~ temperature + branch +
day + operator + driver + bill + pizzas +
discount_customer)
summary(mp)
```

	Estimate	Std. Error	t value	Pr(>|t|)	
(Intercept)	40.42270	2.00446	20.166	< 2e-16	***
temperature	-0.20823	0.02594	-8.027	2.28e-15	***
branchEast	-1.60263	0.42331	-3.786	0.000160	***
branchWest	-0.11912	0.37330	-0.319	0.749708	
dayMonday	-1.15858	0.63300	-1.830	0.067443	.
daySaturday	0.88163	0.50161	1.758	0.079061	.
daySunday	1.01655	0.56103	1.812	0.070238	.
dayThursday	0.78895	0.53006	1.488	0.136895	
dayTuesday	0.79284	0.62538	1.268	0.205117	
dayWednesday	0.25814	0.60651	0.426	0.670468	
operatorMelissa	-0.15791	0.34311	-0.460	0.645435	
driverDomenico	-2.59296	0.73434	-3.531	0.000429	***
driverLuigi	-0.80863	0.58724	-1.377	0.168760	
driverMario	-0.39501	0.43678	-0.904	0.365973	
driverSalvatore	-0.50410	0.43480	-1.159	0.246519	
bill	0.14102	0.01600	8.811	< 2e-16	***

```
pizzas                0.55618     0.11718    4.746 2.31e-06 ***
discount_customer -0.28321        0.36848   -0.769 0.442291
---
Signif. codes:   0 ***   0.001 **   0.01 *   0.05 .   0.1   1

Residual standard error: 5.373 on 1248 degrees of freedom
Multiple R-squared:  0.3178,     Adjusted R-squared:  0.3085
F-statistic:   34.2 on 17 and 1248 DF,  p-value: < 2.2e-16
```

The output shows that lower temperature, higher bills, and more ordered pizzas increase the delivery times. The branch in the East is the fastest, and so is the driver Domenico. While there are differences with respect to day, discount customers and the operator, they are not significant at the 5% level.

(b) The confidence intervals are calculated as: $\hat{\beta}_i \pm t_{n-p-1;1-\alpha/2} \cdot \hat{\sigma}_{\hat{\beta}}$. We know from the model output from (a) that there are 1248 degrees of freedom (1266 observations $-$ 18 estimated coefficients). The respective quantile from the t-distribution is obtained with the qt() function. The coefficients are accessed via the coefficients command (alternatively: mp$coefficients); the variances of the coefficients are either accessed via the diagonal elements of the covariance matrix (diag(vcov(mp))) or the model summary (summary(mp)[[4]][,2])—both of which are laborious. The summary of coefficients, lower confidence limit (lcl) and upper confidence limit (ucl) may be summarized in a matrix, e.g. via merging the individual columns with the cbind command.

```
lcl <- coefficients(mp) - qt(0.975,1248)*sqrt(diag(vcov(mp)))
ucl <- coefficients(mp) + qt(0.975,1248)*sqrt(diag(vcov(mp)))
cbind(coefficients(mp),lcl,ucl)
```

```
                                   lcl          ucl
(Intercept)      40.4227014  36.4902223  44.3551805
temperature      -0.2082256  -0.2591146  -0.1573366
branchEast       -1.6026299  -2.4331162  -0.7721436
branchWest       -0.1191190  -0.8514880   0.6132501
dayMonday        -1.1585828  -2.4004457   0.0832801
...
```

(c) The variance is estimated as the residual sum of squares divided by the degrees of freedom, see also (11.27). Applying the residuals command to the model and using other basic operations yields an estimated variance of 28.86936.

```
sum(residuals(mp)^2)/(mp$df.residual)
```

Taking the square root of the result yields $\sqrt{28.86936} = 5.37$ which is also reported in the model output from (a) under "Residual standard error".

(d) The sum of squares error is defined as $\sum_{i=1}^{n}(y_i - \hat{y}_i)^2$. The total sum of squares is $\sum_{i=1}^{n}(y_i - \bar{y})^2$. This can be easily calculated in R. The goodness of fit is then obtained as $R^2 = 1 - SQ_{Error}/SQ_{Total} = 0.3178$. Dividing SQ_{Error} by $n - p - 1 = 1266 - 17 - 1 = 1248$ and SQ_{Total} by $n - 1 = 1265$ yields $R^2_{adj} = 0.3085$. This corresponds to the model output from (a).

```
SQE <- sum(residuals(mp)^2)
SQT <- sum((time-mean(time))^2)
1-(SQE/SQT)
1-((SQE/1248)/(SQT/1265))
```

(e) Applying `stepAIC` to the fitted model (with option "back" for backward selection) executes the model selection by means of AIC.

```
library(MASS)
stepAIC(mp, direction='back')
```

The output shows that the full model has an AIC of 4275.15. The smallest AIC is achieved by removing the operator variable from the model.

```
Step:   AIC=4275.15
time ~ temperature + branch + day + operator + driver + bill +
    pizzas + discount_customer

                     Df Sum of Sq   RSS    AIC
- operator            1      6.11 36035 4273.4
- discount_customer   1     17.05 36046 4273.8
<none>                            36029 4275.2
- day                 6    448.79 36478 4278.8
- driver              4    363.91 36393 4279.9
- branch              2    511.10 36540 4289.0
- pizzas              1    650.39 36679 4295.8
- temperature         1   1860.36 37889 4336.9
- bill                1   2241.30 38270 4349.6
```

The reduced model has an AIC of 4273.37. Removing the discount customer variable from the model yields an improved AIC (4272.0 < 4273.37).

```
Step:   AIC=4273.37
time ~ temperature + branch + day + driver + bill + pizzas +
    discount_customer

                     Df Sum of Sq   RSS    AIC
```

```
- discount_customer  1      17.57 36053 4272.0
<none>                             36035 4273.4
- day                6     452.00 36487 4277.1
- driver             4     364.61 36400 4278.1
- branch             2     508.57 36544 4287.1
- pizzas             1     649.54 36685 4294.0
- temperature        1    1869.98 37905 4335.4
- bill               1    2236.19 38271 4347.6
```

The model selection procedure stops here as removing any variable would only increase the AIC, not decrease it.

```
Step:   AIC=4271.98
time ~ temperature + branch + day + driver + bill + pizzas

               Df Sum of Sq   RSS    AIC
<none>                       36053 4272.0
- day           6    455.62 36508 4275.9
- driver        4    368.18 36421 4276.8
- branch        2    513.17 36566 4285.9
- pizzas        1    657.07 36710 4292.8
- temperature   1   1878.24 37931 4334.3
- bill          1   2228.88 38282 4345.9
```

The final model, based on backward selection with AIC, includes day, driver, branch, number of pizzas ordered, temperature and bill.

(f) Fitting the linear model with the variables obtained from (e) and obtaining the summary of it yields an R_{adj}^2 of 0.3092.

```
mps <- lm(time ~ temperature + branch + day
+ driver + bill + pizzas)
summary(mps)
```

This is only marginally higher than the goodness of fit from the full model (0.3092 > 0.3085). While the selected model is better than the model with all variables, both with respect to AIC and R_{adj}^2, the results are very close and remind us of the possible instability of applying automated model selection procedures.

(g) Both the normality assumption and heteroscedasticity can be checked by applying plot() to the model. From the many graphs provided we concentrate on the second and third of them:

```
plot(mps, which=2)
plot(mps, which=3)
```

(a) QQ-plot   (b) Fitted values vs. residuals

Fig. B.25 Checking the model assumptions

Figure B.25a shows that the residuals are approximately normally distributed because the dots lie approximately on the bisecting line. There are some smaller deviations at the tails but they are not severe. The plot of the fitted values versus the square root of the absolute values of the residuals shows no pattern, it is a random plot (Fig. B.25b). There seems to be no heteroscedasticity.

(h) Not all variables identified in (e) represent necessarily a "cause" for delayed or improved delivery time. It makes sense to speculate that *because* many pizzas are being delivered (and need to be made!) the delivery time increases. There might also be reasons why a certain driver is improving the delivery time: maybe he doesn't care about red lights. This could be investigated further given the results of the model above. However, high temperature doesn't *cause* the delivery time to be shorter; likely it is the other way around: the temperature is hotter because the delivery time is shorter. However, all of these considerations remain speculation. If there is a significant association, we know that given an accepted error (e.g. 5%), values of **x** are higher when values of **y** are higher. This is useful but it doesn't say whether **x** caused **y** or vice versa, unless we make additional assumptions and have additional knowledge (see Chap. 14).

(i) To check whether it is worth to add a polynomial, we simply add the squared temperature to the model. To make *R* understand that we apply a transformation we need to use I().

```
mps2 <- lm(time ~ temperature +
I(temperature^2) + I(temperature^3) + branch
+ day + driver + bill + pizzas)
summary(mps2)
```

```
                    Estimate Std. Error t value Pr(>|t|)
(Intercept)       -18.954965  8.795301  -2.155  0.03134 *
temperature         1.736692  0.282453   6.149 1.05e-09 ***
I(temperature^2)   -0.015544  0.002247  -6.917 7.36e-12 ***
branchEast         -1.429772  0.416107  -3.436  0.00061 ***
...
```

It can be seen that the null hypothesis $H_0 : \beta_{temp^2} = 0$ is rejected. This indicates that it is worthwile to assume a quadratic relationship between temperature and delivery time.

(j) The prediction can be obtained by the predict command as follows:

```
predict(mps,pizza[1266,])
```

The prediction is 36.5 min. and therefore 0.8 min. higher than the real delivery time.

Chapter 12

Solution to Exercise 12.1

(a) The odds ratio is the cross-product

$$OR = \frac{80 \cdot 60}{20 \cdot 40} = 6.$$

This means the chance of a successful treatment is 6 times higher for treatment *B* compared to treatment *A*.

(b) The estimated coefficients are identical to those derived in Example 12.0.1:

$$\hat{\beta}_0 = -0.4055,$$
$$\hat{\beta}_1 = 1.7918.$$

Exponentiating β_1 yields exp(1.7918), which gives us (approximately) the odds ratio $OR = 6$. There are at least two options to obtain these results in *R*:

```
# Solution 1: 'glm' can handle summarized successes
# and failures directly
treatment <- c('B', 'A')
df <- data.frame( rbind(c(80,20), c(40,60)), treatment)
colnames(df) <- c('y1', 'y0', 'treatment')

m.logit.1 <- glm(cbind(y1,y0)~treatment,
family=binomial, data=df)
summary(m.logit.1)
```

```
# Solution 2: expand the data by using weights
df <- data.frame( rbind( cbind(1,'B',80),
cbind(0,'B',20), cbind(1,'A',40), cbind(0,'A', 60)))
colnames(df) <- c('y', 'treatment', 'weights')
df$weights <- as.numeric(df$weights)
df$y <- as.numeric(df$y)
m.logit.2 <- glm(y~treatment, family=binomial,
weights=weights, data=df)
summary(m.logit.2)
```

Solution to Exercise 12.2

(a) We see that the chances of a long delivery are $\exp(0.011009) \approx 1.011$ times higher for the operator Melissa compared to the operator Laura (which is the reference category). The coefficient (0.011009) is not significantly different from zero $(S = -\log_2(0.625) \approx 0.68)$. Thus we cannot find a significant difference between the delivery times of the two operators. The delivery times of pizzas from the branch in the East are $\exp(-0.199329) \approx 0.82$ times shorter compared to those of the reference branch (i.e. the one in the Centre). The delivery times of pizzas from the Western branch are only marginally shorter $(\exp(0.027212) \approx 0.97$ times) compared to the central branch. Overall, each additionally ordered pizza increases the delivery time by a factor of $\exp(0.076823) \approx 1.08$.

(b) We can calculate the 95% confidence interval as follows:

$$0.076823 \pm \underbrace{z_{1-\alpha/2}}_{1.959964} \times 0.008087 \approx [0.061, 0.093].$$

(c) The exponentiated coefficient for the number of pizzas is $\exp(0.051539) = 1.05289$. It follows that with each additional pizza the chances of a long delivery thus increase by approximately 1.05 times. This is however only true for the reference category, i.e. the branch in the Centre. For the branch in the East we have to evaluate $\exp(0.051539 + 0.058587) = 1.116418$. The association between number of pizzas and delivery time is thus stronger in the East compared to the Centre. Similarly, we obtain for the branch in the West $\exp(0.051539 + 0.011414) = 11.064976$.

Solution to Exercise 12.3

(a) We use

```
theatre$Theatre.bin <- ifelse(theatre$Theatre > 150,1,0)
attach(theatre)
m.t1 <- glm(Theatre.bin~Age+Sex+Income+Culture+Theatre_ly,
family='binomial')
```

The model summary yields the following:

```
Coefficients:
               Estimate Std. Error z value Pr(>|z|)
(Intercept) -8.512305    0.845365 -10.069  < 2e-16  ***
Age          0.013594    0.007274   1.869  0.06164  .
Sex          0.634535    0.196549   3.228  0.00124  **
Income       0.039834    0.007830   5.087 3.63e-07  ***
Culture      0.016468    0.002177   7.564 3.90e-14  ***
Theatre_ly   0.002134    0.004219   0.506  0.61299
```

This means that the chances of spending a large amount of money on theatre
visits are greater for increasing age, female visitors, a larger spending on culture
in general and higher theatre expenditures in the previous year.

(b) To generate the plot, we use `predict` with both `type='response'` and
`type='link'` and then sort the predictions with respect to the η_i-values:

```
predictions <- data.frame(cbind(predict(m.t1,
type='link'), predict(m.t1, type='response') ))
colnames(predictions) <- c('eta', 'p')
predictions <- predictions[order(predictions$eta),]
plot(predictions, type='l')
```

This yields the following plot, which looks like the logistic link function:

(c) The AIC can be extracted from a model as follows:

```
m.t.all2 <-glm(Theatre.bin ~ Age+Sex, family=binomial)
AIC.all2[1] <- m.t.all2$aic
```

Looking at all models with two variables leads to the following results:

Age	Sex	Income	Culture	AIC
X	X			844.8151
X		X		815.0674
X			X	794.7708
	X	X		833.2634
	X		X	776.5759
		X	X	751.2023

We see that the best model with two variables is the model with the variables Income and Culture. If we repeat the exercise for all models with three variables we get:

Age	Sex	Income	Culture	AIC
X	X	X		811.2256
X	X		X	769.5578
X		X	X	749.8926
	X	X	X	743.5074

In this case, the best model is the model with the variables Sex, Income and Culture.

(d) In *R*, we need to load the library generalhoslem and use logitgof as follows:

```
m.t2 <- glm(Theatre.bin~Age+Sex+Income+Culture,
family=binomial)
gof.5 <- logitgof(Theatre.bin, predict(m.t2,
type='response'), g=5)
gof.10 <- logitgof(Theatre.bin, predict(m.t2,
type='response'), g=10)
gof.20 <- logitgof(Theatre.bin, predict(m.t2,
type='response'), g=20)
print(gof.5); print(gof.10); print(gof.20)
```

We indeed get different results for the χ^2-statistic when varying the number of classes. Since the tables of observed and expected frequencies have different dimensions, we cannot compare χ^2-statistics of tables of different dimension. However, the p-values are similar in all 3 cases. For $g = 10$ groups, we get

```
X-squared = 4.334, df = 8, p-value = 0.8258
```

For $g = 20$ groups, we get

```
X-squared = 10.962, df = 18, p-value = 0.896
```

and for $g = 5$ groups we get

```
X-squared = 0.99494, df = 3, p-value = 0.8025
```

(e) We can create the entries for the confusion matrix as follows:

```
table(Theatre.bin)    # absolute frequencies
# collect predictions and true y's & order them
pred.result <- data.frame(Theatre.bin,
predict(m.t2, type='response') )
colnames(pred.result) <- c('y', 'p')
pred.result <- pred.result[order(pred.result$p),]

c <- 0.5
# all cases with p<=c are classified as 0
pred.0 <- pred.result[(pred.result$p<=c), ]
pred.1 <- pred.result[(pred.result$p>c), ]

table(pred.0$y)    # predicted 0's
table(pred.1$y)    # predicted 1's
```

For $c = 0.5$, this yields:

	Predicted		
True outcome y	$y = 1$	$y = 0$	total
1	117	122	239
0	54	406	460

The accuracy (proportion of correctly classified cases) is therefore $(117 + 406)/699 = 0.748$, i.e. 74.8% of the observations are correctly classified by the model using a threshold of $c = 0.5$.

As another example, consider $c = 0.35$ (which is around the proportion of 1's of the response variable Theatre.bin). We then get

	Predicted		
True outcome y	$y = 1$	$y = 0$	total
1	169	70	239
0	126	334	460

Here the accuracy is $(169 + 334)/699 = 0.7196$, i.e. 71.96% of the observations are correctly classified by the model using a threshold of $c = 0.35$. As one can see, this threshold classifies more 1's correctly as 1's (169 compared to 117 for the threshold of $c = 0.5$), but on the other hand fewer 0's are classified correctly (334 instead of 406). There is always a trade-off between correctly predicting 0's and 1's, and we cannot simultaneously maximize both of them.

Chapter 13

Solution to Exercise 13.1

The variance of $\bar{y}$ under SRSWOR and SRSWR is given by

$$\text{Var}(\bar{y}_{WOR}) = \frac{N - \frac{n}{3}}{N \cdot \frac{n}{3}} S^2 \quad \text{and} \quad \text{Var}(\bar{y}_{WR}) = \frac{N - 1}{N \cdot \frac{2n}{3}} S^2,$$

respectively. Then,

$$\frac{\text{Var}(\bar{y}_{WOR})}{\text{Var}(\bar{y}_{WR})} = \frac{N - \frac{n}{3}}{N - 1} \cdot \frac{\frac{Nn}{3}}{\frac{2Nn}{3}} = \frac{2(3N - n)}{3(N - 1)}.$$

Solution to Exercise 13.2

We store the data in a data vector as follows:

```
amounts <- c(60.80, 41.59, 61.01, 59.72, …, 50.29, 39.16)
```

(a) The total number of samples that can be chosen is 75394027566:

```
choose(60, 10)
[1] 75394027566
```

(b) The sampling units 64.41, 35.20, 49.22, 39.16, 63.33, 40.90, 50.22, 31.85, 33.02, and 49.62 are selected as follows:

```
set.seed(4321)
amt_wor <- sample(amounts, size = 10, replace = FALSE)
amt_wor
[1] 64.41 35.20 49.22 39.16 63.33 40.90 50.22 31.85
33.02 49.62
```

(c) The estimate of the average spending is 45.693 €:

```
meanamtwor <- mean(amt_wor)
meanamtwor
[1] 45.693
```

An estimate of its variance is 11.4605 €², which is obtained as follows:

```
N<-60;    n<-length(amt_wor)
var_amt_wor <- ((N- n)/(N*n))*var(amt_wor)
var_amt_wor
[1] 11.4605
```

The 95% confidence interval is (38.03484 €, 53.35116 €). The lower limit is obtained as follows:

```
mean(amt_wor) - qt(0.975, n-1)
*sqrt(((N-n)/(length(amt_wor)*N))*var(amt_wor))
[1] 38.03484
```

The upper limit can be calculated accordingly.

(d) An estimate of the total spending of the customers, and an estimate of its variance, are 2741.58 € and 41257.81 €², respectively:

```
tot_wor <- N*meanamtwor; tot_wor
[1] 2741.58

tot_var_wor <- N*N*var_amt_wor; tot_var_wor
[1] 41257.81
```

Solution to Exercise 13.3

(a) The total number of samples that can be chosen by SRSWR is 6.046618×10^{17}, which is obtained as follows:

```
sample_wor <- 60^10
sample_wor
[1] 6.046618e+17.
```

(b) The sampling units 50.34, 31.64, 49.22, 58.86, 27.64, 33.02, 44.31, 47.29, 27.02, 49.44, 31.64, 59.72, 50.34, 42.79, and 58.86 are selected as follows:

```
set.seed(1234)
amt_wr <- sample(amounts, size = 15, replace = TRUE)
amt_wr
[1] 50.34 31.64 49.22 58.86 27.64 33.02 44.31 47.29
27.02 49.44 31.64 59.72 50.34 42.79 58.86
```

(c) The estimate of the average spending is 44.142 €:

```
meanamtwr <- mean(amt_wr)
meanamtwr
[1] 44.142,
```

An estimate of its variance is 6.500508 €^{2} :

```
N <- 60;   n <- length(amt_wr)
var_amt_wr <- ((N- n)/(N*n))*var(amt_wr)
var_amt_wr
[1] 6.500508.
```

The 95% confidence interval is (37.82767 €, 50.45633 €), where the lower limit is obtained as follows:

```
lcl_amt_wr <- mean(amt_wr) - qt(0.975,
n-1)*sqrt(var(amt_wr)/n)
lcl_amt_wr
[1] 37.82767,
```

The upper limit can be calculated accordingly.

(d) An estimate of the total spending and an estimate of its variance are 2648.52 € and 23401.83 €^{2}, respectively:

```
tot_wr <- N*meanamtwr; tot_wr
[1] 2648.52
tot_var_wr < - N*N*var_amt_wr; tot_var_wr
[1] 23401.83
```

Solution to Exercise 13.4

We store the data in a vector vector as follows
```
marks <- c(29.1, 69.8, 68.7, ..., 39.5, 81.2)
```

(a) The following sampling units, whose values are greater than 30, are drawn by SRSWOR with the help of the logical operator and using the `sample()` command as:

```
set.seed(1234)
markswor <- sample(marks[marks>30], size=10,
replace=FALSE)
markswor
[1] 41.2 41.3 37.5 64.3 75.5 71.2 71.7 39.5
88.9 40.7
```

(b) The estimate of the average score among the students securing more than 30 points is 57.18:

```
meanmarkswor <- mean(markswor)
meanmarkswor
[1] 57.18
```

The variance estimate is 29.14556, which is obtained as follows:

```
N <- 50;   n <- length(markswor)
var_marks_wor <- ((N- n)/(N*n))*var(markswor)
var_marks_wor
[1] 29.14556
```

The 95% confidence interval is (44.96738, 69.39262). The lower limit can be calculated as follows:

```
lcl_marks_wor <- mean(markswor) - qt(0.975, n-1) *
sqrt((N-n)/(n*N)*var(markswor))
lcl_marks_wor
[1] 44.96738
```

The upper limit can be calculated accordingly.

(c) Among the students securing more than 30 points, an estimate of the total amount of points is 2859 and an estimate of its variance is 72863.91:

```
tot_wor <- N*meanmarkswor; tot_wor
[1] 2859
tot_var_wor <- N*N*var_marks_wor; tot_var_wor
[1] 72863.91
```

Solution to Exercise 13.5

(a) The following sampling units, whose values are less than 60, are drawn by SRSWR:

```
set.seed(1234)
markswr <- sample(marks[marks<60], size =
15, replace = T)
markswr
[1] 56.5 44.9 43.4 32.7 23.7 38.6 43.4 42.9
56.5 42.6 20.8 41.3 44.9 42.9 23.7
```

(b) The estimate of the average mark, among students securing less than 60 points, is 39.92:

```
meanmarkswr <- mean(markswr)
meanmarkswr
[1] 39.92
```

The estimated variance is 2.306573:

```
N <- 50;   n <- length(markswr)
se_marks_wr <- sqrt(((N-n)/(N*n))*var(markswr))
se_marks_wr
[1] 2.306573.
```

The 95% confidence interval is (34.97289, 44.86711). The lower limit can be obtained as follows:

```
lcl_marks_wr <- mean(markswr) - qt(0.975,
n-1) * sqrt(((N-n)/(n*N))*var(markswr))
lcl_marks_wr
[1] 34.97289
```

The upper limit can be calculated accordingly.

(c) Among those students securing less than 60 points, an estimate of the total count is 1996 and an estimate of its variance is 13300.7:

```
tot_wr <- N*meanmarkswr; tot_wr
[1] 1996
tot_var_wr <- N*N*se_marks_wr*se_marks_wr;
tot_var_wr
[1] 13300.7
```

Solution to Exercise 13.6   We can store the data in a numeric vector where 'good' is represented by '1' and 'bad' by '0'.:

```
taste <- c(1, 0, 1, 1, 0, 0, 0, 1, …, 1, 1, 0, 0, 1, 0)
```

(a) A total of 6 sampling units are selected into the sample which report 'good'. To obtain this solution, we can simply add up the 1's among the selected sample:

```
set.seed(1234)
taste_wor <- sample(taste, size=15, replace=FALSE)
taste_wor
[1] 1 0 0 1 1 0 0 0 1 1 1 0 0 0 0

sum(taste_wor)
[1] 6
```

(b) The estimate of the proportion of people responding 'good' is 0.4:

```
meantastewor <- mean(taste_wor)
meantastewor
[1] 0.4
```

Its standard error is 0.1095445:

```
N <- 50;   n <- length(taste_wor)
setastewor <- sqrt(((N - n)/(N*(n - 1))) *
mean(taste_wor) * (1 - mean(taste_wor)))
setastewor
[1] 0.1095445
```

Solution to Exercise 13.7

(a) A total of 3 sampling units who answer 'bad' are selected into the sample. Similar to Exercise 13.6a we obtain the solution as follows:

```
set.seed(4321)
taste_wr <- sample(taste, size=10, replace=TRUE)
taste_wr
[1] 1 0 1 1 1 1 0 1 1 0

sum(taste_wr == 0)
[1] 3
```

(b) The estimate of the proportion of people who respond with 'bad' is 0.3 and it is obtained by subtracting the proportion of 'good' from 1 as follows:

```
meantastewr <- 1 - mean(taste_wr)
meantastewr
[1] 0.3
```

The estimated standard error is 0.1527525:

```
N <- 50;    n <- length(taste_wr)
setastewr <- sqrt((mean(taste_wr) *
(1-mean(taste_wr)))/(n-1))
setastewr
[1] 0.1527525
```

Solution to Exercise 13.8

The given data on hours of deep sleep and hours spent on the internet can be stored in the following vectors:

```
child <- c(1:10)
internet <- c(0.3,2.2,0.5,0.7,1.0,1.8,3.0,0.2,2.3,2.2)
sleep <- c(5.8,4.4,6.5,5.8,5.6,5.0,4.8,6.0,6.1,5.9)
internet_sleep_data <- data.frame(child, internet, sleep)
```

(a) We first write a function which returns the correlation coefficient between internet and sleep:

```
corr_internet_sleep_data <- function(d, i){
internet_sleep_data = d[i,]
return(cor(internet_sleep_data$internet,
internet_sleep_data$sleep))
}
```

We can now use the boot command and library to facilitate the bootstrapping:

```
set.seed(1234)
library(boot)
corrboot <- boot(data = internet_sleep_data, statistic =
corr_internet_sleep_data, R = 5000)
corrboot
```

The output of the object corrboot contains the following:

```
ORDINARY NONPARAMETRIC BOOTSTRAP


Bootstrap Statistics :
     original      bias    std. error
t1* -0.5884074 0.02638304   0.2301186
```

Thus, the bootstrap correlation coefficient is estimated as 0.02638304 by using the command mean(corrboot$t) - corrboot$t0. The bootstrap standard error is 0.2301186, as shown in the output.

(b) The bootstrap confidence intervals of the correlation coefficient based on the normal and percentile methods are obtained as $(-1.0658, -0.1638)$ and $(-0.9210, -0.0142)$, respectively. The results are based on the command `boot.ci(boot.out=corrboot, type=c('norm', 'perc'))` which returns the following output:

```
BOOTSTRAP CONFIDENCE INTERVAL CALCULATIONS
Based on 5000 bootstrap replicates


Intervals :
  Level        Normal                Percentile
  95%      (-1.0658, -0.1638 )   (-0.9210, -0.0142 )
Calculations and Intervals on Original Scale
```

Solution to Exercise 13.9

We first capture the data as follows:

```
child <- c(1:10)
sleep <- c(5.8, 4.4, 6.5, 5.8, 5.6, 5.0, 4.8, 6.0, 6.1, 5.9)
sleep_data <- data.frame(child, sleep)
```

(a) We then define a function that calculates the coefficient of variation:

```
sleep_CV <- function(d, i){
d2 <- d[i,]
return(sqrt(var(d2$sleep))/mean(d2$sleep))
}
```

We can now use the `boot` command and library to facilitate the bootstrapping:

```
set.seed(1234)
cvboot <- boot(data = sleep_data, statistic
= sleep_CV, R=5000)
cvboot
```

The output contains the following:

```
ORDINARY NONPARAMETRIC BOOTSTRAP


Bootstrap Statistics :
```

```
     original         bias      std. error
t1* 0.1166835  -0.008124909    0.02501536
```

The bootstrap bias of the coefficient of variation is estimated as -0.008125 and the bootstrap standard error is 0.025015.

(b) The bootstrap confidence intervals of the coefficient of variation based on the normal and percentile methods are obtained as $(0.0758, 0.1738)$ and $(0.0521, 0.1513)$, respectively:

```
boot.ci(boot.out = cvboot, type = c('norm',
'perc'))
```

```
BOOTSTRAP CONFIDENCE INTERVAL CALCULATIONS
Based on 5000 bootstrap replicates

Intervals :
  Level        Normal              Percentile
  95%    ( 0.0758,   0.1738 )    ( 0.0521,   0.1513 )
Calculations and Intervals on Original Scale
```

Chapter 14

Solution to Exercise 14.1

(a) Let Y denote the weight gain after 1 year and A and B refer to the two possible treatment strategies (T); then, a meaningful estimand could compare the expected weight gain under each of the two strategies and evaluate their difference:

$$E(Y^A) - E(Y^B).$$

As weight gain is a metric and not a binary variable, looking at differences is more meaningful then evaluating ratios. Of course, in some instances comparing counterfactual medians instead of expectations may be meaningful too.

(b) The phrase "more likely" indicates that a ratio of probabilities, i.e. risks, are of interest. Thus, the causal risk ratio is the right estimand:

$$P(Y^A = 1)/P(Y^B = 1).$$

(c) In the given question, two interventions are relevant: both regular exercise (T_1) and smoking (T_2). We can thus write the estimand as

$$E(Y^{T_1=1, T_2=0}).$$

(d) In this research question, both the intervention (i.e. drug dose) and the outcome
    (i.e. recovery time) are metric variables. We are therefore interested in the coun-
    terfactual expected recovery time for all possible doses. Because a drug dose
    cannot be negative, we may write the estimand as

$$E(Y^t) \quad \forall t > 0.$$

(e) The binary outcome is "reemployment" and the intervention is the additional
    training program. The keyword "chance" often indicates that an odds ratio is
    of interest because comparing chances between interventions relates to an odds
    ratio. It is an odds ratio conditional on age and whether the odds ratio differs
    between older ($X = 1$) and younger ($X = 0$) job-seekers is the question being
    asked:

$$\frac{P(Y^1=1|X=0)/P(Y^1=0|X=0)}{P(Y^0=1|X=0)/P(Y^0=0|X=0)} - \frac{P(Y^1=1|X=1)/P(Y^1=0|X=1)}{P(Y^0=1|X=1)/P(Y^0=0|X=1)}$$

Solution to Exercise 14.2

The smallest valid adjustment sets are:

(a) $\{X_1, X_2\}$, $\{X_1, X_3\}$
(b) $\{X_2\}$
(c) $\{X_3\}$, $\{X_4\}$ (note that X_1 is a collider)
(d) $\{\emptyset\}$ (there is no back-door path, no adjustment is needed)
(e) $\{X_4, X_5\}$
    The path $T \leftarrow X_4 \rightarrow X_3 \rightarrow Y$ should be closed through X_4 because X_3 is both
    a descendant of a mediator and a collider. The same applies to the back-door
    path $T \leftarrow X_5 \rightarrow X_6 \rightarrow Y$: X_5 needs to be used to close the path, rather than
    X_6.

Solution to Exercise 14.3

(a)

$$
\begin{aligned}
P(Y^1 = 1) &= \sum_x P(Y = 1|T = 1, X = x) \times P(X = x) \\
&= P(Y = 1|T = 1, X = 0) \times P(X = 0) \\
&\quad + P(Y = 1|T = 1, X = 1) \times P(X = 1) \\
&= \frac{1}{3} \times \frac{5}{8} + \frac{1}{2} \times \frac{3}{8} \\
&\approx 0.396 \, .
\end{aligned}
$$

$$P(Y^0 = 1) = \sum_x P(Y = 1 | T = 0, X = x) \times P(X = x)$$

$$= P(Y = 1 | T = 0, X = 0) \times P(X = 0)$$

$$+ P(Y = 1 | T = 0, X = 1) \times P(X = 1)$$

$$= \frac{1}{2} \times \frac{5}{8} + 0 \times \frac{3}{8}$$

$$\approx 0.3125.$$

$$\theta_{MRR} = P(Y^1 = 1) / P(Y^0 = 1) = \frac{0.396}{0.3125} = 1.267.$$

(b) In R, the same result can be achieved as follows:

```
X <- c(0,0,0,0,0,1,1,1)
T <- c(1,1,0,0,1,1,0,1)
Y <- c(0,1,0,1,0,1,0,0)

m1 <- glm(Y~T+X+T:X, family=binomial)
# Set T=1 und T=0
newdata1 <- as.data.frame(cbind(Y,T,X))
newdata1$T <- c(rep(1,8))
newdata0 <- as.data.frame(cbind(Y,T,X))
newdata0$T <- c(rep(0,8))
# MRR, i.e. E(Y^1)/E(Y^0)
mean(predict(m1, type='response', newdata=newdata1))/
mean(predict(m1, type='response', newdata=newdata0))
```

Solution to Exercise 14.4

(a)

$$E(Y^1) = \sum_{x=0}^{1} E(Y | T = 1, X = x) \times P(X = x)$$

$$= E(Y | T = 1, X = 0) \times P(X = 0)$$

$$+ E(Y | T = 1, X = 1) \times P(X = 1)$$

$$= (3445.15 - 281.06) \cdot 0.562 + (3445.15 - 281.06 - 71.2) \cdot 0.438$$

$$= 3132.904.$$

$$E(Y^0) = \sum_{x=0}^{1} E(Y | T = 0, X = x) \times P(X = x)$$

$$= E(Y | T = 0, X = 0) \times P(X = 0)$$

$$+\mathrm{E}(Y|T = 0, X = 1) \times P(X = 1)$$
$$= 3445.15 \cdot 0.562 + (3445.15 - 71.2) \cdot 0.438 = 3413.964 \,.$$

Thus, the ATE is $132.904 - 3413.964 = -281.06$.

(b) The estimated average treatment effects of both approaches are identical: -281.06 grams. This is because the expected difference is a collapsible effect measure (as the risk difference). In the given example, no mediators or colliders were included, only one confounder and thus the respective parameter estimate of the linear regression model can be interpreted as the marginal average treatment effect, which is often obtained through the g-formula.

(c) First, it is advantageous to set a seed for reproducibility, load the `boot` package and convert the strings of the treatment and confounder variables into binary integers:

```
set.seed(01112021)
library(boot)
smoke <- as.integer(mbsmoke=='smoker')
fb <- as.integer(fbaby=='Yes')
```

Then, we can proceed as explained in Sect. 13.6.6 and use the `boot` package by writing a function which calculates g-formula estimates given the data:

```
ATE.gformula <- function(d,i){
boot.data = d[i,]
Y <- boot.data$bweight; T <- boot.data$smoke;
X <- boot.data$fb
model <- lm(Y~T+X, data=boot.data)
newdata1 <- as.data.frame(cbind(Y,T,X))
newdata1$T <- c(rep(1,length(T)))
newdata0 <- as.data.frame(cbind(Y,T,X))
newdata0$T <- c(rep(0,length(T)))
mean(predict(model,newdata=newdata1))-
mean(predict(model,newdata=newdata0))
}

ATE.boot <- boot(data=as.data.frame(cbind(fb,smoke,bweight)),
statistic=ATE.gformula,R=1000)
boot.ci(ATE.boot,type='perc')
```

The result for the 95% confidence interval is:

```
BOOTSTRAP CONFIDENCE INTERVAL CALCULATIONS
Based on 1000 bootstrap replicates
```

```
Intervals :
Level      Percentile
95%     (-265.9, -196.0 )
```

(d) At first glance, one could argue that the intervention of "smoking" during pregnancy is not well-defined. There might be substantial differences between mothers classified under "smoking at all", i.e. the category may contain both mothers who smoke 1 single cigarette during a 9-moth period and mothers who smoke > 20 cigarettes daily. Different levels of smoking may affect the birth weight differently. As the variable msmoke captures more information than the variable mbsmoke, it may be better to work with this variable. However, a cross table between the two variables reveals the following relationship:

	mbsmoke	
msmoke	nonsmoker	smoker
0 cigarettes daily	3778	0
1-5 cigarettes daily	0	200
11 or more cigarettes daily	0	327
6-10 cigarettes daily	0	337

This means "smoker" is defined as someone smoking at least 1 cigarette a day. With this additional information, it seems reasonable to argue that consistency is met.

(e) For conditional exchangeability to be achieved, all potential confounders would have to be measured and included in the model. It is obvious that this is not the case in the given example. For example, the mother's diet may affect both the desire to smoke and the baby's birth weight, but is not captured in the data set. Thus, a causal effect interpretation is not warranted in the given example.

Technical Appendix

C

C.1   More Details on Chap. 3

Proof of equation (3.27).

$$\tilde{s}^2 = \frac{1}{n}\sum_{j=1}^{k}\sum_{x_i \in K_j}(x_i - \bar{x})^2 = \frac{1}{n}\sum_{j=1}^{k}\sum_{x_i \in K_j}(x_i - \bar{x}_j + \bar{x}_j - \bar{x})^2$$

$$= \underbrace{\frac{1}{n}\sum_{j=1}^{k}\sum_{x_i \in K_j}(x_i - \bar{x}_j)^2}_{[i]} + \underbrace{\frac{1}{n}\sum_{j=1}^{k}\sum_{x_i \in K_j}(\bar{x}_j - \bar{x})^2}_{[ii]}$$

$$+ \underbrace{\frac{2}{n}\sum_{j=1}^{k}\sum_{x_i \in K_j}(x_i - \bar{x}_j)(\bar{x}_j - \bar{x})}_{[iii]}$$

We obtain the following expressions for $[i]$–$[iii]$:

$$[i] = \frac{1}{n}\sum_{j=1}^{k}n_j\frac{1}{n_j}\sum_{x_i \in K_j}(x_i - \bar{x}_j)^2 = \frac{1}{n}\sum_{j=1}^{k}n_j\tilde{s}_j^2 \, ,$$

$$[ii] = \frac{1}{n}\sum_{j=1}^{k}n_j(\bar{x}_j - \bar{x})^2 \, ,$$

$$[iii] = \frac{2}{n}\sum_{j=1}^{k}(\bar{x}_j - \bar{x})\sum_{x_i \in K_j}(x_i - \bar{x}_j) = \frac{2}{n}\sum_{j=1}^{k}(\bar{x}_j - \bar{x})\,0 = 0.$$

© Springer Nature Switzerland AG 2022
C. Heumann et al., *Introduction to Statistics and Data Analysis*,
https://doi.org/10.1007/978-3-031-11833-3

Since $[i]$ is the within-class variance and $[ii]$ is the between-class variance, equation (3.27) holds.

C.2   More Details on Chap. 7

Proof of Theorem 7.2.3. Consider the interval $(x_0 - \delta, x_0]$ with $\delta \geq 0$. From (7.12) it follows that $P(x_0 - \delta < X \leq x_0) = F(x_0) - F(x_0 - \delta)$ and therefore

$$
\begin{aligned}
P(X = x_0) &= \lim_{\delta \to 0} P(x_0 - \delta < X \leq x_0) \\
&= \lim_{\delta \to 0} [F(x_0) - F(x_0 - \delta)] \\
&= F(x_0) - F(x_0) = 0 .
\end{aligned}
$$

Proof of Theorem 7.3.1.

$$
\begin{aligned}
\mathrm{Var}(X) &\overset{(7.17)}{=} E(X - \mu)^2 \\
&= E(X^2 - 2\mu X + \mu^2) \\
&\overset{(7.28-7.31)}{=} E(X^2) - 2\mu E(X) + E(\mu^2) \\
&= E(X^2) - 2\mu^2 + \mu^2 \\
&= E(X^2) - [E(X)]^2 .
\end{aligned}
$$

Proof of Theorem 7.4.1.

We define a discrete random variable Y as

$$
Y = \begin{cases} 0 & \text{if } |X - \mu| < c \\ c^2 & \text{if } |X - \mu| \geq c . \end{cases} \tag{C.1}
$$

The respective probabilities are $P(|X - \mu| < c) = p_1$ and $P(|X - \mu| \geq c) = p_2$. The definition of Y in (C.1) implies that

$$
Y \leq |X - \mu|^2
$$

since for $|X - \mu|^2 < c^2$ Y takes the value $y_1 = 0$ and therefore $Y \leq |X - \mu|^2$. If $|X - \mu|^2 \geq c^2$ Y takes the value $y_2 = c^2$, and therefore $Y \leq |X - \mu|^2$. Using this knowledge we obtain

$$
E(Y) \leq E(X - \mu)^2 = \mathrm{Var}(X) .
$$

However, for Y we also have

$$
E(Y) = 0 \cdot p_1 + c^2 \cdot p_2 = c^2 P(|X - \mu| \geq c)
$$

which means that we can summarize the above findings in the following inequality:

$$c^2 P(|X - \mu| \geq c) \leq \text{Var}(X).$$

This equates to Tschebyschev's inequality. Using $P(\bar{A}) = 1 - P(A)$, i.e.

$$P(|X - \mu| \geq c) = 1 - P(|X - \mu| < c),$$

we obtain the second formula of Tschebyschev's inequality:

$$P(|X - \mu| < c) \geq 1 - \frac{\text{Var}(X)}{c^2}.$$

Proof of rule (7.30). For discrete variables, we have

$$\text{E}(a + bX) \overset{(7.16)}{=} \sum_i (a + bx_i) p_i = \sum a p_i + b x_i p_i = a \sum_i p_i + b \sum_i x_i p_i.$$

Since $\sum_i p_i = 1$ and $\sum_i x_i p_i = \text{E}(X)$, we obtain $\text{E}(a + bX) = a + b\,\text{E}(X)$, which is rule (7.30). In the continuous case, we have

$$\text{E}(a + bX) = \int_{-\infty}^{\infty} (a + bx) f(x) dx = \int_{-\infty}^{\infty} (af(x)dx + bxf(x)dx)$$

$$= a \int_{-\infty}^{\infty} f(x) dx + b \int_{-\infty}^{\infty} x f(x) dx = a + b\,\text{E}(X).$$

Proof of rule (7.33). Using $\text{Var}(X) = \text{E}(X^2) - \text{E}(X)^2$, we can write the variance of bX as

$$\text{Var}(bX) = \text{E}([bX]^2) + \text{E}(bX)^2.$$

Using (7.29), we get $\text{E}([bX]^2) = b^2 \text{E}(X^2)$ and $\text{E}(bX)^2 = (b\,\text{E}(X))^2$. Therefore

$$\text{Var}(bX) = b^2 (\text{E}(X^2) - \text{E}(X)^2) = b^2 \text{Var}(X).$$

Proof of $\rho = 1$ for a perfect linear relationship. If $Y = aX + b$ with $a \neq 0$, we get

$$\text{Cov}(X, Y) = \text{E}[(X - \mu_X)(Y - \mu_Y)]$$
$$= a\,\text{E}[(X - \mu_X)(X - \mu_X)]$$

because $\text{Cov}(aX + b, cY + d) = ac\,\text{Cov}(X, Y)$, see Sect. 7.5. Then

$$\text{Cov}(X, Y) = a\,\text{Var}(X),$$

$$\text{Var}(Y) \overset{(7.33)}{=} a^2 \text{Var}(X),$$

and therefore

$$\rho(X, Y) = \frac{a \operatorname{Var}(X)}{\sqrt{a^2 \operatorname{Var}(X) \operatorname{Var}(X)}} = \frac{a}{|a|} = 1$$

if $a > 0$. Similarly, if $Y = aX + b$ with $a < 0$ we get $\rho(X, Y) = -1$.

C.3   More Details on Chap. 8

Theorem of Large Numbers. To explain the Theorem of Large Numbers, we first need to first define **stochastic convergence**.

Definition C.3.1 A sequence of random variables, $(X_n)_{n \in \mathbb{N}}$, converges stochastically to 0, if for any $\epsilon > 0$

$$\lim_{n \to \infty} P(|X_n| > \epsilon) = 0 \tag{C.2}$$

holds.

This is equivalent to $\lim_{n \to \infty} P(|X_n| \leq \epsilon) = 1$.

Theorem C.3.1 (Theorem of Large Numbers) *Consider n i.i.d. random variables* $X_1, X_2, \ldots, X_n$ *with* $\mathrm{E}(X_i) = \mu$, $\operatorname{Var}(X_i) = \sigma^2$ *and* $\bar{X}_n = \frac{1}{n} \sum_{i=1}^{n} X_i$. *It holds that*

$$\lim_{n \to \infty} P(|\bar{X}_n - \mu| < c) = 1, \quad \forall\; c \geq 0. \tag{C.3}$$

This implies that $\bar{X}_n$ converges stochastically to μ. As another motivation, recall Definition 7.6.1 where we define random variables $X_1, X_2, \ldots, X_n$ to be i.i.d. (independently identically distributed) if all X_i follow the same distribution and are independent of each other. Under this assumption we showed in (7.36) that $\operatorname{Var}(\bar{X}) = \frac{\sigma^2}{n}$. It follows that the larger n, the smaller the variance. If we apply Tschebyschev's inequality (Theorem 7.4.1) to $\bar{X}$, we get the following equation for $(\bar{X}_n - \mu)_{n \in \mathbb{N}}$:

$$P(|\bar{X}_n - \mu| < c) \geq 1 - \frac{\operatorname{Var}(\bar{X}_n)}{c^2} = 1 - \frac{\sigma^2}{nc^2}. \tag{C.4}$$

This means that for each $c \geq 0$, the right-hand side of the above equation tends to 1 as $n \to \infty$ which gives a similar interpretation as the Theorem of Large Numbers.

Central Limit Theorem. Let X_i $(i = 1, 2, \ldots, n)$ be n i.i.d. random variables with $\mathrm{E}(X_i) = \mu$ and $\operatorname{Var}(X_i) = \sigma^2$. If we consider the sum $\sum_{i=1}^{n} X_i$, we obtain $\mathrm{E}(\sum_{i=1}^{n} X_i) = n\mu$ and $\operatorname{Var}(\sum_{i=1}^{n} X_i) = n\sigma^2$. If we want to standardize $\sum_{i=1}^{n} X_i$ we can use Theorem 7.3.2 to obtain

$$Y_n = \frac{\sum_{i=1}^{n} X_i - n\mu}{\sqrt{n\sigma^2}}, \tag{C.5}$$

i.e. it holds that $E(Y_n) = 0$ and $Var(Y_n) = 1$.

Theorem C.3.2 (Central Limit Theorem) *Let X_i ($i = 1, 2, \ldots, n$) be n i.i.d. random variables with $E(X_i) = \mu$ and $Var(X_i) = \sigma^2$. Y_n denotes the standardized sum of X_i, $i = 1, 2, \ldots, n$. The CDF of Y_n is*

$$\lim_{n \to \infty} P(Y_n \leq y) = \Phi(y), \quad \forall \ y,$$

where $\Phi(y)$ denotes the CDF of the standard normal distribution.

This theorem tells us that Y_n is approximately standard normal distributed if n is large, i.e.

$$Y_n \sim N(0, 1) \quad \text{as} \quad n \to \infty.$$

This is equivalent to saying that $\sum_{i=1}^{n} X_i$ is approximately $N(n\mu, n\sigma^2)$ distributed if n is large, i.e.

$$\sum_{i=1}^{n} X_i \sim N(n\mu, n\sigma^2). \tag{C.6}$$

As a consequence $\bar{X}_n = \frac{1}{n} \sum_{i=1}^{n} X_i$ is $N(\mu, \frac{\sigma^2}{n})$ distributed for large n, i.e.

$$\bar{X}_n \sim N(\mu, \frac{\sigma^2}{n}).$$

In summary, the Theorem of Large Numbers tells us that $\bar{X}_n$ converges stochastically to μ, whereas the Central Limit Theorem tells us that $\bar{X}_n$ converges to a $N(\mu, \frac{\sigma^2}{n})$-distribution as n tends to infinity.

PDF of the χ^2 -distribution. The PDF of the χ^2-distribution, with n degrees of freedom, is defined as

$$f(x) = \begin{cases} \frac{x^{n/2-1} \exp(-x/2)}{\Gamma(n/2) 2^{n/2}} & \text{if} \ x > 0 \\ 0 & \text{otherwise.} \end{cases} \tag{C.7}$$

Note that $\Gamma(n)$ is the Gamma function, defined as $\Gamma(n) = (n-1)!$ for positive integers and $\Gamma(x) = \int_0^\infty t^{x-1} \exp(-t) dt$ otherwise.

PDF of the t-distribution. The PDF of the t-distribution, with n degrees of freedom, is defined as

$$f(x) = \frac{\Gamma(\frac{n+1}{2})}{\Gamma(n/2)\sqrt{n\pi}} \left(1 + \frac{x^2}{n}\right)^{-(n+1)/2} \quad -\infty < x < \infty . \tag{C.8}$$

Note that $\Gamma(n)$ is the Gamma function, defined as $\Gamma(n) = (n-1)!$ for positive integers and $\Gamma(x) = \int_0^\infty t^{x-1} \exp(-t)dt$ otherwise.

PDF of the F-distribution. The PDF of the F-distribution, with n and m degrees of freedom respectively, is defined as

$$f(x) = \frac{\Gamma(\frac{n+m}{2})(\frac{n}{m})^{n/2}x^{n/2-1}}{\Gamma(\frac{n}{2})\Gamma(\frac{m}{2})(1 + \frac{nx}{m})^{(n+m)/2}} \quad , x > 0 \quad . \tag{C.9}$$

The PDF of the F distribution with m and n degrees of freedom can be derived by interchanging the roles of m and n.

C.4   More Details on Chap. 9

Another Example of Sufficiency. Let $X_1, X_2, \ldots, X_n$ be a random sample from $N(\mu, \sigma^2)$ where μ and σ^2 are both unknown. We attempt to find a sufficient statistic for μ and σ^2.

$$f(x_1, x_2 \ldots, x_n; \mu, \sigma^2) = \left(\frac{1}{\sqrt{2\pi}}\right)^n \exp\left[-\frac{1}{2}\sum_{i=1}^n \frac{(x_i - \mu)^2}{\sigma^2}\right]$$

$$= \left(\frac{1}{\sqrt{2\pi}\sigma}\right)^n \exp\left[-\frac{1}{2\sigma^2}\left(\sum_{i=1}^n x_i^2 - 2\mu\sum_{i=1}^n x_i + n\mu^2\right)\right] .$$

Here the joint density depends on $x_1, x_2, \ldots, x_n$ through two statistics $t_1(x_1, x_2, \ldots, x_n) = \sum_{i=1}^n x_i$ and $t_2(x_1, x_2, \ldots, x_n) = \sum_{i=1}^n x_i^2$ with $h(x_1, x_2, \ldots, x_n) = 1$. Thus $T_1 = \sum_{i=1}^n X_i$ and $T_2 = \sum_{i=1}^n X_i^2$ are jointly sufficient for μ and σ^2. Therefore, $\bar{X}$ and $S^2 = \frac{1}{n-1}\sum_{i=1}^n (X_i - \bar{X})^2$ are jointly sufficient for μ and σ^2 as they are a one-to-one function of T_1 and T_2. They are the maximum likelihood estimates for μ and σ^2.

C.5   More Details on Chap. 10

Exact Test of Fisher. Similar to the approximate two-sample Binomial test in Sect. 10.4.2, we assume two samples following a Binomial distribution with parameters (n_1, p_1) and (n_2, p_2) respectively.

$$\mathbf{X} = (X_1, X_2, \ldots, X_{n_1}), \quad X_i \sim B(1; p_1)$$
$$\mathbf{Y} = (Y_1, Y_2, \ldots, Y_{n_2}), \quad Y_i \sim B(1; p_2) .$$

For the sums of these random variables, we get:

$$X = \sum_{i=1}^{n_1} X_i \sim B(n_1; p_1), \quad Y = \sum_{i=1}^{n_2} Y_i \sim B(n_2; p_2).$$

Let $Z = X + Y$. The Exact Test of Fisher uses the fact that the row marginal frequencies n_1 and n_2 in the following table

	Success	Failure	Total
Population A	X	$n_1 - X$	n_1
Population B	$Z - X = Y$	$n_2 - (Z - X)$	n_2
Total	Z	$(n_1 + n_2 - Z)$	$n_1 + n_2$

are fixed by the sample sizes n_1 and n_2. Conditional on the total number of successes $Z = z$ (i.e. the column margins are assumed to be fixed), the only remaining random variable is X (since the other three entries of the table are then determined by the realization x of X and the margins). Under $H_0 : p_1 = p_2 = p$, it can be shown that

$$X \sim H(n, n_1, z),$$

i.e.

$$P(X = x | Z = z) = \frac{\binom{n_1}{x}\binom{n-n_1}{z-x}}{\binom{n}{z}}.$$

The proof is straightforward using the idea of conditioning on $Z = z$ and the assumption of independence of the two samples:

$$P(X = x | Z = z) = \frac{P(X = x, Z = z)}{P(Z = z)} = \frac{P(X = x, Y = z - x)}{P(Z = z)}$$

$$= \frac{P(X = x)P(Y = z - x)}{P(Z = z)} = \frac{\binom{n_1}{x}p^x(1-p)^{n_1-x}\binom{n_2}{z-x}p^{z-x}(1-p)^{n_2-(z-x)}}{\binom{n}{z}p^z(1-p)^{n-z}}$$

$$= \frac{\binom{n_1}{x}\binom{n_2}{z-x}}{\binom{n}{z}} = \frac{\binom{n_1}{x}\binom{n-n_1}{z-x}}{\binom{n}{z}}.$$

Note that in the equation above we use the fact that under H_0, $Z = X + Y$ is $B(n, p)$ distributed; see the additivity theorem for the Binomial distribution, i.e. Theorem 8.1.1.

Example C.5.1 Consider two competing lotteries A and B. Say we buy 10 tickets from each lottery and test the hypothesis of equal winning probabilities. The data can be summarized in a 2×2 table:

	Winning	Not Winning	Total
Lottery A	1	24	25
Lottery B	7	18	25
Total	8	42	50

In R, we can use the command `fisher.test` to perform the test. Using the example data and $H_1 : p_1 \neq p_2$, we get

```
ft <- matrix(nrow=2,ncol=2,data=cbind(c(1,7),
c(24,18)))
fisher.test(x=ft)
```

with output

```
Fisher's Exact Test for Count Data

data:    fisher.table
p-value = 0.0488
alternative hypothesis: true odds ratio is not equal to 1
95 percent confidence interval:
 0.002289885 0.992114690
sample estimates:
odds ratio
 0.1114886
```

For the example data and $\alpha = 0.05$, the null hypothesis is rejected, since the p-value is lower than α. For the calculation of the p-value, the one-sided and two-sided cases have to be distinguished. The idea is that while fixing the margins at the observed values (i.e. 25, 25, 8, 42), we have to calculate the sum of the probabilities of all tables which have lower probability than the observed table. In R, one can use the functions `dhyper` and `phyper` for calculating (cumulative) probabilities. For example, $P(X = 1|Z = 8)$ can be calculated as

```
dhyper(1,25,25,8)
```

```
[1] 0.02238402
```

A more extreme table than the observed one is

0	25	25
8	17	25
8	42	50

with probability $P(X = 0) = \text{dhyper}(0,25,25,8) = 0.002$, which is lower than $P(X = 1)$. The sum is $0.0224 + 0.002 = 0.0244$ which is the (left) one-sided p-value. In this case (not generally true!), the two-sided p-value is simply two times the one-sided value, i.e. $2 \cdot 0.0244 = 0.0488$.

Remark C.5.1 The two-sided version of the Exact Test of Fisher can also be used as a test of independence of two binary variables. It is equivalent to the test of equality of two proportions, see Example 10.8.2.

One-sample χ^2 -Test for testing hypothesis about the variance. We assume a normal population, i.e., $X \sim N(\mu, \sigma^2)$ and an i.i.d. sample $(X_1, X_2, \ldots, X_n)$ distributed as X. We only introduce the test for two-sided hypotheses

$$H_0 : \sigma^2 = \sigma_0^2$$

versus

$$H_1 : \sigma^2 \neq \sigma_0^2 \,.$$

The test statistic

$$T(\mathbf{X}) = \frac{(n-1)S_X^2}{\sigma_0^2}$$

follows a χ_{n-1}^2-distribution under H_0. The critical region is constructed by taking the $\alpha/2$- and $(1 - \alpha/2)$-quantile as critical values, i.e. H_0 is rejected, if

$$t(x) < c_{n-1;\alpha/2}$$

or if

$$t(x) > c_{n-1;1-\alpha/2} \,,$$

where $c_{n-1;\alpha/2}$ and $c_{n-1;1-\alpha/2}$ are the desired quantiles of a χ^2-distribution. In R, the test can be called by the `sigma.test` function in the `TeachingDemos` library or the `varTest` function in library `EnvStats`. Both functions also return a confidence interval for the desired confidence level. Note that the test is biased. An unbiased level α test would not take $\alpha/2$ at the tails but two different tail probabilities α_1 and α_2 with $\alpha_1 + \alpha_2 = \alpha$.

F-Test for comparing two variances. Comparing variances can be of interest when comparing the variability, i.e. the "precision" of two industrial processes; or when comparing two medical treatments with respect to their reliability. Consider two populations characterized by two independent random variables X and Y which follow normal distributions:

$$X \sim N(\mu_X, \sigma_X^2), \quad Y \sim N(\mu_Y, \sigma_Y^2).$$

For now, we distinguish the following two hypotheses:

$$H_0 : \sigma_X^2 = \sigma_Y^2 \text{ versus } H_1 : \sigma_X^2 \neq \sigma_Y^2, \quad \text{two-sided}$$
$$H_0 : \sigma_X^2 \leq \sigma_Y^2 \text{ versus } H_1 : \sigma_X^2 > \sigma_Y^2, \quad \text{one-sided.}$$

The third hypothesis with $H_1 : \sigma_X^2 < \sigma_Y^2$, is similar to the second hypothesis where X and Y are replaced with each other.

Test Statistic

Let $(X_1, X_2, \ldots, X_{n_1})$ and $(Y_1, Y_2, \ldots, Y_{n_2})$ be two independent random samples of size n_1 and n_2. The test statistic is defined as the ratio of the two sample variances

$$T(\mathbf{X}, \mathbf{Y}) = \frac{S_X^2}{S_Y^2}, \tag{C.10}$$

which is, under the null hypothesis, F-distributed with $n_1 - 1$ and $n_2 - 1$ degrees of freedom, see also Sect. 8.3.3.

Critical Region

Two-sided case. The motivation behind the construction of the critical region for the two-sided case, $H_0: \sigma_X^2 = \sigma_Y^2$ verses $H_1: \sigma_X^2 \neq \sigma_Y^2$, is that if the null hypothesis is true (i.e. the two variances are equal) then the test statistic (C.10) would be 1; also, $T(\mathbf{X}, \mathbf{Y}) > 0$. Therefore very small (but positive) and very large values of $T(\mathbf{X}, \mathbf{Y})$ should lead to a rejection of H_0. The critical region can then be written as $K = [0, k_1) \cup (k_2, \infty)$, where k_1 and k_2 are critical values such that

$$P(T(\mathbf{X}, \mathbf{Y}) < k_1 | H_0) = \alpha/2$$
$$P(T(\mathbf{X}, \mathbf{Y}) > k_2 | H_0) = \alpha/2.$$

Here k_1 and k_2 can be calculated from the quantile function of the F-distribution as

$$k_1 = f_{n_1-1, n_2-1, \alpha/2}, \quad k_2 = f_{n_1-1, n_2-1, 1-\alpha/2}.$$

Example C.5.2 Let $n_1 = 50$, $n_2 = 60$ and $\alpha = 0.05$. Using the qf command in R we can determine the critical values as:

```
qf(q=0.025, df1=50-1, df2=60-1)
qf(q=0.975, df1=50-1, df2=60-1)
```

The results are $k_1 = 0.5778867$ and $k_2 = 1.706867$.

Remark C.5.2 There is the following relationship between quantiles of the F-distribution:

$$f_{n_1-1;n_2-1;\alpha/2} = \frac{1}{f_{n_2-1;n_1-1;1-\alpha/2}}\,.$$

One-sided case. In the one-sided case, the alternative hypothesis is always formulated in such a way that the variance in the numerator is greater than the variance in the denominator. The hypotheses are $H_0\colon \sigma_X^2 \le \sigma_Y^2$ versus $H_1\colon \sigma_X^2 > \sigma_Y^2$ or $H_0\colon \sigma_X^2/\sigma_Y^2 \le 1$ versus $H_1\colon \sigma_X^2/\sigma_Y^2 > 1$. This means it does not matter whether $H_1 : \sigma_X^2 > \sigma_Y^2$ or $H_1 : \sigma_X^2 < \sigma_Y^2$; by constructing the test statistic in the correct way, we implicitly specify the hypothesis. The critical region K consists of the largest values of $T(\mathbf{X}, \mathbf{Y})$, i.e. $K = (k, \infty)$, where k has to fulfil the condition

$$P(T(\mathbf{X}, \mathbf{Y}) > k | H_0) = \alpha\,.$$

The resulting critical value is denoted by $k = f_{n_1-1;n_2-1;1-\alpha}$.

Observed Test Statistic

Using the sample variances, the realized test statistic t is calculated as:

$$t(x, y) = \frac{s_x^2}{s_y^2}, \quad s_x^2 = \frac{1}{n_1 - 1}\sum_{i=1}^{n_1}(x_i - \bar{x})^2, \quad s_y^2 = \frac{1}{n_2 - 1}\sum_{i=1}^{n_2}(y_i - \bar{y})^2$$

Test Decisions

Case	H_0	H_1	Reject H_0, if
(a)	$\sigma_X = \sigma_Y$	$\sigma_X \ne \sigma_Y$	$t(x, y) > f_{n_1-1;n_2-1;1-\alpha/2}$ or
			$t(x, y) < f_{n_1-1;n_2-1;\alpha/2}$
(b)	$\sigma_X \le \sigma_Y$	$\sigma_X > \sigma_Y$	$t(x, y) > f_{n_1-1;n_2-1;1-\alpha}$

Remark C.5.3 We have tacitly assumed that the expected values μ_X and μ_y are unknown and have to be estimated. However, this happens rarely, if ever, in practice. When estimating the expected values by the arithmetic means, it would be appropriate to increase the degrees of freedom from $n_1 - 1$ to n_1 and $n_2 - 1$ to n_2. Interestingly, standard software will not handle this case correctly.

Example C.5.3 A company is putting baked beans into cans. Two independent machines at two sites are used. The filling weights are assumed to be normally distributed with mean 1000 g. It is speculated that one machine is more precise than the other. Two samples of the two machines give the following results:

Sample	n	$\bar{x}$	s^2
X	20	1000.49	72.38
Y	25	1000.26	45.42

With $\alpha = 0.1$ and the hypotheses

$$H_0 : \sigma_X^2 = \sigma_Y^2 \quad \text{versus} \quad H_1 : \sigma_X^2 \neq \sigma_Y^2,$$

we get the following quantiles

```
qf(0.05, 20-1, 25-1)
[1] 0.4730049
qf(0.95, 20-1, 25-1)
[1] 2.039858
```

The observed test statistic is

$$t = \frac{72.38}{45.42} = 1.59.$$

Therefore H_0 is not rejected, since $k_1 \leq t \leq k_2$. We can't reject the hypothesis of equal variability of the two machines. In R, the F-test can be used using the command `var.test`.

Remark C.5.4 For the t-test, we remarked that the assumption of normality is not crucial because the test statistic is approximately normally distributed, even for moderate sample sizes. However, the F-test relies heavily on the assumption of normality. This is why alternative tests are often used, for example the Levene test.

C.6   More Details on Chap. 11

Obtaining the least squares estimates in the linear model.
The function $S(a, b)$ describes our optimization problem of minimizing the residual sum of squares:

$$S(a, b) = \sum_{i=1}^{n} e_i^2 = \sum_{i=1}^{n} (y_i - a - bx_i)^2 \,.$$

Minimizing $S(a, b)$ is achieved using the principle of maxima and minima which involves taking the partial derivatives of $S(a, b)$ with respect to both a and b and setting them equal to 0. The partial derivatives are

$$\frac{\partial}{\partial a} S(a, b) = \sum_{i=1}^{n} \frac{\partial}{\partial a} (y_i - a - bx_i)^2 = -2 \sum_{i=1}^{n} (y_i - a - bx_i) \,, \quad \text{(C.11)}$$

$$\frac{\partial}{\partial b} S(a, b) = \sum_{i=1}^{n} \frac{\partial}{\partial b} (y_i - a - bx_i)^2 = -2 \sum_{i=1}^{n} (y_i - a - bx_i) x_i \,. \quad \text{(C.12)}$$

Now we set (C.11) and (C.12) equal to zero respectively:

$$\begin{array}{lll} \text{(I)} & \sum_{i=1}^{n} (y_i - \hat{a} - \hat{b}x_i) = & 0 \,, \\ \text{(II)} & \sum_{i=1}^{n} (y_i - \hat{a} - \hat{b}x_i) x_i = & 0 \,. \end{array}$$

This equates to

$$\begin{array}{lll} \text{(I')} & n\hat{a} + \hat{b} \sum_{i=1}^{n} x_i & = \sum_{i=1}^{n} y_i \,, \\ \text{(II')} & \hat{a} \sum_{i=1}^{n} x_i + \hat{b} \sum_{i=1}^{n} x_i^2 & = \sum_{i=1}^{n} x_i y_i \,. \end{array}$$

Multiplying (I') by $\frac{1}{n}$ yields

$$\hat{a} + \hat{b}\bar{x} = \bar{y}$$

which gives us the solution for a:

$$\hat{a} = \bar{y} - \hat{b}\bar{x} \,.$$

Putting this solution into (II') gives us

$$(\bar{y} - \hat{b}\bar{x}) \sum_{i=1}^{n} x_i + \hat{b} \sum_{i=1}^{n} x_i^2 = \sum_{i=1}^{n} x_i y_i \,.$$

Using $\sum_{i=1}^{n} x_i = n\bar{x}$ leads to

$$\hat{b}\left(\sum_{i=1}^{n} x_i^2 - n\bar{x}^2\right) = \sum_{i=1}^{n} x_i y_i - n\bar{x}\bar{y}.$$

If we use

$$\sum_{i=1}^{n} x_i^2 - n\bar{x}^2 = \sum_{i=1}^{n} (x_i - \bar{x})^2 = S_{xx}$$

and

$$\sum_{i=1}^{n} x_i y_i - n\bar{x}\bar{y} = \sum_{i=1}^{n} (x_i - \bar{x})(y_i - \bar{y}) = S_{xy},$$

we eventually obtain the least squares estimate of b:

$$\hat{b} S_{xx} = S_{xy}$$
$$\hat{b} = \frac{S_{xy}}{S_{xx}} = \frac{\sum_{i=1}^{n} (x_i - \bar{x})(y_i - \bar{y})}{\sum_{i=1}^{n} (x_i - \bar{x})^2}.$$

Remark C.6.1 To show that the above solutions really relate to a minimum, and not to a maximum, we would need to look at all the second order partial derivatives of $S(a, b)$ and prove that the bordered Hessian matrix containing these derivatives is always positive definite. We omit this proof however.

Variance decomposition. We start with the following equation:

$$y_i - \hat{y}_i = (y_i - \bar{y}) - (\hat{y}_i - \bar{y}).$$

If we square both sides we obtain

$$\sum_{i=1}^{n} (y_i - \hat{y}_i)^2 = \sum_{i=1}^{n} (y_i - \bar{y})^2 + \sum_{i=1}^{n} (\hat{y}_i - \bar{y})^2 - 2\sum_{i=1}^{n} (y_i - \bar{y})(\hat{y}_i - \bar{y}).$$

The last term on the right-hand side is

$$\sum_{i=1}^{n} (y_i - \bar{y})(\hat{y}_i - \bar{y}) \overset{(11.8)}{=} \sum_{i=1}^{n} (y_i - \bar{y})\hat{b}(x_i - \bar{x})$$

$$= \hat{b} S_{xy} \overset{(11.6)}{=} \hat{b}^2 S_{xx} \overset{(11.8)}{=} \sum_{i=1}^{n} (\hat{y}_i - \bar{y})^2.$$

We therefore obtain

$$\sum_{i=1}^{n}(y_i - \hat{y}_i)^2 = \sum_{i=1}^{n}(y_i - \bar{y})^2 - \sum_{i=1}^{n}(\hat{y}_i - \bar{y})^2,$$

which equates to

$$\sum_{i=1}^{n}(y_i - \bar{y})^2 = \sum_{i=1}^{n}(\hat{y}_i - \bar{y})^2 + \sum_{i=1}^{n}(y_i - \hat{y}_i)^2.$$

The relation between R^2 and r.

$$SQ_{\text{Residual}} = \sum_{i=1}^{n}(y_i - (\hat{a} + \hat{b}x_i))^2 \stackrel{(11.8)}{=} \sum_{i=1}^{n}[(y_i - \bar{y}) - \hat{b}(x_i - \bar{x})]^2$$

$$= S_{yy} + \hat{b}^2 S_{xx} - 2\hat{b}S_{xy}$$

$$= S_{yy} - \hat{b}^2 S_{xx} = S_{yy} - \frac{(S_{xy})^2}{S_{xx}}$$

$$SQ_{\text{Regression}} = S_{yy} - SQ_{\text{Residual}} = \frac{(S_{xy})^2}{S_{xx}}.$$

We therefore obtain

$$R^2 = \frac{SQ_{\text{Regression}}}{S_{yy}} = \frac{(S_{xy})^2}{S_{xx}S_{yy}} = r^2.$$

The least squares estimators are unbiased.

$$\mathrm{E}(\hat{\beta}) = \mathrm{E}((\mathbf{X}'\mathbf{X})^{-1}\mathbf{X}'\mathbf{y})$$

Given that $\mathbf{X}$ in the model is assumed to be fixed (i.e. non-stochastic and not following any distribution), we obtain

$$\mathrm{E}(\hat{\beta}) = (\mathbf{X}'\mathbf{X})^{-1}\mathbf{X}'\mathrm{E}(\mathbf{y}).$$

Since $\mathrm{E}(\epsilon) = \mathbf{0}$ it follows that $\mathrm{E}\,\mathbf{y} = \mathbf{X}\beta$ and therefore

$$\mathrm{E}(\hat{\beta}) = (\mathbf{X}'\mathbf{X})^{-1}\mathbf{X}'\mathbf{X}\beta = \beta.$$

How to obtain the variance of the least squares estimator. With the same arguments as above (i.e $\mathbf{X}$ is fixed and non-stochastic) and applying the rule $\mathrm{Var}(bX) = b^2\,\mathrm{Var}(X)$ from the scalar case to matrices we obtain:

$$\mathrm{Var}(\hat{\beta}) = \mathrm{Var}((\mathbf{X}'\mathbf{X})^{-1}\mathbf{X}'\mathbf{y}) = (\mathbf{X}'\mathbf{X})^{-1}\mathbf{X}'\,\mathrm{Var}(\mathbf{y})\mathbf{X}(\mathbf{X}'\mathbf{X})^{-1} = \sigma^2(\mathbf{X}'\mathbf{X})^{-1}.$$

Maximum likelihood estimation in the linear model. The linear model follows a normal distribution:

$$\mathbf{y} = \mathbf{X}\beta + \epsilon \sim N(\mathbf{X}\beta, \sigma^2\mathbf{I}).$$

Therefore the likelihood function of $\mathbf{y}$ also follows a normal distribution:

$$L(\beta, \sigma^2) = (2\pi\sigma^2)^{-n/2} \exp\left\{-\frac{1}{2\sigma^2}(\mathbf{y} - \mathbf{X}\beta)'(\mathbf{y} - \mathbf{X}\beta)\right\}.$$

The log-likelihood function is given by

$$l(\beta, \sigma^2) = -\frac{n}{2}\ln(2\pi\sigma^2) - \frac{1}{2\sigma^2}(\mathbf{y} - \mathbf{X}\beta)'(\mathbf{y} - \mathbf{X}\beta).$$

To obtain the maximum likelihood estimates of β and σ^2, one needs to obtain the maxima of the above function using the principle of maxima and minima that involves setting the partial derivatives equal to zero and finding the solution:

$$\frac{\partial l}{\partial \beta} = \frac{1}{2\sigma^2}2\mathbf{X}'(\mathbf{y} - \mathbf{X}\beta) = \mathbf{0},$$

$$\frac{\partial l}{\partial \sigma^2} = -\frac{n}{2\sigma^2} + \frac{1}{2(\sigma^2)^2}(\mathbf{y} - \mathbf{X}\beta)'(\mathbf{y} - \mathbf{X}\beta) = 0.$$

We therefore have

$$\mathbf{X}'\mathbf{X}\hat{\beta} = \mathbf{X}'\mathbf{y}, \text{ or } \hat{\beta} = (\mathbf{X}'\mathbf{X})^{-1}\mathbf{X}'\mathbf{y}$$

$$\hat{\sigma}^2 = \frac{1}{n}(\mathbf{y} - \mathbf{X}\hat{\beta})'(\mathbf{y} - \mathbf{X}\hat{\beta})$$

which give us the ML estimates of β and σ^2 in the linear regression model. Here, we also need to check the Hessian matrix of second order partial derivatives to show that we really found a minimum and not a maximum. We omit this proof however.

C.7   More Details on Chap. 12

Maximum likelihood estimation in the logistic regression model. Using the individual probabilities p_i, the likelihood function of all n observations y_i, $i = 1, \ldots, n$ is

$$L(p_1, \ldots, p_n|y_1, \ldots, y_n) = \prod_{i=1}^{n} L_i = \prod_{i=1}^{n} p_i^{y_i}(1 - p_i)^{(1-y_i)}.$$

Note that for each observation y_i, a Bernoulli likelihood $L_i = p_i^{y_i} (1 - p_i)^{(1-y_i)}$ is used, so that $L_i = p_i$ if $y_i = 1$ and $L_i = 1 - p_i$ if $y_i = 0$. Using (12.1), the likelihood can be written as a function of the coefficient vector $\beta = (\beta_0, \beta_1, \ldots, \beta_p)'$:

$$L(\beta) = \prod_{i=1}^{n} L_i(\beta) \tag{C.13}$$

$$= \prod_{i=1}^{n} \left[\frac{\exp(\mathbf{x}_i'\beta)}{1 + \exp(\mathbf{x}_i'\beta)} \right]^{y_i} \left[\frac{1}{1 + \exp(\mathbf{x}_i'\beta)} \right]^{(1-y_i)} , \tag{C.14}$$

where $P(y_i = 0|\mathbf{x}_i) = 1/(1 + \exp(\mathbf{x}_i'\beta))$. The likelihood $L(\beta)$ is maximized with respect to β. Usually, the natural logarithm of this function is maximized because the product then becomes a sum of terms, which is easier to maximize. For example, by taking the partial derivatives with respect to β_j and setting these derivatives to zero. The maximization does not, in general, lead to an explicit solution as in the case of linear regression (i.e. no formula exists for the estimator $\hat{\beta}$ of β) but a solution has to be found by some iterative method. Popular numerical iterative methods for obtaining the solution include the Newton–Raphson algorithm or Fisher scoring.

The maximum-likelihood estimates have several desirable properties (see also Sect. 9.3.1): they are consistent, asymptotically unbiased and asymptotically normally distributed. This means that

$$\hat{\beta} \sim N(\beta, \mathrm{Cov}(\beta)) , \tag{C.15}$$

where $\mathrm{Cov}(\beta)$ is the covariance (matrix) of $\hat{\beta}$ which in turn is estimated by replacing it by the maximum-likelihood estimator $\hat{\beta}$, i.e. $\widehat{\mathrm{Cov}}(\hat{\beta})$. The square roots of the diagonal elements are used as estimated standard errors. The z-values are the parameter estimates $\hat{\beta}_j$, $j = 1, 2, \ldots, p$, divided by the estimated standard errors and the two-sided p-values are then calculated under the two-sided null hypotheses $H_0 : \beta_j = 0$ using an $N(0, 1)$-distribution.

The AIC criterion is also properly defined for GLMs when the estimation is based on maximum-likelihood. In this case we get

$$\mathrm{AIC} = -2 \log L(\hat{\beta}) + 2(p + 1) , \tag{C.16}$$

i.e., the AIC is twice the negative log likelihood, evaluated at the maximum-likelihood estimator $\hat{\beta}$ plus twice the number of estimated parameters $(p + 1)$ (the intercept also has to be estimated, that is why it is "$p + 1$"). Models with more parameters are penalized by the AIC. The general interpretation is: "the lower the AIC, the better". An increase of parameters is therefore only be considered to be advantageous if the log-likelihood is substantially increased (or equivalently, the negative log-likelihood decreased, respectively).

C.8   More Details on Chap. 13

Probabilities of Selection of Samples and Sampling Units.

(i) If n units are selected by SRSWOR from a population of size N, the total number
of possible samples that can be chosen is $\binom{N}{n}$. Therefore, the probability of
selecting any one of these samples is $1/\binom{N}{n}$. This can be proven as follows:
Note that a unit can be selected into the sample in any of the n draws. Let u_j be
the jth unit in the population. This unit can be selected in the sample either at the
first draw, or the second draw, . . ., or the nth draw (but not at multiple draws). Let
$P_i(j)$ denote the probability of selection of u_j at the ith draw, $i = 1, 2, \ldots, n$,
$j = 1, 2, \ldots, N$. Then

$$P_i(j) = P_1(j) + P_2(j) + \cdots + P_n(j)$$
$$= \frac{1}{N} + \frac{1}{N} + \cdots + \frac{1}{N} \quad (n \text{ times})$$
$$= \frac{n}{N}.$$

Now if $u_1, u_2, \ldots, u_n$ are the n units selected in the sample, then the probability
of their selection is

$$P(u_1, u_2, \ldots, u_n) = P(u_1) \cdot P(u_2) \cdot \ldots \cdot P(u_n).$$

Note that when the second sampling unit is selected, there are $(n-1)$ sampling
units remaining to be selected in the sample from the population of $(N-1)$
sampling units. Similarly, when the third sampling unit is to be selected, there
are $(n-2)$ sampling units left to be selected in the sample from the population
of $(N-2)$ sampling units and so on. If $P(u_1) = \frac{n}{N}$, then

$$P(u_2) = \frac{n-1}{N-1}, \ldots, P(u_n) = \frac{1}{N-n+1}.$$

Thus

$$P(u_1, u_2, \ldots, u_n) = \frac{n}{N} \cdot \frac{n-1}{N-1} \cdot \ldots \cdot \frac{1}{N-n+1} = \frac{1}{\binom{N}{n}}.$$

(ii) The probability of selection of a unit under SRSWOR is $1/N$. This can be proven
as follows: let A_l denote an event where a particular, say jth unit u_j is not selected
at the lth draw. Then,

$$P(\text{selection of } u_j \text{ at } kth \text{ draw})$$
$$= P(A_1 \cap A_2 \cap \ldots \cap A_{k-1} \cap \bar{A}_k)$$
$$= P(A_1) \cdot P(A_2|A_1) \cdot P(A_3|A_1, A_2) \cdots P(A_{k-1}|A_1, A_2, \ldots, A_{k-2})$$
$$\cdot P(\bar{A}_k|A_1, A_2 \ldots A_{k-1})$$
$$= \left(1 - \frac{1}{N}\right) \cdot \left(1 - \frac{1}{N-1}\right) \cdot \left(1 - \frac{1}{N-2}\right)$$
$$\cdots \left(1 - \frac{1}{N-k+2}\right) \frac{1}{N-k+1}$$
$$= \frac{1}{N}.$$

Proofs for the variance under SRSWOR and SRSWR, respectively. Assuming that each observation has variance σ^2, we first consider the variance of $\bar{y}$ as

$$
\begin{aligned}
\text{Var}(\bar{y}_{WR}) &= \text{E}(\bar{y} - \bar{Y})^2 \\
&= \text{E}\left[\frac{1}{n}\sum_{i=1}^{n}(y_i - \bar{Y})\right]^2 \\
&= \text{E}\left[\frac{1}{n^2}\sum_{i=1}^{n}(y_i - \bar{Y})^2 + \frac{1}{n^2}\sum_{i=1}^{n}\sum_{i'(\neq i)=1}^{n}(y_i - \bar{Y})(y_{i'} - \bar{Y})\right] \\
&= \frac{1}{n^2}\sum_{i=1}^{n}\text{E}(y_i - \bar{Y})^2 + \frac{1}{n^2}\sum_{i=1}^{n}\sum_{i'(\neq i)=1}^{n}\text{E}(y_i - \bar{Y})(y_{i'} - \bar{Y}) \\
&= \frac{1}{n^2}\sum_{i=1}^{n}\sigma^2 + \frac{K}{n^2} \\
&= \frac{N-1}{Nn}S^2 + \frac{K}{n^2}
\end{aligned}
\tag{C.17}
$$

where $K = \sum_{i=1}^{n}\sum_{i'(\neq i)}^{n}\text{E}(y_i - \bar{Y})(y_{i'} - \bar{Y})$. Now we find K under the setups of SRSWOR and SRSWR.

(i) Variance for SRSWOR:

Under SRSWOR, consider

$$\text{E}(y_i - \bar{Y})(y_{i'} - \bar{Y}) = \frac{1}{N(N-1)}\sum_{j=1}^{N}\sum_{j'(\neq j)=1}^{N}(y_j - \bar{Y})(y_{j'} - \bar{Y}).$$

Since

$$\left[\sum_{j=1}^{N} (y_j - \bar{Y}) \right]^2 = \sum_{j=1}^{N} (y_j - \bar{Y})^2 + \sum_{j=1}^{N} \sum_{j'(\neq j)=1}^{N} (y_j - \bar{Y})(y_{j'} - \bar{Y})$$

$$0 = (N-1)S^2 + \sum_{j=1}^{N} \sum_{j'(\neq j)=1}^{N} (y_j - \bar{Y})(y_{j'} - \bar{Y}).$$

Hence

$$E(y_i - \bar{Y})(y_{i'} - \bar{Y}) = \frac{1}{N(N-1)} \sum_{j=1}^{N} \sum_{j'(\neq j)=1}^{N} (y_j - \bar{Y})(y_{j'} - \bar{Y})$$

$$= \frac{1}{N(N-1)} [-(N-1)S^2]$$

$$= -\frac{S^2}{N}.$$

Thus, $K = \sum_{i=1}^{n} \sum_{i'(\neq i)}^{n} E(y_i - \bar{Y})(y_{i'} - \bar{Y}) = -n(n-1)\frac{S^2}{N}$ and substituting the value of K in (C.17), the variance of $\bar{y}$ under SRSWOR is obtained as

$$\text{Var}(\bar{y}_{WOR}) = \frac{N-1}{Nn}S^2 - \frac{n(n-1)}{n^2}S^2.$$

$$= \frac{N-n}{Nn}S^2. \tag{C.18}$$

(ii) Next, under SRSWR, consider

$$K = \sum_{i=1}^{n} \sum_{i'(\neq i)}^{n} E(y_i - \bar{Y})(y_{i'} - \bar{Y})$$

$$= \sum_{i=1}^{n} \sum_{i'(\neq i)}^{n} E(y_i - \bar{Y})E(y_{i'} - \bar{Y})$$

$$= 0$$

because the ith and i'th draws $(i \neq i')$ are independent. Thus, the variance of $\bar{y}$ under SRSWR is

$$\text{Var}(\bar{y}_{WR}) = \frac{N-1}{Nn}S^2 = \frac{\sigma^2}{n}. \tag{C.19}$$

Estimation of the variance of the sample mean. Consider

$$s^2 = \frac{1}{n-1} \sum_{i=1}^{n} (y_i - \bar{y})^2$$

$$= \frac{1}{n-1} \sum_{i=1}^{n} \left[(y_i - \bar{Y}) - (\bar{y} - \bar{Y}) \right]^2$$

$$= \frac{1}{n-1} \sum_{i=1}^{n} \left[(y_i - \bar{Y})^2 - n(\bar{y} - \bar{Y})^2 \right].$$

Then

$$E(s^2) = \frac{1}{n-1} \sum_{i=1}^{n} \left[E(y_i - \bar{Y})^2 - n E(\bar{y} - \bar{Y})^2 \right]$$

$$= \frac{1}{n-1} \left[\sum_{i=1}^{n} \text{Var}(y_i) - n \, \text{Var}(\bar{y}) \right].$$

$$= \frac{1}{n-1} \left[n\sigma^2 - n \, \text{Var}(\bar{y}) \right]. \tag{C.20}$$

Now we try to substitute the value of $\text{Var}(\bar{y})$ under SRSWOR and SRSWR in (C.20). First we consider the case of SRSWOR,

$$E(s^2) = \frac{1}{n-1} \left[n\sigma^2 - n \, \text{Var}(\bar{y}_{WOR}) \right] \tag{C.21}$$

$$= \frac{n}{n-1} \left[\sigma^2 - \frac{N-n}{Nn} S^2 \right]$$

$$= \frac{n}{n-1} \left[\frac{N-1}{N} S^2 - \frac{N-n}{Nn} S^2 \right]$$

$$= S^2. \tag{C.22}$$

Next we consider the case of SRSWR:

$$E(s^2) = \frac{1}{n-1} \left[n\sigma^2 - n \, \text{Var}(\bar{y}_{WR}) \right] \tag{C.23}$$

$$= \frac{n}{n-1} \left[\sigma^2 - \frac{\sigma^2}{n} \right]$$

$$= \sigma^2. \tag{C.24}$$

Hence

$$E(s^2) = \begin{cases} S^2 & \text{in SRSWOR} \\ \sigma^2 & \text{in SRSWR}. \end{cases}$$

This completes the proof of (13.3).

C.9 Distribution Tables

Table C.1 CDF values for the standard normal distribution, $\Phi(z)$

z	0.00	0.01	0.02	0.03	0.04
0.0	0.500000	0.503989	0.507978	0.511966	0.515953
0.1	0.539828	0.543795	0.547758	0.551717	0.555670
0.2	0.579260	0.583166	0.587064	0.590954	0.594835
0.3	0.617911	0.621720	0.625516	0.629300	0.633072
0.4	0.655422	0.659097	0.662757	0.666402	0.670031
0.5	0.691462	0.694974	0.698468	0.701944	0.705401
0.6	0.725747	0.729069	0.732371	0.735653	0.738914
0.7	0.758036	0.761148	0.764238	0.767305	0.770350
0.8	0.788145	0.791030	0.793892	0.796731	0.799546
0.9	0.815940	0.818589	0.821214	0.823814	0.826391
1.0	0.841345	0.843752	0.846136	0.848495	0.850830
1.1	0.864334	0.866500	0.868643	0.870762	0.872857
1.2	0.884930	0.886861	0.888768	0.890651	0.892512
1.3	0.903200	0.904902	0.906582	0.908241	0.909877
1.4	0.919243	0.920730	0.922196	0.923641	0.925066
1.5	0.933193	0.934478	0.935745	0.936992	0.938220
1.6	0.945201	0.946301	0.947384	0.948449	0.949497
1.7	0.955435	0.956367	0.957284	0.958185	0.959070
1.8	0.964070	0.964852	0.965620	0.966375	0.967116
1.9	0.971283	0.971933	0.972571	0.973197	0.973810
2.0	0.977250	0.977784	0.978308	0.978822	0.979325
2.1	0.982136	0.982571	0.982997	0.983414	0.983823
2.2	0.986097	0.986447	0.986791	0.987126	0.987455
2.3	0.989276	0.989556	0.989830	0.990097	0.990358
2.4	0.991802	0.992024	0.992240	0.992451	0.992656
2.5	0.993790	0.993963	0.994132	0.994297	0.994457
2.6	0.995339	0.995473	0.995604	0.995731	0.995855
2.7	0.996533	0.996636	0.996736	0.996833	0.996928
2.8	0.997445	0.997523	0.997599	0.997673	0.997744
2.9	0.998134	0.998193	0.998250	0.998305	0.998359
3.0	0.998650	0.998694	0.998736	0.998777	0.998817

(continued)

Table C.1 (continued)

z	0.05	0.06	0.07	0.08	0.09
0.0	0.519939	0.523922	0.527903	0.531881	0.535856
0.1	0.559618	0.563559	0.567495	0.571424	0.575345
0.2	0.598706	0.602568	0.606420	0.610261	0.614092
0.3	0.636831	0.640576	0.644309	0.648027	0.651732
0.4	0.673645	0.677242	0.680822	0.684386	0.687933
0.5	0.708840	0.712260	0.715661	0.719043	0.722405
0.6	0.742154	0.745373	0.748571	0.751748	0.754903
0.7	0.773373	0.776373	0.779350	0.782305	0.785236
0.8	0.802337	0.805105	0.807850	0.810570	0.813267
0.9	0.828944	0.831472	0.833977	0.836457	0.838913
1.0	0.853141	0.855428	0.857690	0.859929	0.862143
1.1	0.874928	0.876976	0.879000	0.881000	0.882977
1.2	0.894350	0.896165	0.897958	0.899727	0.901475
1.3	0.911492	0.913085	0.914657	0.916207	0.917736
1.4	0.926471	0.927855	0.929219	0.930563	0.931888
1.5	0.939429	0.940620	0.941792	0.942947	0.944083
1.6	0.950529	0.951543	0.952540	0.953521	0.954486
1.7	0.959941	0.960796	0.961636	0.962462	0.963273
1.8	0.967843	0.968557	0.969258	0.969946	0.970621
1.9	0.974412	0.975002	0.975581	0.976148	0.976705
2.0	0.979818	0.980301	0.980774	0.981237	0.981691
2.1	0.984222	0.984614	0.984997	0.985371	0.985738
2.2	0.987776	0.988089	0.988396	0.988696	0.988989
2.3	0.990613	0.990863	0.991106	0.991344	0.991576
2.4	0.992857	0.993053	0.993244	0.993431	0.993613
2.5	0.994614	0.994766	0.994915	0.995060	0.995201
2.6	0.995975	0.996093	0.996207	0.996319	0.996427
2.7	0.997020	0.997110	0.997197	0.997282	0.997365
2.8	0.997814	0.997882	0.997948	0.998012	0.998074
2.9	0.998411	0.998462	0.998511	0.998559	0.998605
3.0	0.998856	0.998893	0.998930	0.998965	0.998999

These values can also be obtained in R by using the `pnorm(p)` command.

Table C.2 $(1 - \alpha)$ quantiles for the t-distribution

df	$1 - \alpha$			
	0.95	0.975	0.99	0.995
1	6.3138	12.706	31.821	63.657
2	2.9200	4.3027	6.9646	9.9248
3	2.3534	3.1824	4.5407	5.8409
4	2.1318	2.7764	3.7469	4.6041
5	2.0150	2.5706	3.3649	4.0321
6	1.9432	2.4469	3.1427	3.7074
7	1.8946	2.3646	2.9980	3.4995
8	1.8595	2.3060	2.8965	3.3554
9	1.8331	2.2622	2.8214	3.2498
10	1.8125	2.2281	2.7638	3.1693
11	1.7959	2.2010	2.7181	3.1058
12	1.7823	2.1788	2.6810	3.0545
13	1.7709	2.1604	2.6503	3.0123
14	1.7613	2.1448	2.6245	2.9768
15	1.7531	2.1314	2.6025	2.9467
16	1.7459	2.1199	2.5835	2.9208
17	1.7396	2.1098	2.5669	2.8982
18	1.7341	2.1009	2.5524	2.8784
19	1.7291	2.0930	2.5395	2.8609
20	1.7247	2.0860	2.5280	2.8453
30	1.6973	2.0423	2.4573	2.7500
40	1.6839	2.0211	2.4233	2.7045
50	1.6759	2.0086	2.4033	2.6778
60	1.6706	2.0003	2.3901	2.6603
70	1.6669	1.9944	2.3808	2.6479
80	1.6641	1.9901	2.3739	2.6387
90	1.6620	1.9867	2.3685	2.6316
100	1.6602	1.9840	2.3642	2.6259
200	1.6525	1.9719	2.3451	2.6006
300	1.6499	1.9679	2.3388	2.5923
400	1.6487	1.9659	2.3357	2.5882
500	1.6479	1.9647	2.3338	2.5857

These values can also be obtained in R using the `qt(p,df)` command.

Table C.3 $(1 - \alpha)$ quantiles of the χ^2 distribution

df	$1 - \alpha$					
	0.01	0.025	0.05	0.95	0.975	0.99
1	0.0001	0.001	0.004	3.84	5.02	6.62
2	0.020	0.051	0.103	5.99	7.38	9.21
3	0.115	0.216	0.352	7.81	9.35	11.3
4	0.297	0.484	0.711	9.49	11.1	13.3
5	0.554	0.831	1.15	11.1	12.8	15.1
6	0.872	1.24	1.64	12.6	14.4	16.8
7	1.24	1.69	2.17	14.1	16.0	18.5
8	1.65	2.18	2.73	15.5	17.5	20.1
9	2.09	2.70	3.33	16.9	19.0	21.7
10	2.56	3.25	3.94	18.3	20.5	23.2
11	3.05	3.82	4.57	19.7	21.9	24.7
12	3.57	4.40	5.23	21.0	23.3	26.2
13	4.11	5.01	5.89	22.4	24.7	27.7
14	4.66	5.63	6.57	23.7	26.1	29.1
15	5.23	6.26	7.26	25.0	27.5	30.6
16	5.81	6.91	7.96	26.3	28.8	32.0
17	6.41	7.56	8.67	27.6	30.2	33.4
18	7.01	8.23	9.39	28.9	31.5	34.8
19	7.63	8.91	10.1	30.1	32.9	36.2
20	8.26	9.59	10.9	31.4	34.2	37.6
25	11.5	13.1	14.6	37.7	40.6	44.3
30	15.0	16.8	18.5	43.8	47.0	50.9
40	22.2	24.4	26.5	55.8	59.3	63.7
50	29.7	32.4	34.8	67.5	71.4	76.2
60	37.5	40.5	43.2	79.1	83.3	88.4
70	45.4	48.8	51.7	90.5	95.0	100.4
80	53.5	57.2	60.4	101.9	106.6	112.3
90	61.8	65.6	69.1	113.1	118.1	124.1
100	70.1	74.2	77.9	124.3	129.6	135.8

These values can also be obtained in R using the `qchisq(p,df)` command.

Quantiles of the F-distribution. These quantiles can be obtained in R using the `qf(p,df1,df2)` command.

D.1   Descriptive Data Analysis

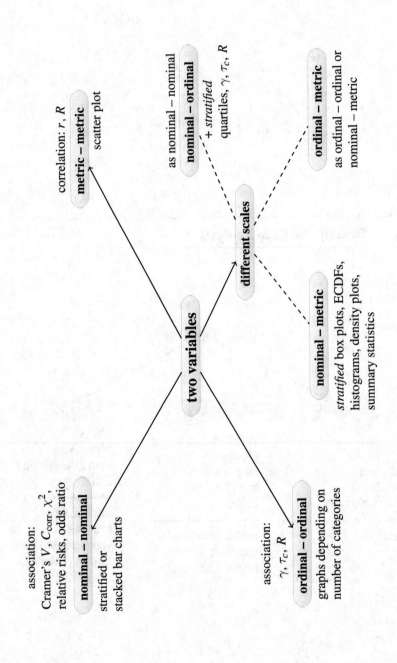

association:
Cramer's V, C_{corr}, χ^2,
relative risks, odds ratio

nominal – nominal

stratified or
stacked bar charts

correlation: r, R

metric – metric

scatter plot

as nominal – nominal

nominal – ordinal

+ *stratified*
quartiles, γ, τ_c, R

ordinal – metric

as ordinal – ordinal or
nominal – metric

different scales

nominal – metric

stratified box plots, ECDFs,
histograms, density plots,
summary statistics

two variables

association:
γ, τ_c, R

ordinal – ordinal

graphs depending on
number of categories

D.2 Summary of Tests for Metric and Ordinal Variables

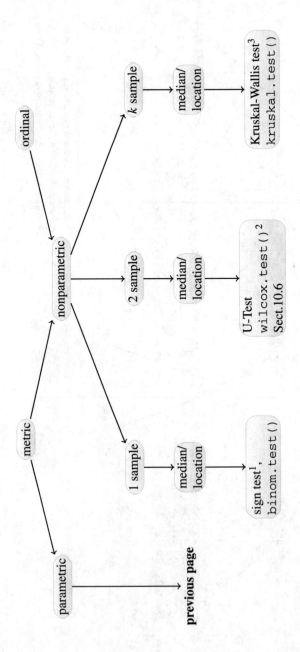

¹ not explained in this book; alternative: Mood's median test
² use option paired=TRUE for dependent data
³ not explained in this book; use Friedman test for dependent data

D.3 Summary of Tests for Nominal Variables

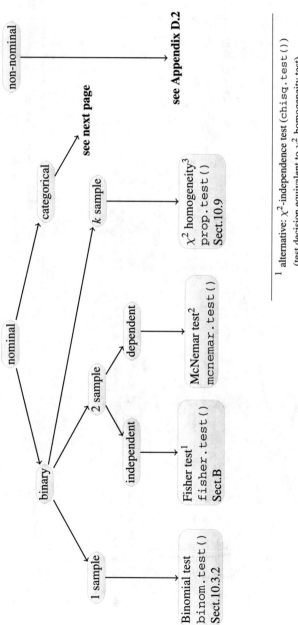

1 alternative: χ^2-independence test (chisq.test()) (test decision equivalent to χ^2-homogeneity test)
2 not explained in this book
3 test decision equivalent to χ^2-independence test

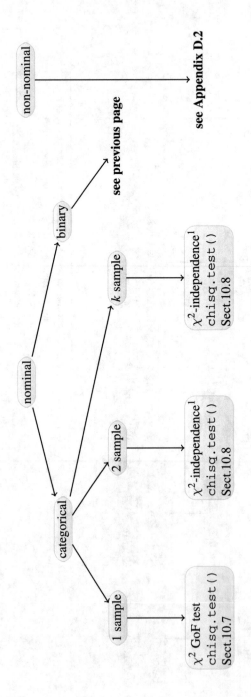

^{1} test decision equivalent to χ^2-homogeneity test (prop.test)

References

Adler, J. (2012). *R in a nutshell*. Boston: O'Reilly.

Albert, J., & Rizzo, M. (2012). *R by example*. New York: Springer.

Casella, G., & Berger, R. (2002). *Statistical inference*.

Chow, S., Wang, H., & Shao, J. (2007). *Sample size calculations in clinical research*. Chapman & Hall.

Cole, S. R., Platt, R. W., Schisterman, E. F., Chu, H., Westreich, D., Richardson, D., & Poole, C. (2010). Illustrating bias due to conditioning on a collider. *International Journal of Epidemiology, 39*(2), 417–20.

Crawley, M. (2013). *The R book*. Wiley.

Dalgaard, P. (2008). *Introductory statistics with R*. Springer.

Efron, B., & Tibshirani, R. (2014). *An introduction to the bootstrap*. Chapman and Hall/CRC.

Everitt, B., & Hothorn, T. (2011). *An introduction to applied multivariate analysis with R*. New York: Springer.

Groves, R., Fowler, F., Couper, M., Lepkowski, J., Singer, E., & Tourangeau, R. (2009). *Survey methodology*. Wiley series in survey methodology. Hoboken, N.J.: Wiley.

Hernan, M., & Robins, J. (2021). *Causal inference*. Boca Raton: Chapman & Hall/CRC.

Hosmer, D. W., & Lemeshow, S. (2013). *Applied logistic regression*. New York: Wiley.

Hyndman, R. J., & Fan, Y. (1996). Sample quantiles in statistical packages. *American Statistician, 50*, 361–365.

Luque-Fernandez, M. A., Schomaker, M., Rachet, B., & Schnitzer, M. E. (2018). Targeted maximum likelihood estimation for a binary treatment: A tutorial. *Statistics in Medicine, 37*(16), 2530–2546.

R Core Team (2021). *R: A Language and Environment for Statistical Computing*, R Foundation for Statistical Computing, Vienna, Austria. http://www.R-project.org/

Rafi, Z., & Greenland, S. (2020). Semantic and cognitive tools to aid statistical science: Replace confidence and significance by compatibility and surprise. *BMC Medical Research Methodology, 20*(1), 244.

Rao, P. (2000). *Sampling methodologies with applications*. Chapman and Hall/CRC.

Young, G., & Smith, R. (2005). *Essentials of statistical inference*. Cambridge University Press.

© Springer Nature Switzerland AG 2022

C. Heumann et al., *Introduction to Statistics and Data Analysis*,

https://doi.org/10.1007/978-3-031-11833-3

Index

© Springer Nature Switzerland AG 2022
C. Heumann et al., *Introduction to Statistics and Data Analysis*,
https://doi.org/10.1007/978-3-031-11833-3

Printed in the United States
by Baker & Taylor Publisher Services